Mössbauer Spectroscopy Applied to Magnetism and Materials Science

Volume 2

MODERN INORGANIC CHEMISTRY

Series Editor: John P. Fackler, Jr., *Texas A&M University*

Recent volumes in the series:

CARBON-FUNCTIONAL ORGANOSILICON COMPOUNDS
Edited by Václav Chvalovský and Jon M. Bellama

COOPERATIVE PHENOMENA IN JAHN–TELLER CRYSTALS
Michael D. Kaplan and Benjamin G. Vekhter

GAS PHASE INORGANIC CHEMISTRY
Edited by David H. Russell

HOMOGENEOUS CATALYSIS WITH METAL PHOSPHINE COMPLEXES
Edited by Louis H. Pignolet

INORGANOMETALLIC CHEMISTRY
Edited by Thomas P. Fehlner

THE JAHN–TELLER EFFECT AND VIBRONIC INTERACTIONS IN MODERN CHEMISTRY
I. B. Bersuker

METAL COMPLEXES IN AQUEOUS SOLUTIONS
Arthur E. Martell and Robert D. Hancock

MÖSSBAUER SPECTROSCOPY APPLIED TO INORGANIC CHEMISTRY
Volumes 1 and 2 • Edited by Gary J. Long
Volume 3 • Edited by Gary J. Long and Fernande Grandjean

MÖSSBAUER SPECTROSCOPY APPLIED TO MAGNETISM AND MATERIALS SCIENCE
Volumes 1 and 2 • Edited by Gary J. Long and Fernande Grandjean

ORGANOMETALLIC CHEMISTRY OF THE TRANSITION ELEMENTS
Florian P. Pruchnik
Translated from Polish by Stan A. Duraj

PHOTOCHEMISTRY AND PHOTOPHYSICS OF METAL COMPLEXES
D. M. Roundhill

A Continuation Order Plan is available for this series. A continuation order will bring delivery of each new volume immediately upon publication. Volumes are billed only upon actual shipment. For further information please contact the publisher.

Mössbauer Spectroscopy Applied to Magnetism and Materials Science

Volume 2

Edited by

Gary J. Long

University of Missouri–Rolla
Rolla, Missouri

and

Fernande Grandjean

University of Liège
Sart-Tilman, Belgium

Springer Science+Business Media, LLC

Library of Congress Cataloging-in-Publication Data

On file

ISBN 978-1-4899-1765-2 ISBN 978-1-4899-1763-8 (eBook)
DOI 10.1007/978-1-4899-1763-8

Originally published by Plenum Press, New York in 1996
Softcover reprint of the hardcover 1st edition 1996

10 9 8 7 6 5 4 3 2 1

Contributors

S. J. Campbell • School of Physics, University College, University of New South Wales, Australian Defence Force Academy, Canberra, ACT 2600, Australia

Susanna Carbonin • Dipartimento di Mineralogia e Petrologia, Università di Padova, I-35100 Padova, Italy

Desmond C. Cook • Department of Physics, Old Dominion University, Norfolk, VA 23529, USA

Antonio Della Giusta • Dipartimento di Mineralogia e Petrologia, Università di Padova, I-35100 Padova, Italy

Richard G. Grant • Department of Physics, Hollins College, Roanoke, VA 24020, USA

C. E. Johnson • Department of Physics, University of Liverpool, Liverpool L69 3BX, UK

J. A. Johnson • School of the Built Environment, Liverpool John Moores University, Liverpool, UK

W. A. Kaczmarek • Department of Applied Mathematics, Research School of Physical Sciences and Engineering, The Australian National University, Canberra ACT 0200, Australia

W. Meisel • Institut für Anorganische und Analytische Chemie, Johannes Gütenberg-Universität Mainz, Staudinger Weg 9, D-55099 Mainz, Germany

Q. A. Pankhurst • Department of Physics and Astronomy, University College London, London WC1E 6BT, England

Moshe P. Pasternak • School of Physics and Astronomy, Tel Aviv University, Ramat Aviv, Tel Aviv 69978, Israel

D. G. Rancourt • Department of Physics, University of Ottawa, Ottawa, Ontario, K1N 6N5, Canada

Umberto Russo • Dipartimento di Chimica Inorganica, Metallorganica ed Analitica, Università di Padova, Via Loredan 4, I-35131 Padova, Italy

Ch. Sauer • Institut für Festkörperforschung, Forschungszentrum Jülich GmbH, D-52425 Jülich, Germany

Zbigniew M. Stadnik • Department of Physics, University of Ottawa, Ottawa, Ontario, K1N 6N5, Canada

R. Dean Taylor • Los Alamos National Laboratory, P. O. Box 1663 MS-K764, Los Alamos, NM 87545, USA

G. Vogl • Institut für Festkörperphysik der Universität Wien, Strudlhofgasse 4, A-1090 Wien, Austria

K. F. E. Williams • Department of Physics, University of Liverpool, Liverpool L69 3BX, UK

Preface

This book represents Volume 2 in a series on the use of Mössbauer spectroscopy in the study of magnetism and materials. However, the perceptive reader will notice some differences from Volume 1. Specifically, in order to market the book at a more affordable price for most universities and research laboratories, the book has been prepared in camera-ready format. The editors and the authors agreed to do this because there is a demand for such a book in the Mössbauer community. This format has placed an extra burden on the editors and the authors and we hope we have overcome all the difficulties generated by the transfer of files between different computers. In order to make the book more attractive to materials scientists who are not experts in Mössbauer spectroscopy, this volume is particularly oriented towards the study of materials by Mössbauer spectroscopy and related complementary techniques, such as neutron scattering and a variety of surface scattering techniques.

The authors of this volume can be proud of the high quality professional effort they have devoted to clearly presenting their specific topics. As a result we very much enjoyed working with the authors on this volume. We hope that their effort will help to educate the next generation of Mössbauer effect spectroscopists, a generation which will face the challenge of maintaining equally high scientific and professional standards in their research work.

We would like to thank each of the authors for providing lively, timely, and instructive chapters within a tight schedule. An editorial project such as this volume requires adherence to a schedule and, at this time, we are pleased to note that the project is only three weeks beyond the schedule proposed several years ago. We would also like to thank several colleagues for their help in the preparation of this volume, specifically Prof. H. P. Leighly, Jr., who carefully reviewed several of the chapters, and our young colleagues, D. Hautot, B. Tilkens, and D. Vandormael, who greatly assisted in the preparation of the camera-ready manuscript. Our younger colleagues are far more adept than we are at dealing with the new personal computers and their complex software. Finally, we would like to thank our professional colleagues, Jürgen Buschow, André Hautot, Dimitri Hautot, Olivier Isnard, Bill James, Kanishka Marasinghe, Dave Middleton, Sanjay Mishra, Allan Pringle, Denis Vandormael, and Bill Yelon, for providing a stimulating environment for our scientific work over the past several years.

Rolla, Missouri

Gary J. Long
Fernande Grandjean

Contents

1. Surface and Thin Film Analysis by Mössbauer Spectroscopy and Related Techniques

W. Meisel

2. Magnetic and Structural Properties of Real Metal Layer Interfaces

Ch. Sauer

3. Iron-based Amorphous Ribbons and Wires

Q. A. Pankhurst

4. Diffusion Studies by Mössbauer Spectroscopy

G. Vogl

5. Analytical Methods for Mössbauer Spectral Analysis of Complex Materials

D. G. Rancourt

6. Quasicrystalline Materials

Zbigniew M. Stadnik

Mössbauer Spectroscopy Applied to Magnetism and Materials Science

Volume 2

1

Surface and Thin Film Analysis by Mössbauer Spectroscopy and Related Techniques

W. Meisel

1. PROBLEMS OF SURFACE AND THIN FILM ANALYSIS

The main trend in current high-tech development is miniaturization. This holds for complex electronic devices, such as microprocessors, as well as for coatings, such as Langmuir-Blodgett coatings, optical components, such as fiber couplers, chemical devices, such as microreactors, mechanical components, such as micromotors, and many others. Development and production of components with dimensions in the micron and submicron range require the availability of appropriate analytical tools for chemical, physical, and morphological analyses. In close interaction with the development of microcomponents, many, more and more efficient, analytical methods have been developed, many of which are "surface and thin layer methods." For brevity, they are often described by acronyms, and, to date, more than a hundred different acronyms are in use in this field. A selection of these acronyms is given in Table 1.

It is not the intention of this chapter, to cover the entire field of surface analysis, nor to give a general introduction to the broad variety of methods and techniques used in the field. Rather this chapter is directed to Mössbauer spectroscopists and points out some special problems in surface analytical applications and the need for interconnections with related surface methods. At this time, there is no textbook available covering the entire field of surface analysis. All relevant monographs, even those with the most generalized titles, focus on only a very few of the possible surface methods. The interested reader should consult some of these useful references.[1-6]

In the following discussion only solid surfaces will be considered, and the surrounding medium could be a vacuum, a gas, a liquid or, in the case of interfaces, a solid. There are only a very few methods, which record equipotential surfaces or surfaces of equal charge density, "mathematical surfaces" in the form of a two dimensional array. The methods of most practical interest are rather concerned with thin surface or interface layers, i.e., with objects exhibiting a small but finite in-depth extension.

W. Meisel • Institut für Anorganische und Analytische Chemie, Johannes Gutenberg-Universität Mainz, Staudinger Weg 9, D-55099 Mainz, Germany.

Mössbauer Spectroscopy Applied to Magnetism and Materials Science, Volume 2, edited by Gary J. Long and Fernande Grandjean. Plenum Press, New York, 1996.

Table 1. Selection of Acronyms Used in Surface Science.

Acronym	Technique	Acronym	Technique
AES	Auger Electron Spectroscopy	MD	Magnetic Dichroism
AFM	Atomic Force Microscopy	MDAD	Magnetic Dichrosim in Angular Distribution of Photoelectrons
APS	Appearance Potential Spectroscopy	MEIS	Medium Energy Ion Scattering
ARISS	Angle-Resolved Ion Scattering Spectroscopy	MIES	Metastable Induced Electron Spectroscopy
ARUPS	Angle Resolved UV-Induced Photoelectron Spectroscopy	MOKE	Magneto-optical Kerr Effect
ARXPS	Angle Resolved X-ray Induced Photoelectron Spectroscopy	MSHG	Magnetization Induced Second Harmonic Generation
ASPID	Adsorption Spectroscopy by Polariton Induced Desorption	MXPS	Monochromatized X-ray Induced Photoelectron Spectroscopy
BIAES	Bremsstrahlung Induced Auger Electron Spectroscopy	NEXAFS	Near Edge X-ray Absorption Fine Structure
CDAD	Circular Dichroism in Angular Distribution of Photoelectrons	NISXW	Normal Incidence Standing X-ray Wavefield
CEMS	Conversion Electron Mössbauer Spectroscopy	NRA	Nuclear Reaction Analysis
DCEMS	Depth Selective Conversion Electron Mössbauer Spectroscopy	PD	Photoelectron Diffraction
DLEED	Diffuse Low Energy Electron Diffraction	PEED	Photo-emission Electron Microscopy
EAPFS	Electron Appearance Potential Fine Structure	PIXE	Proton Induced X-ray Emission
EDX	Energy Dispersive X-ray Detection	PNR	Polarized Neutron Reflectometry
EXELFS	Extended X-ray Energy Loss Fine Structure	PSD	Photon Stimulated Desorption
EELS	Electron Energy Loss Spectroscopy	RAIRS	Reflection Absorption Infrared Spectroscopy
ESD	Electron Stimulated Desorption	RAS	Reflectance Anisotropy Spectroscopy
EMP	Electron Microprobe	RBS	Rutherford Backscattering Spectroscopy
EPMA	Electron Probe Micro-Analysis	REELS	Reflected Electron Energy Loss Spectroscopy
ESCA	Electron Spectroscopy for Chemical Analyses	REM	Reflection Electron Microscopy
ESD	Electron Stimulated Desorption	REMPI	Resonantly Enhanced Multi-Phonon Ionization
EXAFS	Extended X-ray Abs. Fine Structure	RHEED	Reflected High Energy Electron Diffraction
FABMS	Fast Atom Bombardment Mass Spectroscopy	SAM	Scanning Auger Microprobe
FEEM	Field Emission Electron Microscopy	SAXS	Small Angle X-ray Scattering
FIM	Field Ion Microscopy	SDA	Sputter Depth Analysis
GAXRD	Glancing Angle X-ray Diffraction	SEM	Scanning Electron Microprobe
GDOS	Glow Discharge Optical Spectroscopy	SERS	Surface Enhanced Raman Scattering
GSMS	γ-ray Scattering Mössbauer Spectroscopy	SEXAFS	Surface Extended X-ray Absorption Fine Structure
HAS	Helium Atom Scattering	SIMS	Secondary Ion Mass Spectroscopy
HEIS	High-Energy Ion Scattering	SMOKE	Surface Magneto-optical Kerr Effect
HIID	Heavy Ion-Induced Desorption	SNMS	Secondary Neutrals Mass Spectroscopy
HREELS	High Resolution Electron Energy Loss Spectroscopy	SPLEED	Spin-Polarized Low Energy Electron Diffraction
IP	Inverse Photoemission	SPLEEM	Spin-Polarized Low Energy Electron Microscopy
IBSCA	Ion Beam Spectrochemical Analysis	SPMDS	Spin-Polarized Metastable Deexcitation Spectroscopy
IRAS	Infrared Reflection-Absorption Spectroscopy	STM	Scanning Tunneling Microscopy
ICEMS	Integral Conversion Electron Mössbauer Spectroscopy	SXAPS	Soft X-ray Appearance Potential Spectroscopy
IMMA	Ion Microprobe Mass Analysis	SXPS	Soft X-ray Photoelectron Spectroscopy
IPES	Inverse Photo-Electron Spectroscopy	TDS	Thermal Desorption Spectroscopy
ISS	Ion Scattering Spectroscopy	TEM	Transmission Electron Microscopy
IRRS	Infrared Reflection Spectroscopy	TPD	Temperature Programmed Desorption
LAMMA	Laser Microprobe Mass Analysis	UPS	UV-Induced Photoelectron Spectroscopy
LDAD	Linear Dichroism in Angular Distribution of Photoelectrons	VI	Visual Inspection
LEED	Low Energy Electron Diffraction	WFS	Work Function Spectroscopy
LEF	Laser Excited Fluorescence	XANES	X-ray Absorption Near Edge Spectroscopy
LEIS	Low Energy Ion Scattering	XPD	X-ray Photoelectron Diffraction
MBE	Molecular Beam Epitaxy	XPS	X-ray Induced Photoelectron Spectroscopy
MBRS	Molecular Beam Relaxation Spectroscopy		

A basic characteristic of a surface is the loss by a surface atom of one or more of its bulk near neighbors. A simple surface model is that of "dangling bonds," see Fig. 1, in which a surface is assumed to be produced by breaking an extended solid while leaving the surface atoms unchanged, i.e., free valences are left at the surface, dangling like arms searching for a partner to form a new bond. This model illustrates the high reactivity of a surface but does not reflect its real structure, a structure which results from a rearrangement of the surface electron density distribution and the surface atom or ion positions in order to minimize the surface free energy. A simplified model of a reconstructed surface is shown schematically in Fig. 2. A consequence of the energetic changes resulting from surface reconstruction is a reduction in the work function, as is shown in Table 2, of the different faces of a silicon crystal.[7]

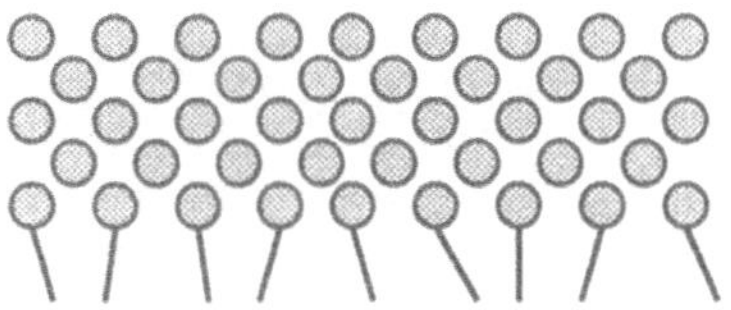

Figure 1. The concept of "dangling bonds."

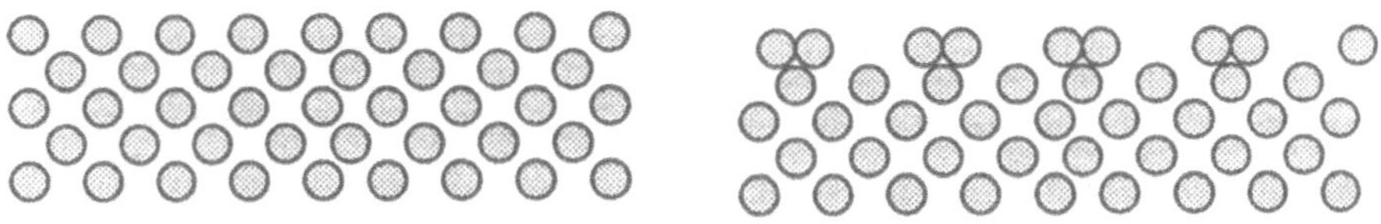

Figure 2. A schematic of an unchanged surface, left, and a reconstructed surface, right.

It is reasonable to assume that a truncated crystal exhibits plane faces with edges, corners, holes, and more complex defects. Sometimes, in Mössbauer spectral investigations of surfaces, the model shown in Fig. 3a has been used to discuss the different Debye-Waller factors of Mössbauer atoms, circles in Fig. 3a, at different surface positions. Unfortunately, this is an unrealistic model. A free surface reduces its energy by adsorption of atoms from the surrounding medium. The sticking coefficient, giving the probability that an incident atom will be fixed at the surface, is rather high and, for most cases of interest, is close to unity. Therefore, long before a Mössbauer active atom, the large circles in Fig. 3, can be deposited, the substrate is already covered by an adsorbate, see Fig. 3b. After site exchange with the adsorbate, the final surface may be described more realistically as that shown in Fig. 3c.

A pure surface can be prepared by sputter cleaning or by breaking a bulk sample in the fracture stage of an ultrahigh vacuum chamber. For a sticking coefficient of unity, the pressure dependence of the coverage time of nitrogen is shown in Fig. 4. Consequently, for "clean" experiments taking a few minutes or hours, a vacuum in the range of 10^{-10} to 10^{-12} mb must be provided. This is a very difficult condition for Mössbauer spectral experiments which normally extend over hours and even days. The situation is even more difficult if only a partial coverage of a few percent of a monolayer can be tolerated. Therefore, conclusions drawn from Mössbauer spectra concerning a finite or vanishing Debye-Waller factor for free atoms on a surface, for instance the right atom in Fig. 3a, are not very reliable. Such atoms are, in the least, embedded within an adsorbate layer.

Table 2. The Work Function of Reconstructed Silicon Faces[a]

Initial face	Reconstructed	Φ, eV
Si {001}	2 × 1	4.85
Si {111}	7 × 7	4.6
Si {111}	2 × 1	4.83

[a]Data obtained from ref. 7.

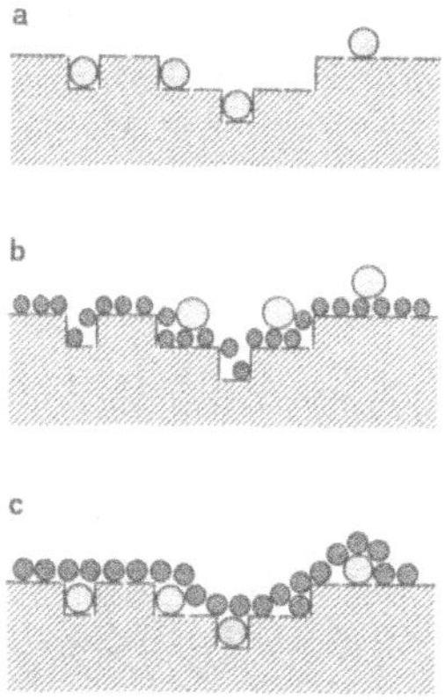

Figure 3. A simple, upper, and more realistic model, lower, for atoms located on a surface.

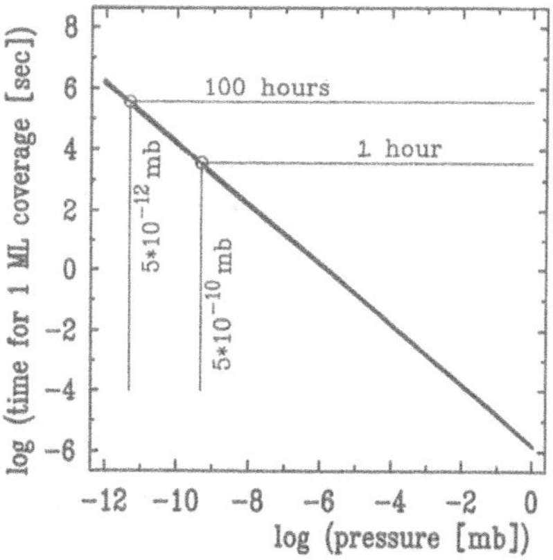

Figure 4. The pressure dependence of the coverage time of one monolayer of nitrogen.

2. SOME PARAMETERS OF EXPERIMENTAL INTEREST

2. 1. Exposure

A classical "high-vacuum" is understood to be a pressure of the order of 10^{-6} Torr. Therefore, the unit of gas exposure of a sample is defined as one Langmuir, L, the exposure of a sample to a gas or vapor at a pressure of 10^{-6} Torr or 1.33×10^{-6} mb for one second. For nitrogen and a sticking coefficient of unity, see Fig. 4, one Langmuir corresponds to the formation of one monolayer within one second, i.e., a coverage of approximately 100 percent is obtained within one second.

2. 2. Basic Principle of Surface Analysis

The majority of the methods applied to surface and thin film analysis operate according to the scheme shown in Fig. 5. In this approach, a primary beam is directed towards the sample to be analyzed. This beam may consist of particles or electromagnetic radiation. The particles can be neutral, positively, or negatively charged, e.g., atoms, ions, electrons, or positrons, which can be unpolarized or spin polarized, and the beam may be focused, scanned, or varied by its angle of incidence or particle energy. The secondary radiation, re-emitted, reflected, or scattered by the sample, is recorded by a particle or electromagnetic radiation detector, in which the intensity may be recorded versus mass, charge, potential,

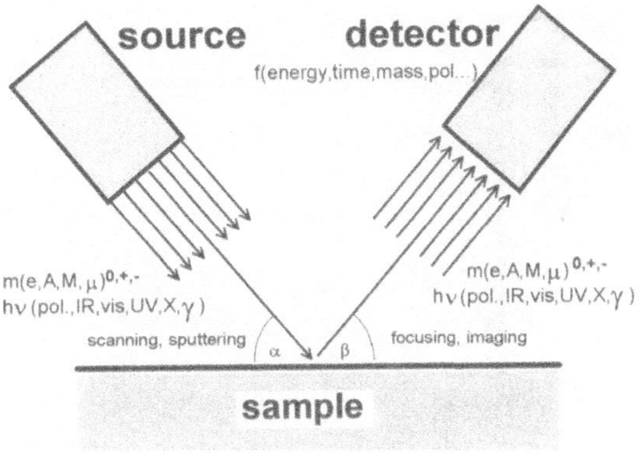

Figure 5. A generalized schematic for various methods of surface analysis.

frequency, energy, angle, or polarization, under non-focusing, focusing, or imaging conditions.

An inspection of Table 1 reveals that there are only a few methods which cannot be characterized easily by the scheme shown in Fig. 5, for instance STM, AFM, the Kelvin cell, and some electrochemical devices cannot be characterized by this scheme.

2. 3. Information Depth

Among the surface sensitive methods, there are characteristic differences in their depth sensitivity, differences which are determined by the penetration power of the primary radiation, e.g., as in SEM-EDX, or by that of the secondary radiation, e.g., as in XPS or CEMS. Regardless of whether the attenuation occurs via absorption, elastic collisions, or inelastic collisions, the depth attenuation usually follows the exponential law,

$$I = I_0 \exp(-x/\lambda), \qquad (1)$$

where $1/\lambda = N\sigma$, N is the number of absorbing or scattering centers per unit volume, and σ is the corresponding cross section.

For practical applications, the most important parameter is the "information depth," i.e., the maximum depth within a sample from which it is possible to obtain information by a respective method. This means that the information depth depends on the best signal to noise ratio detectable by a specific method under given experimental conditions, i.e., it depends on the detector resolution and the background signal as well as on the form, size, and composition of the sample. For example, if the smallest signal detectable in a spectrum is one percent of the most intense signal in the same spectrum, then the actual information depth is 4 to 5 × λ, where λ is the average value determined by taking into account the actual composition of the entire sampled surface layer. Fig. 6 shows a comparison of the typical information depth for several methods.

For electromagnetic radiation above 100 eV, the information depth is often assumed to increase with increasing energy or decreasing wavelength. For example, a larger information depth is expected for scattered γ-rays than for the softer secondary x-rays observed in conversion electron Mössbauer spectroscopy. Fig. 7 shows that this assumption may be misleading. The position of the absorption edge may result in a strong attenuation and, consequently, to a reduced information depth for γ-rays as compared with x-rays, as is illustrated in Fig. 7. For the 5 atomic percent content of iron in corundum, see the bottom of Fig. 7a, the attenuation difference is, as expected, large, whereas for the reduced iron content

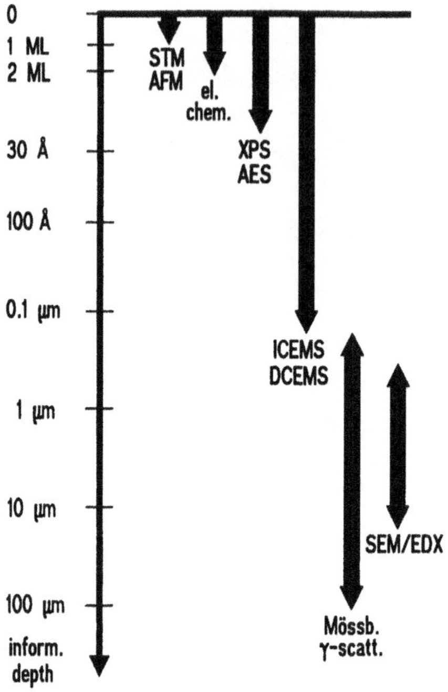

Figure 6. The typical information depth for several surface sensitive techniques.

in iron stearate it is smaller. The attenuation difference decreases even more for magnetite, $Fe_{50}Ni_{50}$, and α-iron, as is shown in the upper parts of Fig. 7a. Fig. 7b indicates that the situation is inverted in antimony(III) oxide, in which the attenuation of the γ-radiation is much stronger than that of the x-rays. The lower part of Fig. 7b shows the spectrum of the electromagnetic radiation re-emitted from both the iron-57 and the antimony-121 nuclei after Mössbauer excitation.

2. 4. Small Particles

Sometimes, surface sensitive methods are also used to investigate surface layers on small particles. However, this applicability is limited by the size of the particles relative to the information depth. If the thickness of the surface layer is d and the diameter of a spherical particle is D, then β, the ratio of the volume of the surface layer to the volume of the spherical particle is given by

$$\beta = 6\, d/D - 12\, d^2/D^2 + 8\, d^3/D^3. \quad (2)$$

In this equation, if D is much greater than d, then β is approximately 6d/D. The thickness of a monolayer, ML, can be estimated from the lattice constant and is less than or equal to c. For example, it should be noted that ML = c/2 for a bcc structure, see Fig. 8. If the surface contribution to a spectrum is 10 percent of the bulk contribution, β equals 0.1, and from the above equation D_{10} = 60d which is approximately 8.6 nm, when d is 0.143 nm. This is in the size range where most magnetically ordered iron alloys and compounds become superparamagnetic at room temperature. Consequently, a surface versus bulk composition study of superparamagnetic iron particles by conversion electron Mössbauer spectroscopy, whose information depth is approximately 250 nm, is not appropriate.

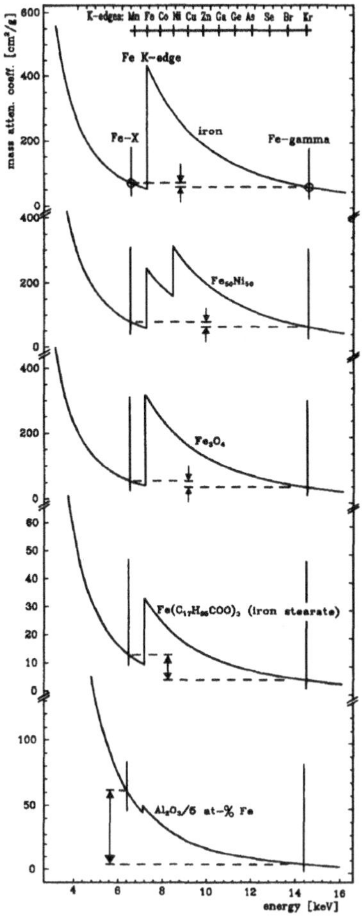

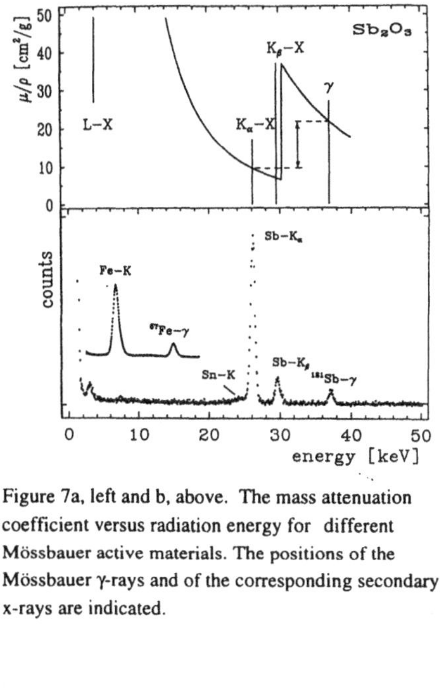

Figure 7a, left and b, above. The mass attenuation coefficient versus radiation energy for different Mössbauer active materials. The positions of the Mössbauer γ-rays and of the corresponding secondary x-rays are indicated.

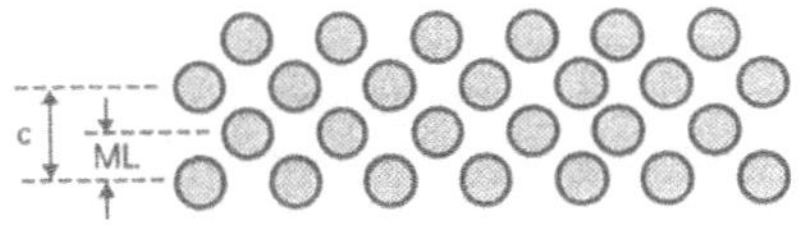

Figure 8. The definition of a monolayer in a bcc structure.

3. SOME SELECTED METHODS OF SURFACE ANALYSIS

In the previous section the general principles of the majority of the surface sensitive methods have been described. However, a few of these methods, which need to be discussed in more detail, will be covered in this section. The specific selection of these methods is more or less fortuitous except that they are often used in combination with Mössbauer spectroscopy. Only the basic principles are described, even though all these methods exhibit a broad variety of experimental techniques and options.

3. 1. Visual Inspection

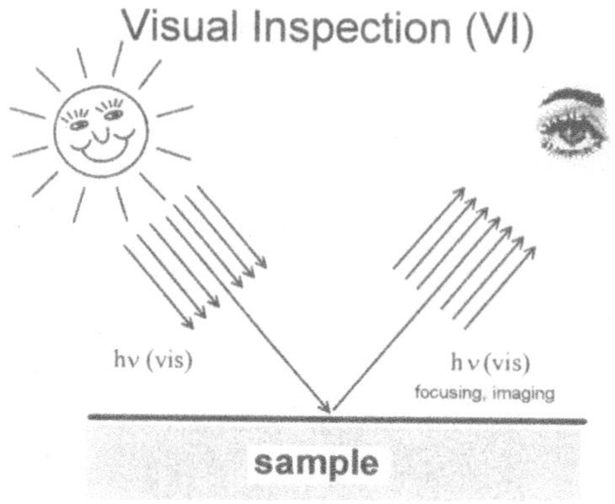

Figure 9. A schematic representation of visual inspection.

Of the senses by which a man perceives his surroundings, sight yields the highest information density. The faculty of vision, and also, somehow, the sense of touch concerns only surfaces. In nearly all analytical work, visual inspection, see Fig. 9, is the first investigative step, even though we often do so unconsciously. It is not necessary to discuss herein the excellent performance of sight with respect to imaging and energy resolution. The high surface sensitivity, however, should be pointed out. Visible light has a very small information depth in most materials and often can also show interference effects. Therefore, under favorable circumstances, such as for flat liquid surfaces, polished materials, or single crystals, even a monomolecular layer can be detected by visual inspection. An example is the observation of a single Langmuir-Blodgett film layer on a flat substrate.

3. 2. Auger Electron Spectroscopy

An electron beam focused onto a sample produces holes in the atomic shells by impact, holes which are subsequently filled by electrons from higher shells. The energy released by this process either creates electromagnetic radiation or, often with a rather high probability, causes the emission of an additional electron. The energy of this emitted Auger electron does not depend on the energy of the primary electrons of the exciting beam, but rather its energy is characteristic of the atom, especially when core electrons are involved. As a consequence, there are always three electrons involved in the Auger process. Thus, Fe KL_1L_2 describes an Auger electron emitted from the iron L_2 shell [2p] as the consequence of an initial hole in the iron K shell [1s], a hole which was filled by an electron from the iron L_1 shell [2s]. Because the resulting Auger lines appear on the very intense but smooth background of the inelastically scattered electrons, it is usual to record the first derivative instead of the direct intensity of the electron energy spectrum. Of course, an electron energy analyzer is the decisive part of an Auger electron spectrometer, and in Fig. 10 it is a constitutive part of the detector. It should be noted that the actual information depth of an Auger line depends on the energy of the respective electrons. In Fig. 6, a typical commercial Auger electron spectrometer for energies up to 2000 eV has been assumed. The energy of an AXY-Auger electron emitted from an atom of atomic number Z can be estimated with the approximation,

$$E_{AXY} = E_A(Z) - 0.5[E_X(Z) + E_X(Z+1)] - 0.5[E_Y(Z) + E_Y(Z+1)] - \Phi, \quad (3)$$

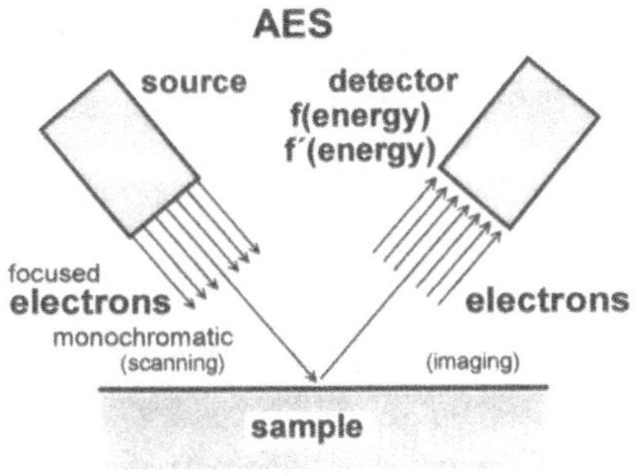

Figure 10. A schematic representation of Auger electron spectroscopy.

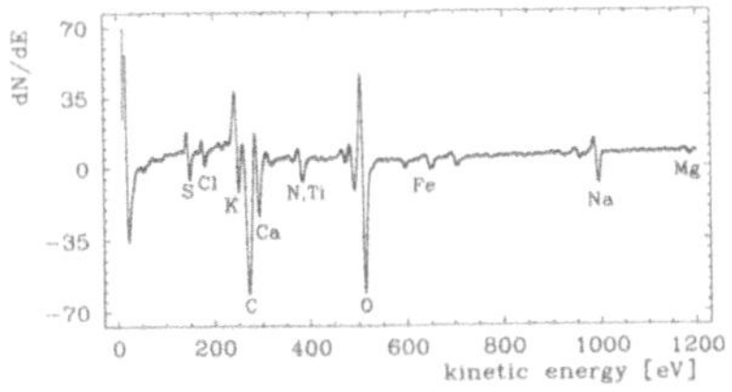

Figure 11. The Auger electron spectrum of a worn forging tool.

where E_i is the binding energy in the respective shell and Φ is the work function difference, the difference between the sample and the spectrometer. A typical Auger electron spectrum of a worn surface region of a TiN coated cold forging tool, a hardened alloy steel, tested under industrial load, is shown in Fig. 11.[8]

The normal Auger electron spectrum permits a local analysis of a surface spot on which the primary electron beam is focused. In contrast, the primary beam can be scanned over a distinct surface area in the technique known as scanning Auger microprobe analysis. If the electron energy analyzer is fixed to an Auger line characteristic of a special element, the lateral distribution of this element can be recorded synchronously by a monitor and an appropriate data memory. This option is called "element mapping." In the same way, the detection of all secondary electrons without energy selection provides useful information about surface morphology and is helpful in choosing the appropriate spots for the analyses of surface details.

3. 3. X-ray Induced Photoelectron Spectroscopy

Photoelectrons are produced by directing a more or less monochromatic x-ray beam onto a sample. The electrons escaping from the surface are selected with respect to their kinetic energy by means of an energy analyzer and are usually recorded as an intensity which is dependent on the binding energy, BE, which is given by

$$BE = h\nu - E_{kin} - \Phi, \tag{4}$$

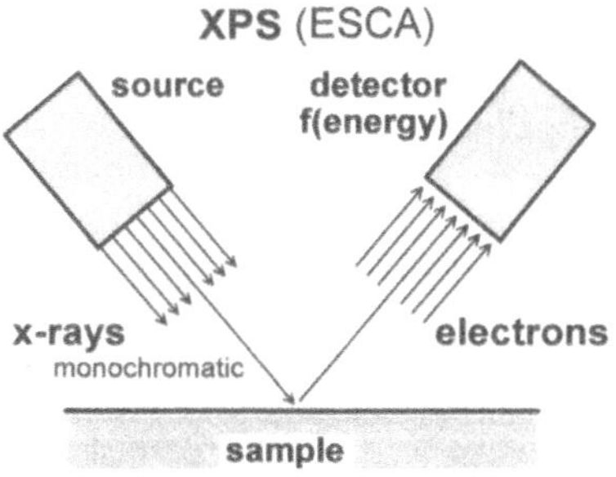

Figure 12. A schematic representation of x-ray induced photoelectron spectroscopy.

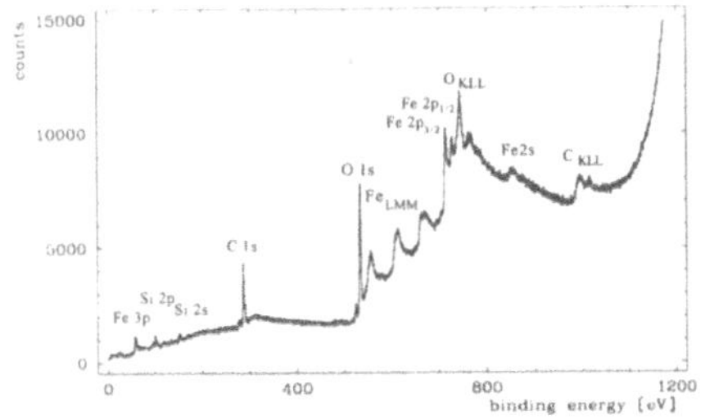

Figure 13. The x-ray induced photoelectron spectrum of two monolayers of ferric oxide deposited on a silicon wafer.

where hν is the energy of the primary x-rays, E_{kin} is the kinetic energy of the emitted electrons, and Φ is the work function. The information depth is determined by the inelastic mean free path, λ. Electrons which lost some energy through inelastic collisions on their way to the surface contribute to the background. Therefore, the background exhibits steps at the positions of x-ray induced photoelectron spectral lines. This has to be taken into account in quantitative evaluations. The principle of x-ray induced photoelectron spectroscopy is shown in Fig. 12, and an example of the photoelectron spectrum of an ultrathin layer, consisting of about two monolayers of iron(III) oxide deposited by a modified Langmuir-Blodgett technique onto the surface of a silicon wafer, is shown in Fig. 13. The fact that only very weak silicon 2s and 2p signals, originating from the substrate, are visible in the low binding energy region, i.e., at high kinetic energies, demonstrates the shallow information depth, and consequently the high surface selectivity provided by x-ray induced photoelectron spectroscopy. The lines visible in Fig. 13 may also contain a substructure due to different chemical bonds. The oxygen 1s x-ray induced photoelectron signal, recorded from a ternary optical glass, is shown in Fig. 14. The signal is decomposed into components belonging to two species of bridging and to one species of non-bridging oxygens present in the glassy network. The bridging oxygens interconnect network formers with each other, whereas the non-bridging oxygens connect the network modifiers with network formers.

3. 4. Extending the Information Depth

If the layers to be investigated are thicker than the normal information depth of the method in use, there are some optional techniques available to analyze the deeper lying layers or interfaces.

3. 4. 1. Sputtering Techniques

A simple but rough method for extending the information depth is to remove sequentially, slice by slice, parts of the thick layer. This can be done mechanically or, usually more precisely, by directing a powerful scanned ion beam onto the surface. Upon

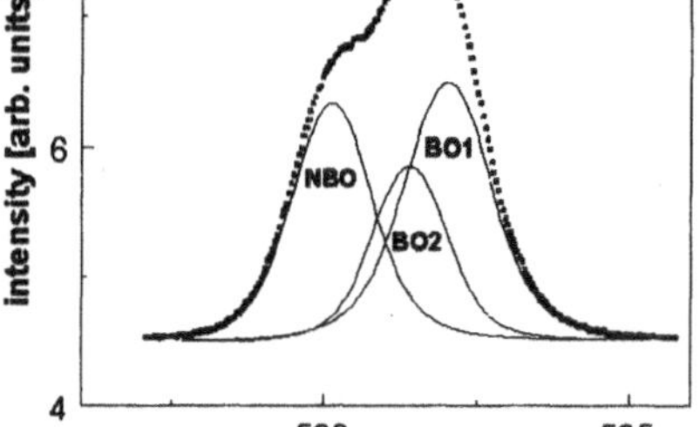

Figure 14. The x-ray induced photoelectron spectrum of the oxygen 1s signal of a glass sample showing the deconvolution into components due to bridging, BO, and non-bridging, NBO, oxygen sites.

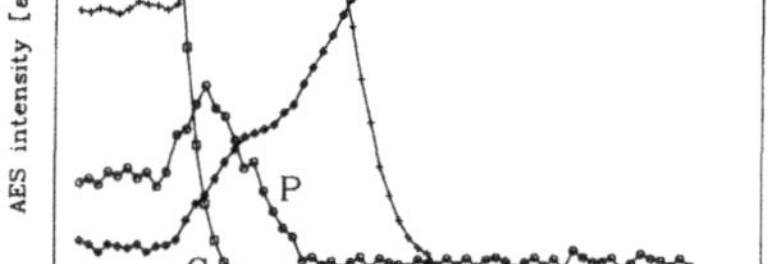

Figure 15. The Auger sputter depth profile of a steel surface passivated in a phosphate buffer solution.

impact atoms are removed from the surface, a procedure sometimes also called local evaporation or ion etching. All these methods, however, involve the risk of changing the surface states, surface structure, and surface morphology. During the etching process or intermittently, the actual surface can be analyzed by an appropriate method and, in this way, a depth profile can be obtained. An example of an Auger electron spectral depth profile, obtained by argon ion etching, is shown in Fig. 15. It was recorded on a steel sample electrochemically treated at an anodic potential, U_{NHE}, of 1.62 V for one hour in a phosphate buffer solution. It shows the outermost adsorbed layer containing carbon, followed by a phosphate layer, and, at the metal interface, by an oxidic layer.

3. 4. 2. Bremsstrahlung Induced Auger Electron Spectroscopy

The information depth depends on the initial energy of the respective particles. For example, in x-ray induced photoelectron spectroscopy the photoelectrons with higher kinetic energy, i.e., with lower binding energy, exhibit a larger inelastic mean free path. The electrons producing the x-rays by impinging onto the anode of the x-ray tube, however, should be much more energetic than the x-ray line in order to be able to excite the x-rays with a sufficiently high efficiency. For example, it is usual to utilize the K_α line of magnesium, with $h\nu$ = 1253.6 eV, and to excite it by electrons of up to 25 keV. Auger electrons, however, are produced not only by the narrow x-ray line, but also by the entire x-ray background of bremsstrahlung ranging up to the anode voltage of the x-ray tube. Therefore, the high energy bremsstrahlung induced Auger lines produced in this way can be used for analyses with much larger information depths, provided an appropriate energy analyzer is available. Furthermore, in this energy range, Auger lines of light elements, e.g., of adsorbed contaminants, do not exist and consequently do not overlap with the Auger signals of higher-Z elements. For example, the titanium LMM Auger lines at 387 and 418 eV change with the titanium oxidation state and can overlap with the 379 eV KLL line of adsorbed nitrogen. The upper plot in Fig. 16 shows the bremsstrahlung induced Auger electron spectrum in the range of 2 to 10 keV and the lower plot shows the first derivative of this spectrum, a spectrum which was recorded from a steel sample covered by a 25 nm layer of titanium. Whilst the LMM signals of iron are not visible in a "normal" Auger electron spectrum because of the insufficient information depth provided by 700 eV electrons, the KLL and KLM signals of iron and titanium are clearly observed in the energy range of ca. 4 to 6 keV, an energy which permits an information depth of ca. 50 nm. The signals are well resolved and not disturbed by contaminated atom signals. The argon in the sample originates from implantation during the sputter deposition of the titanium layer.[9] The shift and broadening of the iron signals are characteristic of the thickness of the titanium layer which the iron Auger electrons have to penetrate during their escape to the surface of the sample.

The lower part of Fig. 16 shows the first derivative of the bremsstrahlung induced Auger electron spectrum for comparison with the usual representation of a "normal" Auger spectrum. It demonstrates the resolution which can be obtained in such a spectrum. The Auger lines in the upper part of Fig. 16 show characteristic differences in their lineshapes. The titanium lines are relatively narrow, whereas the iron lines exhibit very broad wings on their low energy side. This is due to the different layer thicknesses. Because the titanium layer is thin, the titanium Auger electrons do not loose much of their initial energy on their way to the surface. The iron substrate, however, is "infinitely" thick and, consequently, there are electrons escaping from all depths, i.e., the energy losses are distributed over the entire energy range. This effect is of great importance not only for electron spectroscopies, such as x-ray induced photoelectron spectroscopy and Auger electron spectroscopy, but also in conversion electron Mössbauer spectroscopy. Therefore, another example, dealing with iron containing layers, will also be discussed.

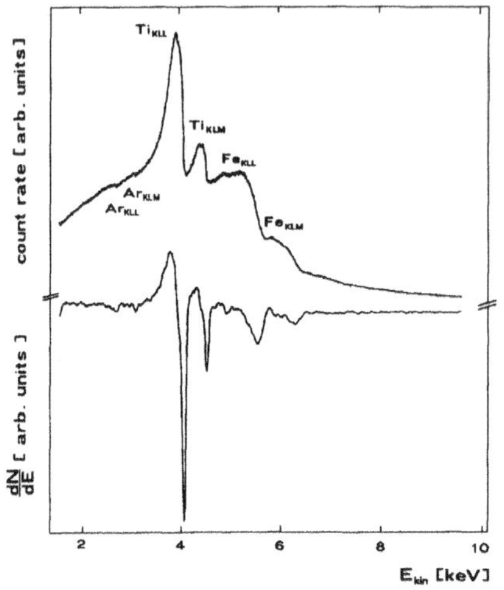

Figure 16. The bremsstrahlung induced Auger electron spectrum of steel coated with 25 nm of titanium, upper plot, and its first derivative, lower plot. Reproduced with permission from ref. 9.

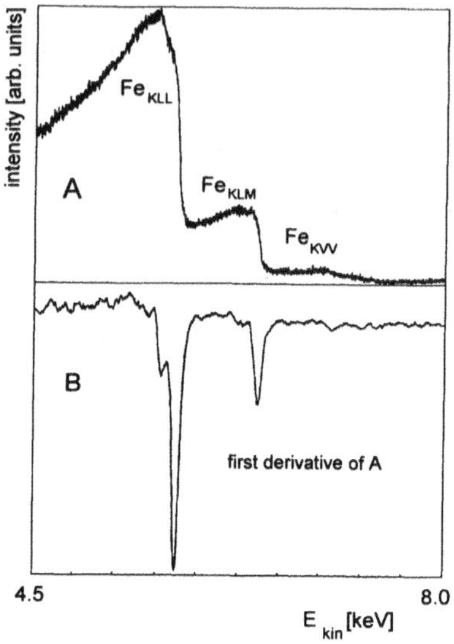

Figure 17. The bremsstrahlung induced Auger electron spectrum of "thick" pure iron, A, and the first derivative of the spectrum, B.

If an iron containing phase on the surface of a sample is thick, the energy distribution of the emitted electrons extends broadly on the low energy side of the energy axis. In contrast, a thin layer of iron permits narrow energy distributions. This is exemplified in Figs. 17 and 18. Curve A in Fig. 17 is a segment of the bremsstrahlung induced Auger electron spectrum of 0.1 mm thick 99.995 percent pure iron. The flat tails left of the Auger energies reflect the energy losses of the electrons created in deeper lying layers; on the high energy side the slopes are rather sharp. This accounts for the good energy resolution shown in curve B of Fig. 17, which is the smoothed first derivative of curve A. As a counterpart to Fig. 17, Fig. 18 shows the analogous spectrum of approximately one monolayer of iron oxide deposited on the polished surface of a silicon wafer. Curve A in Fig. 18 is the as-recorded spectrum induced by bremsstrahlung of up to 15 keV, and curve B is the same spectrum after removal of the background radiation, including the substrate contribution, and after expansion of the count scale. Spectrum B shows well resolved lines with a negligible broadening by energy losses and provides evidence that even details within the iron KLL lines are visible.

3. 5. Scanning Electron Microprobe with an Energy Dispersive X-ray Detector

The possibility of precisely focussing and deflecting electron beams may be utilized in scanning electron microprobe analysis. The secondary electrons produced by the scanning primary beam and emitted from the surface permit the production of an amplified picture of the surface morphology. An example is shown in Fig. 19. The lower picture shows the horizontal boundary between an iron oxide layer, which is less than a monolayer thick, deposited by a Langmuir-Blodgett technique, and the blank silicon substrate in the upper

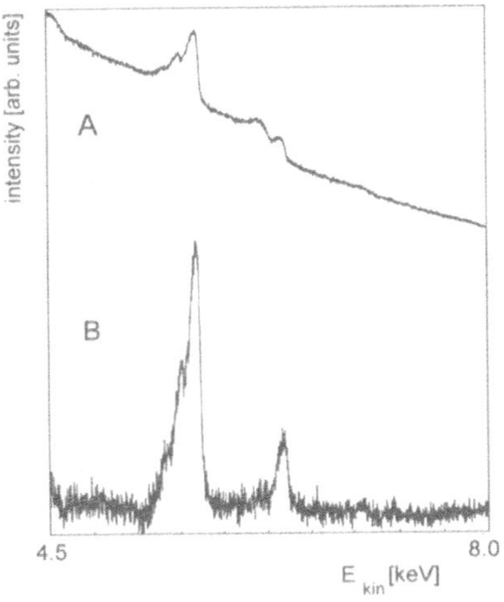

Figure 18. The bremsstrahlung induced Auger electron spectrum of one monolayer of iron oxide on a silicon wafer, A, and the spectrum after background correction, B.

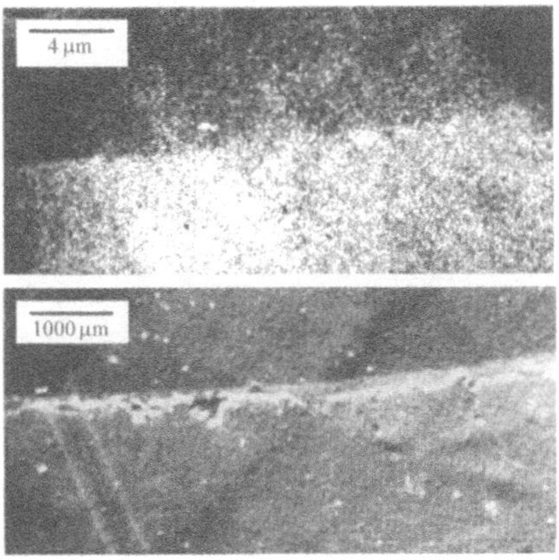

Figure 19. The scanning electron microprobe picture of an ultrathin layer of iron oxide on a silicon wafer at two different magnifications.

part of the lower picture. The shadowed stripes are due to Kikushi type interference. The upper picture shows a small section of this boundary under high magnification. Although the boundary is rather ragged on the "macroscopic" scale of the lower picture, on the microscopic scale the layer appears to be rather smooth, with a straight and well-defined boundary.

The lateral resolution of a scanning electron microprobe can be better than 100 nm. The primary electrons, with energies typically in the range of 10 to 60 keV, however, also induce x-ray fluorescence. With an energy dispersive silicon or germanium x-ray detector a distinction between different elements is possible. In this way, an elemental analysis of precisely located surface spots can be performed, or, if the electron beam is scanned over a

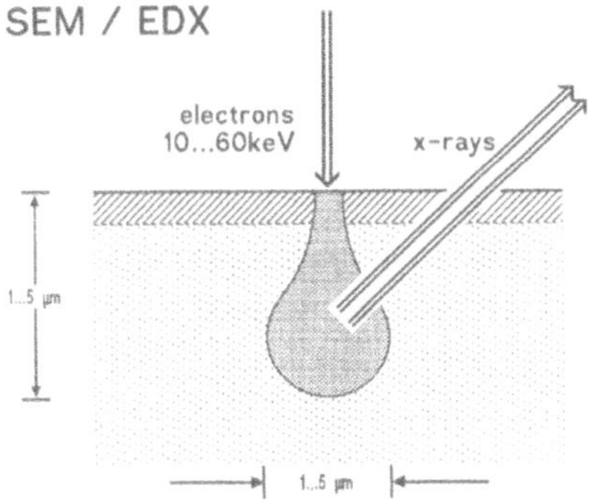

Figure 20. The bubble-like volume sampled in a scanning electron microprobe with energy dispersive electron analysis.

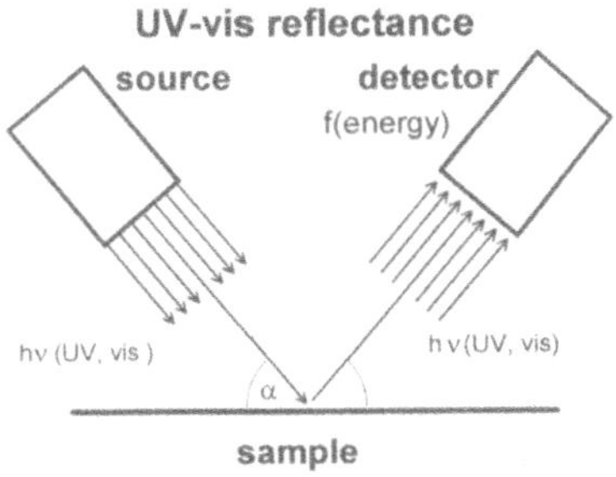

Figure 21. A schematic representation of UV-visible reflectance measurements.

well defined region, the lateral distribution of selected elements, the so called "element map" can be obtained. The local resolution, however, is reduced by the formation of an electron collision cloud which is produced when the primary electrons are scattered by collisions with the components of the sample. This scattering broadens the sampled surface area and this broadening is even stronger below the surface. Fig. 20 shows schematically the shape of the sampled volume. Because the x-rays produced are practically unattenuated by the thin sampled layer, the information depth is given by the energy of the primary beam. The scattering of the electrons, however, limits the lateral resolution of element mapping to the micrometer region, as is shown in Fig. 20.

3. 6. UV-Visible Reflectance

The reflection of an electromagnetic wave is determined by the complex refractive index of the reflecting material. Therefore, reflection provides information about the physical properties of a surface. This is, of course, the basis for the visual inspection mentioned in Section 3. 1. In optics, the reflectance is the ratio of the intensity of the reflected radiation to the intensity of the incident radiation. If absorption can be neglected, the refractive index is real and reflectance measurements can be performed easily, as is shown in Fig. 21. If the reflectance of a multilayer system is recorded versus wavelength, both the refractive indices and the thicknesses can be determined for the respective layers.[10] In particular, this method is advantageous for the characterization of layers on optical glasses, layers which can be formed by coating, evaporation, leaching, or corrosion. Reflectance studies are a useful supplementary method to other surface techniques and yield the "macroscopic" parameters of thickness and refractive index, n_i. The reflectance information depth extends from ca. 20 nm to micrometers.

Fig. 22 shows a typical series of reflectance measurements, recorded from the surface of a polished commercial lead glass as a function of the leaching time in an acid solution. The determination of the details of changes in glass composition, for instance, the determination

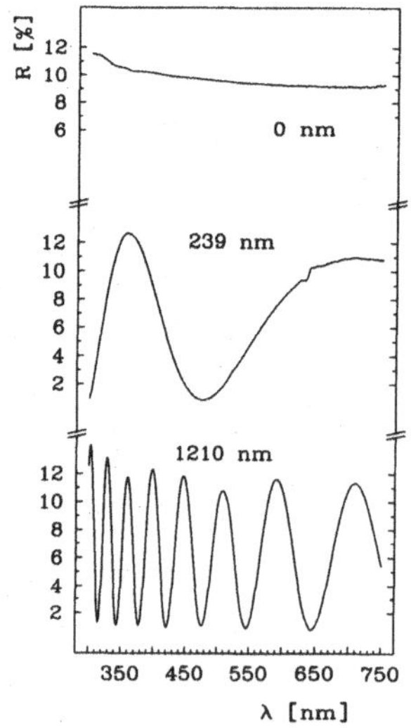

Figure 22. Reflectance measurements on leached glass surfaces.

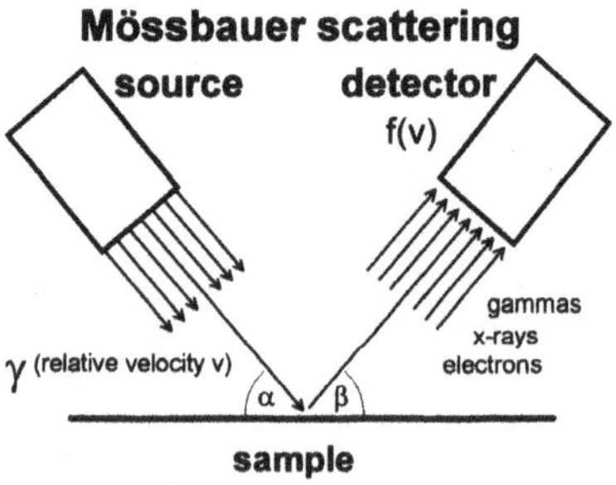

Figure 23. A schematic representation of Mössbauer scattering experiments.

of the small concentration of network modifying cations in the surface layers after leaching, requires the application of other surface techniques, techniques which may exhibit quite a different information depth.

3. 7. *Surface Sensitive Mössbauer Spectroscopy*

The classical and most often used form of Mössbauer spectroscopy is absorption spectroscopy, i.e., a technique in which a spectrum is recorded as the transmitted intensity of the Mössbauer γ-rays as a function of the relative velocity, v, between the radiation source and the absorber. The velocity, v, is an energy shift parameter resulting from the Doppler effect. The "Mössbauer effect" is a resonant excitation of a nuclear state, which after its mean lifetime, τ, relaxes and emits its excitation energy via several different kinds of radiative processes involving γ-rays, x-rays, and/or electrons. These secondary radiative processes can be utilized for surface investigations, in which the information depth depends on the nature of the recorded radiation, see Section 2. 3. The geometry used for such Mössbauer surface studies is shown in Fig. 23. The nuclear relaxation occurs either by re-emission of a γ-ray or by internal conversion. The internal conversion results in a cascade of different radiations and, therefore, the total number of emitted electromagnetic quanta and electrons is larger than the number of the initially excited nuclei. Table 3 gives the emission probability[11-14] for the different components of the secondary radiation for three different Mössbauer isotopes. The relative percentage of each radiation given in this table is obtained by assuming that the number of initially excited nuclear states is one hundred percent.

Table 3. Energy and Relative Intensity of Re-emitted Radiations after Population of the Metastable Nuclear Mössbauer Levels of Three Mössbauer Isotopes.

	Iron-57		Tin-119		Antimony-121	
Radiation	E, keV	%	E, keV	%	E, keV	%
Re-emitted γ-rays	14.41	9	23.87	16.56	37.15	8.2
K-conversion electrons	7.30	81	-	0	6.66	78.9
L_1-conversion electrons	13.56	9	19.41	65.7	32.45	9.5
L_2-conversion electrons	13.69		19.71	2.7	32.77	0.9
L_3-conversion electrons	13.70		19.94	0.7	33.01	0.4
M_1-conversion electrons	14.32	1	22.99	12.9	36.21	1.9
M_2-conversion electrons	14.36		23.11	1.3	36.34	0.3
M_3-conversion electrons	14.36		23.16	0.3	36.38	0.1
$K_{\alpha 1,2}$-x-rays	6.40	21	25.27	0	26.3	56.2
$K_{\beta 1,2}$-x-rays	7.06	3	28.5	0	29.7	12.4
KLL-Auger electrons	5.4	44	20-21	0	21-22	10.3
KLM-Auger electrons	6.2	13	24-25	0	25-26	
L-x-rays	0.7	3	3.4	2.6	3.8	3.7
LMM-Auger electrons[a]	0.7	33	2.6-3.4	66.6	2.8-3.6	92.2

[a]Including the Coster-Kronig transitions.

3. 7. 1. Mössbauer γ–ray Scattering

The relatively high penetrating power of the typical Mössbauer γ-rays requires rather thick layers for investigation. A typical application is the analysis of the phase composition of the corrosion products on a rusty material, an analysis which provides insight into the degradation history of the corroded material. An analysis of corrosion layers by γ–ray scattering is shown in Fig. 24. A rusty steel sample was cooled to 80 K in order to identify the corrosion products from its low temperature Mössbauer spectrum. The sample was embedded in styrofoam insulation and the radiation had to penetrate the styrofoam walls both before and after scattering at the sample. The spectrum shows that the main corrosion products on the steel are α-FeOOH and γ-FeOOH, and that no significant amount of β-FeOOH is present. A similar analysis by x-ray diffraction did not give a useful result because of the high content of small particles which appear to be amorphous with respect to x-ray diffraction. The low-temperature measurement distinguishes decisively between the different FeOOH modifications. This example demonstrates that, when the radiation has to penetrate layers of some thickness in order to reach the sample and/or the detector, γ-ray scattering may be utilized advantageously.

Another application of Mössbauer γ–ray scattering is the *in-situ* analysis of layers existing at a well-defined electrochemical potential in an electrolyte, a situation in which transmission experiments are not possible. The experimental arrangement for *in situ* studies is shown schematically in Fig. 25. This experimental arrangement can be used for the investigation of the electrolytic deposition of coatings. For surface passivation studies, however, it is often not advisable because of the thinness of the layers formed. In a similar fashion, γ-ray scattering is also useful to study processes occuring below a coating, for example, rusting below a corrosion protection coating.

It should be mentioned that, in iron-57 Mössbauer γ-ray scattering experiments, a very effective radiation shielding between the source and the detector is essential. As shown in Fig. 25, 50 mm of lead is typically required. During the radioactive decay of cobalt-57, there

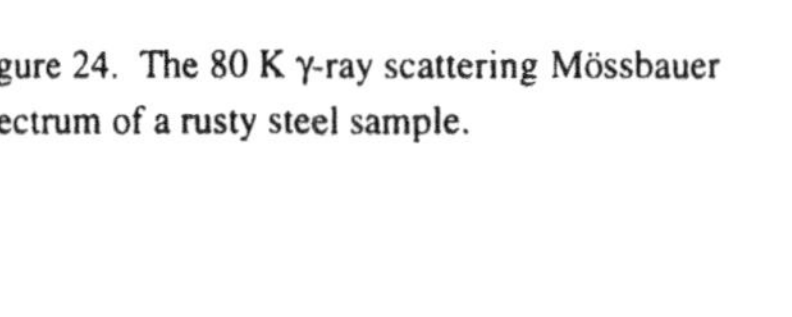

Figure 24. The 80 K γ-ray scattering Mössbauer spectrum of a rusty steel sample.

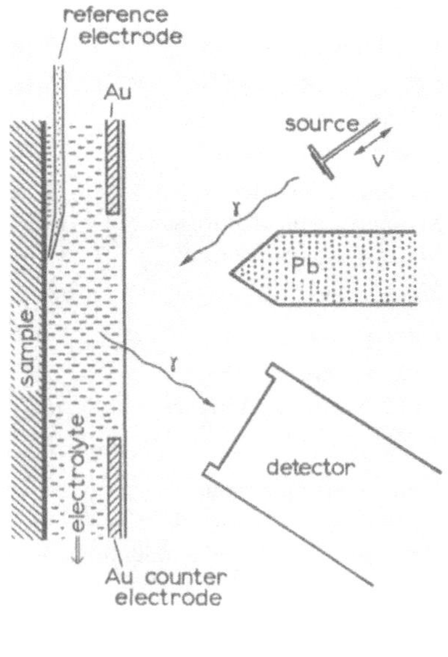

Figure 25. A schematic of an *in situ* γ-ray Mössbauer scattering experiment in an electrochemical cell.

is a γ-ray emitted in the MeV energy range with a relative intensity of about 0.16 percent. This low intensity γ-ray does not play any role in transmission or conversion electron Mössbauer spectroscopy, but in γ-ray scattering experiments it competes with the low intensity of the resonantly scattered 14.4 keV γ-rays.

It is worthwhile mentioning that Mössbauer γ-ray scattering is a valuable method for studying processes occuring on the timescale of the lifetime of the resonantly excited metastable Mössbauer nucleus. In these cases, selective excitation,[15,16] is applied by using a constant relative velocity corresponding to only one component of a split spectrum. This technique can be used to ascertain the details of relaxation processes,[16,17] or to perform Mössbauer-NMR double resonance, or MONDOR, experiments,[18] or perturbed angular correlation of Mössbauer scattering, or PADMOS, experiments.[19]

3. 7. 2. Grazing Incidence Mössbauer Spectroscopy

The surface sensitivity of Mössbauer γ-ray scattering can be improved for thin layers if the angle of the incident radiation, α, see Fig. 23, is very small and, at most, a few milliradians.[20,21] Below a critical angle of incidence, total external reflection occurs. Under these circumstances, the shape of the Mössbauer pattern differs from the usual shape and a more complex fitting procedure is advisable.[22] The major difficulties with such experiments are the required flatness, over a large area, of the sample and the appropriate collimation of the incident beam. Even when Soller slits are used for collimation, the γ-ray intensity reduction by collimation leads to rather long counting times and/or requires an isotopic enrichment of the Mössbauer nuclide in the sample.

3. 7. 3. Mössbauer Surface Studies with Secondary X-rays

In addition to the re-emitted Mössbauer γ-rays, the secondary x-rays can be used for Mössbauer surface studies. Their usefulness, however, is restricted for two reasons. First, there is always an intense background of the same x-rays excited by non-resonant processes, a background produced by the intense high-energy components of the γ-ray radiation via photoexcitation processes. Second, in most cases, the information depth of the x-rays is not significantly smaller than that of the resonantly scattered γ-rays, as has already been discussed in Section 2. 3. Thus, for intensity reasons, a simultaneous detection of both the γ-rays and the x-rays is often advisable.

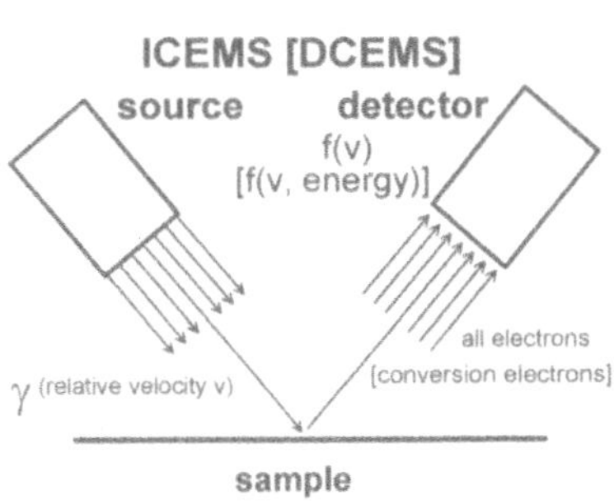

Figure 26. A schematic representation of conversion electron Mössbauer spectroscopy.

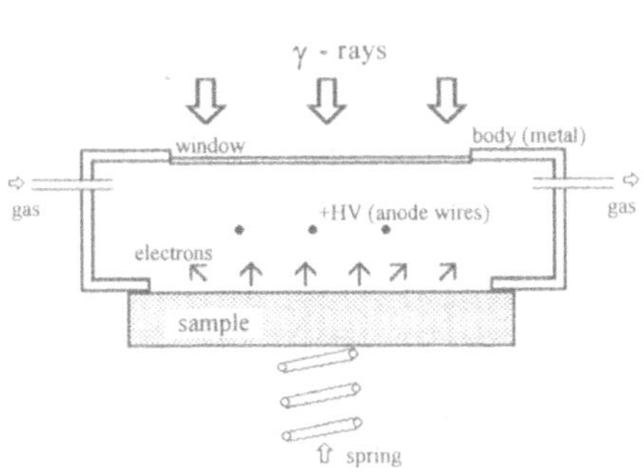

Figure 27. The operating principles of a gas-flow integral conversion electron Mössbauer spectroscopy detector.

3. 7. 4. Integral Conversion Electron Mössbauer Spectroscopy

Conversion electron Mössbauer spectroscopy, CEMS, a schematic of which is shown in Fig. 26, is the most useful technique for surface studies by Mössbauer spectroscopy. Within a decade after the discovery of the Mössbauer effect, it was realized that conversion electron Mössbauer spectroscopy was an important surface sensitive technique. For the majority of the Mössbauer nuclides, the conversion coefficient, α_{tot}, is of the order of ten and, thus, the re-emitted conversion electrons provide a higher sensitivity for the detection of resonant scattering events than does detection of the re-emitted γ-rays or x-rays. In addition, the information depth, which, for example, is about 300 nm for the K-conversion electron in iron, is very appropriate for surface analyses.

Bonchev et al.[23,24] were the first to study in detail the surface sensitivity of conversion electron Mössbauer spectroscopy. Although they performed their experiments with tin-119, later work focussed on iron-57 studies.[25,26] Indeed, by the late 1970s, conversion electron Mössbauer spectroscopy was widespread and commonly used as a method for surface and thin-film analysis. Currently, the majority of the conversion electron Mössbauer spectral investigations are performed with iron-57, followed by tin-119, antimony-121, some of the rare earth elements, and a very few other elements.

Because the mean path of conversion electrons in gases is of the order of a few millimeters, there are only two possible means of detection. Either the electrons are detected within this millimeter range, or the sample under investigation is mounted in a vacuum chamber. The latter option involves the risk of changing the sample surface through the release of some of the surface constituents, such as water. The former option requires the use of a "counting gas" in an ionization chamber, an avalanche detector, or a proportional counter, a gas which may also affect the surface of the sample. *In vacuo*, channeltrons and scintillators[27] are appropriate detectors. Unfortunately, although channel plates exhibit a large effective area, their higher background count rate limits their usefulness. Detection *in vacuo* by channeltrons or similar detectors does not result in the formation of a charge avalanche by gas multiplication and, therefore, no significant charging of an insulating sample occurs. Charging effects in gas-flow detectors, however, only very rarely lead to serious problems.[28] A typical gas-flow detector is shown in Fig. 27. Its efficiency is high if a mixture of helium, with 3 to 5 percent of methane quench gas, and a sample to anode distance of 1 to 3 mm are used.[29] Gas-flow counters with argon exhibit a non-negligible sensitivity for the primary γ-ray radiation, resulting in an enhanced background count rate. With the detectors described herein, all electrons re-emitted from the sample, independent of their escape energy above a certain threshold energy, are detected. Thus, this technique is called integral conversion electron Mössbauer spectroscopy, or ICEMS.

By using a detector of the type shown in Fig. 27, the integral conversion electron Mössbauer spectrum of a phosphate optical glass, containing 1.4 atomic percent of natural abundance iron, was measured and is shown in Fig. 28. This spectrum demonstrates both the applicability of such detectors for insulating samples as well as their high sensitivity. For the latter it should be noted that leaching had reduced the amount of iron at the surface of the glass.[30] A comparison with a transmission spectrum, see Fig. 28 right, clearly indicates an altered iron(II) to iron(III) ratio in the surface layer.

3. 7. 5. Depth Selective Conversion Electron Mössbauer Spectroscopy

A conversion electron is emitted from an atom with a well defined initial energy. However, on its way to the surface some of its kinetic energy may be lost through inelastic collisions. The extent of this energy loss depends on the escape distance, the escape direction, and the average composition of the material. Therefore, if only conversion electrons within a narrow energy window, a window set at some specific value below the initial energy, are detected, the resulting Mössbauer spectrum preferentially samples atoms

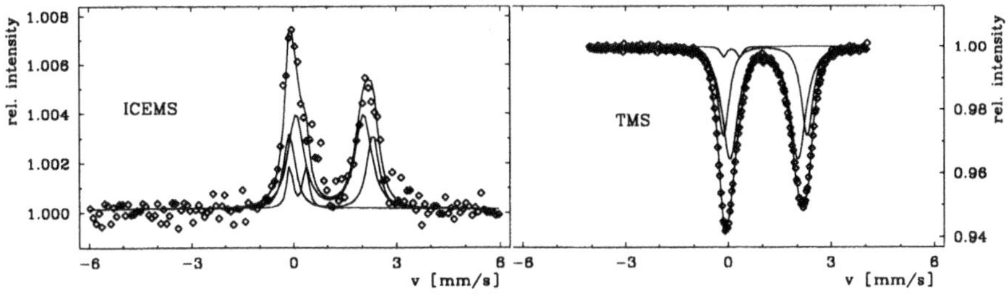

Figure 28. Integral conversion electron Mössbauer spectrum, left, and the transmission Mössbauer spectrum, right, of a phosphate glass containing 1.4 atomic percent iron. Reproduced with permission from ref. 30.

located in a layer at a specific distance below the surface. This technique is known as depth selective conversion electron Mössbauer spectroscopy, or DCEMS. Consequently, under certain conditions, a Mössbauer spectroscopic phase analysis can be obtained from different layers in a material without destruction of the original surface. The depth limit is given by the information depth of the conversion electrons detected, which for iron-57 is 250 to 300 nm in a medium equivalent to α-iron. The depth resolution depends on several factors and is, as a rule, in the range of 10 to 50 nm. For depth selective conversion electron Mössbauer spectral studies, an efficient energy selector with a high brightness or wide field is necessary. The energy resolution should be better than three percent.

Two depth selective conversion electron Mössbauer spectra of a thin layer of α-iron evaporated onto the surface of austenitic steel are shown in Fig. 29. The spectra consist of a singlet from the austenitic steel, a sextet from α-iron, and iron(II) and iron(III) doublets. As Fig. 29 reveals, the relative intensities of the sextet and the iron(II) doublet are reduced in the spectrum corresponding to an energy loss of 250 eV. The most reliable interpretation of the intensities of the components in the two spectra indicates that the thin layer of α-iron is covered by an iron(II) oxide layer, whereas the iron(III) is located between the steel substrate and the evaporated α-iron layer. Thus, the iron(III) belongs to an oxide layer which was formed on the substrate before the evaporation of the α-iron.

The major problem in depth selective conversion electron Mössbauer spectral studies is intensity. The use of an energy analyzer means that only a very small portion of the electrons which escape from the surface are recorded to form a Mössbauer spectrum. Further, for a

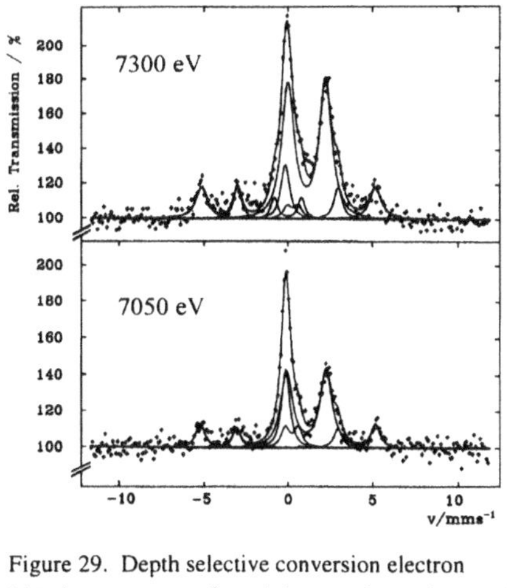

Figure 29. Depth selective conversion electron Mössbauer spectra of a stainless steel sample coated with a thin layer of α-iron made with enriched iron-57.

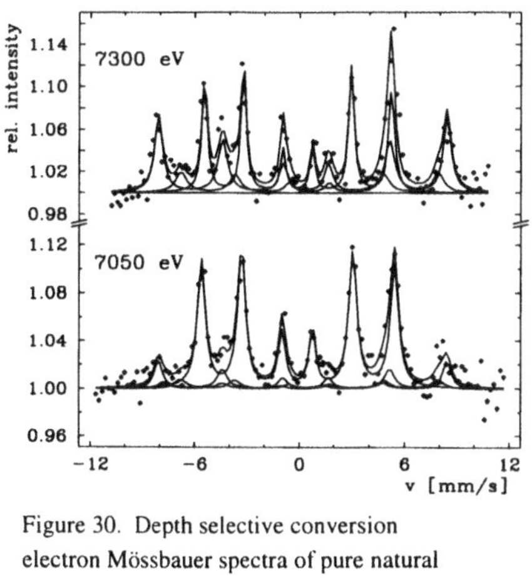

Figure 30. Depth selective conversion electron Mössbauer spectra of pure natural abundance iron oxidized in air at 270°C.

complete depth profile analysis, a series of two or more spectra is generally required. Therefore, if the different phases in the surface layers contain unenriched iron, the study of a single sample may take days or even weeks. Of course, stronger Mössbauer sources, sometimes above 200 mCi, shorten the duration of an experiment, but such strong sources are expensive. However, the assumption that depth selective conversion electron Mössbauer spectroscopy is a fruitful method only if layers enriched in the Mössbauer isotope are analyzed, is incorrect.

Fig. 30 demonstrates that depth selective conversion electron Mössbauer spectroscopy is often useful even for non-sophisticated studies. A sample of 99.995 percent pure natural abundance iron was heated in air at 270°C for 20 min. The upper spectrum in Fig. 30 was obtained with a 2.6 percent resolution energy selector set at 7300 eV, i.e., at the initial energy of the iron-57 K-conversion electrons. In the lower spectrum shown in Fig. 30, only electrons which have lost 250 eV of their initial energy during escape were detected. A simple interpretation of the results is easy. The spectrum shows two sextets characteristic of magnetite with a considerable deficit in iron, i.e., $Fe_{3-x}O_4$ with $0 < x < 0.33$. There is also a well defined iron(II) doublet which, because of its large quadrupole splitting, is characteristic of superparamagnetic magnetite rather than of wüstite, FeO. The lower spectrum in Fig. 30 shows the same oxidation products, but shows a much higher relative intensity for the sextet belonging to α-iron. Therefore, the surface layer thickness should be, at most, of the order of 20 nm. A more precise determination of the layer thickness requires a consideration of the absolute subspectra intensities in both spectra, normalized to identical conditions during measurement, see below. The depth selective conversion electron Mössbauer spectra shown in Fig. 30 were measured in order to check for the formation of FeO during the oxidation process. From the results it is evident that no significant amount of FeO is formed between the outer oxide layer and the bulk α-iron.

3. 7. 6. Quantitative Conversion Electron Mössbauer Spectroscopy

Although the qualitative determination of the components of a surface layer from a conversion electron Mössbauer spectrum is similar to that of transmission Mössbauer spectroscopy, and therefore is a standard technique, the quantitative determination of the thickness of layers and sublayers requires some additional caution. The electron flux escaping from a sample surface depends on the attenuating effect of the material and the thickness of the phases penetrated by the electrons during their escape to the surface. Fortunately, the influence on the resonantly attenuated intensity of the primary γ-ray is often negligible. The attenuation processes are of a stochastic nature, and a calculation of the attenuation is a rather complex problem. A solution to this difficult problem has been attempted by several authors using Monte-Carlo calculations. Without describing the details of the calculations, two approaches will be mentioned.

In order to calculate layer thicknesses from integral conversion electron Mössbauer spectra, a method which has been developed by Salvat et al.[31,32] requires a knowledge of the line positions and the relative intensities of all the components constituting the phases of the different layers under study. It is supposed that the layers are laterally homogeneous and that their sequence is known. In addition, all densities, stoichiometric compositions, and f-factors are required. Similarly, Liljequist and Ismail[33] have developed a method to calculate layer thicknesses from depth selective conversion electron Mössbauer spectra. A specific number of such spectra, the number depending on the number of unknown parameters, must be recorded, each with different energy windows and otherwise under identical experimental conditions. From all these spectra, the *absolute* intensities of the subspectra belonging to the different phases, have to be derived. The layer sequence can be determined only by trial and error, but the layers need not be laterally homogeneous. In both cases the method allows for inert adsorbed layers, layers without Mössbauer active nuclei, and for an infinitely thick active or inert backing material.

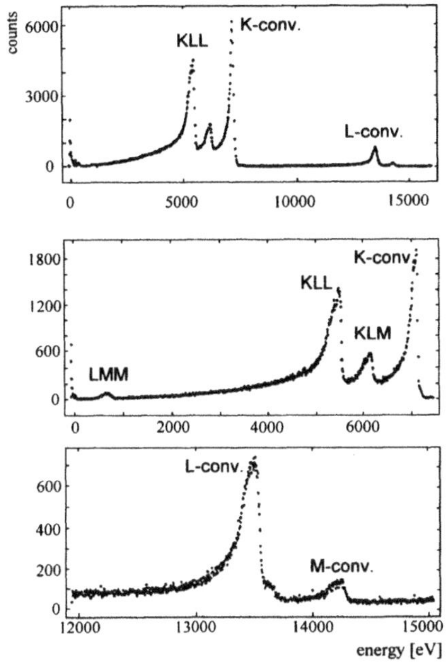

Figure 31, left. The energy spectrum of the electrons emitted by the isomeric iron-57 nuclear excited state, top, and two expanded regions of interest, center, and bottom. Reproduced with permission from ref. 39.

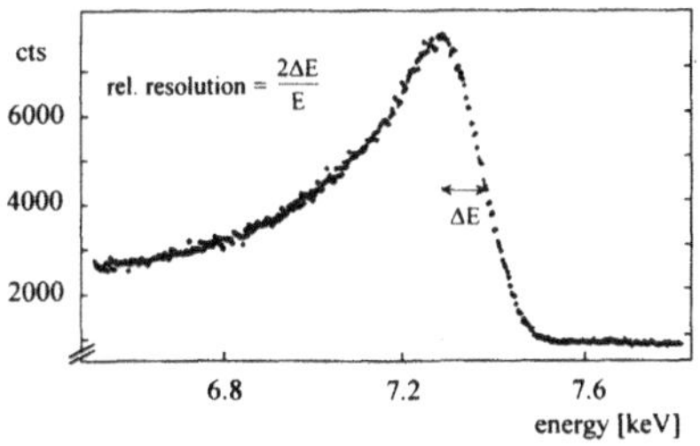

Figure 32. The edge of the iron-57 K-conversion electron spectrum which may be used for energy analyzer calibration.

3. 7. 7. Some Peculiarities of Conversion Electron Mössbauer Spectroscopy

Occasionally, in some papers and presentations, misinterpretation or over-interpretation of depth selective or integral conversion electron Mössbauer spectral data has occured. Although this does not occur very often, some comments seem to be appropriate to prevent this technique from gaining a poor reputation.

The initial energy of the K-conversion electrons is very well defined, and their energy distribution is narrower than the resolution of the energy selectors used in depth selective conversion electron Mössbauer spectroscopy. In contrast, the successive Auger processes are of higher complexity and yield groups of poorly resolved, broadened lines. In general, the electron emitting atoms are not located in the outermost monolayer of a surface, see for example Fig. 3, and, therefore, the electrons loose some energy during escape to the surface. In the energy distribution spectrum, this leads to a tail at the low-energy side of the initial electron energy. This holds true for both conversion electrons and Auger electrons.

The energy spectrum of the electrons emitted from a small amount of a cobalt-57 solution sprayed onto the surface of a gold support is shown in Fig. 31. In spite of the very carefully prepared lateral distribution of the cobalt-57, the observed tails show evidence of some non-vanishing in-depth energy losses. The right half of the high-energy edge of the K-conversion peak, shown in detail in Fig. 32, can be used to calibrate the energy axis and to determine the energy resolution of the energy selector. Both parameters are necessary for the correct evaluation of depth selective conversion electron Mössbauer spectra.

It is much more difficult to record an electron energy spectrum, such as that shown in Fig. 31, if the radioactive material is available only with a very low specific activity. Then practically all electrons have to penetrate on average a non-negligible thickness of the matrix material, and only a rather small fraction of all electrons reach the surface. This generally occurs for tin-121, the parent nuclide in antimony-121 Mössbauer effect sources. Therefore, the electron energy spectrum covering the K-conversion and the LXY-Auger regions, as shown in Fig. 33, could be measured only with poor statistics.

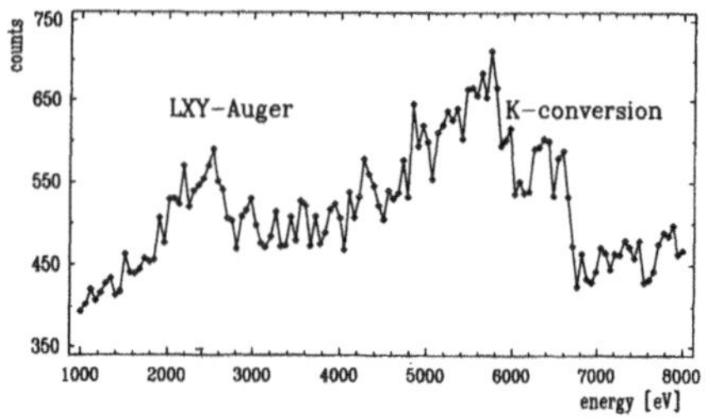

Figure 33. The energy spectrum of the Auger and K-conversion electrons emitted by the isomeric antimony-121 nuclear excited state. Reproduced with permission from ref. 41.

From Fig. 31, an important conclusion is evident. For a reliable thickness determination from a depth selective conversion electron Mössbauer spectrum which has used the K-conversion electrons of iron-57, only the energy range between the K-conversion electron energy and the KLM-Auger electron energy is available, i.e., a range from 6.4 to 7.3 keV. This narrows the useful range of energy losses to 900 eV. If larger losses from thick layers, are observed, the complex influence of the KLM and KLL-Auger electrons have to be considered. This makes the layer thickness determinations discussed above much more complex and less reliable.

The actual electron energy spectrum of a sample investigated by conversion electron Mössbauer spectroscopy is quite different from the spectrum shown in Fig. 31. The energy spectrum recorded from "infinitely" thick α-iron is shown in Fig. 34. It demonstrates that the low energy tails extend continuously over several keV with a strong overlap of the conversion and Auger electrons. The middle spectrum in Fig. 34 shows the energy distribution in the absence of any Mössbauer effect resonances, whereas the upper spectrum was recorded with one of the outermost Mössbauer sextet lines *on-resonance*. The lower curve in Fig. 34, which is the difference between the *on-resonance* and *off-resonance* spectra, exhibits a maximum below 7 keV, whereas the other curves exhibit a maximum close to 7.3 keV. This demonstrates that the average contribution of the Mössbauer spectral line intensities originates from iron atoms located below the surface and that the resonance is extended over the entire energy range shown in Fig. 34. Further, as expected, the Auger electrons contribute to the Mössbauer signal, which leads to another point of interest.

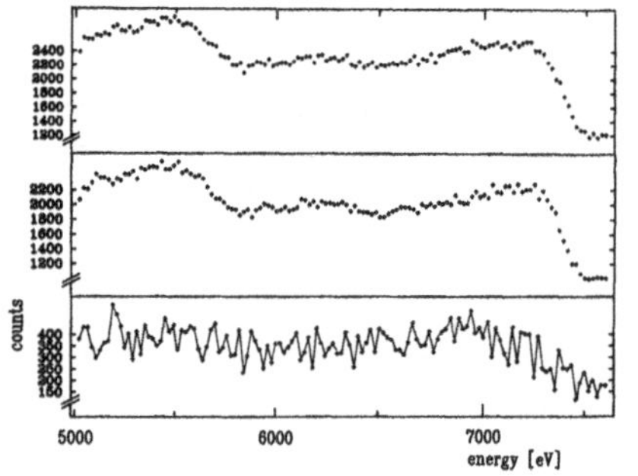

Figure 34. The electron energy spectrum of a natural iron sample measured by iron-57 conversion electron Mössbauer spectroscopy on resonance, top, off-resonance, center, and the difference, bottom. Reproduced with permission from ref. 39.

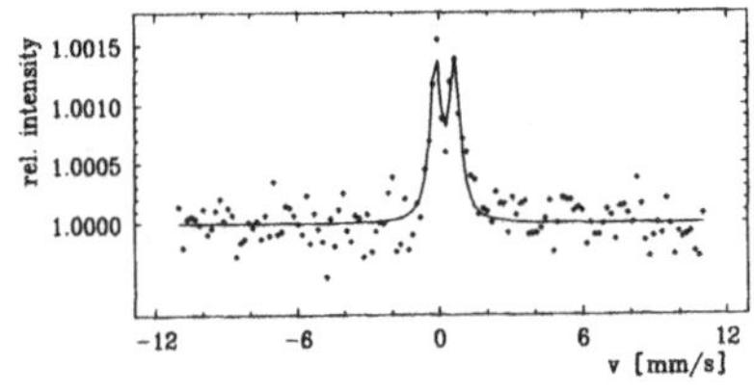

Figure 35. The conversion electron Mössbauer spectrum of a sample coated with less than one monolayer of iron oxide.

There has been a proposal to utilize the 600 to 700 eV LMM Auger electrons to investigate extremely thin iron containing layers, for example, oxidic layers on iron-based alloys. It was found, however, that in this energy region the observed Mössbauer subspectra are characteristic of the bulk material. Unfortunately, the consequent interpretation that ultrathin surface layers exhibit a vanishingly small Debye-Waller f-factor is wrong. This would contradict the phenomena discussed in Section 1. In fact, high f-factors are observed even for a surface coverage of less than a monolayer. This is demonstrated in Fig. 35, which shows an integral conversion electron Mössbauer spectrum obtained from an ultrathin layer of iron oxide. The oxide thickness was about 50 percent of a single monolayer of $^{57}FeOOH$, prepared by a special treatment of Langmuir-Blodgett layers.[34]

The intensity of the LMM Auger electrons, created as a direct consequence of the Mössbauer after-effects is very low, as can be seen in Fig. 31. The signal from the bulk material is due to both the tail extending from the K-conversion and KLL/KLM-Auger electrons down to this low energy region, and to the secondary electrons excited by photoexcitation processes and the Auger effect in the layer close to the surface by γ-rays and conversion and Auger electrons emitted from lower lying layers after Mössbauer excitation. The latter is the major contribution. These problems are discussed in reference 35 and the references given therein. In addition, there are a remarkable number of secondary electrons emitted with energies below 50 eV. It is known that the mean free path is rather large for low energy electrons if their kinetic energies are lower than the binding energies of the valence electrons of the surface atoms. All these factors contribute to the high surface sensitivity realized by integral conversion electron Mössbauer spectroscopy.

Some integral conversion electron Mössbauer spectrometer detectors are reported to exhibit energy resolution, i.e., the output pulse height correlates with the energy of the detected electron. In this case, the setting of a pulse height window could correlate to a depth selectivity as in depth selective conversion electron Mössbauer spectroscopy. However, such an energy resolution, if really present, is of limited value. In practice, the calibration of the pulse height energy axis in units of the electron energy has often been found to depend on the sample under investigation. This implies some difficulties in choosing an appropriate window setting.

In iron-57 conversion electron Mössbauer spectroscopy the appearance of one or two peaks in the pulse height distribution should not automatically be associated with the conversion and direct Auger electrons of iron-57. Some problems can be illustrated by a discussion of Fig. 36. By using an electrostatic energy analyzer, the energy distribution of electrons emitted from two samples during irradiation from a commercial cobalt-57 in rhodium Mössbauer source were analyzed. The upper spectrum in Fig. 36 corresponds to a steel sample covered with a titanium layer, whereas the lower spectrum corresponds to a steel sample covered with an ca. 25 nm layer of TiN.[36] Above 14.4 keV, a high electron count intensity is recorded, an intensity which is due to Auger and photoelectrons excited directly or indirectly by the intense 122 and 136 keV γ-ray radiation of the source. Below 14.4 keV, this contribution becomes less important as compared with the electrons excited by the 14.4 keV γ-ray radiation. Down to about 12 keV, there is a region in which the photoelectron lines appear, lines which are well known from the usual x-ray induced photoelectron spectra, i.e., the ESCA lines in Fig. 36, or the electrons with binding energies of zero to 2 keV. At 9 to 10 keV, the titanium 1s photoelectrons produce a well defined signal, a signal which may cause an additional peak in the pulse height spectrum measured by a Mössbauer conversion electron detector. However, this peak is not useful for obtaining a conversion electron Mössbauer spectrum, because it contains only a very small portion of Mössbauer resonant events resulting from resonantly re-emitted γ-rays, secondary γ-rays which may be created deep inside the sample. This titanium 1s peak could become a more prominent peak for thicker titanium layers or for a titanium containing substrate. Similar photopeaks are to be expected at their respective positions, if a sample contains other elements.

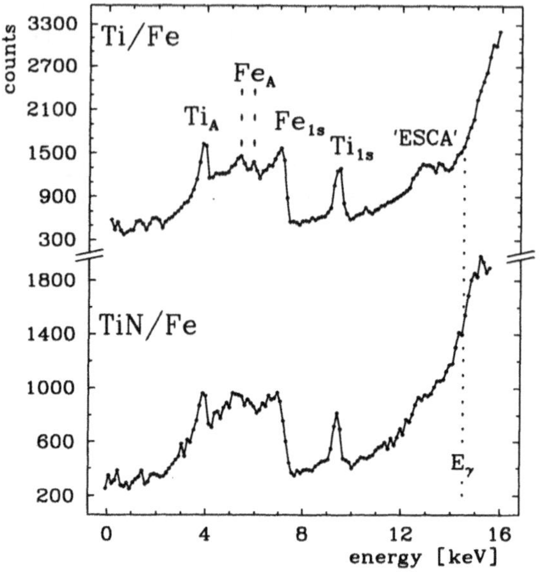

Figure 36. The energy spectra of the electrons emitted by a cobalt-57 irradiated steel sample coated with 25 nm of titanium, top, and TiN, bottom.

The iron 1s peak contains both the 1s photoelectrons from all iron atoms excited by the 14.4 keV radiation, and the K-conversion electrons from the iron-57 nuclei. While the former intense component does not contribute to the Mössbauer spectrum, the latter is the resonant component utilized in conversion electron Mössbauer spectroscopy and both components are excited by the 14.4 keV γ-ray radiation. The iron KLM and KLL-Auger lines, Fe_A in Fig. 36, are a consequence of the total iron 1s electron emission and, hence, contain both a non-resonant and a Mössbauer effect resonant component. The titanium KXY-Auger electrons, Ti_A in Fig. 36, are the consequence of the emission of the titanium 1s photoelectrons. However, titanium Auger electrons can also be produced with a rather high yield by the iron 1s photoelectrons and conversion electrons, because their energy is well above the titanium 1s binding energy. Therefore, the Ti_A peak is expected to contain a non-resonant as well as a resonant component. This fact, which also applies to other elements, complicates layer thickness determinations by conversion electron Mössbauer spectroscopy.

By comparing the two spectra in Fig. 36, it is evident that the intensity of the titanium peaks is reduced in the presence of nitrogen. Although there is no peak due to nitrogen in this energy region, the effect of the nitrogen concentration is a weakening of the titanium peaks. In addition, for the same thickness there is somewhat less titanium in the TiN layer.

4. LAYER GROWTH

In addition to information about the composition of layers, a further very important parameter provided by surface sensitive methods is the layer thickness. A knowledge of the time dependence of the formation of a layer, i.e., the layer growth rate, provides insight into the mechanism of layer formation. In the following sections, some types of layer growth important from a practical point of view, are summarized.[37,38]

4. 1. In-plane Layer Growth

For in plane layer growth it is assumed that the layer thickness increases homogeneously with no lateral differences.

Simple parabolic layer growth. Simple parabolic growth occurs according to the expression

$$y^2 = at + b, \tag{5}$$

where y is the layer thickness and t is the time. Simple parabolic layer growth is characteristic of growth processes which are controlled by the diffusion of a species through the layer. From this equation it is clear that the growth rate decreases with increasing layer thickness, i.e., the layer as a whole gradually hinders the progress of growth by increasing the diffusion path length. Many oxidation processes can be approximated by this equation which is known as Wagner's law. Thus, the parabolic growth requires a high migration rate through the layer caused by a high ionic mobility at a given temperature.

Inverse logarithmic layer growth. If layer growth is controlled by migration, but occurs at temperatures producing a very low mobility, then the growth can be approximated by

$$1/y = a \ln[b(t - t_o) + 1] + 1/y_o. \tag{6}$$

Cubic layer growth. In the intermediate case, between the conditions of the previous two growth laws, the growth can be approximated by a cubic law,

$$y^3 = at + b. \tag{7}$$

Linear layer growth. If the rate of the layer formation is fully determined by an interfacial reaction, then linear growth occurs according to the expression

$$y = at + b. \tag{8}$$

This condition is fulfilled if the migration rate through the already formed layer is much faster than the rate of the reaction forming the layer components. This reaction can take place at the surface, at the layer-substrate interface, or at an interface within the layer.

Mixed parabolic layer growth. This type of growth is expected if growth is partially controlled by an interface reaction and partially controlled by migration, i.e., both rates are competitive. Then the rate is approximated by

$$y^2 + ay = bt. \tag{9}$$

Logarithmic layer growth. If the layer growth is fully controlled by a flux of electrons and no ion transport is involved, then the growth rate is independent of temperature and should follow the expression

$$y = a \ln(1 + bt). \tag{10}$$

4. 2. Island Layer Growth

If layer formation is not laterally homogeneous, it may proceed via the growth of islands.

S-shaped island layer growth. In this case islands are formed around a seed on the surface and extend laterally. If the formation of new seeds is slow as compared with the island growth, the island growth by itself is the rate controlling step. This means that, at a high seed formation energy and for fast lateral growth, the layer growth rate can be approximated by

$$dy/dt = a[1 - \exp(-bt^n)]. \tag{11}$$

Logarithmic island layer growth. This case occurs when a partial coverage of a surface is possible by a clustering of vacancies or defects. In this case, a logarithmic behavior is expected according to the expression

$$y = a \ln(bt^n + c), \qquad (12)$$

where n has been reported to be 0.5 or 1.

4. 3. Local Transport Layer Growth

If the components which are needed to form a layer cannot migrate freely but are restricted to limited transport paths, the mechanism of layer growth will be affected. Such transport paths are, for example, the grain boundaries in metals.

Asymptotic layer growth. An asymptotic growth is expected if reaction products are formed in such a way that they progressively obstruct the transport paths. Such autoblocking leads to a growth which asymptotically reaches a final maximum thickness according to the expression

$$y = a[1 - \exp(-bt)]. \qquad (13)$$

Logarithmic layer growth. If there are two different processes leading to surface layer formation by two migrating species and/or by two different types of transport paths, one of the species may obstruct the path of the other species. For such interblocking, the growth should proceed in a fashion similar to Equation 10 and will be given by the expression

$$y = a \ln(1 + bt). \qquad (14)$$

Some normalized growth curves are shown in Fig. 37. To determine which mechanism of layer growth yielded a given surface film, the different expressions described above, and the constants, a and b, involved have to be obtained from the experimental data. From Fig. 37 it is evident that some curves do not differ enough to permit a distinction between different formation mechanisms. However, a distinction between asymptotic and linear layer growth and the other types of growth is often possible. Fortunately, this distinction covers many important applications. An example is the growth and action of passive layers on metals under defined electrochemical potentials.[39]

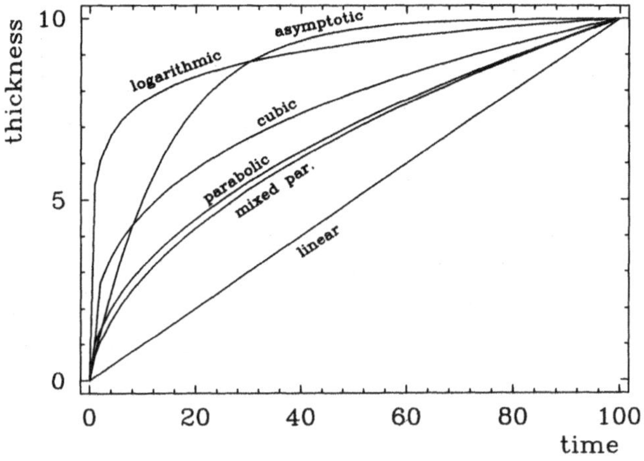

Figure 37. The thickness versus time for different layer growth processes, normalized to y(100)=10 in arbitrary units.

4. 4. Layer Growth Approximations

The layer growth expressions given in Sections 4. 1 to 4. 3 are formulated for practical reasons. According to reference 40, some of the expressions can be derived from basic principles. The net rate of transport for a species, i, which can be atoms, ions, or defects in a solid, in the presence of a concentration gradient and under the influence of an electric potential gradient, is given by

$$v_i = \lambda_i c_i \frac{kT}{h} \exp\left(-\frac{E_{a,i}}{kT}\right)\left[\exp\left(-\alpha\lambda_i z_i \frac{F}{RT}\frac{dU}{dx}\right) - \left(1 + \lambda_i \frac{d\ln c_i}{dx}\right)\exp\left((1-\alpha)\lambda_i z_i \frac{F}{RT}\frac{dU}{dx}\right)\right], \quad (15)$$

where v_i is the migration rate, F is the Faraday constant, $E_{a,i}$ is the activation energy, c_i is the concentration of the migrating species, z_i is the charge, λ_i is the distance between the equilibrium positions, U is the electrostatic potential, x is the distance from the initial substrate surface, y is the layer thickness, α is a symmetry factor, and V is the molar volume. Then, the total rate of transport for all species is the summation of v_i.

By assuming that

$$\frac{dU}{dx} \approx \frac{\Delta U}{y} = \frac{\Delta G}{Fy},$$

i.e., a homogeneous electric field, and that

$$\frac{dc}{dx} \approx \frac{\Delta c}{y},$$

i.e., a homogeneous concentration gradient, that α is approximately 1/2, and that

$$v_{total} = (1/V)\,(dy/dt),$$

one finds the following approximation for the growth rate of a layer,

$$\frac{dy}{dt} = \sum_i K_i c_i(y)\left[\exp\left(-\frac{\lambda_i z_i \Delta G}{2yRT}\right) - \left(1 + \lambda_i \frac{\Delta c_i}{y c_i(y)}\right)\exp\left(\frac{\lambda_i z_i \Delta G}{2yRT}\right)\right], \quad (16)$$

with

$$K_i = \frac{V\lambda_i kT}{h}\exp\left(-\frac{E_{a,i}}{RT}\right).$$

Some special cases should be considered. First, if dy/dt is independent of y, the thickness increases linearly with time such that

$$y = at + b. \quad (17)$$

Second, for a thick layer at high temperature,

$$yT > 0.65\,\lambda z \Delta G/R,$$

for which a ten percent approximation is

$$\frac{dy}{dt} = \frac{1}{y}\sum_i K_i \lambda_i c_i(y)\left[\frac{z_i \Delta G}{RT} + \frac{\Delta c_i}{c_i(y)}\right] = \frac{a}{y},$$

and the thickness increases according to the parabolic law

$$y^2 = 2at + b. \quad (18)$$

Third, if the layer is thin and the temperature is low, and

$$yT < 0.43\ \lambda z\Delta G/R,$$

for which a ten percent approximation is

$$\frac{dy}{dt} = \sum_i K_i c_i(y) \exp\left(-\lambda_i z_i \frac{\Delta G}{2yRT}\right),$$

then

$$\frac{1}{y} = \sum_i \frac{4.606RT}{\lambda_i z_i \Delta G}\left[\ln\frac{K_i c_i(y)}{y} + \ln\frac{t}{t_0}\right] = a\ln[b(t - t_0) + 1] + 1/y_0 \quad (19)$$

is the inverse logarithmic law. Finally, in the intermediate region

$$0.43\ \lambda z\Delta G/R < yT < 0.58\ \lambda z\Delta G/R$$

an approximation is

$$\frac{dy}{dt} = \frac{A}{y^2}\sum_i K_i c_i(y)\left(\frac{\lambda_i z_i \Delta G}{2RT}\right)^2,$$

which leads to the cubic law,

$$y^3 = at + b. \quad (20)$$

5. CONCLUSIONS

As this chapter illustrates, Mössbauer spectroscopy is an excellent technique for informative surface and thin film studies. Some comments concerning special aspects of such analyses and some hints to avoid misinterpretation of spectral data have been presented. In particular it seems evident that, for a detailed characterization of surface layers, a combination of conversion electron Mössbauer spectral results with those obtained by other methods is generally both necessary and useful. There are two main features which make the combination of conversion electron Mössbauer spectrroscopy with other techniques quite necessary. First, conversion electron Mössbauer spectroscopy is possible for only a few elements, and information about other elements constituting the layer, can be derived only indirectly from Mössbauer spectral studies. Second, the range of information depth available from conversion electron Mössbauer spectroscopy is fixed by the initial energy of the conversion electrons and cannot be varied beyond given upper and lower limits. The sensitivity of conversion electron Mössbauer spectroscopy is, however, very high and is often good enough to detect a fraction of a monolayer. There is no question that conversion electron Mössbuer spectroscopy, if applicable, is an outstanding method for qualitative and quantitative phase analyses and, in some important fields of surface research, such as materials science, corrosion studies, and catalytic research, it cannot be replaced by other methods.

A typical example[30] of a complex application of surface sensitive methods can be described in a few words. In Fig. 28, the difference in the iron(II) to iron(III) ratio for iron ions on the surface of a glass, as compared to the ratio in the bulk is evident. X-ray induced photoelectron spectoscopy can be used to show that there is a leaching of iron and other cations from the surface region upon polishing. Structural changes in the glassy network can also be derived from the shape of some of the signals obtained by this technique, as is shown in Fig. 14. Through optical reflectance measurements, similar to those shown in Fig. 22, the thickness, refractive index, and homogeneity of the leached layer can be determined.

Altogether, a rather detailed picture of the structure of the leached layer and the altered role of the iron ions as either network modifiers or network formers in this layer can be determined.

REFERENCES

1. L. C. Feldman and J. W. Mayer, *Fundamentals of Surface and Thin Film Analysis*, North Holland/Elsevier Science Publ. Co., New York (1986).
2. G. A. Somorjai, *Introduction to Surface Chemistry and Catalysis*, Wiley, New York (1994); G. Ertl and J. Küppers, *Low Energy Electrons and Surface Chemistry*, VCH Verlagsanstalt, Weinheim (1985).
3. J. M. Walls, *Methods of Surface Analysis*, Cambridge University Press, Cambridge (1989).
4. A. W. Adamson, *Physical Chemistry of Surfaces*, Wiley, New York (1990).
5. D. Briggs and M. P. Seah, "Practical Surface Analysis," in *Auger and X-ray Photoelectron Spectroscopy*, Vol. 1, Wiley, Chichester (1990).
6. S. R. Morrison, *The Chemical Physics of Surfaces*, 2nd ed., Plenum Press, New York (1990).
7. W. Mönch, in *Chemistry and Physics of Solid Surfaces*,Vol. V,. R. Vanselow and R.Howe, eds., p. 516, Springer Series in Chemical Physics, Vol. 35, Springer Verlag, Berlin (1984).
8. B. Navinsek, D. Hanzel, and W. Meisel, *Vacuum* **43**, 325-330 (1992).
9. W. Meisel and D. Hanzel, *Surface Interface Anal.* **18**, 284-288 (1992).
10. E. Delano and R. J. Pegis, *Progress in Optics* **7**, 66 (1967).
11. F. Rösel, H. M. Fries, K. Alder, and H. C. Pauli, *Atomic Nucl. Data Tables* **21**, 91 (1978).
12. I. M. Band, M. B. Trzhaskovskaya, and M. A. Listengarten, *Atomic Nucl. Data Tables* **18**, 433-457 (1976).
13. M. O. Krause, *J. Phys. Chem. Ref. Data* **8**, 307 (1979).
14. D. Barb and W. Meisel, eds., *Grundlagen und Anwendungen der Mössbauer-Spektroskopie*, Akademie-Verlag, Berlin (1980).
15. W. Meisel, *Mber. Dt. Akad. Wiss. Berlin* **11**, H 5/6, 355-58 (1969).
16. W. Meisel, *Proceedings of the International Mössbauer Conference*, Bratislava 1973, **1**, 200-213 (1975).
17. W. Meisel, *Proceedings of the International Conference on Mössbauer Spectroscopy*, Bucharest, D. Barb and D. Tarina, eds., Vol. 2, p. 71-94 (1977).
18. W. Meisel, in *Modern Physics in Chemistry*, E. Fluck and V. I. Goldanski, eds., Vol. 1, p. 237-267, Academic Press, London (1976).
19. W. Meisel and L. Keszthelyi, *Hyp. Interact.* **3**, 314-322 (1977).
20. P. L. Crouse and C. M. Stander, *J. Phys. Chem. Solids* **49**, 1145-1151 (1988).
21. J. C. Frost, B. C. C. Cowie, S. N. Chapman, and J. F. Marshall, *Appl. Phys. Lett.* **47**, 581-583 (1985).
22. S. M. Irkaev, M. A. Andreeva, and V. G. Semenov, *Nucl. Instr. Meth. B* **74**, 554 and 564 (1993).
23. Zv. Bonchev, A. Jordanov, and A. Minkova, *Nucl. Instr. Meth.* **70**, 36 (1969).
24. Zv. Bonchev, A. Jordanov, and A. Minkova, in *Proceedings of the International Conference on the Applications of the Mössbauer Effect*, I. Dézsi, L. Cser, D. L. Nagy, and I. Vincze, eds., p. 333-337, Akadémiai Kiado, Budapest (1971).
25. K. R. Swanson and J. J. Spijkerman, *J. Appl. Phys.* **41**, 3155 (1970).
26. B. J. Tatarchuk and J. A. Dumesic, in *Chemistry and Physics of Solid Surfaces, V*, R. Vanselow and R. Howe, eds., p. 65-109, Springer, Berlin (1984).

27. Zs. Kajcsos, W. Meisel, P. Griesbach, P. Gütlich, Ch. Sauer, R. Kurz, K. Hildebrand, R. Albrecht, and M. A. C. Ligtenberg, *Nucl. Instr. Meth. B* **93,** 505-515 (1994).
28. W. Meisel, *Hyp. Interact.* **92**, 1213-1220 (1994).
29. D. Hanzel, P. Griesbach, W. Meisel, and P. Gütlich, *Hyp. Interact.* **71**, 1441-1444 (1992).
30. W. Meisel, D. Sprenger, and P. Gütlich, *Surface Interface Anal.* **22**, 267-270 (1994).
31. F. Salvat and J. Parellada, *Nucl. Instr. Meth. B* **1**, 70-84 (1984).
32. F. Salvat, R. Mayol, J. D. Martinez, and J. Parellada, *Nucl. Instr. Meth. B* **6**, 547-550 (1985).
33. D. Liljequist and M. Ismail, *Phys. Rev. B* **31**, 4131-4136 (1985).
34. T. Faldum, W. Meisel, and P. Gütlich, *Hyp. Interact.* **92**, 1263-1269 (1994).
35. G. Klingelhöfer and W. Meisel, *Hyp. Interact.* **57**, 1911-1918 (1990).
36. D. Hanzel, W. Meisel, D. Hanzel, and P. Gütlich, *Hyp. Interact.* **69**, 807-810 (1991).
37. U. R. Evans, *An Introduction to Metallic Corrosion*, Edward Arnold Publ., London (1981).
38. U. R. Evans, *The Corrosion and Oxidation of Metals: Scientific Principles and Practical Applications,* Edward Arnold Publ., London (1971).
39. W. Meisel, *Hyp. Interact.* **45**, 73-90 (1989).
40. T. Hurlen, in *The Corrosion and Oxidation of Metals*, 2nd Supplementary Volume, U. R. Evans, ed., p. 37-42, Edward Arnold Publ., London (1976).
41. W. Meisel, *Appl. Phys.* **A59**, 381-384 (1994).

2

Magnetic and Structural Properties of Real Metal Layer Interfaces

Ch. Sauer

1. INTRODUCTION

The magnetic properties of bulk metals, for instance, their magnetic moments and anisotropies, have been extensively studied in the past and are now well understood in great detail for most homogeneous metals and alloys. In contrast, the magnetic properties of ultrathin film interfaces and surfaces, which usually differ from the bulk properties, have generated growing new interest over the last decade. This rapid growth in interest has resulted because the preparation of thin films with well defined compositions and structures, by molecular beam epitaxy in an ultra high vacuum or by sputtering techniques, has shown enormous progress over the past few years. As a result it has even been possible to prepare thin films of materials with crystal structures deviating from the usual crystal structure of the bulk material. Examples are thin films of fcc-iron or bcc-cobalt. Simultaneously with the growth in preparative methods, many subtle new experimental methods for the analysis of the various properties of thin films and interfaces have been developed. In addition, modern band structure calculations have provided a new and powerful tool for the theoretical treatment of the magnetic properties of thin films and interfaces.

An interface is the transition zone between two thin films, A and B, differing in chemical composition or crystal structure. For an ideal interface this zone consists of two atomically flat monolayers, one belonging to film A and the other to film B. Whereas most theoretical models and calculations are based on such ideal interfaces, real interfaces show significant interdiffusion, roughness, dislocations, or strains. These imperfections strongly influence the magnetic properties of thin film structures, such as their magnetic moments, magnetic anisotropies, interlayer magnetic coupling, or magnetoresistance. Several sensitive analytical methods have been developed in recent years, based mainly on spectroscopy with unpolarized or polarized photoelectrons. But, due to the strong absorption of low energy photoelectrons, these techniques are suitable for the analysis of, at most, a few monolayers at the film surface. For the analysis of internal interfaces hidden within a film structure, nuclear resonance techniques, such as the Mössbauer effect, nuclear magnetic resonance, and per-

Ch. Sauer • Institut für Festkörperforschung, Forschungszentrum Jülich GmbH, D-52425 Jülich, Germany.

Mössbauer Spectroscopy Applied to Magnetism and Materials Science, Volume 2, edited by Gary J. Long and Fernande Grandjean. Plenum Press, New York, 1996.

turbed angular correlation, have been developed and extensively improved in order to provide the necessary sensitivity. This chapter will focus on the magnetic and structural properties of real interfaces in metallic thin layer systems as investigated by using these nuclear techniques. In addition, high resolution transmission electron microscopy and x-ray scattering will be covered, but in less detail. Recent reviews dealing with the properties of thin magnetic films have been published by Shinjo,[1] Gradmann,[2] Sauer et al.,[3] and Falicov,[4] whereas the conditions determining the film growth are described by Gautier et al.[5]

2. EXPERIMENTAL TECHNIQUES OF INTERFACE ANALYSIS

Numerous experimental methods are now available for investigating both the chemical and structural quality of thin films and their electronic and magnetic properties. Most of these methods are applied in-situ, under an ultra high vacuum, and probe, at most, the upper surface layers. Powerful tools for such structural analysis are low energy electron diffraction,[6] reflection high energy electron diffraction[7] and, particularly, their respective spot profile analysis versions, often referred to as SPA-LEED[8] and SPA-RHEED,[9] respectively. These techniques provide very detailed information on the surface topology of freshly prepared films and the results can be compared with the images obtained by scanning tunneling microscopy[10] or atomic force microscopy,[11,12] which often give atomic resolution. Chemical analysis of the surface can be performed using Auger electron spectroscopy[13] or x-ray photoelectron spectroscopy.[14] By using surface extended x-ray absorption fine structure techniques[15] the local symmetry of adsorbate atoms and the adsorbate-substrate bond lengths can be analysed. The electronic properties, e.g., the band structure, can be studied by applying x-ray photoelectron spectroscopy or polarized electron methods. The various methods based on polarized electrons provide powerful tools for the investigation of surface magnetism.[16] In Chapter 1 of this volume these methods are treated in more detail. However, as mentioned above, for the study of internal interfaces only a few techniques are available so far, techniques which will be discussed briefly in the following sections.

2. 1. High Resolution Transmission Electron Microscopy

The most favorable way to analyze the crystalline structure of thin films and interfaces would be a direct visualization of the atomic arrangement. High resolution transmission electron microscopy, using electron energies of several 100 keV, in combination with high-performance computing and image processing techniques, now offers the resolution, about 0.2 nm, necessary for atomic imaging.[17] The structural quality, i.e., the roughness of interfaces, and the crystalline structure of the two films in contact at the interface, and even lattice defects, such as dislocations, can be observed and analysed from high resolution transmission electron microscopy images. However, the preparation of samples suitable for high resolution imaging is subtle and time consuming. Usually a thin cross section is cut from the thin film system and thinned to about 5 to 10 nm by grinding, followed by ion milling. Fig. 1 illustrates an example of a high resolution transmission electron microscopy image of two Fe/Gd film structures.[18] It is clearly apparent from this figure that the 10 nm thick gadolinium film grows first as an amorphous material on the GaAs substrate and then switches, after about 5 nm, to its usual hcp structure. An iron film evaporated on this polycrystalline gadolinium film grows in either an amorphous or a bcc polycrystalline form, depending upon whether the iron film thickness is below or above, see Fig. 1a and b, a critical thickness of 2.3 nm, respectively.[18]

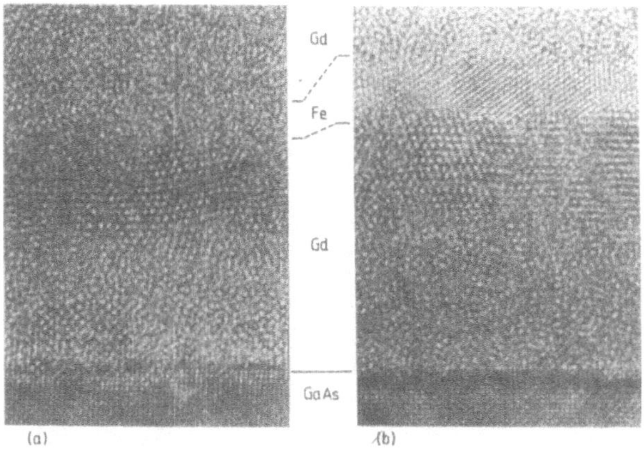

Figure 1. Cross-section high resolution transmission electron microscopy images of two GaAs(110)/10 nm Gd/Fe/3 nm Gd/10 nm Al films with iron film thicknesses of 1.6 nm, a, and 3.2 nm, b. The iron interlayer is amorphous in a and bcc-nanocrystalline in b.[18]

2. 2. Nuclear Techniques

Common to all nuclear techniques, such as conversion electron Mössbauer spectroscopy, nuclear magnetic resonance, and perturbed angular correlation, is the deposition, as local probes, of suitable nuclei at definite positions at the surface or interface of a thin film. The nuclear transition under observation must be sensitive to the local magnetic and structural properties in the vicinity of the tracer atom via the hyperfine interactions. Usually these hyperfine interactions are of short range, extending at most to the second or third neighbor shell. The nuclear transition within the probe atom is detected by its emission of a γ-ray in perturbed angular correlation methods, or a conversion electron of several keV energy in conversion electron Mössbauer spectroscopy, or by the absorption or emission of a radiofrequency signal in NMR spectroscopy. Due to the larger penetration depth, compared with photoelectrons, the probing depth of these methods can be extended to as much as 1000 nm below the surface, and as a result, they permit the study of internal interfaces.

2. 2. 1. Conversion Electron Mössbauer Spectroscopy

Conversion electron Mössbauer spectroscopy[19] is an experimental modification of conventional Mössbauer spectroscopy, see also Chapter 1, which is usually performed in transmission geometry. In a Mössbauer effect experiment with the prominent and for conversion electron Mössbauer spectroscopy the most suitable iron-57 nucleus, a radioactive source containing cobalt-57 decays by electron capture to the 14.4 keV excited state of iron-57 which has a lifetime of 97.7 ns. The 14.4 keV γ-rays created by the deexcitation of the iron-57 nuclear excited state in the source can be resonantly absorbed by an absorber which also contains iron-57, which are excited to their 14.4 keV state. The experimental detection of the resonant absorption process can be performed either in standard transmission geometry, by measuring the intensity reduction of the γ-rays, due to their absorption or, in a scattering geometry, by detecting the conversion electrons, as is shown in Fig. 2. The second version is called conversion electron Mössbauer spectroscopy and is possible because the excited iron-57 nucleus in the absorber can decay via re-emission of a 14.4 keV γ-ray, or,

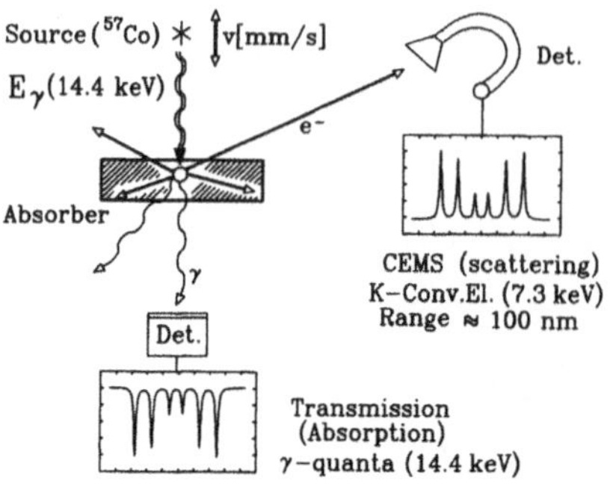

Figure 2. The fundamental design of the transmission and conversion electron Mössbauer spectroscopy, CEMS, experiments. The spectra represent the magnetic sextet observed for α-iron.

alternatively, via a radiationless energy transfer to its core electrons, preferentially the K-electrons, the electrons closest to the nucleus. Then, an iron K-conversion electron is ejected with a kinetic energy, $E_{kin} = E_\gamma - E_{K\text{-binding}} = 14.4 - 7.1 = 7.3$ keV. Of course, L and M-conversion electrons can also be emitted with, however, a much smaller probability.

Because of the relatively low energy of the 7.3 keV K-conversion electrons, the characteristic range of iron-57 conversion electron Mössbauer spectroscopy investigations is usually restricted to a sub-surface region of, at most, 100 nm. Even though the count rates are much lower, such experiments often are superior in thin film research as compared to transmission Mössbauer spectral experiments because of the improved signal to background noise ratio.[3] A compilation[20] of Mössbauer nuclei used for conversion electron Mössbauer spectroscopy applications clearly indicates that iron-57 is the most sensitive tracer isotope. For example, the limit, detectable by conversion electron Mössbauer spectroscopy amounts to 10^{13} to 10^{14} atoms/cm^2 for iron-57, 10^{15} atoms/cm^2 for tin-119, and 10^{16} atoms/cm^2 for europium-151. For iron-57 this detection limit corresponds to about 0.1 monolayers of iron metal. This permits the investigation of as little as one interface, whereas for transmission geometry multilayer samples with many interfaces are always required.

Of course, the same information is obtained from conversion electron Mössbauer spectroscopy and transmission Mössbauer spectroscopy. This includes the isomer shift, which provides insight into the valence state of the Mössbauer effect atom or on charge transfer, the quadrupole splitting, which reflects the local symmetry of neighboring atoms and the charge distibution around the Mössbauer effect atom, the magnetic hyperfine field, which reflects the magnetic properties of the film, and the relative line intensities of a magnetic Mössbauer spectrum, which gives the orientation of the magnetization within the film, e.g., whether the magnetization lies in or out of the plane of the thin film. The last two parameters are most important for the study of magnetic thin film systems and, especially the magnetic hyperfine field can often serve as a fingerprint for local atomic configurations at interfaces.

The decay of a resonantly excited Mössbauer effect nucleus in an absorber film is quite a complex process and yields a variety of Auger electrons and x-rays of different energies. By setting an energy window for counting those electrons or x-rays, one is able to derive the depth at which the Mössbauer spectral absorption process occurs relative to the film surface. This technique, which uses an electron energy analyzer, is called depth selective conversion electron Mössbauer spectroscopy and is often applied in-situ to experiments in which the film preparation and the spectral measurements are performed in the same ultra high vacuum system. The interested reader should consult reference 3 for a compilation of the various types of these energy analyzers. In contrast, integral conversion electron Mössbauer spectroscopy collects all electrons and x-rays without any energy discrimination. For such integral conversion electron Mössbauer spectroscopy experiments, simple gas flow proportional counters, with the Mössbauer effect absorber as its window, are widely used because of their large 2π solid angle, which yields a high counting effficiency. By means of a channeltron electron detector, integral conversion electron Mössbauer spectroscopy can be carried out in an ultra high vacuum. Depth selective measurements are also possible by integral conversion electron Mössbauer spectroscopy by inserting approximately two monolayers of iron-57 at a definite position in an iron film, a film which otherwise contains only the non-Mössbauer active iron-56 isotope. This probe layer technique makes possible a depth resolution as precise as one monolayer in favorable cases. A detailed analysis[21] of the merits and demerits of integral conversion electron Mössbauer spectroscopy and depth selective conversion electron Mössbauer spectroscopy is also presented in Chapter 1.

2. 2. 2. Nuclear Magnetic Resonance Spectroscopy

Nuclear magnetic resonance, or NMR, is an early nuclear resonance method used for the determination of solid state properties,[22] a method which is widely applied in physics, chemistry, and biology. NMR work based on the copper-63 and 65 and cobalt-59 isotopes has become a particularly useful way to study interfaces, e.g., interfaces in the Co/Cu multilayers, which reveal the presence of a large magnetoresistance. NMR is based on the idea, that the nuclear magnetic moments are partly oriented in a magnetic field, yielding a nuclear magnetization which can be resonantly destroyed by a high-frequency radiowave, which can be continuous or pulsed. In the latter case, the so-called spin-echo method, the nuclear magnetization is first rotated 90° by a radio-frequency pulse and the resulting transverse magnetization is allowed to decay due to magnetic inhomogeneities. After a time, τ, all spin orientations are reversed by a 180° pulse and, after a time, 2τ, all spins are again in phase leading to an echo signal which can be detected with a pick-up coil. The spin-echo method has some advantages for inhomogeneous magnetic samples.

The information obtained from NMR measurements are essentially the same as those obtained from conversion electron Mössbauer spectroscopy, i.e., the magnetic hyperfine fields and quadrupole interactions. However, the energy resolution of NMR is much better than that of conversion electron Mössbauer spectroscopy because the resonance frequencies in NMR can be measured with high precision. Hence, it is usually no problem to resolve the various atomic neighbor configurations ocurring at interfaces in magnetic film structures, structures which appear as satellite lines beside the central resonance line stemming from the bulk of the film. These neighbor configurations differ in their magnetic hyperfine fields mainly because of the polarization of the conduction electrons by the magnetic neighbor atoms. In contrast, iron-57 conversion electron Mössbauer spectroscopy has a superior sensitivity. In NMR usually a few monolayers of the NMR isotope are required in order to obtain a detectable signal. So far, most NMR studies of thin films have dealt with either a detailed analysis of the atomic intermixing and the roughness at the interfaces, see for example reference 23 which deals with studies of Co/X multilayers, where X is Cr, Fe, Ni, or Cu, or with the correlation of magnetoresistance properties, see for example reference 24.

2. 2. 3. Perturbed Angular Correlation

A technique much more sensitive and rapid than conversion electron Mössbauer spectroscopy and NMR is available through perturbed γ-γ angular correlation. This technique requires only about 10^{10} radioactive tracer atoms, a number which corresponds to ca. 10^{-4} of a monolayer. Thus, this method is very attractive for surface and interfacial studies and the reader should consult reference 25 for further details and a review of the technique. The most prominent nucleus used to date for such applications is indium-111, which decays via electron capture and a 172 to 247 keV γ-ray cascade to cadmium-111, see Fig. 3. The basis of the method is the measurement of the anisotropy of the γ_2-emission from

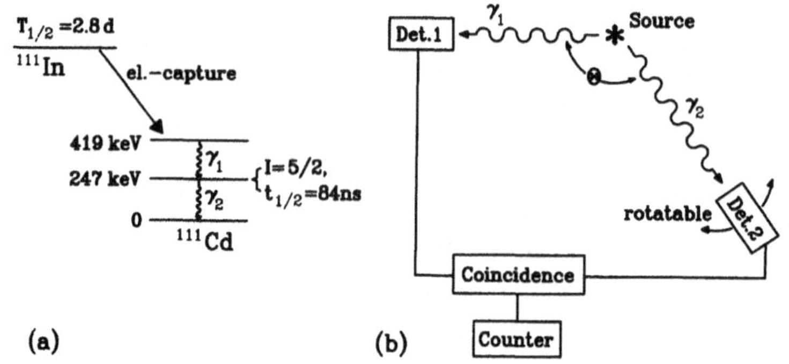

Figure 3. The nuclear level scheme of cadmium-111, a, and the fundamental design of the perturbed angular correlation experiment, b.

the 247 keV intermediate nuclear state by using the preceding 172 keV γ_1-emission as a reference direction. By registering γ_1 and γ_2 by delayed coincidence and then using γ_1 as a start and γ_2 as a stop signal, one can record the anisotropy pattern by the rotation of detector 2 in Fig. 3. In the presence of a magnetic hyperfine field or an electric quadrupole interaction the γ-γ angular correlation becomes time dependent and, as a consequence, the measured count rate will oscillate as a function of the delay time. From the oscillation period information about the magnetic hyperfine interaction, the quadrupole interaction, and the orientation of the magnetization[26] within the film, can be obtained in a fashion similar to that used in conversion electron Mössbauer spectroscopy. However, because for perturbed angular correlations no Debye-Waller factor restrictions exist, perturbed angular correlation investigations can be extended up to the melting point of the sample.

Cadmium-111 is a non-magnetic tracer atom, and thus magnetic interactions can be probed only via the transferred hyperfine interactions. So far, most applications of indium-111 perturbed angular correlations use the electric quadrupole interaction for the determination of the environment of the indium atom. The quadrupole interaction serves as a fingerprint and can provide detailed information about the local probe position, e.g., whether it is located at a step, a corner, or a hole defect.[25] Further applications comprise diffusion of adsorbates, surface roughening, structural phase transitions on stepped metal surfaces, as well as compound formation at reactive metal/metal interfaces.[25] Recently there has been some success in using palladium-100 as a proper perturbed angular correlation probe.[27] For magnetic interface investigations palladium-100 is particularly useful because its nuclear magnetic moment is large, whereas its quadrupole moment is small, a combination which leads to negligible effects from the electric field gradients. The general problem with perturbed angular correlation in thin film studies is the complex sample preparation technique required when a radioactive material has to be deposited in ultra high vacuum at definite lattice positions. Of course, by performing source experiments with radioactive cobalt-57 as a probe atom, instead of iron-57, conversion electron Mössbauer spectroscopy would have the same sensitivity and as short a measuring time as perturbed angular correlation studies.

2. 3. X-ray Scattering

All x-ray techniques, either absorption or scattering,[28] become surface sensitive at grazing incidence angles. At small angles of incidence of less than 0.5° total reflection occurs, and the x-rays penetrate 1 to 5 nm into the film.[29] From x-ray reflection the density and thickness of thin films, as well as the roughness of the surfaces and the internal interfaces, can be determined.[30] The experiments can be carried out in the laboratory by means of high intensity x-ray sources with rotating anodes or, preferably, with synchrotron radiation.

Near an absorption edge the index of refraction becomes complex and is given by

$$n = (1 - \delta) - i\beta,$$

where δ is proportional to $\lambda^2\Sigma\rho_i Z_i/A_i$, which is approximately 10^{-5}, and the absorption part, β, is proportional to $\lambda^2\Sigma\rho_i/A_i$, which is approximately 10^{-6}, at an x-ray energy of 10 keV.[31] Thus, for x-rays the index of refraction is close to one and depends on the wavelength, λ, of the x-rays, the density, ρ_i, the atomic weight, A_i, and the charge number, Z_i, of atom i in the sample. At surfaces and interfaces separating j films with different atomic species, the x-rays are scattered with different angles, Θ_j, and reflectivities, r_j, see Fig. 4a. This leads to interference effects and, therefore, to oscillations of the total reflectivity, R, as a function of the angle of incidence, Θ_1. This is demonstrated in Fig. 4b for a Ta_2O_5 film on a silicon substrate[32] for which both the surface and interface were fairly smooth. From the oscillation period, the thickness of the Ta_2O_5 film was obtained with an error of 1 to 2 percent, independently from the optical parameters. With increasing roughness of the interface these oscillations are damped, see Fig. 4c. Based on model calculations the *vertical* roughness, i.e., the averaged *amplitude* of the roughness, can be derived from measurements of the reflectivity of the specular beam.[30] Furthermore, the *lateral* roughness can also be determined by analyzing the intensity of the diffuse scattering *around* the specular beam.[33] In this way, x-ray scattering can provide detailed structural information about the interfaces. Neutron scattering[33] at grazing incidence gives basically the same information, but often with a different contrast between the elements as compared with x-rays, a difference which makes other combinations of elements in a film structure accessible. By using polarized neutrons, magnetic as well as chemical interface profiles can be probed. For neutron spectroscopy, however, more than one interface is typically required, which means that multilayer samples are necessary. Furthermore, the samples must be a few square centimeters in area, a size which often is not easy to prepare with sufficient lateral homogeneity.

2. 4. Rutherford Backscattering

Rutherford backscattering[34] is based on the backscattering of helium ions of typically 1 to 5 MeV from thin films. This technique yields a chemical depth analysis up to the penetration depth of the helium ions which is typically about one micrometer. The helium ions hitting the atoms of the sample are backscattered with different kinetic energies and detected by a suitable energy selective detector. The backscattering energy increases with the atomic mass and decreases with the penetration depth. Through a comparison of computer simulations[35] with the experimental results, the chemical depth profile and the thickness of the individual films in a multilayer system, can be determined. Usually films thicker than 50 nm are required for Rutherford backscattering studies.

Another important application of Rutherford backscattering is channelling experiments with epitaxial films. If the helium ions hit a single crystal sample exactly along the direction of an open channel within an *ideal* crystal lattice, they will penetrate the sample without perturbation and will not be backscattered. In a *non-ideal* crystal lattice, however, the ions are scattered at lattice imperfections resulting from interfaces, impurity atoms, dislocations, or strains, leading to a non-zero backscattering rate. On changing the angle of incidence by a few degrees around the channel direction, the backscattering rate will pass through a minimum at the exact parallel orientation. The lower the minimum yield, i.e. the lower the count rate at this optimum angle, the more perfect is the crystal lattice. Films of high epitaxial quality are characterized by a ratio of the in to off channel counting rates, R, of less

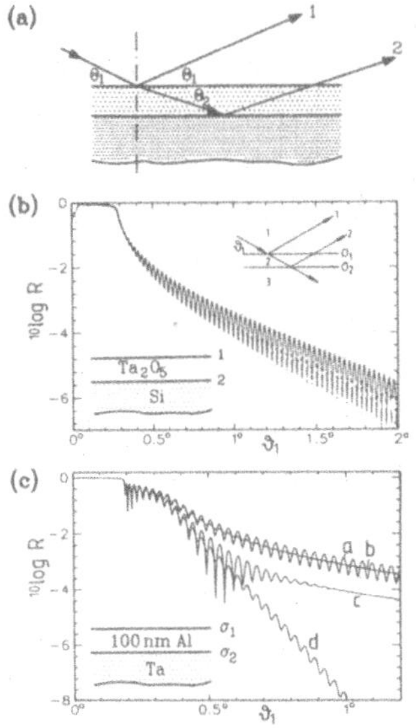

Figure 4. At surfaces and interfaces , a, x-rays are totally reflected for small angles of incidence, $\Theta_1 < 0.5°$. The 9.8 keV x-ray reflectivity, b, at a Ta_2O_5 film deposited on a silicon substrate with a smooth surface and interface. The oscillations arise from the interference of the beams reflected at the surface and interface, respectively. A simulation, c, of the 9.8 keV x-ray reflectivity at a 100 nm thick aluminum film, grown on a tantalum substrate, for three different values of the interface roughnesses, σ_1, at the air-aluminum surface, and σ_2, at the aluminum-tantalum interface. Roughness dampens the oscillation amplitude.[30]

than ten percent. An example[36] of a Rutherford backscattering spectrum and of a strain analysis by channeling experiments of an epitaxial EuS/SrS multilayer structure, is shown in Fig. 5. The individual EuS and SrS films are clearly resolved. From the angular shift of the minima, shown in Fig. 5b, the elastic dilatation and compression of the EuS and SrS layers, due to the misfit of the lattice parameters, could be determined. Based on detectors with improved energy resolution, Rutherford backscattering has recently been extended to lower ion energies of a few 100 keV, making feasible experiments on thinner films and on surface regions.[37]

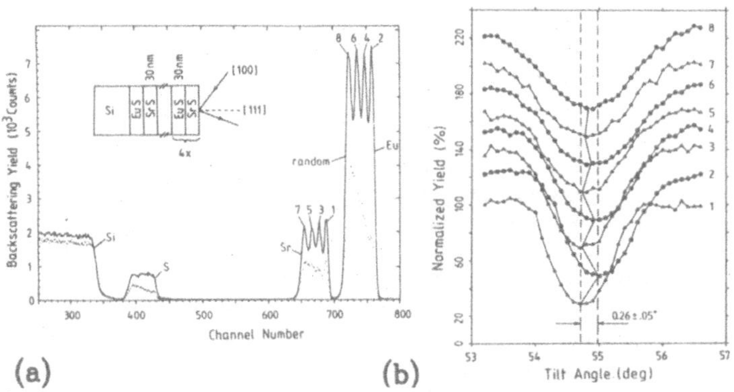

Figure 5. The random, solid line, and [100] aligned, dotted line, Rutherford backscattering spectra, a, for a EuS/SrS multilayer grown on a Si(111) substrate and obtained using 3.37 MeV He^{2+} ions.[36] The signals from Si, S, Eu, and Sr are clearly separated. The angular scans of the normalized backscattering yield from the individual EuS and SrS layers, b. From the observed angular shift the strain due to the lattice mismatch can be evaluated.

3. MULTILAYER AND INTERFACE MAGNETIC PHENOMENA

The magnetic behavior near a surface or interface may differ in many aspects from that of the bulk material. At an interface the translational symmetry of the bulk is broken. The reason for this break can be a different electronic band structure and/or a different arrangement of atoms and charges on both sides of the interface. New and interesting magnetic phenomena arise from the existence of interfaces as, for example, the modification of magnetic moments and hyperfine fields, changes in exchange interactions and spin wave excitations, or the appearance of magnetic interface anisotropies.

3. 1. Magnetic Moments and Hyperfine Fields: Theory and Experiment

It is well known that in metallic systems the magnetic moments are sensitive to their local environment. Consequently, the magnetic moments are expected to change at surfaces and interfaces. One can argue simply that, due to the missing neighbor atoms in one half of space, the electronic bands will become narrower at a surface, leading to an enhancement of the magnetic moments. As a rule, there exists a linear relationship between magnetic moments and the hyperfine fields in the *bulk*. Similarly, one would expect a modification of the hyperfine fields at interfaces, but as will be shown below, in this case the linear relationship is usually violated.

In recent years ab-initio band structure calculation methods have made enormous progress and are now able to calculate, and even predict, relevant magnetic quantities such as the magnetic moments, hyperfine fields, or, in a few cases, magnetic anisotropies for magnetic thin films, surfaces, and interfacial regions, for reviews see refs. 38, 39, 40, 41, 42 The theory is based on a full potential linearized augmented plane wave, or FLAPW, band structure calculation[43] and yields the local spin density from first principles. From the spin density, the ground state properties at zero temperature, i.e., the magnetic moments per atom, the hyperfine fields, the isomer shifts, the electric field gradients, and the type of magnetic order, can be derived.[44] The results for the hyperfine field are usually split into two contributions, $B_{hf} = B_{core} + B_{ce}$, due to the core polarization and the conduction electron polarization, respectively. Because the relevant contributions to the hyperfine interactions arise from the wave functions close to the nucleus, where the Coulomb potential is very large, relativistic effects become important and have to be taken into account. The calculations assume a thin film model structure containing the interface, which is infinitely extended in the film plane and symmetrically built up in a perpendicular direction. The total thickness of these model films is restricted by computer capacity and currently is about 11 to 15 monolayers. By calculating the magnetic properties for the individual monolayers, a profile across the interface can be obtained.

Freeman and his coworkers have investigated theoretically a wealth of different metal/metal interfaces, calculating magnetic moments and hyperfine fields on both sides of the interface. The results covering mainly 3d/3d, 3d/4d, and 3d/5d interfaces, as well as different crystal orientations, are reviewed in reference 45. Depending on the system, for example, the magnetic moment of iron is enhanced at interfaces compared with the bulk, whereas, in most cases, the iron hyperfine field is reduced. This effect is explained[46] by the field contribution, B_{ce}, of the conduction electron polarization changing sign, from negative in the bulk layers to positive in the interface layer, because of the broken translational symmetry. Meanwhile, other research groups have also performed band structure calculations, as a few examples will show. Fernando et al.[47] treated fcc-iron on Cu (001),Ostroukhov et al.[48] the free Fe(100) and Ni(100) surface, and Coehoorn[49] has calculated the magnetic hyperfine fields of bcc-Fe, fcc-Co, hcp-Co, and fcc-Ni. Blügel studied the magnetism of 4d-metal monolayers on Ag(001), Au(001), and Pd(001) surfaces[50] and found, in part large magnetic moments and ferromagnetic ordering for such two-dimensional arrangements. Even some 5d-metals should order ferromagnetically[51] for a film thickness of one monolayer.

Conversion electron Mössbauer spectroscopy now offers, together with the recent advances in epitaxial film preparation by ultra high vacuum-molecular beam epitaxy the opportunity to measure the magnetic hyperfine fields locally at and near iron interfaces or even surfaces. Therefore, theoretical predictions can now be checked experimentally. Conversion electron Mössbauer spectroscopy measurements at a free iron surface are extremely difficult because of the contamination of the surface by residual gases, even in an ultra high vacuum of 10^{-11} mbar. This is especially a problem for Mössbauer spectral work in which one to several days are often required to obtain a single Mössbauer spectrum. Gradmann and his coworkers are, so far, the only ones to have performed a successful experiment.[52] By using the iron-57 probe layer technique, see Section 2. 2. 1, with a sample structure of W(110)/10ML-^{56}Fe/1-3ML-^{57}Fe/dML-^{56}Fe/vacuum, where d ranged from zero to ten monolayers, ML, they achieved monolayer resolution and could measure the hyperfine field as a function of the distance of the layer from the iron-56/vacuum surface. For the experiment with the iron-57 probe layer deposited directly at the free surface they overcame the contamination problem by measuring the spectrum several times, for 8 hours each. The contamination, which increased from spectrum to spectrum, could be eliminated by extrapolation of the evaluated hyperfine field data to zero time. Comparison with theory for Fe(110) surfaces[53] gave fairly good quantitative agreement for the reduction of the hyperfine field in the surface layer and the enhancement in the first sub-surface layer. The quadrupole interaction due to the broken symmetry, which was observed only in the surface layer, vanished during the course of the experiments because of gradual covering by residual gases.[54]

The iron-57 probe layer technique was suggested and first applied by Tyson et al.,[55] who measured the profile of the hyperfine fields across an iron film at 4.2 K and room temperature and found remarkable deviations from the bulk value. Since these early studies, many interfaces in Fe/M systems have been studied with this technique. By means of transmission Mössbauer effect studies, Shinjo and his coworkers investigated polycrystalline Fe/X multilayers, where X is Mg,[56] Sn,[57] Sb, MgO, MnF_2,[58] Dy,[59] Nd,[60] and V,[61] containing many interfaces which were prepared on thin mylar films as substrates. The hyperfine fields at the Fe/Ag interface in epitaxial film structures with (001) and (110) orientations were studied by several groups.[62,63,64] Detailed conversion electron Mössbauer spectroscopy studies of the W/Fe and W/Fe/Ag systems, epitaxially grown mainly on the (110) orientation, were carried out by Gradmann and his coworkers. The interested reader should consult the review of this work for a complete compilation of the systems studied.[65] Such investigations, on samples with well defined epitaxial structure, permit a comparison between the experimentally and theoretically obtained hyperfine fields. For the W(110)/Fe interface, which reveals a strong decrease of the hyperfine field in the iron interface layer, good quantitative agreement with theory[66] was observed. For the Fe(110)/Ag interface, which reveals only a slightly enhanced hyperfine field in the iron interface layer, the results contradict the early theoretical results of Onishi et al.,[46] for Fe(001)/Ag, but are in good agreement with recent calculations by Handschuh[67] for Fe(001)/Ag. When performing such comparisons between theory and experiment, one must remember that the calculations are based on ideal flat interfaces, whereas the experimental results must involve more or less imperfect, rough interfaces, a situation which usually leads to a reduction of interfacial effects.

Fairly good agreement between conversion electron Mössbauer spectroscopy experimental results and theory was obtained for the Fe(110)/Cr and Fe(001)/Cr systems.[67,68] However, recent measurements[69] on W(110)/Cr(110)/Fe(110)/Cr(110) indicated that *two* different hyperfine fields may exist for the iron interface layer. The origin of these two fields was explained by two magnetically inequivalent sites due to the antiferromagnetic order in the chromium film. The distance dependence of the hyperfine fields near the interfaces in

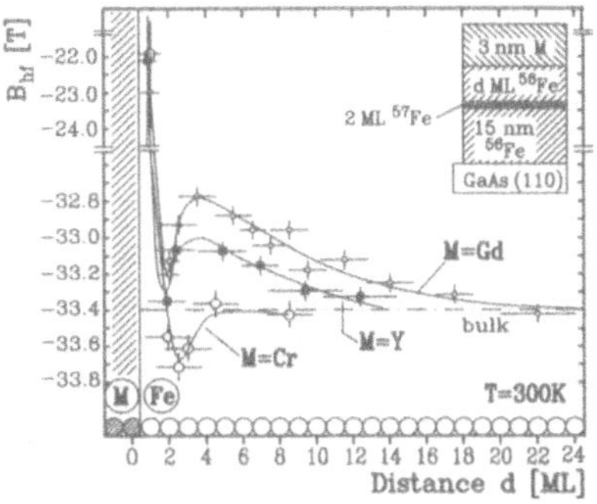

Figure 6. The distance dependence with respect to the interface of the magnetic hyperfine field, B_{hf} at 300 K in the Fe/M film structures, where M is Cr, Y, and Gd.[70]

Fe/M film structures, where M is a 3d, 4d, or 5d metal,[70] shows, in many cases at least, in which the iron and M couple *antiferromagnetically*, similar features, see Fig. 6. A strong reduction of the hyperfine field in the first iron monolayer at the interface is followed by a peak-like enhancement in the second and third iron monolayer. Then the bulk value of the hyperfine field is approached at different rates for the 3d, 4d, and 5d metals. The different range of the perturbation induced by the interface seems to be correlated to the increasing hybridization between the iron-3d and the 3d, 4d, and 5d wave functions of the neighboring metal film.[71] In the Fe/Gd system a surprisingly long interaction range was observed, a range which was explained by a polarization of the iron conduction electrons by the large gadolinium 4f moment of $7\mu_B$ [70] An interesting system is the Fe/Pd interface at which iron and palladium couple *ferromagnetically*. In this case a large enhancement of the hyperfine field of about 12 percent, as compared with a typical ca. 3 percent increase, was observed in the second iron monolayer, see Fig. 7, followed by an oscillating approach to the bulk value.[71] The oscillation with distance could be fit with a RKKY-like function with a wavelength described well by a wave vector of the iron Fermi surface. In addition to this phenomenological description, ab-initio full potential linearized augmented plane wave band structure calculations gave an excellent, even quantitative agreement with the conversion electron Mössbauer spectroscopy results.[67,72] A conversion electron Mössbauer spectroscopy

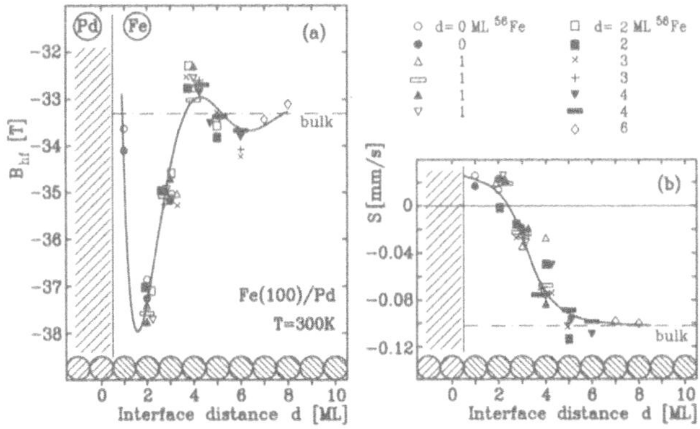

Figure 7. The distance dependence[71] relative to the Fe(100)/Pd interface of the magnetic hyperfine field, B_{hf}, at 300 K, a, in which the solid line represents a fit based on an RKKY-like function, and the isomer shift, b, in which the solid line is simply a guide to the eye.

of polycrystalline Fe/Pd multilayers with reduced resolution and less pronounced effects was also carried out by Li et al.[73]

3. 2. Hyperfine Fields in Pseudomorphic Structures

As it is well known, iron can transform from its usual bcc crystal structure into other metastable phases, such as fcc γ-iron at high temperatures, or hcp ε-iron at high pressures. Another procedure to stabilize such unusual phases is now available through pseudomorphic growth of thin films of a few monolayers thickness on suitable substrates. This growth becomes possible if the total energy of the various phases does not differ by very much. Maurer et al.[74] could grow epitaxial hcp-iron in Fe/Ru(0001) *superlattices* with sharp interfaces in which the hcp-iron structure was maintained for up to ten iron monolayers. It was concluded that two iron monolayers at the Fe/Ru interface are non-magnetic, and that the rest of the iron layers were in a high-spin state with a magnetic moment of 2.1 μ_B. Although the usual high pressure hcp ε-iron phase is non-magnetic, this surprisingly large magnetic moment was explained by the expansion of the iron lattice enforced by the pseudomorphic growth on ruthenium. However, in contrast to this work, Korecki et al.[75] recently found for Fe/Ru(0001) *bilayers*, that, after the first pseudomorphic hcp iron monolayer on Ru(0001), iron grows in a three-dimensional bcc island mode. Band structure calculations for Fe(001) and Fe(110) covered with one monolayer of ruthenium are presented in references 67 and 76, respectively.

Another prominent and thoroughly investigated example is fcc-iron on Cu(001). There was confusion on the results for a long time because all types of magnetic order, paramagnetic, ferromagnetic, and antiferromagnetic, were observed by conversion electron Mössbauer spectroscopy.[77] The variety of experimental results, which were obviously related to the sample preparation,[78,79] found an explanation through theoretical band structure calculations.[80] The resulting magnetic phase diagram for the ground state of fcc-iron shows a strong dependence of the magnetic properties on the Fe-Fe atomic distance at a critical Wigner-Seitz radius of ca. 2.71 a.u. and contains paramagnetic and ferromagnetic phases, as well as a high-spin and low-spin branch within an antiferromagnetic phase.[81,82] The review paper by Gradmann[2] treats the fcc-Fe/Cu problem in detail. Other examples of unusual pseudomorphic structures are bcc-cobalt grown on GaAs(001),[83] fcc-Co/Cu, which shows large magnetoresistance effects,[84] and bcc-nickel on Fe(001).[85]

3. 3. Temperature Dependence of Hyperfine Fields and Magnetic Interactions

Usually the band structure calculations of interfacial properties refer to the magnetic ground state at zero temperature, and give, in most cases, reliable results. The situation is not as satisfactory for the calculation of magnetic properties, such as magnetization or critical ordering temperatures, of band magnets at finite temperatures. A short review of the history of these calculations is given in ref. 2. Hasegawa[86] and Hubbard[87] independently proposed a finite temperature theory using the functional integral method within a static approximation. This is a mean field theory which regards the 3d-metal system as an ensemble of local magnetic moments. This method has also been extended to surface related problems.[88] By using a generalized Hubbard model, Nolting et al. calculated the temperature dependence of the magnetization and the Curie temperature of nickel.[89] This model was extended and applied to iron as well as nickel,[90] and to the localized magnet, gadolinium.[91] For interfaces a direct experimental verification of these theories is difficult because magnetization measurements can hardly provide monolayer resolution. However, by means of the iron-57 probe layer technique, conversion electron Mössbauer spectroscopy is able to determine layer by layer, near surfaces and interfaces, the temperature dependence of the magnetization from that of the hyperfine field, B_{hf}. Although the linear relationship between B_{hf} and the magnetic moment is lost at interfaces, it is strictly preserved for the *temperature dependences* of B_{hf} on the moment.[64]

In the low temperature regime, for $T < T_C/3$, the magnetization is determined by spin wave excitations and follows the Bloch $T^{3/2}$ spin-wave law,

$$\Delta M/M(0) = (M(0) - M(T))/M(0) = \Delta B_{hf}/B_{hf}(0) = bT^{3/2}.$$

Whereas this $T^{3/2}$ law is found to usually hold for thin magnetic films, including the surface or interface regions, the spin wave parameter, b, differs at interfaces as compared with the bulk. Rado[92] and Mills et al.[93] argued that the spin wave parameter, b_s, at the surface should be related to the bulk value by a factor of two, i.e., $b_s/b_{bulk} = 2$. Recently Mathon[94] suggested that the exchange coupling can be weakened close to an interface, leading to a b-value larger than twice the bulk value, b_{bulk}.. Furthermore, this interface b-value should be mainly related, reciprocally, to the interlayer exchange $J_\perp$ locally coupling two adjacent monolayers. Conversion electron Mössbauer spectroscopy can measure a profile of the spin wave parameter, b, across the film from the surface/interface to the center or bulk material. Some examples of such experiments are shown in Fig. 8, which gives the b-parameter in an iron film near the interface in the epitaxial Fe/M systems, where M is Cr, Y, or Gd.[70] The b-parameters follow an exponential distance law,

$$\Delta b = b(d) - b_{bulk} = \alpha \exp(-d/\delta),$$

where the distance d from the interface is given in iron monolayers and $b_{bulk} = (5.4 \pm 0.2) \times 10^{-6}\ K^{-3/2}$.[64,70] The range parameter, δ, increases from chromium to gadolinium, namely, $\delta_{Fe/Cr} = 0.7$, $\delta_{Fe/Y} = 1.3$, and $\delta_{Fe/Gd} = 2.8$ monolayers, reflecting the increasing interaction range. An exponential decay of these b-parameters was also observed by Gradmann et al.[95] on various Fe/W systems. In their paper a compilation of all b-parameters measured to date is presented. By plotting the spin wave parameter, in units of nm, as a function of the d-band width of the contacting film, a fairly linear relationship is obtained.[71] This indicates that the hybridization between the Fe 3d and the 3d, 4d, and 5d wave functions of the contact film should play an important role in determining the range of the magnetic interaction at the interface.

Another important aspect is the Curie temperature of thin magnetic films, monolayers and interfaces. Generally, the Curie temperature, T_C, of thin films decreases with reduced film thickness.[96] Related experiments have been performed, for instance on Ni[97] and Gd[98] thin films. Gradmann and coworkers performed a skillful study of the Curie temperature of

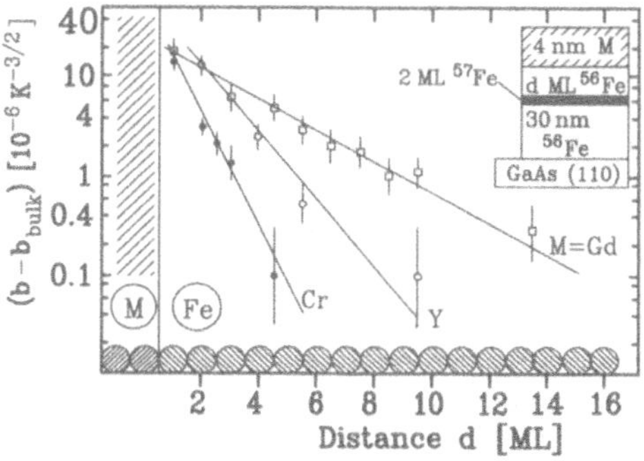

Figure 8. The deviation of the spin wave parameter, b, from the iron bulk value as a function of the distance from the Fe/M interface, where M is Cr, Y, and Gd.[68,70]

one iron monolayer on W(110) coated with Ag, including a spot profile analysis of low energy electron diffraction of the structure of the iron monolayer, which yielded a Curie temperature of 282 K for the Ag covered sample and of ca. 210 K for the uncoated sample. For detailed results, as well as for a survey of other Curie temperature studies on thin films the reader should consult reference 2.

3. 4. Interface Anisotropy

As a consequence of the predominantly dominating shape anisotropy in thin ferromagnetic films, the orientation of the spontaneous magnetization usually lies parallel to the film plane. Then, for homogeneously magnetized samples, the orientation within the film plane is determined mainly by the crystalline anisotropy. In very thin films of a few monolayer thickness, however, the surface or interface anisotropy arising from the break of the translational symmetry also becomes apparent. This anisotropy can even rotate the easy axis normal to the film plane. For 3d ferromagnets these interface anisotropy fields are quite large and can amount to several Tesla.[99,100] Finally, the superposition of all the contributions to the anisotropy, including strain effects at the interface, determines the orientation of the magnetization within the film.[101]

The surface anisotropy is inversely proportional to the film thickness, d, and can be determined from a series of similarly prepared films with different thickness.[99] Ferromagnetic resonance and sensitive magnetic studies, such as torsion balance, Faraday balance, vibrating sample magnetometer, SQUID magnetometer, and magnetooptical Kerr effect studies, permit a quantitative determination and separation of the various anisotropy contributions.[102,103,104] Conversion electron Mössbauer spectroscopy and perturbed angular correlation studies are also able to determine the orientation of the magnetization in thin films through the use of a proper experimental geometry, e.g., with perpendicular or nearly grazing incidence of the γ-rays relative to the film plane, in which case this information can be obtained uniquely from the relative line intensities of the conversion electron Mössbauer spectrum, see Section 2. 2. 1. Especially attractive for possible technical applications as high density magnetic recording media, are film structures having perpendicular magnetization. This can be achieved, for example, with Co/X, where X is Pd, Pt, Ir, or Au[105] or Co/Ru[106,107] multilayers, provided that the cobalt film thickness is less than ca. 1.5 nm. The same effect occurs for Fe(001)/Ag(001)[108] and Fe/Mg[56] systems for an iron thickness of about two monolayers. The out-of-plane magnetization is explained by the interface anisotropy overcoming the other anisotropy contributions for a sufficiently small film thickness, together with strain effects due to the lattice mismatch. Additionally, theory is also able to treat the interface anisotropy from first principles.[109,110] A nice example, which also illustrates the possibility of tailoring modern material for a particular purpose, is the prediction of perpendicular magnetization in the fcc (111) multilayers, $(1\text{ML-Co}/2\text{ML-Ni})_n$ by band structure calculations,[111] and their subsequent experimental verification.[112]

3. 5. Iron/Rare-Earth Film Structures

Just as for the TM-RE alloys, where TM is a transition metal, such as iron or cobalt and RE is a rare-earth metal, the layered TM/RE structures reveal a variety of interesting features. Among them are the existence of a compensation temperature for the total magnetization, as well as a perpendicular anisotropy, a feature which is of interest for magnetooptical data storage applications. As a rule, TM and heavy RE metals couple antiferromagnetically leading to a layered ferrimagnetic structure. Then, at a certain "compensation temperature" the magnetizations of the RE and TM layers can compensate each other, resulting in a zero total magnetization. This effect is similar to the behavior of amorphous TM-RE alloy films. For sputtered Co/Tb[113] and ultra high vacuum-evaporated Fe/Tb[114] multilayers, it is well known that the compensation temperatures agree quantitatively for layered and mixed alloy films of nominally equivalent composition. This finding holds for iron film thicknesses of less than ca. 3 nm and indicates a long range RE-TM exchange interaction.

Because of the half-filled 4f shell of gadolinium, the layered Fe/Gd system has attracted broad interest in recent years.[115,116] Because of the antiferromagnetic Fe-Gd coupling a compensation temperature is observed,[117] a temperature which varies depending on the iron and gadolinium film thicknesses. The system has a complex magnetic phase diagram with a "twisted spin" phase which was suggested theoretically by Camley and coworkers,[118] and was confirmed experimentally by neutron scattering[119] and Mössbauer effect[120] studies. Fe/Tb is another system which was thoroughly studied by the Mössbauer effect because of the occurrence of perpendicular anisotropy. There seems to be no distinct difference in the magnetic properties between sputtered and ultra high vacuum-evaporated samples. The magnetization makes an angle with the normal to the film plane which varies in the range of 30 to 90°, depending on temperature and thickness of the iron and terbium layers.[121] From a recent conversion electron Mössbauer spectral study of iron films grown in ultra high vacuum on a Tb(0001) single crystal and coated with terbium, Scholz et al.[122] concluded that the out-of-plane magnetization is caused by the interfacial anisotropy. A structural and magnetic phase diagram of sputtered Fe/Tb multilayers was proposed by Teillet et al.[123] By using conversion electron Mössbauer spectroscopy, Shinjo and his coworkers have recently studied[124] the Fe/RE multilayer systems, where RE covered all the rare earth elements between lanthanum and lutetium. They found out-of-plane anisotropy for Pr, Nd, Eu, Dy, and Tb.

3. 6. Magnetic Interlayer Coupling and Giant Magnetoresistance

In 1986 Grünberg et al.[125] observed, by Brillouin light scattering, an antiferromagnetic coupling between two iron films separated by only a few monolayers of a chromium film. Later, Parkin et al.[126] discovered that this coupling oscillates from ferromagnetic to antiferromagnetic alignment upon continuously increasing the thickness of the non-magnetic interlayer. Such exchange coupled multilayers very soon became particularly interesting

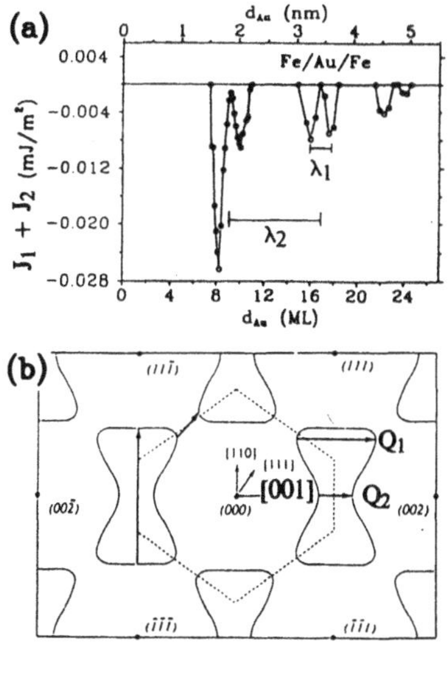

Figure 9. The magnetic interlayer coupling, a, in a Fe(100)/Au/Fe trilayer as a function of the gold interlayer thickness for two oscillations with wavelengths λ_1 and λ_2 and the Fermi surface of gold, b. The wave vectors Q_1 and Q_2 in the [001] direction are inversely proportional to λ_1 and λ_2, respectively.[134]

because of their "giant magnetoresistance," or GMR, which was discovered first in Fe/Cr layered structures,[127,128] and will soon be used commercially in miniaturized magnetic sensors.[129] For a recent survey of this work the reader should consult reference 130.

Generally, there seems to be agreement[131] that interlayer magnetic coupling is based on an oscillating RKKY-type exchange interaction mediated by the conduction electrons of the non-magnetic interlayer. This interpretation is confirmed experimentally by the observation of a superposition of two oscillation periods in the thickness dependence of the coupling parameter, J_1, namely a short period with approximately two monolayer wavelength and a longer period. For example, this coupling was found for Fe/Cr[132] and Fe/Au[133] as well as for many other systems. In fact, the observed wavelength, λ_i, of the oscillation, i, is related quantitatively to a vector, Q_i, on the Fermi surface of the interlayer material by the relationship, $\lambda_i = 2\pi/Q_i$, see refererence 134 and Fig. 9. Phenomenologically, the interlayer coupling can be described by an exchange parameter, J_1, which oscillates with the interlayer thickness and can therefore be positive and yield ferromagnetic coupling, or negative and yield antiferromagnetic coupling. Additionally a 90° coupling was also observed,[135] a coupling which tends to orient the two ferromagnetic films at 90° to each other. This coupling is described by a parameter, J_2, which is usually small and dominant only near the zero crossing points of J_1. Whereas J_1 is thought to arise from a RKKY-type interaction and to be related to the shape of the Fermi surface, J_2 seems to have its origin in the roughness of the interfaces separating the layers, for example, the iron and chromium layers. Various models based on this idea have been suggested.[136,137] The coupling parameters, J_i, can be determined using the magnetooptical Kerr effect, ferromagnetic resonance, or Brillouin light scattering.

Although giant magnetoresistance effects were recently also discovered in the related clustered alloys, such as Cu-Co and Co-Ag alloys,[138] the layered films that are antiferromagnetically coupled, will probably be more important for technical applications because of their high sensitivity in weak magnetic fields.[139] The giant magnetoresistance results from the strong enhancement of the resistance for antiferromagnetic alignment of a magnetic film structure as compared with ferromagnetic alignment. The ferromagnetic alignment can be caused by a magnetic field, e.g., by the stray field of the magnetic bits in a data storage medium. As an example, Fig. 10 shows the giant magnetoresistance ratio, i.e., the ratio between the resistances measured in a field and in a zero-field, for Fe/Cr/Fe(110) trilayers,[128] Fig. 10a, and Co/Cu(111) multilayers,[140] Fig. 10b. Apparently, the giant magnetoresistance also shows oscillations as a function of the spacer layer thickness and follows exactly the oscillations of the interlayer coupling, see Fig. 10b. The giant magnetoresistance is generally ascribed to the different scattering probability of the spin-up and the spin-down conduction electrons in successive magnetic layers and, particularly at interfaces[130] between antiferromagnetically coupled layers. A necessary condition is that the distance between the layers must be smaller than the mean free path of the electrons. Giant magnetoresistance, as well as magnetic interlayer coupling, depend on the roughness of the interface. Rough interfaces wipe out the details of the interlayer coupling, for example, the short two monolayer oscillations. Giant magnetoresistance depends strongly on the interface roughness,[141,142,143] which can sometimes be increased,[144] but also reduced.[145] Here, nuclear methods, such as conversion electron Mössbauer spectroscopy and NMR are able to provide information on the epitaxial quality of the interfaces.

4. STRUCTURAL PROPERTIES OF INTERFACES

Some magnetic properties of magnetic thin film structures are related primarily to the sharpness and roughness of the interfaces. The sharpness of the interface is determined by local interdiffusion leading to an alloy-like zone of a finite width, whereas the roughness is

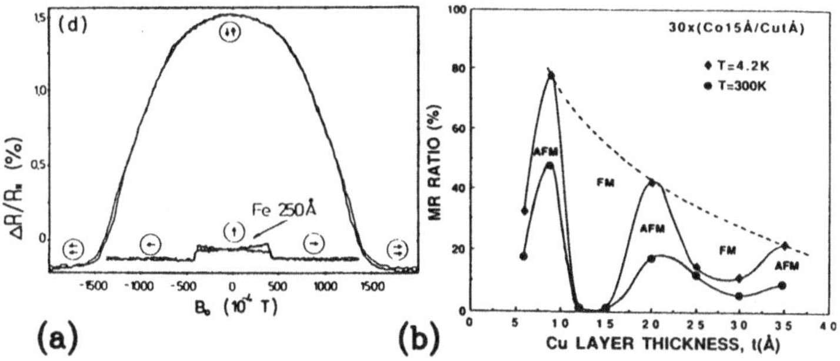

Figure 10. The magnetoresistance ratio, $\Delta R/R_{||} = (R - R_{||})/R_{||}$, of a Fe/Cr/Fe(110) trilayer with antiferromagnetic coupling, a. The ordinary anisotropic magnetoresistance effect of a 25 nm thick iron film is shown for comparison.[128] The oscillatory variation of the magnetoresistance ratio as a function of the copper layer thickness in (1.5 nm Co)/Cu(111) multilayers, b. The thickness ranges with ferromagnetic and antiferromagnetic coupling are indicated.[140]

related to the probability and distribution of steps of one or a few atoms height and to a characteristic mean extension of the steps. Of course, the differences cannot actually be distinguished for really sharp interfaces, which may be defined as two monolayer thick interfaces. The hyperfine field, as measured by conversion electron Mössbauer spectroscopy or NMR, is very sensitive to the local atomic configuration which is determined predominantly by the nearest and next nearest neighbor atoms. Therefore, the value of the hyperfine field can serve as a fingerprint of the chemical environment of the probe atom.

4. 1. Interdiffusion

With a few exceptions, the iron-57 hyperfine field is reduced on substitution of iron neighbor atoms by other elements. This reduction permits the determination of the interdiffusion profile at interfaces with the iron-57 probe layer technique, see Section 2. 2. 1. In the Mössbauer spectra additional subspectra show up with hyperfine field values smaller than the iron bulk value, fields corresponding to lattice sites near a foreign atom. For iron atoms diffused into the adjacent film the corresponding subspectrum usually collapses to a broadened single line which can be easily identified. Such investigations have been performed over the past ten years on Fe/V multilayers with iron-57 conversion electron Mössbauer spectroscopy[146,147] and with vanadium-51 NMR,[148] respectively, revealing an interface width of about three monolayers. A simple and straightforward estimation of the interface width can be made by comparing the area of the interface related subspectra to the area of the bulk related subspectrum. As a few recent examples of interdiffusion studies we mention conversion electron Mössbauer spectroscopy studies of Fe/Co,[149] Fe/Pt,[150] and Fe/Rh[151] multilayers.

4. 2. Interface Roughness

The effect of interface roughness on the shape of Mössbauer spectra is demonstrated in Fig. 11. A sharp six line bulk spectrum of α-iron is obtained if the iron-57 probe layer is deposited seven monolayers from the Fe/Cr interface, see Fig. 11c. On approaching the interface, Figs. 11a and b, an additional complex structure appears at the expense of the bulk component.[152] From these subspectra information about both interface roughness and interdiffusion can be obtained. An example of such a roughness analysis, on several epitaxial Fe/Cr film structures grown on MgO(001) and GaAs(001) substrates at different substrate

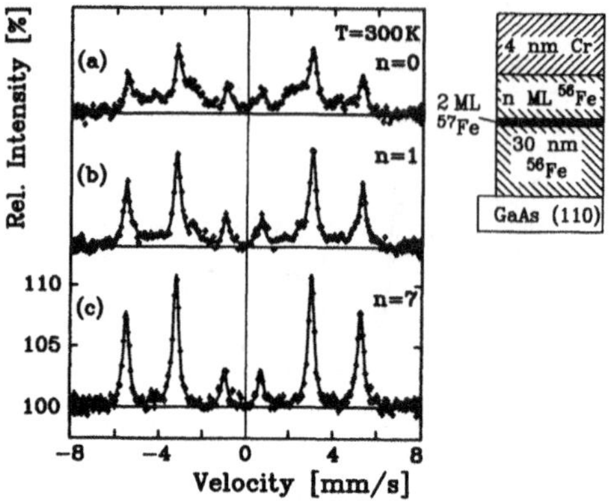

Figure 11. The conversion electron Mössbauer spectra of Fe(110)/Cr film structures measured at distances of zero, one and seven monolayers from the Fe/Cr interface by means of the iron-57 probe layer technique.[152]

temperatures, is given in reference 153 and is shown in Fig. 12. An Fe/Cr/Fe(001) trilayer was grown on a GaAs(001) substrate with a 150 nm thick silver buffer layer, see Fig. 12b. The interface related structure, revealed in the conversion electron Mössbauer spectrum, Fig. 12a, could be fit rather well by assuming a superposition of ten magnetically split six line subspectra. By applying the line sharpening technique of Afanas'ev and Tsymbal,[154] the fit procedure could be performed with enough resolution and reliability to yield a discrete distribution of hyperfine fields, as is shown in Fig. 12c.

In order to derive a roughness pattern for the Fe/Cr interface, the following model was used. By first assuming, that the roughness consists only of steps and islands of one monolayer height and, second, by taking into account nearest and next nearest neighbor interactions, then there exist only ten different Fe-Cr neighbor configurations at an interface in a crystalline (100) oriented bcc lattice. From the analysis of many Fe(001)/Cr interfaces it was found empirically, that the hyperfine field of these ten configurations can be described by the equation

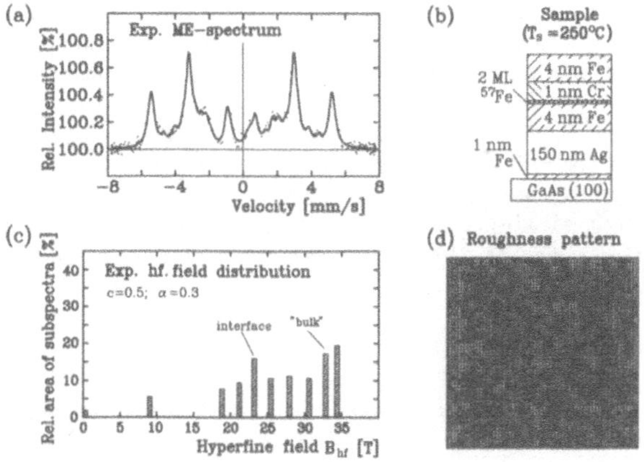

Figure 12. The conversion electron Mössbauer spectrum, a, of a Fe(100)/Cr film with the structure sketched in b, and the hyperfine field distribution, c, as derived from this spectrum. The Fe/Cr interface roughness pattern obtained by comparison of the experimental hyperfine field distribution with model calculations, d.[153]

$$B_{hf} = B_{bulk} + n_1 B_1 + n_2 B_2 + \Delta B,$$

where B_{bulk} is −33.3 T, the bulk value of the hyperfine field at 300 K, and B_1 is +2.5 T and B_2 is +2.05 T, the contributions to the field of the n_1 near neighbor chromium atoms and the n_2 next nearest neighbor atoms around the iron-57 probe atom, respectively. For a consistent description of the experimental hyperfine fields, an additional constant contribution, ΔB of −1.75 T had to be introduced, a contribution which was explained qualitatively as the field enhancement arising at interfaces due to the break of the symmetry, see Section 3. 1. In order to obtain the distribution of the Fe-Cr configurations, it was assumed that the iron and chromium films are separated by only one monolayer, which represented a two-dimensional Fe-Cr alloy. This alloy-like monolayer is characterized by two parameters, the concentration, c, of the iron-57 atoms and a short range order parameter, a, which is the probability of finding an iron near neighbor atom at a specific iron site. Statistical mixture means that a is zero and complete separation means that a is one. Based on this model various hyperfine field distributions, with different c and a values, can be calculated and compared with the experimental hyperfine field distribution. By following this procedure, a roughness pattern, see Fig. 12d, which represents a top-view of the alloy-like interface monolayer, was derived. As its principal result, this study showed that this fine grained roughness pattern, on a scale of 1 to 2 nm, does not change very much with different sample preparation conditions, as long as the samples still have good epitaxial quality. Apparently, there is an intermixing of iron and chromium atoms within one to two monolayers of the interface even in high quality films. In addition to such experiments, first principle calculations of the magnetic moments, but unfortunately not of the hyperfine fields, at rough Fe/Cr interfaces have been carried out by Hasegawa[155] and Coehoorn.[141]

Similar results, namely an intermixture, on a small scale, were reported on high quality epitaxial Fe(001)/Ag(001) structures.[156,157] In the latter work of Schurer et al., the isomer shift, rather than the hyperfine field, was used as a fingerprint for distinguishing the different atomic configurations. As has already been mentioned, NMR is more sensitive than conversion electron Mössbauer spectroscopy and even makes possible a strain analysis of the interface region. A nice example of a detailed NMR study of the atomic topology at the interfaces in fcc-Co/Cu multilayers was reported by de Gronckel et al.[158] Their conclusion was that the Co/Cu interface also consists of a mixed monolayer. Naturally, in addition to the hyperfine fields, the magnetic moments and magnetic anisotropies are modified at the step positions of the rough surfaces and interfaces. This has been investigated with Fe(110) films grown on W(110)[159] substrates and Fe/Cr[160] systems.

4. 3. Amorphous Growth of Iron

By preparing thin films by ultra high vacuum evaporation one usually obtains polycrystalline samples with some degree of texture or even, in some cases, epitaxial films. However, by choosing the appropiate preparation conditions, i.e., substrate, substrate temperature, and evaporation rate, one can form and stabilize unusual crystal structures. For example, iron metal can seldom be produced in an amorphous state because of its strong tendency toward recrystallisation, even at low temperature. But its preparation becomes possible, by evaporating an iron film up to a critical thickness on a rare earth substrate, as it was observed first by Shinjo and his colleagues, with Fe/Dy[59] and Fe/Nd[60] multilayers.

A conversion electron Mössbauer spectroscopy and high resolution electron microscopy study of Gd/Fe/Gd *trilayers*[18] revealed that iron grows in an amorphous phase on hcp-gadolinium up to a critical thickness, d_{cr}, of 2.3 nm. Above this thickness the iron film crystallizes in the usual bcc structure, a crystallization that is accompanied by the simultaneous recrystallisation of the 2.3 nm thick initially deposited iron substrate film. This thickness dependent crystallisation was found to be quite sharp, as is shown in Fig. 13. On increasing the iron film thickness up to 2.3 nm only an amorphous conversion electron

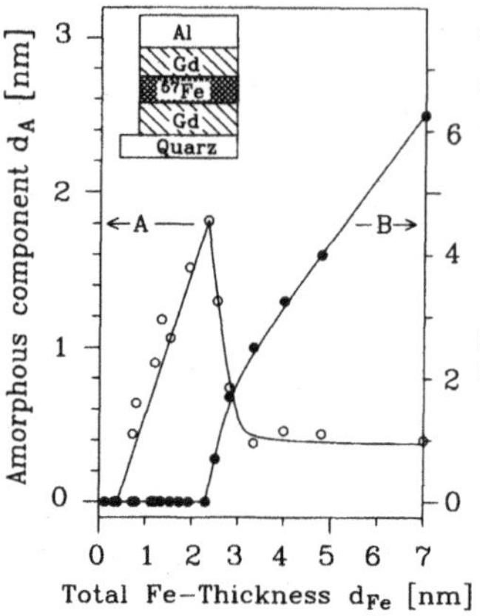

Figure 13. The relative contributions of the amorphous, A, and bcc-crystalline, B phases of the conversion electron Mössbauer spectra of the Gd/Fe film structures converted to partial iron thicknesses, d_A and d_B, respectively, as a function of the total iron interlayer thickness, d_{Fe}.[18]

Mössbauer spectrum was observed. However, starting at a thickness of ca. 2.3 nm a crystalline phase appeared at the expense of the amorphous phase. Above ca. 3.0 nm the iron film is completely polycrystalline except for a resistant interfacial zone which is non-magnetic at 300 K, an interface zone of one to two monolayer thickness which may consist of an Fe-Gd mixture. The amorphous growth mode occurs also without the gadolinium coating film, i.e., with a Gd/Fe *bilayer.*[18] Therefore the amorphous growth seems not to be a multilayer effect, but is rather solely a property of the Gd/Fe interface. Apparently, the large differences in crystal structure, lattice parameters, atomic sizes, and surface energies, between iron and the hcp rare-earth are responsible for the amorphous growth. In the high resolution electron microcroscopy images, shown in Fig. 1, the thickness dependent amorphous/polycrystalline iron structure can be visualized directly. The amorphous growth mode was also observed for other Fe/RE systems, namely Fe/Ce[161] and Fe/Tb,[123] with a critical thickness of ca. 2.2 nm, and for Fe/Mg[56] with a critical thickness of ca. 1.2 nm.

In order to obtain information on the magnetic properties of amorphous iron, a-iron, to date only extrapolations x equal to zero of the data on amorphous $Fe_{1-x}M_x$, where M is Zr or Y, were available, an extrapolation which suggests a reduced magnetic moment and spin

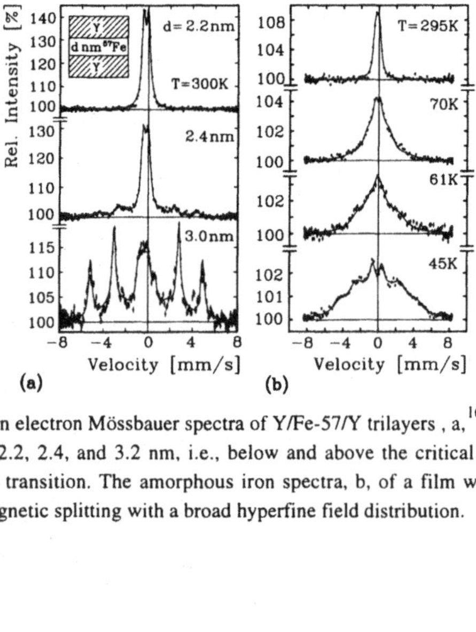

Figure 14. The conversion electron Mössbauer spectra of Y/Fe-57/Y trilayers , a,[163] at 300 K for samples with iron thicknesses, d_{Fe} of 2.2, 2.4, and 3.2 nm, i.e., below and above the critical thickness of 2.3 nm of the amorphous to crystalline transition. The amorphous iron spectra, b, of a film with d_{Fe} = 2.2 nm at various temperatures showing magnetic splitting with a broad hyperfine field distribution.

glass properties. The layered system Fe/Y, showing amorphous growth of an iron film up to a d_{cr} of ca. 2.3 nm, offers the chance to measure those properties directly because the results are not influenced by the *non*-magnetic yttrium. The conversion electron Mössbauer spectra of Y/Fe/Y trilayers,[162] as shown in Fig. 14a, illustrate the paramagnetic a-iron phase for d_{Fe} = 2.2 nm, the transition to amorphous/bcc at 2.4 nm, and the magnetic crystalline bcc phase at 3.0 nm. On cooling the amorphous sample with d_{Fe} = 2.2 nm, the spectrum splits magnetically and because of the broadening can only be fit with a distribution of hyperfine fields, as is shown in Fig. 14b. From the temperature dependence of the average hyperfine field, an average magnetic ordering temperature was determined. Magnetization measurements and conversion electron Mössbauer spectra indicated that the ordering should be that of a disordered ferromagnet. When plotted as a function of d_{Fe}, the average ordering temperature approaches saturation close to the critical crystallization thickness of d_{cr} = 2.3 nm, where the influence of the Fe/Y interface is smallest, see Fig. 15b. The saturation value of ca. 80 K for the average magnetic ordering temperature was regarded as an average Curie temperature of amorphous iron. Magnetization measurements on a set of samples with different d_{Fe} values yielded the magnetic moment per iron atom. This moment decreases from the value of bcc-iron at large d_{Fe} thicknesses to ca. 1.2 μ_B directly below d_{cr}, i.e., in the amorphous phase, see Fig. 15a.

Concurrently with the above experimental work, amorphous iron was also studied by modern theoretical methods. By using numerical simulations within an itinerant description of the local magnetic properties, Krey et al.[163] calculated the magnetic moment of amorphous iron as a function of the iron density, i.e., of the mean volume, V, of amorphous iron. They found a moment versus volume relationship similar to the antiferromagnetic branch in the magnetic phase diagram of fcc-iron, which was derived by band structure calculations by Moruzzi et al.[81] For the magnetic ordering of amorphous iron they obtained two nearly energetically equivalent states, a spin glass state and an asperomagnetic state, in which the spins are randomly oriented with a net orientation in one direction. Kakehashi et al.[164] also proposed the existence of a spin glass state in amorphous iron from a finite temperature theory of amorphous transition metals. Finally, in a recent theoretical study, Lorenz et al.[165] suggested that amorphous iron undergoes, with increasing density, a continuous transition from a ferromagnetic to an asperomagnetic, and then to a speromagnetic, spin glass like, state. This spin glass state should exist over a considerable range of densities. Experimentally,[166] a spin glass behavior for amorphous iron is supported by magnetization and conversion electron Mössbauer spectroscopy measurements on sputtered and ultra high vacuum evaporated Fe/Y multilayers.

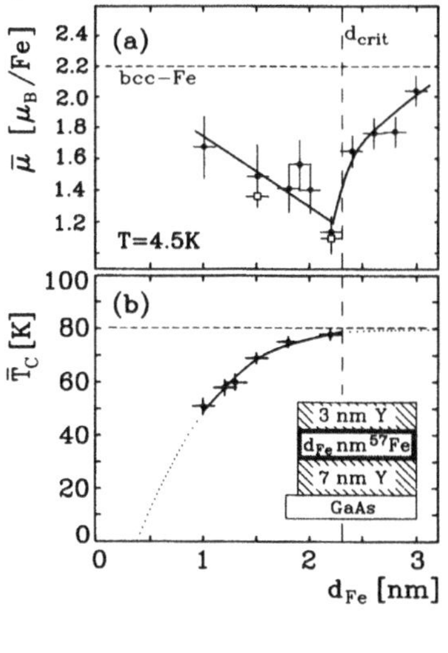

Figure 15. The variation of the mean magnetic moment and the mean Curie temperature of amorphous iron with respect to the iron interlayer thickness, d_{Fe} , in $Y/^{57}Fe/Y$ trilayers.[163]

5. CONCLUSIONS

The application of nuclear techniques to the study of thin magnetic film structures has undergone a strong impetus in the last decade because of both the tremendous growth in interest in magnetic films and the remarkable progress in the preparation and tailoring of multilayer structures. The high detection sensitivity, the monolayer resolution, and the access to internal interfaces in magnetic multilayers, are outstanding features and advantages of these methods.

Unfortunately, these techniques are still restricted to the use of only a few suitable tracer nuclei, such as iron-57 and tin-119 in conversion electron Mössbauer spectroscopy or indium-111 in perturbed angular correlation studies. Of course, by combining conversion electron Mössbauer spectroscopy, NMR, and perturbed angular correlation, the number of employable nuclei is widened. For conversion electron Mössbauer spectroscopy on thin film structures, important progress is expected from the availability, in the near future, of synchotron radiation sources for Mössbauer effect γ-radiation. These new sources will also permit studies using other low energy Mössbauer effect transitions, such as europium-151. The high intensity of synchrotron radiation will reduce the time required to collect a spectrum from the present several hours to several days to perhaps a few minutes. A completely new feature of Mössbauer spectroscopy arises from the small synchrotron radiation beam diameter, which can be reduced even further by focussing. For the first time this offers the advantage of a point analysis, i.e., of *scanning* conversion electron Mössbauer spectral image of the surface of a thin film. In view of these new developments it is expected that, in the near future, nuclear techniques will make valuable new contributions to the study of thin magnetic layered structures.

REFERENCES

1. T. Shinjo, *Surf. Sci. Repts.* **12**, 49 (1991).
2. U. Gradmann, in *Handbook of Magnetic Materials,* K. H. J. Buschow, ed., Vol. 7, p. 1, Elsevier Science Publ., Amsterdam (1993).
3. Ch. Sauer and W. Zinn, in *Magnetic Multilayers*, L. H. Bennett and R. E. Watson, eds., p. 147, World Scientific, Singapore (1994).
4. L. M. Falicov, *Thin Solid Films* **216**, 169 (1992).
5. F. Gautier and D. Stoeffler, *Surf. Sci.* **249**, 265 (1991).
6. M. Henzler, in *Electron Spectroscopy for Surface Analysis,* H. Ibach, ed., p. 119 Springer, Berlin (1977).
7. M. G. Lagally, D. E. Savage, and M. C. Tringides, in *Reflection High-Energy Electron Diffraction and Reflection Electron Imaging of Surfaces,* P. K. Larsen and P. J. Dobson, eds., p. 139, Plenum Press, New York (1988); S. Ino, *ibid.* p. 3.
8. M. Henzler, *Surf. Sci.* **298**, 369 (1993).
9. B. Müller, Ph. D. Thesis, University of Hannover, 1994.
10. A. L. de Loxanne, in *Physical Methods of Chemistry*, B. W. Rossiter and R. C. Baetzold, eds., Vol. IXA, "Investigations of Surfaces and Interfaces - Part A," p. 141, John Wiley, New York (1993).
11. H. J. Güntherodt, in *Scanning Tunneling Microscopy,* Vol. 1, *General Principles and Applications to Clean and Adsorbate Covered Surfaces*, Springer, Berlin (1992).
12. S. M. Hues, R. J. Colton, E. Meyer, and H. J. Güntherodt, *Mater. Res. Bull.* **18**, 41 (1993).
13. D. P. Woodruff and T. A. Delchar, *Modern Techniques of Surface Science*, Chapters 2 and 3, Cambridge Univ. Press, Cambridge (1986).
14. N. H. Turner, see ref. 10, Part-B, p. 123.
15. D. Norman, see ref. 10, Part-B, p. 331.
16. R. Feder, *Spin Polarized Electrons in Surface Physics*, World Scientific, Singapore (1985).

17. J. P. Eberhart, *Structural and Chemical Analysis of Materials*, p. 391, John Wiley, Chichester (1991).
18. J. Landes, Ch. Sauer, B. Kabius, and W. Zinn, *Phys. Rev. B* **44**, 8342 (1991).
19. P. Gütlich, R. Link, and A. Trautwein, *Mössbauer Spectroscopy and Transition Metal Chemistry*, Springer, Berlin (1978); A. Vertes, L. Korecs, and K. Burger, *Mössbauer Sectroscopy*, Elsevier Science Publ., Amsterdam (1979); B. V. Thosar and P. K. Iyengar, *Advances in Mössbauer Spectroscopy*, Elsevier Science Publ., Amsterdam (1983).
20. J. A. Sawicki and B. D. Sawicka, *Hyp. Interact.* **13**, 199 (1983).
21. Zs. Kajcsos, Ch. Sauer, and W. Zinn, *Hyp. Interact.* **51**, 1147 (1989); *Hyp. Interact.* **57**, 1889 (1990); Zs. Kajcsos, Ch. Sauer, W. Zinn, R. Kurz, W. Meyer, and M. A. C. Ligtenberg, *Hyp. Interact.* **58**, 2519 (1990); Zs. Kajcsos, Ch. Sauer, W. Zinn, W. Meisel, H. Spiering, M. Alflen, and P. Gütlich, *Hyp. Interact.* **71**, 1469 (1992); Zs. Kajcsos, W. Meisel, P. Griesbach, P. Gütlich, Ch. Sauer, R. Kurz, K. Hildebrand, R. Albrecht, and M. A. C. Ligtenberg, *Nucl. Instr. Meth. B* **93**, 505 (1994).
22. C. P. Slichter, *Principles of Magnetic Resonance*, Springer Series in Solid State Physics, Vol. 1, Berlin (1978).
23. H. A. M. de Gronckel and W. J. M. de Jonge, see ref. 3, p. 111.
24. T. Thomson, P. C. Riedi, and D. Greig, *Phys. Rev. B* **50**, 10319 (1994).
25. T. Klas, R. Fink, G. Rausch, R. Platzer, J. Voigt, R. Wesche, and G. Schatz, *Surf. Sci.* **216**, 270 (1989); G. Rausch, R. Fink, K. Jacobs, U. Kohl, J. Lohmüller, B. Luckscheiter, R. Platzer, B. U. Runge, U. Wöhrmann, and G. Schatz, *Hyp. Interact.* **78**, 261 (1993); H. Haas, *Hyp. Interact.* **84**, 69 (1994).
26. B. Swinnen, J. Dekoster, G. Langouche, and M. Rots, *J. Magn. Magn. Mat.* **148**, 148 (1995).
27. S. Seeger, H. Haas, and A. Mayer, *Hyp. Interact.* **84**, 211 (1994).
28. C. F. Majkrzak, J. F. Ankner, N. F. Berk, and D. Gibbs, see ref. 3, p. 299.
29. B. Lengeler, in *Photoemission and Absorption Spectroscopy of Solids and Interfaces with Synchrotron Radiation*, M. Campagna and R. Rosei, eds., p. 157, North-Holland, Amsterdam (1990).
30. B. Lengeler, *Adv. X-ray Analysis* **35**, 127 (1992).
31. R. W. James, *The Optical Principles of the Diffraction of X-rays*, Cornell University Press, Ithaca (1967).
32. M. Huppauff, K. Bange, and B. Lengeler, *Thin Solid Films* **230**, 191 (1993).
33. H. Zabel, *Appl. Phys. A* **58**, 159 (1994).
34. D. David, *Surf. Sci. Repts.* **16**, 333 (1992).
35. G. Vizkelethy, *Nucl. Instr. Meth. B* **89**, 122 (1994).
36. S. Mantl, Ch. Buchal, B. Strizker, and B. Saftic, *Nucl. Instr. Meth. B* **15**, 314 (1986).
37. E. Taglauer, *Surf. Sci.* **299-300**, 64 (1994).
38. A. J. Freeman and C. L. Fu, in *Magnetic Properties of Low-Dimensional Systems*, L. M. Falicov and J. L. Moran-Lopez, eds., p. 16, Springer, Berlin (1986).
39. A. J. Freeman and C. L. Fu, *J. Appl. Phys.* **61**, 3356 (1987); A. J. Freeman, C. L. Fu, M. Weinert, and S. Onishi, *Hyp. Interact.* **33**, 53 (1987).
40. V. L. Moruzzi and P. M. Marcus, see ref. 2, p. 97.
41. M. Weinert and S. Blügel, see ref. 3, p. 51.
42. S. Blügel, B. Drittler, R. Zeller, and P. H. Dederichs, *Appl. Phys. A* **49**, 547 (1988).
43. E. Wimmer, H. Krakauer, M. Weinert, and A. J. Freeman, *Phys. Rev. B* **24**, 864 (1981); M. Weinert, E. Wimmer, and A. J. Freeman, *Phys. Rev. B* **26**, 4571 (1982).
44. H. Akai, M. Akai, S. Blügel, B. Drittler, H. Ebert, K. Terakura, R. Zeller, and P. H. Dederichs, *Progr. Theor. Phys.* **101**, 11 (1990).
45. A. J. Freeman and R. Wu, *J. Magn. Magn. Mat.* **100**, 497 (1991).
46. S. Onishi, A. J. Freeman, and M. Weinert, *Phys. Rev. B* **28**, 6741 (1983).
47. G. W. Fernando and B. R. Cooper, *Phys. Rev. B* **38**, 3016 (1988).
48. A. A. Ostroukhov, V. M. Floka, V. T. Cherepin, and V. N. Tomilenko, *J. Magn. Magn. Mat.* **147**, 205 (1995).
49. R. Coehoorn, *J. Magn. Magn. Mat.* (1995), in press.
50. S. Blügel, *Europhys. Lett.* **18**, 257 (1992).
51. S. Blügel, *Phys. Rev. Lett.* **68**, 851 (1992).
52. J. Korecki and U. Gradmann, *Europhys. Lett.* **2**, 651 (1986).
53. A. J. Freeman and C. L. Fu, *J. Appl. Phys.* **61**, 3356 (1987).

54. J. Korecki and U. Gradmann, *Hyp. Interact.* **28**, 931 (1986).
55. J. Tyson, A. H. Owens, and J. C. Walker, *J. Appl. Phys.* **52**, 2487 (1981).
56. K. Kawaguchi, R. Yamamoto, N. Hosoito, T. Shinjo, and T. Takada, *J. Phys. Soc. Japan* **55**, 2375 (1986).
57. T. Shinjo, N. Hosoito, S. Hine, and T. Takada, *Japan J. Appl. Phys.* **19**, L531 (1980).
58. S. Hine, T. Shinjo, and T. Takada, *J. Phys. Soc. Japan* **47**, 767 (1979).
59. K. Yoden, N. Hosoito, K. Kawaguchi, K. Mibu, and T. Shinjo, *Japan J. Appl. Phys.* **27**, 1680 (1988).
60. K. Mibu, N. Hosoito, and T. Shinjo, *Hyp. Interact.* **54**, 831 (1990); N. Hosoito, K. Mibu, T. Shinjo, and W. Keune, *Hyp. Interact.* **68**, 337 (1992); *ibid.*, p. 341.
61. N. Hosoito, K. Kawaguchi, T. Shinjo, T. Takada, and Y. Endoh, *J. Phys. Soc. Japan* **53**, 2659 (1984).
62. R. Droste, G. Stern, and J. C. Walker, *J. Magn. Magn. Mat.* **54-57**, 763 (1986).
63. F. A. Volkening, B. T. Jonker, J. J. Krebs, N. C. Koon, and G. A. Prinz, *J. Appl. Phys.* **63**, 3869 (1988).
64. G. Lugert and G. Bayreuther, *Phys. Rev. B* **38**, 11068 (1988).
65. M. Przybylski, J. Korecki, and U. Gradmann, *Appl. Phys. A* **52**, 33 (1991).
66. S. C. Hong, A. J. Freeman, and C. L. Fu, *Phys. Rev. B* **38**, 12156 (1988).
67. S. Handschuh, Ph. D. Thesis, University of Cologne, 1994; S. Handschuh and S. Blügel, *Phys. Rev. Lett.* submitted (1996).
68. Ch. Sauer, J. Landes, W. Zinn, and H. Ebert, *Mat. Res. Soc. Symp. Proc.* **231**, 153 (1992).
69. J. Zukrowski, G. Liu, and U. Gradmann, *J. Magn. Magn. Mat.* **140-144**, 1977 (1995).
70. J. Landes, Ch. Sauer, S. Dörrer, and W. Zinn, *J. Magn. Magn. Mat.* **113**, 137 (1992).
71. G. Kisters, Ch. Sauer, E. Tsymbal, and W. Zinn, *Hyp. Interact.* **92**, 1285 (1994).
72. H. Nait-Laziz, H. Dreysse, C. Demangeat, and J. Khalifeh, *J. Magn. Magn. Mat.* **148**, 28 (1995).
73. M. Li, X. D. Ma, C. B. Peng, J. G. Zhao, L. M. Mei, Y. H. Liu, B. G. Sheng, and D. S. Dai, *J. Phys.: Condens. Matter* **6**, L785 (1994).
74. M. Maurer, M. Piecuch, M. F. Ravet, J. C. Ousset, J. P. Sanchez, C. Aaron, J. Decoster, D. Raoux, A. de Andres, M. de Santis, A. Fontaine, F. Baudelet, J. L. Rouviere, and B. Dieny, *J. Magn. Magn. Mat.* **93**, 15 (1991).
75. J. Korecki, M. Przybylski, J. Prokop, K. Krop, and P. Auric, *J. Magn. Magn. Mat.* **140-144**, 673 (1995); W. Karas, J. Korecki, and M. Przybylski, *J. Magn. Magn. Mat.* **140-144**, 665 (1995).
76. W. Karas, *Vacuum* **45**, 345 (1994).
77. R. D. Ellerbrook, A. Fuest, A. Schatz, W. Keune, and R. A. Brand, *Phys. Rev. Lett.* **74**, 3053 (1995) and references therein.
78. H. Zillgen, B. Feldmann, and M. Wuttig, *Surf. Sci.* **321**, 32 (1994).
79. C. Liu, E. R. Moog, and S. D. Bader, *Phys. Rev. Lett.* **60**, 2422 (1988).
80. V. L. Moruzzi, P. M. Marcus, K. Schwarz, and P. Mohn, *Phys. Rev. B.* **34**, 1784 (1986).
81. V. L. Moruzzi, P. M. Marcus, and J. Kübler, *Phys. Rev. B* **39**, 6957 (1989).
82. M. Uhl, L. M. Sandratskii, and J. Kübler, *J. Magn. Magn. Mat.* **103**, 314 (1992).
83. G. A. Prinz, *Phys. Rev. Lett.* **54**, 1051 (1985).
84. J. P. Renard, P. Beauvillain, C. Dupas, K. De Lang, P. Veillet, E. Velu, C. Marliere, and D. Renard, *J. Magn. Magn. Mat.* **115**, L147 (1992).
85. J. Zhang, Z.-L. Han, S. Varma, and B. P. Tonner, *Surf. Sci.* **298**, 351(1993).
86. H. Hasegawa, *J. Phys. F: Metal Phys.* **16**, 347 (1986).
87. J. Hubbard, *Phys. Rev. B* **19**, 2626 (1979).
88. H. Hasegawa, *J. Phys. F: Metal Phys.* **17**, 165 (1987); H. Hasegawa and F. Herman, *Phys. Rev. B* **38**, 4863 (1988); H. Hasegawa, *J. Magn. Magn. Mat.* **126**, 384 (1993).
89. W. Nolting, W. Borgiel, V. Dose, and Th. Fauster, *Phys. Rev. B* **40**, 5015 (1989).
90. W. Nolting and A. Vega, *J. Appl. Phys.* **76**, 6691 (1994); W. Nolting, G. Borstel, T. Dambeck, T. Fauster, and A. Q. Vega, *J. Magn. Magn. Mat.* **140-144**, 55 (1995).
91. W. Nolting, T. Dambeck, and G. Borstel, *Z. Physik B, Cond. Matter* **90**, 413 (1993).
92. G. T. Rado, *Bull. Am. Phys. Soc.* **2**, 127 (1957).

93. D. L. Mills and A. A. Maradudin, *J. Phys. Chem. Solids* **28**, 1855 (1967).
94. J. Mathon, *Physica B* **149**, 31 (1988); J. Mathon and S. B. Ahmad, *Phys. Rev. B* **37**, 660 (1988); J. Mathon, *Rept. Progr. Phys.* **51**, 1 (1988).
95. G. Liu and U. Gradmann, *J. Magn. Magn. Mat.* **118**, 99 (1993).
96. G. A. T. Allan, *Phys. Rev. B* **1**, 352 (1970).
97. C. A. Ballentine, R. L. Fink, J. Araya-Pochet, and J. L. Erskine, *Phys. Rev. B* **41**, 2631 (1990); Y. Li, M. Farle, and K. Baberschke, *Phys. Rev. B* **41**, 9596 (1990).
98. M. Farle and K. Baberschke, *Phys. Rev. Lett.* **58**, 511 (1987).
99. U. Gradmann, *J. Magn. Magn. Mat.* **54-57**, 733 (1986).
100. H. Fritzsche, J. Kohlhepp, H. J. Elmers, and U. Gradmann, *Phys. Rev. B* **49**, 15665 (1994); H. Fritzsche, H. J. Elmers, and U. Gradmann, *J. Magn. Magn. Mat.* **135**, 343 (1994).
101. M. T. Johnson, R. Jungblut, P. J. Kelly, and F. J. A. den Broeder, *J. Magn. Magn. Mat.* **148**, 118 (1995).
102. H. J. Elmers and U. Gradmann, *Appl. Phys. A* **51**, 255 (1990).
103. G. T. Rado, *J. Magn. Magn. Mat.* **104-107**, 1679 (1992).
104. T. Kaneyoshi, *J. Phys.: Condens. Matter* **3**, 4497 (1991).
105. F. J. A. den Broeder, W. Hoving, and P. J. H. Bloemen, *J. Magn. Magn. Mat.* **93**, 562 (1991).
106. H. W. van Kesteren, F. J. A. den Broeder, P. J. H. Bloemen, E. A. M. van Alphen, and W. J. M. de Jonge, *J. Magn. Magn. Mat.* **102**, L9 (1991).
107. A. Dinia, K. Ounadjela, A. Arbaoui, G. Suran, D. Muller, and P. Panissod, *J. Magn. Magn. Mat.* **104-107**, 1871 (1992).
108. B. Heinrich, K. B. Urquhart, A. S. Arrot, J. F. Cochran, K. Myrtle, and S. T. Purcell, *Phys. Rev. Lett.* **59**, 1756 (1987); F. A. Volkening, B. T. Jonker, J. J. Krebs, N. C. Koon, and G. A. Prinz, *J. Appl. Phys.* **63**, 3869 (1988); R. Cabanel, P. Etienne, S. Lequien, A. Barthelemy, and A. Fert, *J. Appl. Phys.* **67**, 5409 (1990); Y. Suzuki, H. Kikuchi, T. Katayama, and S. Yoshida, *J. Appl. Phys.* **67**, 5394 (1990).
109. D. S. Wang, R. Wu, and A. J. Freeman, *Phys. Rev. B* **47**, 14932 (1993); D. S. Wang, R. Q. Wu, L. P. Zhong, and A. J. Freeman, *J. Magn. Magn. Mat.* **140-144**, 643 (1995).
110. E. du Tremolet de Lacheisserie and O. F. K. McGrath, *J. Magn. Magn. Mat.* **147**, 160 (1995).
111. G. H. O. Daalderop, P. J. Kelly, and F. J. A. den Broeder, *Phys. Rev. Lett.* **68**, 682 (1992).
112. F. J. A. den Broeder, E. Janssen, A. Mud, and J. M. Kerkhof, *J. Magn. Magn. Mat.* **126**, 563 (1993).
113. L. Ertl, G. Endl, and H. Hoffmann, *J. Magn. Magn. Mat.* **113**, 227 (1992).
114. J. Tappert, J. Jungermann, B. Scholz, and R. A. Brand, *J. Appl. Phys.* **76**, 6293 (1994).
115. H. Fujimori, Y. Kamiguchi, and Y. Hayakawa, *J. Appl. Phys.* **67**, 5716 (1990).
116. M. Nawate, H. Kiriake, K. Doi, S. Honda, and T. Kusuda, *J. Magn. Magn. Mat.* **104-107**, 1861 (1992).
117. T. Morishi, Y. Togami, and K. Tsushima, *J. Phys. Soc. Japan* **54**, 37 (1985).
118. J. G. Le Page and R. E. Camley, *Phys. Rev. Lett.* **65**, 1152 (1990) and references therein.
119. M. Löwenhaupt, W. Hahn, Y. Y. Huang, and G. P. Felcher, *J. Magn. Magn. Mat.* **121**, 173 (1993).
120. M. Sajieddine, Ph. Bauer, K. Cherifi, C. Dufour, G. Marchal, and R. E. Camley, *Phys. Rev. B* **49**, 8815 (1994).
121. K. Cherifi, C. Dufour, M. Piecuch, A. Bruson, Ph. Bauer, G. Marchal, and Ph. Mangin, *J. Magn. Magn. Mat.* **93**, 609 (1991).
122. B. Scholz, R. A. Brand, and W. Keune, *Phys. Rev. B* **50**, 2537 (1994).
123. J. Teillet, A. Fnidiki, F. Richomme, P. Boher, and Ph. Houdy, *J. Magn. Magn. Mat.* **123**, 359 (1993); F. Richomme, J. Teillet, P. Auric, P. Veillet, A. Fnidiki, Ph. Houdy, and P. Boher, *J. Magn. Magn. Mat.* **140-144**, 627 (1995).
124. K. Mibu, N. Hosoito, and T. Shinjo, *J. Magn. Magn. Mat.* **126**, 343 (1993).
125. P. Grünberg, R. Schreiber, Y. Pang, M. B. Brodsky, and H. Sowers, *Phys. Rev. Lett.* **57**, 2442 (1986).
126. S. S. P. Parkin, N. More, and K. P. Roche, *Phys. Rev. Lett.* **64**, 2304 (1990).

127. M. N. Baibich, J. M. Broto, A. Fert, F. Nguyen Van Dau, F. Petroff, P. Etienne, A. Friederich, and J. Chazelas, *Phys. Rev. Lett.* **61**, 2472 (1988).
128. G. Binasch, P. Grünberg, F. Saurenbach, and W. Zinn, *Phys. Rev. B* **39**, 4828 (1989).
129. M. Pardavi-Horvarth, see ref. 3, p. 1 and p. 355.
130. A. Fert, P. Grünberg, A. Barthelemy, F. Petroff, and W. Zinn, *J. Magn. Magn. Mat.* **140-144**, 1 (1995) and references therein.
131. Y. Yafet, see ref. 3, p. 19.
132. J. Unguris, R. J. Celotta, and D. T. Pierce, *Phys. Rev. Lett.* **67**, 140 (1991).
133. A. Fuss, S. Demokritov, P. Grünberg, and W. Zinn, *J. Magn. Magn. Mat.* **103**, L221 (1992).
134. Q. Leng, V. Cross, R. Schäfer, A. Fuss, P. Grünberg, and W. Zinn, *J. Magn. Magn. Mat.* **126**, 367 (1993).
135. M. Rührig, R. Schäfer, A. Hubert, R. Mosler, J. A. Wolf, S. Demokritov, and P. Grünberg, *Physica Stat. Sol. A* **125**, 635 (1991).
136. J. C. Slonczewski, *Phys. Rev. Lett.* **67**, 3172 (1991); J. C. Slonczewski, *J. Appl. Phys.* **73**, 5957 (1993).
137. S. Demokritov, E. Tsymbal, P. Grünberg, W. Zinn, and I. K. Schuller, *Phys. Rev. B* **49**, 720 (1994).
138. L. B. Steren, R. Morel, A. Barthelemy, F. Petroff, A. Fert, P. Holody, R. Loloee, and P. A. Schroeder, *J. Magn. Magn. Mat.* **140-144**, 495 (1995); B. Rodmacq, G. Palumbo, and P. Gérard, *J. Magn. Magn. Mat.* **118**, L11 (1993).
139. W. Folkerts, J. C. S. Kools, T. G. S. M. Rijks, R. Coehoorn, M. C. de Nooijer, G. H. J. Somers, J. J. M. Ruigrok, and L. Postma, *IEEE Trans. Magn.* **30**, 3813 (1994).
140. D. H. Mosca, F. Petroff, A. Fert, P. A. Schroeder, W. P. Pratt, and R. Loloee, *J. Magn. Magn. Mat.* **94**, L1 (1991); S. S. P. Parkin, R. Bhadra, and K. P. Roche, *Phys. Rev. Lett.* **66**, 2152 (1991).
141. R. Coehoorn, *J. Magn. Magn. Mat.* **151**, 341 (1995).
142. P. Belien, R. Schad, C. D. Potter, G. Verbanck, V. V. Moshchalkov, and Y. Bruynseraede, *Phys. Rev. B* **50**, 9957 (1994).
143. Z. J. Yang and M. R. Scheinfein, *Phys. Rev. B* **52**, 4263 (1995).
144. E. E. Fullerton, D. M. Kelly, J. Guimpel, I. K. Schuller, and Y. Bruynseraede, *Phys. Rev. Lett.* **68**, 859 (1992).
145. T. Thomson, P. C. Riedi, and D. Greig, *Phys. Rev. B* **50**, 10319 (1994).
146. N. Hosoito, K. Kawaguchi, T. Shinjo, T. Takada, and Y. Endoh, *J. Phys. Soc. Japan* **53**, 2659 (1984).
147. N. K. Jaggi, L. H. Schwarz, H. K. Wong, and J. B. Ketterson, *J. Magn. Magn. Mat.* **49**, 1 (1985).
148. K. Takanashi, H. Yasuoka, K. Kawaguchi, N. Hosoito, and T. Shinjo, *J. Phys. Soc. Japan* **53**, 4315 (1984).
149. T. Okuyama and T. Shinjo, *J. Phys. Soc. Jaan.* **61**, 286 (1992).
150. E. Devlin, V. Psycharis, A. Kostikas, A. Simopoulos, D. Niarchos, A. Jankowski, T. Tsakalakos, H. Wan, and G. Hadjipanayis, *J. Magn. Magn. Mat.* **120**, 236 (1993).
151. K. Hanisch, W. Keune, R. A. Brand, C. Binek, and W. Kleemann, *J. Appl. Phys.* **76**, 6528 (1994).
152. J. Landes, Ch. Sauer, R. A. Brand, W. Zinn, S. Mantl, and Z. Kajcsos, *J. Magn. Magn. Mat.* **86**, 71 (1990).
153. F. Klinkhammer, Ch. Sauer, E. Tsymbal, S. Handschuh, Q. Leng, and W. Zinn, *J. Magn. Magn. Mat.* (1996), in press.
154. A. M. Afanas'ev and E. Yu. Tsymbal, *Hyp. Interact.* **62**, 325 (1990).
155. H. Hasegawa, *J. Phys.: Condensed Matter* **4**, 169 (1992).
156. D. J. Keavney, M. D. Wieczorek, D. F. Storm, and J. C. Walker, *J. Magn. Magn. Mat.* **121**, 49 (1993).
157. P. J. Schurer, Z. Celinski, and B. Heinrich, *J. Appl. Phys.* **75**, 5583 (1994); *Phys. Rev. B* **48**, 2577 (1993); *Phys. Rev. B* **51**, 2506 (1995).
158. H. A. M. de Gronckel, K. Kopinga, W. J. M. de Jonge, P. Panissod, J. P. Schille, and F. J. A. den Broeder, *Phys. Rev. B* **44**, 9100 (1991).
159. M. Albrecht, U. Gradmann, T. Furubayashi, and W. A. Harrison, *Europhys. Lett.* **20**, 65 (1992); M. Albrecht, T. Furubayashi, M. Przybylski, J. Korecki, and U. Gradmann, *J. Magn. Magn. Mat.* **113**, 207 (1992).

160. S. Miethaner and G. Bayreuther, *J. Magn. Magn. Mat.* **148**, 42 (1995).
161. J. Thiele, F. Klose, A. Schurian, O. Schulte, W. Felsch, and O. Bremert, *J. Magn. Magn. Mat.* **119**, 141 (1993).
162. S. Handschuh, J. Landes, U. Köbler, Ch. Sauer, G. Kisters, A. Fuss, and W. Zinn, *J. Magn. Magn. Mat.* **119**, 254 (1993).
163. U. Krey, S. Krompiewski, and U. Krauss, *J. Magn. Magn. Mat.* **86**, 85 (1990); U. Krauss and U. Krey, *J. Magn. Magn. Mat.* **98**, L1 (1991); U. Krey, U. Krauss, and S. Krompiewski, *J. Magn. Magn. Mat.* **103**, 37 (1992).
164. Y. Kakehashi and M. Yu, *Phys. Rev. B* **50**, 6189 (1994).
165. R. Lorenz and J. Hafner, *J. Magn. Magn. Mat.* **139**, 209 (1995).
166. E. Arenholz and G. Bayreuther, private communication.

3

Iron-based Amorphous Ribbons and Wires

Q. A. Pankhurst

1. INTRODUCTION

Amorphous metallic alloys may be prepared by rapid solidification, from the gas or liquid phase to achieve a solid state in which the atoms are frozen in a configuration resembling that of a liquid. They have no long range atomic order: other than atomic nearest-neighbour correlations, they are completely without structure. As well as being of intrinsic scientific interest, this amorphous structure results in a variety of magnetic, electrical, and mechanical properties which are of fundamental and technological importance. These include very low magnetic coercivities, high electrical resistivities, and high tensile strengths. Amongst the many compositional variations of amorphous alloys that have been explored over the last twenty years, some have been optimised with respect to one or another of these physical properties. These include alloys with very small hysteresis loss, alloys with a very high magnetostrictive strain response, and alloys with a coefficient of thermal expansion equal to zero. Given the boundless possibilities with regard to the composition, preparation conditions, and post-production treatment of amorphous metallic alloys, there are almost certainly more unique and useful materials that have not yet been discovered.

To date there are perhaps three distinguishable classes of amorphous alloys that are particularly important from a magnetic perspective: (i) the transition metal - metalloid alloys, (ii) the rare earth - transition metal alloys, and (iii) the transition metal - early transition metal alloys. The transition metal - metalloid alloys are soft magnetic alloys, with low coercivities, low hysteresis loss and high permeability, making them attractive in e.g., power transformer applications. Typical alloys in this class contain about 80 percent iron, cobalt, or nickel, with the metalloid remainder comprising boron, carbon, silicon, phosphorus, or aluminium. The rare earth - transition metal alloys have low saturation magnetisation and high perpendicular anisotropy, and have been studied extensively as possible magnetic recording materials. Alloys of the main series transition metals, such as iron, cobalt, and nickel, and early transition metals, such as scandium, yttrium, zirconium, and hafnium, exhibit soft magnetic properties similar to those of the transition metal - metalloid alloys, but in addition they have unique magnetic properties such as spin-glass or cluster spin-glass behaviour.

Q. A. Pankhurst • Department of Physics and Astronomy, University College London, London WC1E 6BT, England.

Mössbauer Spectroscopy Applied to Magnetism and Materials Science, Volume 2, edited by Gary J. Long and Fernande Grandjean. Plenum Press, New York, 1996.

The full force of modern techniques in materials characterisation and modelling have been brought to bear on studies of amorphous alloys, and as a result our understanding of the physics underlying them is continually expanding. Amongst the techniques utilised in these studies, Mössbauer spectroscopy has played a central role, particularly with regard to investigations of the structural and short range properties of amorphous alloys, and with regard to their atomic scale magnetic behaviour. In this chapter current progress in Mössbauer studies of amorphous alloy magnetism will be reviewed and set within the context of the broader range of experimental and theoretical studies of amorphous alloys. Material in the chapter will focus primarily on those alloys that are now commercially produced in the form of both ribbons and wires. These comprise a subset of the first class of amorphous alloys, namely transition metal - metalloid alloys in which the transition metal is iron.

The remainder of this introductory section provides a review of techniques for preparing amorphous alloys, some examples of commercial applications in which they are used, and a survey of the fundamental research avenues that are currently being explored in order to understand and harness their magnetic properties. In following sections the role of Mössbauer spectroscopy in studies of iron-based amorphous alloys will be outlined, and then expanded upon in relation to the structural and magnetic information that may be obtained. Finally, a few selected topics are introduced to illustrate the utilisation of Mössbauer spectroscopy in amorphous alloy research, and the chapter will conclude with some remarks on future possibilities and areas for special attention.

A number of excellent books and review articles are available both on the general topic of amorphous alloys and their applications,[1-5] and on aspects of the use of Mössbauer spectroscopy in their study.[6-9] These are recommended as complementary reading to this chapter.

1. 1. Preparation Techniques

The formation of an amorphous alloy requires that the constituent alloy atoms be cooled at a rate that is sufficiently high to avoid crystallisation. This rate is of order 10^5 Ks^{-1}. Condensation from the gaseous phase, as in sputtering and vapour deposition, provides an inherently high quench rate, more than 10^8 Ks^{-1}, and therefore allows the formation of systems not obtainable by other routes. However, much of the progress in the field of amorphous magnetic alloys springs from the development of cheap and simple techniques for the continuous formation of amorphous materials from the molten state: first by the 'splat cooling' technique[10,11] in 1960, and then by the 'melt spinning' technique[12] in 1974. In both these techniques, and in other variations, rapid cooling is brought about by maximising the contact area between the melt and the cooling medium. A schematic diagram of the melt-spinning process for the production of both amorphous ribbons and amorphous wires is shown in Fig. 1; further details of the formation process are available in the literature.[13-15]

Amorphous transition metal - metalloid alloys may be formed relatively easily because the addition of solute atoms of a different size and/or chemical character to the host atoms inhibits crystallisation and stabilises the glassy state. The formation of an amorphous system is favoured for compositions at which there is a deep eutectic in the phase diagram, where the melting temperature of the mixture is a minimum. At such concentrations the temperature through which the melt needs to be cooled without inducing crystallisation is minimised. For the transition metal - metalloid alloys there is a eutectic at approximately 20 at. percent metalloid and 80 at. percent transition metal. This is illustrated in Fig. 2 for the case of the iron-boron binary alloy system.

Melt-spinning is not the only method currently being used to prepare amorphous alloys. In large-scale industrial laboratories a continuous melt-extrusion process can produce high quality ribbons several cm in width. More recently there have been a proliferation of studies

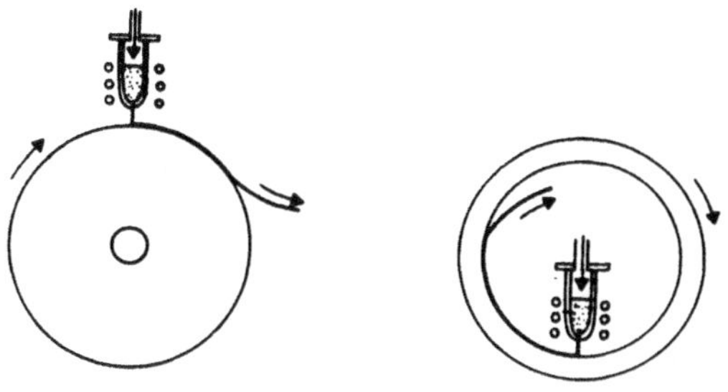

Figure 1. Schematic diagram of the melt-spinning method of formation of continuous lengths of amorphous alloys. Left, the molten alloy is dropped in a continuous stream onto a rapidly rotating cylinder, producing a long ribbon, typically 2 to 3 mm wide and 25 μm thick. Right, the melt falls on the inner surface of a rapidly rotating drum. When the drum is partially filled with water, amorphous wires of circular cross-section, typically 100-150 μm in diameter, are formed.

of amorphisation processes via mechanical alloying, where the high impact velocity and shearing forces generated within a rotating ball-mill can be sufficient to access the amorphous state.[16-19] Another popular route that is being explored is the chemical synthesis of amorphous alloy particles, usually involving the reduction of metal ions with borohydride in aqueous solution and the precipitation of the product.[20-22] In a recent twist to this, a method has been found for encapsulating ultrafine amorphous Fe-B alloy particles in an organic compound,[23] a development which may be important for medical and biological applications. Finally, perhaps the most unusual method yet used for amorphous alloy preparation is that of 'acoustic cavitation': the formation and collapse of bubbles in a liquid subjected to ultrasonic irradiation.[24] Acoustic cavitation induces extreme local heating in otherwise cold liquids, and subsequent cooling rates of order 10^9 Ks^{-1}. Using this technique, and taking as a starting point a solution of $Fe(CO)_5$, a product of nearly pure amorphous iron may be obtained.[25, 26] It remains to be seen what other new materials may be produced by this technique.

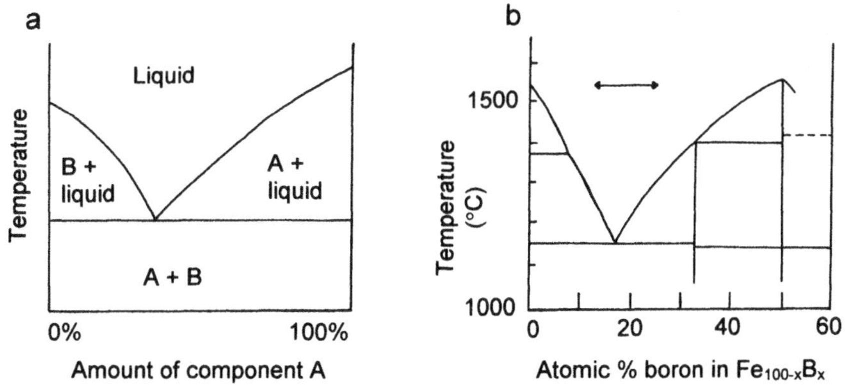

Figure 2. Schematic phase diagram of a simple binary eutectic system, a, and the phase diagram of the Fe-B alloy, showing the eutectic at approximately 18 at. percent boron, and the extent of the glass-forming region, b.

1. 2. Commercial Applications

The commercial applications of amorphous alloys have been surveyed and commented on in many reviews and books.[5, 27-31] These applications arise from the magnetic, electrical, and mechanical properties of the alloys. Some pertinent magnetic properties of a selection of amorphous alloys are illustrated in Table 1, in comparison to those of two well-known crystalline ferromagnetic materials, mumetal and grain-oriented Fe-Si steel.

Table 1. Magnetic Parameters of Amorphous Transition Metal - Metalloid Alloys.[15]

	Coercivity H_c (Am^{-1})	Saturation magnetisation (298 K) I_s (T)	Curie temperature T_c (°C)	Magneto-striction $\lambda_s(\times 10^6)$
Amorphous				
$Fe_{80}B_{20}$ - METGLAS 2605	3.2	1.6	374	30
$Fe_{80}P_{16}B_1C_3$ - METGLAS 2615	4.0	1.5	292	30
$Fe_{40}Ni_{40}P_{14}B_6$ - METGLAS 2826	1.5	0.8	247	11
$Fe_{29}Ni_{49}P_{14}B_4Si_2$ - METGLAS 2826B	0.9	0.4	382	5
Crystalline				
$Ni_{80}Fe_{16}Mo_4$ - Mumetal	2.0	0.8	460	~0
$Fe_{96.8}Si_{3.2}$ - grain-oriented steel	25-40	2.0	730	4

As discussed by Jiles,[30] the main commercial interest in soft ferromagnetic amorphous alloys over the last fifteen years has stemmed from their low coercivities, down to 0.1 Am^{-1}, which is about an order of magnitude smaller than for Si-Fe, coupled with their large permeabilities, of order 10^6, which is about an order of magnitude greater than Si-Fe. These properties are distinctly advantageous for applications in power devices, especially transformers, where the additional factor of the high resistivity of an amorphous alloy, typically 1.5 μΩm as compared with 0.5 μΩm in Si-Fe, results in very low power losses. However, there are also some important disadvantages associated with using the alloys in transformer applications. They have rather low saturation magnetisations, see Table 1, and at high flux densities their core losses become more significant. Also, because the alloys require post-production heat treatment to relieve the large internal stresses and strains due to the casting process, they tend to become brittle, making the task of transformer production technically much harder than with Si-Fe. Despite these set-backs, research continues to be directed towards overcoming the problems with high-power applications, such as identifying means of avoiding embrittlement.[32] Another approach currently being explored is to use ultrathin amorphous ribbons, 3-10 μm thick, for improved transformer performance in the megahertz range.[33]

At the same time there is a great deal of interest in low-current applications, as in pulsed power transformers, and in specialist devices, such as magnetic sensors, magnetostrictive transducers, motors, magnetic recording and reading heads, and security tag systems. The scope and diversity of these various applications is huge. This may be illustrated by considering a selection of papers presented at the latest International Conference on Rapidly Quenched and Metastable Materials in August 1993, which included designs for: high-sensitivity non-contact torque sensors for the control of rotating shafts using amorphous ribbons;[34,35] an improved longitudinal magnetic shielding system comprising a laminate of amorphous alloy sheets and a panel of grain-oriented Fe-Si steel;[36] a quick-response magnetic field sensor using a 200 MHz resonance oscillator and amorphous wire magnetoinductive elements;[37] and an energy transferring system for totally implantable medical devices using amorphous wires.[38] In another novel proposal, amorphous wires are being tested for potential high-density magnetic recording systems based on scanning tunnelling microscopy.[39] Whatever the outcome of these endeavours, there is little doubt that the commercial applications of amorphous alloys will continue to expand and develop in many different areas.

1. 3. Characterisation and Fundamental Research

A wide range of experimental and theoretical approaches have been used to study amorphous alloys. In the context of this chapter, it is worthwhile mentioning a few of these approaches, partly to remind the reader that Mössbauer spectroscopy is only one of many techniques available, but also because the results obtained from Mössbauer studies of amorphous alloys are often closely related to the results of other techniques. It is the complementary use of the information made available from a variety of sources that ultimately leads to better understanding of the magnetism in these materials.

Bulk magnetometry measurements are probably the most direct and basic measurements on any given magnetic alloy. Torque magnetometry experiments enable the measurement of the uniaxial anisotropy constant, K_u, due either to induced magnetic anisotropy, or to the small inherent anisotropy found in melt-spun ribbons as a result of the casting process.[40] High field magnetisation and susceptibility measurements provide information about the approach to ferromagnetic saturation in the alloy, and the effect of this on spin-waves.[41, 42] In recent years some surprising results have been obtained in high field magnetisation studies, with evidence for persistent non-collinear magnetic states in some iron-based alloys in steady-state fields of up to 20 T, and in pulsed fields of up to 120 T.[43] Similar data have been reported from lower field experiments using vibrating sample magnetometry and SQUID magnetometry.[44]

Surface domain imaging is an important method for monitoring the micromagnetic state. A variety of different techniques are currently employed, ranging from the conventional Bitter technique of magnetic particle deposition[45] and magneto-optical Kerr effect imaging,[46] to high resolution studies using scanning electron microscopy[47, 48] and transmission electron microscopy.[49] It is interesting to note that in the high resolution transmission electron microscopy study on $Fe_{78}Si_9B_{13}$ there is evidence for considerable domain sub-structure: major domain walls could be seen, but in addition there were strong components of magnetisation orthogonal to the major domain walls. This implies that the domain structure in amorphous alloys may in general prove to be more complex than was previously supposed.

Magnetostriction studies[31, 50, 51] are another sensitive tool for investigating amorphous alloys, both as a macroscopic phenomenon, and as an indication of local atomic structure, the latter because it originates from the strain dependence of the atomic scale magnetic anisotropy. Field dependent measurements are usually based on either the physical change in sample dimension, quantified by the engineering magnetostriction, λ_e, or by the change in Young's modulus, known as the ΔE effect. Recent developments in this field include a high resolution vibrating reed technique for measuring the ΔE effect in FeSiB amorphous wires.[52]

Neutron and x-ray scattering studies are used for a wide range of structural and magnetic characterisations of amorphous alloys.[53, 54] Recent examples of structural studies using neutrons include the inference of density fluctuations in $Fe_{85}B_{15}$ from the dynamic structure factors in a low momentum transfer regime,[55] structural changes in Co-Si-B glasses as a function of composition,[56] and evidence for induced in-plane structural anisotropy in field annealed ribbons of $Fe_{78}Si_9B_{13}$.[57] Magnetic measurements include small angle neutron scattering experiments on magnetic phase transitions[58] and polarised neutron studies of non-collinear magnetic structures in iron-based alloys.[59, 60]

Theoretical modelling of the atomic structure and the magnetism of amorphous alloys has always been an integral part of their research.[61-63] Current modelling methods include self-consistent calculations based on local spin-density theory,[64, 65] molecular dynamics,[66] density functional theory,[67] and 'muffin-tin' orbital Hamiltonians.[68] The complexity of some of these approaches reflects the degree of sophistication that is required to obtain realistic results. One development that may prove to be fruitful is the application of reverse Monte Carlo simulation techniques to the analysis of neutron diffraction data to obtain both the atomic and the magnetic structure.[69, 70] The benefit of this approach is that no prior knowledge of interatomic or spin-spin interactions is assumed. In recent work the reverse Monte Carlo technique has been used in conjunction with combined neutron, x-ray, and extended x-ray absorption fine structure data to obtain models of Fe-Zr;[71] application of this technique to transition metal - metalloid systems is also in progress.

2. MÖSSBAUER SPECTROSCOPY OF AMORPHOUS ALLOYS

The role of Mössbauer spectroscopy in studies of amorphous ribbons and wires is one of providing specific and detailed information on the atomic scale environments of the Mössbauer effect atoms. The technique cannot be said to be a standard or routine characterisation tool for amorphous alloys. Nevertheless it has proved to be invaluable in a wide range of different contexts, from the practical realm of identifying impurity phases and gauging product quality, to the more esoteric realm of revealing fundamental phenomena of purely scientific interest.

Mössbauer studies of the transition metal - metalloid amorphous alloys have almost exclusively utilised ^{57}Fe Mössbauer spectroscopy. A variety of data collection modes are used: transmission and backscattering, with unpolarised or polarised γ-ray sources. All of these modes will be discussed in this chapter. It may be noted also that there have been many rare earth isotope Mössbauer studies of the transition metal - rare earth amorphous alloys, using e.g., ^{129}I, ^{161}Dy, ^{151}Eu, and ^{174}Yb. These will not be discussed here, but reviews are available in the literature.[8]

To illustrate the type of data that may be obtained from a conventional ^{57}Fe Mössbauer experiment, Fig. 3 shows the room temperature spectra of an amorphous FeSiB alloy and a crystalline ferromagnet, α-iron. As both materials are below their respective Curie temperatures, they have the characteristic six-line patterns of a magnetically ordered material. The broad lines of the FeSiB alloy are typical of amorphous systems and reveal the variable nature of the atomic nearest neighbour environments for the ^{57}Fe atoms. In comparison, the α-iron lines are sharp, as the environment of the ^{57}Fe atoms in the crystalline state is the same for all the atoms.

We may consider the spectrum from the alloy to be comprised of a summation of individual Mössbauer spectra, each one associated with a single iron atom in the alloy, and each one having sharp lines, just as in the α-iron spectrum, and well defined line positions and line intensities. The line positions are determined by the magnitude and sign of the local electric monopole and quadrupole interactions at the iron atom, line positions which are quantified by the Mössbauer parameters, the isomer shift, δ, the quadrupole splitting, Δ, and the electric field gradient asymmetry, η, and by the magnetic hyperfine field at the iron nucleus, H. The line intensities are determined by the angle, θ, between the hyperfine field direction and the γ-ray beam. The resultant spectrum therefore depends on the nature of the probability distribution in atomic environments, and on how this translates into probability distributions in the parameters δ, Δ, η, H and θ. Each of these parameters may be correlated with one or more of the others in some way. The resultant summation of the individual ^{57}Fe subspectra is therefore a complete but complex description of the full range of atomic scale structural and magnetic environments in the alloy.

This is both promising and daunting. The promising part is that the Mössbauer spectrum of an amorphous alloy contains a wealth of information on the local structural and magnetic properties of the system. The daunting prospect is the challenge of reliably extracting data

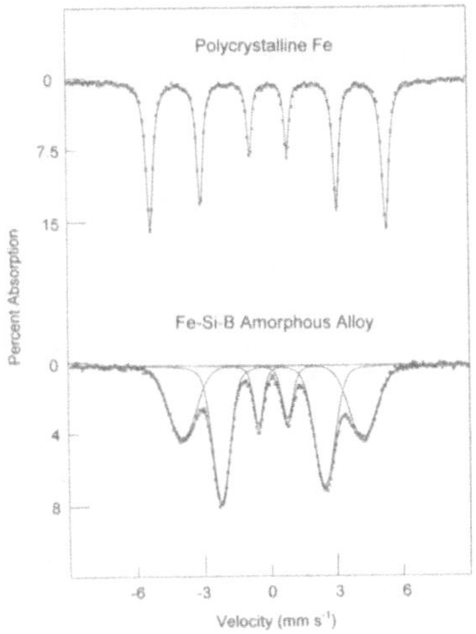

Figure 3. Room temperature ^{57}Fe Mössbauer spectra of ribbons of polycrystalline α-iron and the amorphous alloy $Fe_{78}Si_9B_{13}$ The spectra were recorded in the conventional transmission mode, using an unpolarised ^{57}Co:Rh source.

from what may potentially be a highly complex and correlated set of hyperfine parameters, based on largely featureless spectra such as in Fig. 3. One of the main themes of this chapter will be the exploration of different strategies for rising to this challenge, with the goal of being able to derive unambiguous interpretations of the Mössbauer data. In the remainder of this section we review some of the techniques and concepts that need to be taken into consideration when modelling and interpreting the spectra of amorphous alloys.

2. 1. Quadrupole Splitting

At first sight the amorphous alloy spectrum, shown in Fig. 3, might seem to be purely magnetic. There is no obvious sign of any quadrupole shift: the characteristic shift in the positions of the inner four lines relative to the outer two is not apparent, and the spectrum is roughly symmetric around its centroid. In this respect the spectrum looks like that of α-iron, where the quadrupole splitting is almost exactly zero because of the symmetric arrangement of the bcc iron lattice. However, in fact the quadrupole splitting in $Fe_{78}Si_9B_{13}$ is quite significant, with a mean value of Δ ca. 0.50 mms^{-1}.[72] The reason it is not apparent in the spectrum is that in a six-line Mössbauer spectrum with combined magnetic and electric interactions, the quadrupole shift is scaled by a factor, $(3\cos^2\theta - 1)/2$, where θ is the angle between the principal axis of the local electric field gradient, and the γ-ray beam. In the amorphous alloy, this angle θ is randomly oriented from atom to atom, and the weighted average over a sphere of the scaling factor is zero. In other words, the spectra of the ^{57}Fe atoms each individually contain a quadrupole shift, but when they are all summed together the mean quadrupole shift that is apparent falls to zero.

This is important, partly because it emphasises that caution must be exercised when analysing the spectra, but also because the information contained in the distribution of quadrupole parameters at the iron atoms is needed for comparison with model predictions of, e.g., short range structural ordering.[73] One way to overcome this is to heat the alloy above its Curie temperature, T_c, and measure its paramagnetic spectrum. However, for many amorphous alloys this is impossible because they crystallise close to or below T_c. An alternative is to utilise the magnetic radiofrequency collapse technique to induce a paramagnetic-like state well below T_c.[72, 74, 75] In this technique an alternating magnetic field of sufficient strength and frequency is applied to the sample so that the magnetic splitting of the Mössbauer spectrum falls to zero, revealing only the quadrupole and isomer shift interactions. An example of this is given in Fig. 4, along with a curve showing the composition dependence of quadrupole splitting in some FeSiB alloys. In a comprehensive survey of seventy different iron - metalloid amorphous alloys, Kopcewicz[72] found that the average quadrupole splitting fell in the range 0.45 - 0.61 mms^{-1}, and the average isomer shift was between -0.21 and 0.17 mms^{-1}. Given this, it is clear that quadrupole effects need to be taken into consideration when analysing the spectra of amorphous alloys.

2. 2. Modelling the Mössbauer Spectra of Amorphous Alloys

Many papers have been written on various aspects of analysing the magnetically split ^{57}Fe Mössbauer spectra of amorphous alloys.[77-93] All of these procedures deal in some way with the task of determining the probability distribution in magnetic hyperfine fields - the so-called 'P(H) distribution'. Some of the analysis methods also consider the probability distributions in isomer shift, P(δ), and in quadrupole splitting, P(Δ), and some deal explicitly with the correlations between these three distributions. Broadly speaking, the different analysis strategies fall into three distinguishable groups: (i) lineshape fitting, (ii) model-dependent P(H) fitting, and (iii) model-independent P(H) fitting.

In the lineshape fitting approach the usual Lorentzian line profiles are replaced by Gaussian or pseudo-Lorentzian profiles. This enables the average hyperfine parameters to be found simply and directly, although no quantitative information about the P(H), P(Δ), and P(δ) distributions are obtained. In a review of the efficacy of this approach, Vandenberghe et al.[90] concluded that the so-called 'Pearson VII' line profile, a function that varies continuously between the Lorentzian and Gaussian profiles, gave the best results when fitting the Mössbauer spectra of $Fe_{81}B_{13.5}Si_{3.5}C_2$.

In model-dependent fitting an *a priori* assumption is made regarding the shape of the P(H) distribution, such as Gaussian or binomial, and the spectrum is generated by convoluting this distribution with an elementary Lorentzian sextet. The most widely used variation on this approach is due to Lines and Eibschütz.[86, 89] In this model the spectra are fitted using Voigtian line profiles, i.e., Gaussian distributions of Lorentzian lines, whose positions are constrained to be physically realistic. Experimentally fitted correlations between the isomer shift, quadrupole shift, and hyperfine field are included in the model, so that some information on the P(δ) and P(Δ) distributions is also obtained.

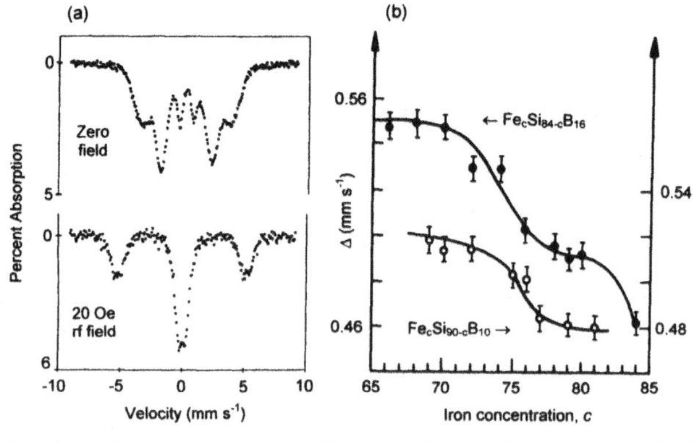

Figure 4. Quadrupole splitting in amorphous alloys revealed by the magnetic radiofrequency collapse technique: (a) Mössbauer spectra of as quenched $Fe_{76}Y_4B_{20}$ recorded in zero field and in a radiofrequency field of 20 Oe at 62 MHz,[76] and (b) variation of the average quadrupole splitting, Δ, versus iron concentration, *c*, of the amorphous alloy series $Fe_cSi_{84-c}B_{16}$ and $Fe_cSi_{90-c}B_{10}$.[72]

Model-independent fitting procedures make no *a priori* assumptions about the shape of the P(H) distribution. The distribution is obtained by fitting the spectrum with a convolution of an elementary sextet, often Lorentzian, and a free or partially constrained P(H) distribution. Parameterisation of the P(H) distribution has developed over the years, beginning in the early 1970s with unconstrained expansions of trigonometric functions[77] or histograms comprising delta functions at discrete hyperfine fields values.[78] The histogram method was later improved by adding constraints forcing all the P(H) values to be positive[80] and requiring the distribution end-points to be zero.[84] In some of the more sophisticated procedures that have been used to date, the elementary sextet lineshape incorporates the effect of the quadrupole interaction at the Fe nucleus.[80, 90] In this way the effect of a large quadrupole interaction, with Δ of ca. 0.5 mms^{-1} and asymmetry η, may be fully accounted for by using a complex lineshape, denoted S^{η}, rather than a simple Lorentzian. The introduction of this level of mathematical rigour can sometimes have a discernible effect on the shape and structure of the P(H) derived from the data,[80] as is illustrated in Fig. 5. However, there are also instances where better fits to the data are obtained, but with virtually no significant change in the P(H) distribution.[90] Perhaps because of this, and the complexity of the S^{η} lineshape in terms of computational analysis, the majority of studies today still use the Lorentzian lineshape in the elementary sextet.

In addition to the different analysis strategies discussed above, some scientists have devised methods to try to simplify the initial data that is being analysed. These include a di-

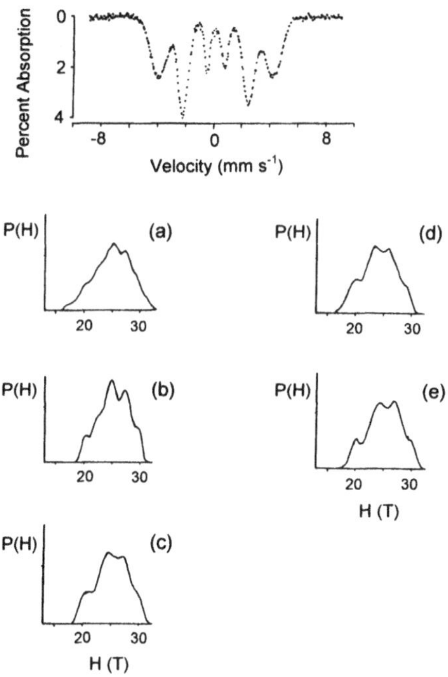

Figure 5. Hyperfine field P(H) distributions obtained by fitting the Mössbauer spectrum of amorphous $Fe_{79.5}Si_{1.5}B_{19}$ with various lineshapes for the elementary sextets: (a) Lorentzians with narrow lines, (b) Lorentzians with broader lines, (c) S^{η} with $\eta = 1$ and $\Delta = +0.58$ mms^{-1}, (d) S^{η} with $\eta = 0$ and $\Delta = +0.58$ mms^{-1}, and (e) S^{η} with $\eta = 0$ and $\Delta = -0.58$ mms^{-1}. Figure adapted from Le Caër and Dubois.[80] Note the variety of shapes evident in the P(H) distributions, compared to the similarity in the quality of fit for each of the models.

rect Fourier deconvolution approach[85] and a rotated sample 'magic angle' recording method designed to remove problems associated with ill-defined magnetic texture.[94] Another idea is the use of 'spectral subtraction,[95] wherein a linear combination of two spectra with different relative intensities of the 2nd and 5th lines are added in such a way as to remove the 1st, 3rd, 4th, and 6th lines, leaving only two lines in the resultant spectrum. This can be particularly helpful in materials where the Mössbauer spectral lines overlap strongly. However, the change in intensity of the 2nd and 5th lines is usually achieved by applying a small magnetic field to the alloy, which should be small enough to avoid perturbing the P(H) distribution. In light of this it has been suggested that unless the material is very soft indeed the moments cannot be reliably assumed to have been equally affected by the applied field, and that therefore the method is not reliable.[82] More recently the spectral subtraction approach has been used in zero field for magnetically textured ribbons, thereby avoiding this potential complication.[96]

Lastly, a note of warning should be given regarding all of the methods discussed above for extracting P(H) distributions from the Mössbauer spectra of amorphous alloys. In almost all cases the elementary spectra that are used are derived on the basis of first-order perturbation theory, under the assumption that the magnitude of the quadrupole interaction is small compared with the magnitude of the magnetic hyperfine interaction. If this prerequisite is not the case, then the results obtained will be unreliable. Uncertainties will also arise with increasing levels of overlap between the spectral components. For example, if the distribution in hyperfine fields is Gaussian, then for large standard deviations, σ_H, the outermost lines, the first and sixth lines, of the smaller field components may overlap with the middle lines, the second and fifth lines, of the larger field components. Le Caër et al.[88] have considered these effects in detail, and derived a 'validity diagram,' shown in Fig. 6, that can be used to gauge the degree to which problems may be anticipated for a given alloy. Most iron rich alloys lie in or near the 'valid' region of the diagram, especially at temperatures well below their Curie points, but some of the mixed transition metal alloys do not.[88]

2. 3. Corrections for Saturation Effects

One final concept that needs to be considered when modelling the spectra of amorphous alloys is that of saturation effects. Strictly speaking, any absorber of finite thickness will exhibit line broadening and attenuation due to the energy-dependent resonant absorption of the γ-rays by the Mössbauer effect nuclei. A simple way to imagine this is to suppose that one has a thick absorber. Incident γ-rays of an energy corresponding to the maximum absorption cross-section of the Mössbauer nuclei in the material will be removed from the beam more frequently than γ-rays with slightly less or more energy. Consequently, nuclei at

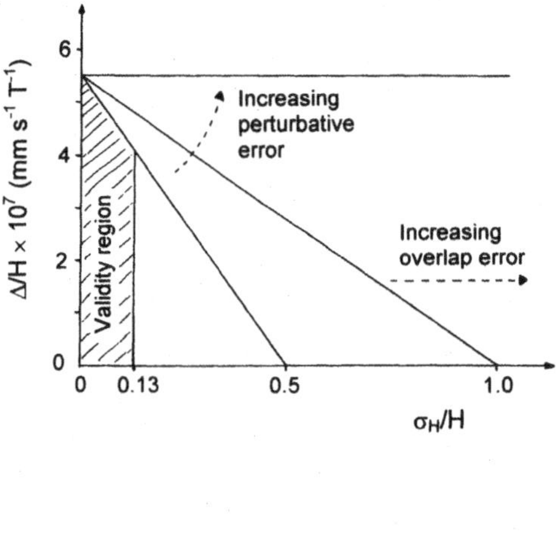

Figure 6. Validity diagram for P(H) distribution calculations, where Δ is the mean quadrupole splitting, H is the mean hyperfine field, and σ_H is the Gaussian standard deviation of the P(H) distribution. Figure adapted from Le Caër and Dubois.[80]

the latter part of the traversal of the γ-rays through the sample may not register as an absorption event simply because no γ-rays of the appropriate energy are left in the beam. The resulting spectrum will contain non-Lorentzian lines, with the relative areas of the less probable nuclear transitions appearing to have an enhanced absorption relative to the more probable transitions.

In practice, however, saturation effects are often negligible. This is the case for modern γ-ray source foils. In ^{57}Fe Mössbauer spectroscopy a general rule of thumb is that a material containing less than ca. 10 mg cm^{-2} of natural iron may usually be regarded as a 'thin' absorber. In iron-based amorphous ribbons, which have densities of ca. 6 g cm^{-3}, this corresponds to a ribbon thickness of up to ca. 17 μm. Although spectra taken from ribbons thicker than this may still be acceptably modelled using the thin absorber approximation because of the intrinsic breadth of the absorption lines,[97] care should still be taken when deciding whether or not to incorporate saturation corrections into the data analysis. In the case of amorphous wires, which are often of diameter ca. 100 μm, saturation effects are significant, and must be accounted for in the analysis.

A number of methods of varying complexity are used to model saturation effects. The most rigorous is the 'transmission integral' method,[98, 99] in which the saturation effects are incorporated into the underlying description of the shape and area of the absorption lines. Another method is to deconvolute the source and absorber characteristics from the measured spectrum, obtaining a 'deblackened' spectrum that corresponds to what one would expect from an ideally thin absorber.[100] Less rigorous, but simpler and faster to compute, is the first-order phenomenological approach of Shirley et al.[101] In this model the effect of moderately thick absorber cross-sections is related to spectral lines that remain Lorentzian in shape, but are broadened and have a reduced area depending on an 'effective thickness' parameter, τ_A. This method has been used successfully for the case of relatively thick $Fe_{78}Si_9B_{13}$ ribbon samples, with physical thickness of 25 to 50 μm, with a modified Lorentzian being used as the elemental lineshape in a Voigtian profile analysis.[96]

3. STRUCTURAL INFORMATION

The most direct measurements of the structure of amorphous alloys are x-ray and neutron scattering experiments, both of which provide radial density functions. These represent one-dimensional projections of the three dimensional arrangement of atoms within the solid. Mössbauer spectroscopy is a technique that provides complementary, and in some cases, extremely sensitive data on the structure of amorphous systems. Other techniques of note include nuclear magnetic resonance, Raman spectroscopy, extended x-ray absorption fine structure, and electron diffraction. In this section some of the areas of structural research in which Mössbauer spectroscopy has made some contribution are reviewed.

3. 1. Short Range Order

One of the most fundamental questions regarding the structure of an amorphous material is whether or not there is a discernible short-range order. Even in models based on Bernal's pioneering concept of an amorphous solid as a dense random packing of hard spheres[102] the atomic structure is not completely random because, for example, a degree of repulsion between metalloid atoms is needed to reproduce the scattering measurements. Other models assume that the short-range order is rather well defined, and similar to that of a related crystalline compound or of a specific structural unit.[103] In this case it is assumed that there is a random network of these structural units, leading to the overall amorphous structure. O'Handley[5] suggests a classification where an amorphous alloy is one in which any short range order is frustrated beyond ca. 10 Å, with two subdivisions: class I having local atomic ordering very like that of a crystalline phase, and class II having a noncrystalline local arrangement, such as one of icosahedral symmetry.

Mössbauer spectroscopy offers two means of distinguishing between different possible structural states, one based on analysis of the distribution of quadrupole splittings, and the other based on correlations between the magnetic and nuclear hyperfine parameters.

Quadrupole splitting distributions contain information about the angular distribution of the local atomic structure. A generalised expression for the probability distribution is given by $P(\Delta) \propto \Delta^{d-1}\exp(-\Delta^2/2\sigma^2)/\sigma^d$, where d is the number of independent degrees of freedom available to the system.[73, 104-106] In materials that are strictly governed by the dense random packing model, it is predicted that $d = 5$. The same expression, but with $d = 2$, has been derived for the amorphous insulator ferric fluoride, a system which exhibits strong short range order.[107] However, it should be noted that different assumptions are made in the two models.[106] Nevertheless, the generalised expression provides a good mathematical approximation for the quadrupole splitting distribution and, when used with $d \leq 5$ as a variable parameter, can give good fits to the experimentally observed function. Although theoretical understanding is not yet complete, it appears that forms with $d < 5$ are consistent with short or medium range order in the system.[106]

Another way to gauge changes in the local atomic structure is to simply monitor the mean quadrupole splitting, Δ_{av}.[72, 76] This parameter gives a good qualitative measure of the short range order, and can reveal any discrete changes in the atomic arrangement. An example of this is the data for the ternary Fe-Si-B system, shown in Fig. 4, in which the abrupt changes in Δ_{av} as a function of composition have been correlated with a structural reconstruction model.[72] According to the model,[108] silicon atoms preferentially replace iron atoms with no boron nearest neighbours. As the concentration of silicon atoms is increased, a limit is reached beyond which a change in the short range structure occurs so that new 'sites' without boron nearest neighbours may be accommodated. These rather abrupt changes in the local atomic arrangements are reflected by significant changes in the mean quadrupole splitting.

The second distinct approach to obtaining structural information from Mössbauer data on amorphous alloys is to consider the magnetic hyperfine parameters. One method of doing this is based on the Lines and Eibschütz model, where 'fluctuation correlation functions' of the type $\mu_N\langle\Delta H\Delta\delta\rangle$ between the fluctuations in hyperfine field, ΔH, and the fluctuations in isomer shift, $\Delta\delta$, can be used to characterise the local structure.[86, 89, 109-111] This is achieved by comparing values for the correlation functions with those obtained from crystalline materials of similar composition. For the case of amorphous Fe-B this has led to the conclusion that a value of $\mu_N\langle\Delta H\Delta\delta\rangle$ that is greater than ca. 0.03 mm^2s^{-2} characterises an orthorhombic short range order of the Fe_3C, cementite, type, while a value of order -0.02 mm^2s^{-2} corresponds to a local structure like that of tetragonal Fe_3B.[110, 111]

3. 2. Crystallisation and Structural Relaxation

Studies of crystallisation mechanisms and processes, although complex, can give inferred information on the structural state of the amorphous alloy. The basic concept is that, under thermal annealing, the short range order present in the as-quenched material will determine which crystalline products will nucleate and grow. The most difficult aspect of this procedure is that, because the crystallisation phase transition is rapid, it is important to ensure that only the initial crystallisation products are obtained. Careful annealing is therefore needed. In a recent example of such a study, $Fe_{84}B_{10}C_6$ samples were subjected to a range of different annealing sequences, and the crystallisation products were determined from their room temperature Mössbauer spectra.[112] Representative spectra from this work are shown in Fig. 7. It was concluded that in amorphous $Fe_{84}B_{10}C_6$ there were two cohabiting short range order states - one like orthorhombic Fe_3B and one like body-centred-tetragonal Fe_3B.[112] This result was confirmed in complementary boron-11 nuclear magnetic

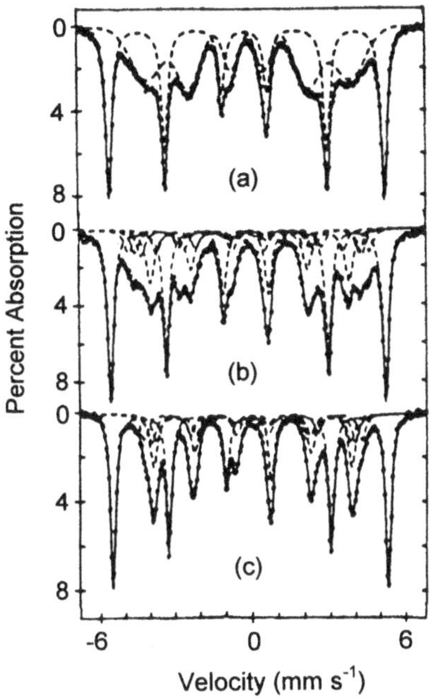

Figure 7. Room temperature Mössbauer spectra of $Fe_{84}B_{10}C_6$ annealed for 10 minutes at (a) 703 K, (b) 743 K, and (c) 868 K.[112] Separate phases are identified as follows: (a) amorphous $Fe_{84}B_{10}C_6$ and crystalline α-iron; (b) α-iron, orthorhombic Fe_3B and Fe_3C, and body-centred-tetragonal Fe_3B; and (c) α-iron, orthorhombic Fe_3B and Fe_3C, and tetragonal Fe_2B.

resonance experiments. At high annealing temperatures the tetragonal Fe_3B phase transforms to the tetragonal Fe_2B phase.

Amorphous to crystalline transformations are also studied in their own right. In melt-spun amorphous $Fe_{78}Si_9B_{13}$ ribbons Mössbauer spectroscopy has provided detailed information on the kinetics of the growth on annealing of α-FeSi and Fe_3B crystallites.[113] This work has also established that the process of crystallisation at the surface of the ribbon differs from that in the bulk, and that furthermore there are differences in the crystallisation at the 'dull' and 'shiny' surfaces of the ribbon. It is postulated that these differences are due to the different amounts of structural and chemical inhomogeneities induced in the alloy surfaces at the quench.

It should also be mentioned that a great deal of interest in the crystallisation mechanism has arisen in the last few years after the discovery of the excellent soft magnetic properties of the nanocrystalline alloy, $Fe_{73.5}Cu_1Nb_3Si_{13.5}B_9$,[114] which is now marketed under the trademark 'FINEMET'. This material is produced by controlled annealing of the as-cast amorphous alloy. Mössbauer studies are currently being used to help to understand the crystallisation kinetics of the system with a view to both understanding the phenomenon, and controlling the characteristics of the product.[115, 116]

Short of full crystallisation, the effect of thermal annealing on the structure of amorphous alloys is to induce either reversible or irreversible structural relaxation.[117] This relaxation is found not only in the region of the glass transition, but also at much lower temperatures, and is the subject of a good deal of fundamental research. Irreversible relaxation occurs in all amorphous alloys, irrespective of composition, and carries an attendant increase in the density of the alloy. Reversible relaxation is strongly dependent on composition and exhibits no change in density. In a recent study of $(Co_{0.75}Fe_{0.25})_{75}Si_{10}B_{15}$, one of those alloys that exhibit both reversible and irreversible relaxation, Mössbauer spectroscopy has provided sufficiently precise measurements of the relaxation effects to enable the separation of the contributions.[117] Structural relaxation effects were evidenced by a narrowing of the P(H) distribution, a splitting of the quadrupole splitting distribution, an increased mean hyperfine field, and a decreased mean isomer shift. It was concluded that the irreversible relaxation was likely to be due to topological short range ordering, with the silicon and boron atoms moving from their random distribution to well defined positions in the material, and that the reversible relaxation corresponded to chemical short range ordering between cobalt and iron atoms.

Another new development is an attempt at correlating annealing-induced changes in Mössbauer parameters with changes in the magnetoelastic Young's modulus, the latter being a commonly used order parameter for characterising structural relaxation.[118] Although detailed correlations have not yet been fully realised, some trends have been identified, so that this may be a promising area for future work.

4. MAGNETIC INFORMATION

In magnetically split Mössbauer spectra of amorphous alloys, the magnitude of the hyperfine splitting reflects the core electron polarisation due to the valence and conduction band electrons, and the dipolar field contributions from neighbouring magnetic atoms. The P(H) distribution of hyperfine fields, evident in the broad lines, results from the variation in local environments experienced by the iron atoms in the alloy. Analysis of the P(H) distribution can therefore provide detailed information about the magnetic state of the alloy. In addition to this, the relative areas of the magnetically split lines reveal the orientation of the iron atomic moments in the alloy. The outer:middle:inner pairs of lines making up the sextet contribution of each individual iron atom contribute absorption areas in the ratio 3:*r*:1 where *r* can vary between 0 and 4, dependent on whether the iron moment is parallel to or perpendicular to the propagation direction of the γ-ray beam. In this way the Mössbauer spectra also contain specific information regarding the distribution of moment directions within the alloy. In this section we consider further the way in which these two facets of the magnetic properties of amorphous alloys may be extracted from the Mössbauer data.

4. 1. Hyperfine Field

A number of review articles have been written which detail the fundamental magnetic properties of amorphous alloys[5, 119] and the relationship between the magnetic hyperfine field and the atomic moment.[7, 120] It has been established that in transition metal - metalloid alloys there is an approximately linear correlation between the mean hyperfine field, as measured in a Mössbauer experiment, and the mean magnetic moment, as measured in a bulk magnetisation experiment.[120] Assuming that this correlation holds for the individual magnetic moments as well as for the mean, the P(H) distribution obtained in Mössbauer experiments can be used in conjunction with magnetisation data to obtain the distribution in moments.

A revealing example of this approach comes from investigations of the temperature dependence of the magnetisation and hyperfine field in iron-based amorphous alloys. It is observed experimentally that the reduced data $m(T) = M_s(T)/M_s(0)$ and $h(T) = H(T)/H(0)$ follow the same curves, and that they are flatter and lie below the Brillouin curve that is found in crystalline materials. This is shown in Fig. 8 for amorphous $Fe_{80}B_{20}$, with data for crystalline α-iron for comparison. There are two likely contributors to this behaviour: (i) variations in exchange interactions and (ii) chemical disorder. Variations in exchange interaction between the magnetic atoms will arise from the structural disorder. If the resulting exchange interactions fluctuate about some mean value so that $J = J_o + \Delta J$, it has been shown that the temperature dependence of *m* will be modified in accordance with the value of a parameter, $\delta^2 = \langle\Delta J^2\rangle/\langle J\rangle^2$.[121] Although this model works extremely well for rare-earth based alloys, it is only moderately successful for the transition metal alloys.[5] This

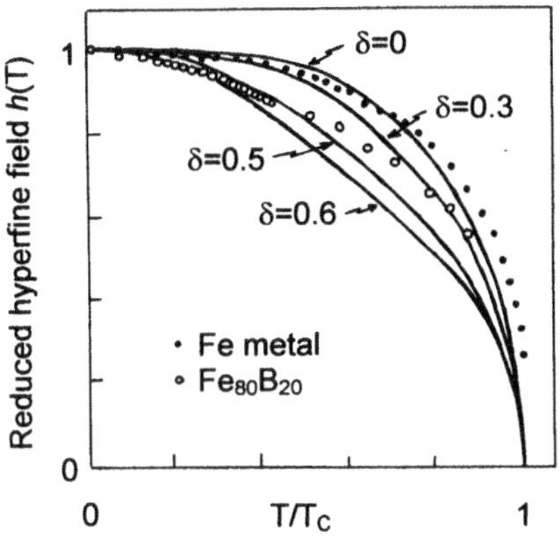

Figure 8. Reduced hyperfine field, $h(T) = H(T)/H(0)$, versus reduced temperature, T/T_c, for amorphous $Fe_{80}B_{20}$ and crystalline α-iron.[5] The theoretical curves are based on a disturbed exchange model, in which the structural disorder introduces variations in exchange interactions between iron atoms. Increasing δ corresponds to increased variations in the exchange.

leads to the idea that a contributing factor could be chemical disorder, where it is supposed that the 3d magnetic states in the transition metal alloys are strongly perturbed by the structural disorder, leading to a variation in moment magnitude from one iron atom to another, as well as a variation in exchange coupling. This can explain the data for the rare-earth alloys, where the 4f magnetic states are only weakly affected by the structural disorder.

Important data related to this problem is revealed by Mössbauer studies. First, it is found that the P(H) distribution is rather broad, supporting the notion of chemical disorder giving rise to a distribution in moment magnitudes.[5] Second, analysis of the P(H) distribution shows that the hyperfine field at each site follows roughly the same curve as a function of temperature.[122] This implies that large fluctuations in exchange interaction cannot be the only contributing factor, because, if they were, there would then be different temperature variations in hyperfine field from the different iron atoms. It is therefore most likely that a combination of exchange variations and chemical disorder are responsible for the flattened magnetisation curves. The chemical disorder effect should be more important at low temperatures where the local environment is most irregular. Such considerations have been incorporated in modified random exchange models, in which the parameter δ is taken to be a function of temperature, $\delta = \delta_o (1 - t^2)$, where $t = T/T_c$.[123, 124] Very good agreement with experiment is found for values of $\delta_o \sim 0.65$, both for the temperature dependent magnetisation curves[7] and for the breadth of the moment distribution.[5] It is also inferred that, at least at low temperatures, many of the iron atom pairs experience either zero or even negative exchange coupling.[5] These are conclusions that could not have been arrived at had it not been for the detail afforded by the Mössbauer data.

4. 2. Moment Distributions

Magnetic texture is observed in virtually all amorphous alloy ribbons and wires. In as cast ribbons and wires there is a high degree of internal stress, with a complex distribution. Given the positive magnetostriction that exists in most iron-based alloys, the resulting distribution in moment directions is almost random. If the sample is stress-relieved by thermal annealing to a temperature below the crystallisation point, then the moments are free to respond to the appropriate demagnetisation fields. In ribbons this results in the moments lying preferentially in the ribbon plane, while in wires the moments lie preferentially along the wire axis. If a macroscopic anisotropy is introduced, as in the field annealing of ribbon samples, then the moments will also lie preferentially along the direction of the induced anisotropy. The moment orientations are also sensitive to the effect of applied stresses, again through the magnetostrictive response, as well as being strongly affected by the application of an external magnetic field.

Given the degree to which the distribution of moment directions in an alloy specimen is indicative of its microscopic mechanical and magnetic state, it is not surprising that magnetic texture is widely studied. Mössbauer spectroscopy is particularly well suited for this purpose, because the data intrinsically contain information on the orientation of every ^{57}Fe atom in the alloy. However, some steps need to be taken to enable the reliable extraction of the moment distribution from a given set of Mössbauer spectra. In early experiments research workers tended to use only one spectrum, with the γ-ray beam incident at 90° to the plane of the sample. This method only provides a single mean value for the angle between the moments and the γ-rays. Later on more sophisticated methods were developed, such as a technique whereby one can generate an effectively texture-free spectrum from the superposition of spectra recorded with the incident γ-ray beam at the 'magic angle' of 54° to the ribbon plane.[125] More recently scientists are using series of spectra recorded in different geometries and fitting the trends in the data as a function of the orientation to the γ-ray beam.[96, 126, 127] This offers better definition in that there are a larger number of data points for the limited number of degrees of freedom associated with a functional form for the moment distribution.

A variety of models have been used to describe the moment distributions. Some are illustrated in Fig. 9. The simplest of all is the uniform moment model, in which it is assumed that all the moments point along the same direction, as if they were all collinear within a single magnetic domain. An extended version of this, involving two uniform moments corresponding to an observed biaxial domain pattern, has been used to model the moment distributions in $Fe_{40}Ni_{40}B_{20}$.[127] Another model has a generalised form for the distribution in terms of an ellipsoidal probability function.[128] This formalism has been extended to one in which the principal axes of the ellipsoidal distribution, and therefore the principal axes of the moment distribution, may be determined directly from the Mössbauer data.[125] Although the ellipsoidal model is intuitively reasonable, at least for as cast ribbons, there are aspects that are not physical in the case of stress relieved or field annealed ribbons. In particular, as long as the principal axis normal to the plane is even slightly non-zero, there will be a finite probability that some moments will be directed in that normal direction. An alternative model which circumvents this is the wedge-shaped distribution, where all the moments are assumed to lie within a region bounded by an in-plane angle and an out-of-plane angle.[129] This model has been found to be useful in field annealed ribbons and in applied field experiments, where there is a well resolved axis in the ribbon, about which the moments are distributed.

In the past few years studies of moment canting in amorphous alloys have necessitated some further development of models of moment distributions. Moment canting refers to a phenomenon which is observed in a number of iron and FeNi based amorphous alloys, in which even in response to large applied fields the moments do not align completely with the applied field. The details of this will be dealt with more fully in the following section. Of relevance here is that two new models of moment distributions have been developed from this research. The first is a double-Gaussian distribution, in which it is assumed that the in-plane and out-of-plane moment distributions follow independent Gaussian distributions.[130] This is a physically realistic model, albeit still a phenomenological one, in keeping with all others previously mentioned. The only *a priori* theoretically determined distribution to date has recently been derived from a 'combed hair' model of moment canting induced by the presence of frustrated exchange interactions.[96] This model predicts truncated inverse-sine distributions both in-plane and out-of-plane.

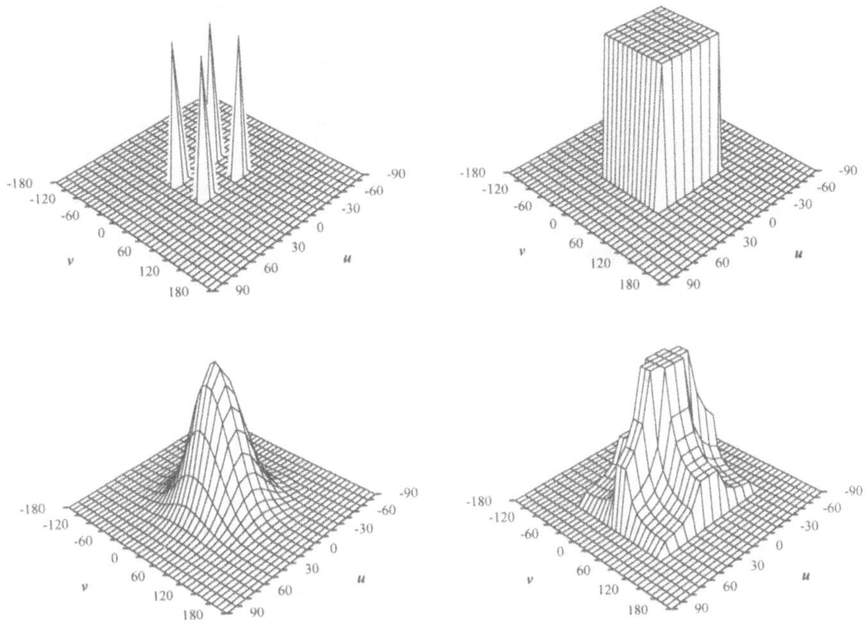

Figure 9. Three-dimensional plots representing the shape of the probability distribution of the moment directions used in some common models: (a) uniform moment, (b) wedge-shaped distribution, (c) double-Gaussian distribution, and (d) truncated inverse-sine distribution. The plot coordinates are the out-of-plane angle, u, the in-plane angle, v, and the probability distribution, $P(u,v)$.

5. SELECTED TOPICS

In this section, we consider some selected topics of current interest in studies of amorphous alloys and wires: (i) the determination of moment distributions in as cast and surface crystallised wires, (ii) the question of residual moment canting in amorphous ribbons in applied fields, and the consequences of this for magnetic models of the short range order, and (iii) the use of polarised Mössbauer spectroscopy for determination of in-plane moment distributions and P(H) distributions in amorphous ribbons.

5. 1. Moment Distributions in Wires

Iron-based amorphous wires, such as $Fe_{77.5}Si_{7.5}B_{15}$, have received considerable attention in the last few years because of a characteristic large Barkhausen jump in their hysteresis loops. This spontaneous magnetisation reversal accounts for about half of the total flux change. The size and abruptness of this effect has aroused interest in using amorphous wires for a variety of applications in non-contact sensor devices. It has been found[52,131,132] that post-production thermal treatment can enhance the magnetic and magnetoelastic properties of the wires, with the determining factors being the growth of surface α-iron crystallites and the associated modification of the magnetic domain structure in the wire. Mössbauer spectroscopy can play a part in identifying and characterising both these features.

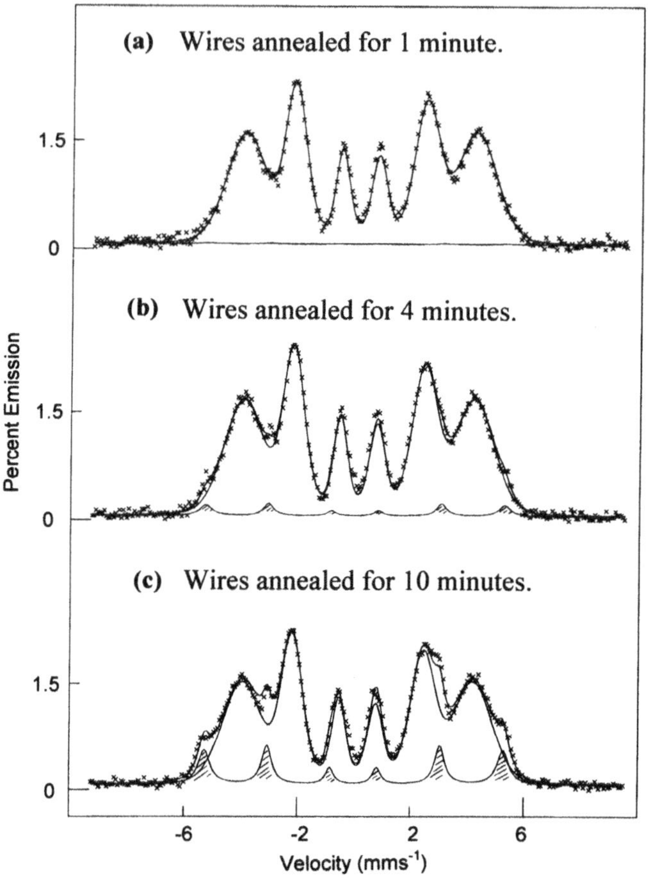

Figure 10. Room temperature conversion electron Mössbauer spectra of $Fe_{77.5}Si_{7.5}B_{15}$ wires annealed at 425°C for (a) 1 minute, (b) 4 minutes, and (c) 10 minutes. After 4 minutes annealing the spectra show the growth of a second phase corresponding to surface crystallites of α-iron.

The growth of surface crystallites may be studied using conversion electron Mössbauer spectroscopy, or CEMS, in which one monitors the backscattered signal from those conversion electrons that are within the ca. 200 nm escape depth of the surface. This method offers highly sensitive measurement of any surface phases, as is illustrated in Fig. 10, which shows the Mössbauer spectra of $Fe_{77.5}Si_{7.5}B_{15}$ wires thermally annealed for 1, 4, and 10 minutes, respectively. Crystallisation in the form of a sharp six-line subspectrum due to α-iron is already visible after 4 minutes and grows to about 20 percent of the spectral area after 10 minutes. The ability to detect such rapid crystallisation is important, especially as in this case x-ray diffraction data was insensitive to any such crystallisation for less than 10 minutes.[133]

Magnetic texture in wires can be resolved using transmission Mössbauer spectroscopy. A particular complication is that the cross-sectional diameter of most commercial wires, which is of the order of 125 μm, is large enough to introduce inhomogeneous thickness effects into the spectra. In the two reported Mössbauer studies of amorphous wires to date, both the transmission integral approach[134] and a weighted phenomenological approach[132, 135] have been used to account for these saturation effects. Le Gal and Varret[134] also used a method for simulating standard densities for resonant absorption by using a reference sample, free of texture and thickness effects, obtained by mechanically crushing the wires. By looking at the magnetic texture in three different as cast Fe-Cr-P-C-Si wires of different diameter, they conclude that the iron moments lie preferentially along the wire axis, but that near the surface of the wire the moments are directed radially outwards. This result is understandable in view of the radial casting stresses that are generated in the quench, competing with an axial demagnetisation field. Data on as cast $Fe_{77.5}Si_{7.5}B_{15}$ wires support this conclusion, and additionally show that, in surface crystallised wires, the moments become preferentially aligned perpendicular to the wire axis.[132, 135] This is understood in terms of compressive stresses from the surface crystallisation giving rise to a sheath of tangentially orientated iron moments surrounding an axially orientated core.[132]

5. 2. Residual Moment Canting

The issue of residual moment canting in amorphous alloys has attracted considerable interest in recent years. A body of evidence has emerged from polarised neutron scattering,[60,136] magnetostriction,[51,137] high field magnetization,[43] and Mössbauer effect[96,127,129,138,139] experiments which supports the conclusion that in many nominally ferromagnetic amorphous alloys the atomic moments are not collinear in the presence of an applied field. The most surprising aspect of this non-collinear state is that it appears to persist in applied fields well in excess of the technical saturation field of the alloys concerned. An example of this is the metallic glass, $Fe_{78}Si_9B_{13}$, which has a technical saturation field at room temperature of order 25 mT, but which shows a continuing increase in magnetization, and therefore a lack of absolute saturation, even in pulsed magnetic fields as high as 120 T.[43] The observation of canting effects in such large fields indicates that the determining factors are related to the strong exchange interactions between magnetic ions, rather than the much weaker anisotropic interactions.

Mössbauer spectroscopy has proved to be invaluable for detailing the applied field dependence of the moment canting. In the latest report on moment canting in $Fe_{78}Si_9B_{13}$ ribbons, Mössbauer effect measurements established that the same degree of non-collinearity, corresponding to a Gaussian distribution with a standard deviation of the order of 11°, was present in a ribbon subjected to an 0.5 T applied field in-plane, and to a 9 T field directed normal to the plane.[96] This result confirms the intrinsic nature of the moment canting, and has led to the development of an *a priori* 'combed hair' model that attributes the canting to the presence of a few antiferromagnetically exchange coupled iron atoms in the alloy.[96] The model predicts that long range perturbations, falling off in a very slow $\exp(-r)/r$ decay law, will surround a pair of antiferromagnetically coupled moments, much like the way combed hair will part and flow around a tangle or knot.

Given that moment canting is attributed to the effect of disturbed exchange interactions, the potential arises for controlling or modifying the degree of frustration present, e.g., through varying preparation conditions, compositions, and post-production treatment. If one could achieve an improved moment collinearity in moderate applied fields, the attendant increase in technical saturation magnetisation could be a significant factor in future device development and manufacture.

5. 3. Polarised Mössbauer Spectroscopy

In other recent work, *polarised* Mössbauer spectroscopy has been used to obtain systematic data on *in plane* moment distributions in amorphous alloy ribbons.[139,140] Although it has been known for 35 years,[141] polarised Mössbauer spectroscopy has only in-

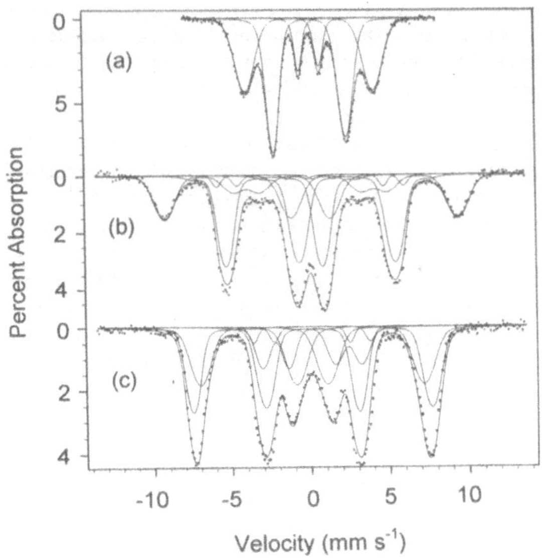

Figure 11. Conventional and polarised Mössbauer spectroscopy of $Fe_{78}Si_9B_{13}$: (a) single line ^{57}Co:Rh source, (b) and (c) polarised six-line ^{57}Co:Fe source.[139] In all cases an in-plane field, B_A = 250 mT, was applied to the $Fe_{78}Si_9B_{13}$ absorber. For (b) and (c) an in-plane field, B_S = 250 mT, was applied to the source also. In (b) $B_A // B_S$ while in (c) $B_A \perp B_S$. The strong difference in the spectra of (b) and (c) indicates the sensitivity of the polarised source technique to in-plane moment distributions.

frequently been applied to studies of amorphous alloys.[142, 143] The new experiments take advantage of modern improvements in source manufacture and computational analysis.

In conventional, unpolarised, ^{57}Fe Mössbauer spectroscopy a single line source, ^{57}Co diffused into a Rh metal foil, is used. In polarised ^{57}Fe Mössbauer spectroscopy a six-line source, ^{57}Co diffused into a polycrystalline α-iron foil, is used. If the α-iron foil is magnetised with an in-plane magnetic field, such that the cobalt and iron atoms are aligned normal to the γ-ray propagation direction, the six emitted γ-rays are plane polarised either parallel or perpendicular to the magnetising field. The relative intensities and the polarisations of the six lines are approximately $3_{//}:4_{\perp}:1_{//}:1_{//}:4_{\perp}:3_{//}$.[141] If the absorber foil is also magnetised normal to the γ-ray direction, the differently polarized γ-rays may be separated. This is illustrated in Fig. 11 for the case of $Fe_{78}Si_9B_{13}$.

The polarised spectra of Fig. 11 demonstrate two important features. First, the direction of the iron moments in the plane of the ribbon has a strong influence on the spectral profile obtained, as evidenced by the differences between Fig. 11(b) and (c). This allows the relatively accurate measurement of in-plane moment directions without the need for multiple dataset analysis, as with unpolarised γ-rays. Second, there is a clear misfit between the theoretical solid lines in Fig. 11 and the experimental data. This misfit appears in the polarised spectra, but not in the unpolarised spectrum. This is revealing, as the same model was used to fit all three spectra.[139] This shows that the polarised spectra are more sensitive to the fine details of the distributions of hyperfine parameters than are unpolarised spectra. This newly recognised feature may prove to be important in future improved modelling of the magnetic and structural states of amorphous alloys.

Further developmental work is progressing on the methodology of polarised Mössbauer spectroscopy. A particular focus in the last few years has been in the preparation of *single-line* polarised sources. The technique employed is one that was established in the 1960s and

1970s, in which a resonant polarising filter preferentially absorbs one of the polarised components of an unpolarised source, leaving only the other polarised component.[144, 145] The latest work has centred on optimising the polarisation of the transmitted γ-ray beam by using ^{57}Fe-enriched polarising filters of $FeBO_3$ and α-iron metal heated to temperatures of order 340 and 850 K, respectively.[146] The heating is necessary to bring the optimal absorption line into coincidence with the γ-rays emitted from the source. Because single line sources are not yet as well polarised as the six-line sources, there is still room for improvement, and in the future they may prove to be of superior quality.

6. CONCLUSIONS

In this chapter we have reviewed the current state of research and development of amorphous alloys and ribbons with a view to both their fundamental properties and their technological application. We have focused in particular on the role that Mössbauer spectroscopy plays in this process. The underlying concepts of Mössbauer spectroscopy for studying amorphous alloys have been discussed, highlighting some of the experimental and analytical techniques needed to deal with hidden quadrupole splitting contributions, distributions in hyperfine parameters, and thickness effects. The structural information embodied in the Mössbauer data has been described, including studies of short range order, crystallisation processes, and structural relaxation. The magnetic information contained in the Mössbauer data was similarly described, concentrating on the relationship between the magnetic hyperfine field and the atomic moment, and on determinations of magnetic texture. Finally a few selected topics of current interest were reviewed: magnetic texture and surface crystallisation in amorphous wires, residual moment canting in applied fields, and the use of polarised Mössbauer spectroscopy as an improved means of studying both texture and hyperfine parameter distributions.

Looking ahead, it seems certain that amorphous alloy research will continue unabated as long as there remains the potential for improved applications and commercially viable devices. The burgeoning work in the field of composite materials may also be of direct interest, especially in the exploration of new systems comprising combinations of ultra-soft and ultra-hard magnetic materials. In structural investigations there are good prospects for more detailed modelling of amorphous alloys. One potential area of future research is in the combined modelling of neutron, x-ray, and Mössbauer experimental data using techniques such as reverse Monte Carlo analysis. Such modelling has been used recently, with good results, for combined neutron and x-ray data;[71] the addition of Mössbauer data would add another dimension to this work. In magnetic studies there is ample opportunity for more detailed and refined measurements on magnetic texture and hyperfine field distributions. Polarised Mössbauer spectroscopy is currently being explored for this purpose. A possible contender for accuracy in such work is the new resonant scattering capability of synchrotron x-ray sources tuned to Mössbauer effect frequencies. Experiments utilising the Mössbauer beamline at the European Synchrotron Research Facility in Grenoble to study magnetic texture in amorphous alloy ribbons are already under way.[147] As well as these new techniques and developments, it is certain that the more standard Mössbauer studies of amorphous alloys will continue as before. Mössbauer spectroscopy is an excellent tool for quality control in the preparation of amorphous materials, because it is very sensitive to both impurities and crystallised phases. This very practical aspect of Mössbauer spectroscopy is central to its use, and underpins its continuing role as one of the defining spectroscopies used for characterising and understanding the magnetism and materials science of amorphous alloys.

ACKNOWLEDGMENTS. The author thanks the UK Engineering and Physical Sciences Research Council for their financial assistance of research into moment canting in amorphous alloys through grants GR/H28073 and GR/J97618. He is also grateful to the University of London Central Research Fund for contributing towards the purchase of a backscattering detector for use in the study of crystallisation in amorphous wires.

REFERENCES

1. H. J. Guntherodt and H. Beck, *Glassy Metals I*, Springer Verlag, Berlin (1981).
2. H. J. Guntherodt and H. Beck, *Glassy Metals II*, Springer Verlag, Berlin (1983).
3. F. E. Luborsky, *Amorphous Metallic Alloys*, Butterworths, London (1983).
4. K. Moorjani and J. M. D. Coey, *Magnetic Glasses*, Elsevier, Amsterdam (1984).
5. R. C. O'Handley, *J. Appl. Phys.* **62**, R15-R49 (1987).
6. U. Gonser, in *Atomic Energy Review Suppl. No. 1: Application of Nuclear Techniques to the Studies of Amorphous Metals*, p. 203-228, Int. Atomic Energy Review Agency, Vienna (1981).
7. A. K. Bhatnagar, *Hyp. Interact.* **24-26**, 637-666 (1985).
8. G. Czjzek, *Hyp. Interact.* **24-26**, 667-680 (1985).
9. G. Longworth, in *Mössbauer Spectroscopy Applied to Inorganic Chemistry, Volume 2*, G. J. Long, ed., p. 289-342, Plenum Press, New York (1987).
10. W. Klement Jr., R. H. Willens, and P. Duwez, *Nature* **187**, 869 (1960).
11. P. Duwez, R. H. Willens, and W. Klement Jr., *J. Appl. Phys.* **31**, 1136 (1960).
12. H. S. Chen and D. E. Polk, *J. Non-Cryst. Solids* **15**, 174 (1974).
13. H. A. Davies, in *Amorphous Metallic Alloys*, F. E. Luborsky, ed., p. 8-25, Butterworths, London (1983).
14. H. H. Liebermann, in *Amorphous Metallic Alloys*, F. E. Luborsky, ed., p. 26-41, Butterworths, London (1983).
15. S. R. Elliott, *Physics of Amorphous Materials*, Wiley, New York (1990).
16. H. Okumura, K. N. Ishihara, P. H. Shingu, H. S. Park, and S. Nasu, *J. Mat. Sci.* **27**, 153-160 (1992).
17. S. J. Campbell, K. D. Jayasuriya, A. Calka, and J. Jing, *Nucl. Instrum. Meth. B* **76**, 81-84 (1993).
18. K. D. Jayasuriya, S. J. Campbell, A. Calka, and J. Jing, *Nucl. Instrum. Meth. B* **76**, 85-88 (1993).
19. J. Balogh, T. Kemeny, I. Vincze, L. Bujdoso, L. Toth, and G. Vincze, *J. Appl. Phys.* **77**, 4997-5003 (1995).
20. J. van Wonterghem, S. Mørup, C. J. W. Koch, S. W. Charles, and S. Wells, *Nature* **322**, 622-623 (1986).
21. I. D. Dragieva, M. S. Slavcheva, and D. T. Buchkov, *J. Less-Common Metals* **117**, 311-317 (1986).
22. S. Linderoth, *Hyp. Interact.* **68**, 107-118 (1991).
23. S. M. Huang and C. Z. Yang, *J. Mat. Sci. Lett.* **13**, 779-783 (1994).
24. L. A. Crum, *Physics Today*, 22-29 (1994).
25. K. S. Suslick, S. B. Choe, A. A. Cichowlas, and M. W. Grinstaff, *Nature* **353**, 414-416 (1991).
26. M. W. Grinstaff, M. B. Salamon, and K. S. Suslick, *Phys. Rev. B* **48**, 269-273 (1993).
27. C. D. Graham Jr. and T. Egami, *J. Magn. Magn. Mater.* **15-18**, 1325-1330 (1980).
28. F. E. Luborsky, in *Amorphous Metallic Alloys*, F. E. Luborsky, ed., p. 360-380, Butterworths, London (1983).
29. D. Raskin and C. H. Smith, in *Amorphous Metallic Alloys*, F. E. Luborsky, ed., p. 381-400, Butterworths, London (1983).
30. D. Jiles, *Introduction to Magnetism and Magnetic Materials*, Chapman & Hall, London (1991).
31. M. R. J. Gibbs, *Physics World*, 40-44 (1994).
32. K. Shibuya, *Mat. Sci. Eng. A* **182**, 1435-1439 (1994).
33. T. Sawa and K. Nakagawa, *Mat. Sci. Eng. A* **182**, 902-905 (1994).
34. F. Hirose, M. Takagi, H. Mori, and T. Imura, *Mat. Sci. Eng. A* **182**, 1359-1362 (1994).
35. H. Hase, R. Shoji, and M. Wakamiya, *Mat. Sci. Eng. A* **182**, 1378-1382 (1994).

36. Y. Okazaki and H. Mogi, *Mat. Sci. Eng. A* **182**, 1374-1377 (1994).
37. Y. Yoshida, K. Mohri, J. Yamasaki, T. Uchiyama, and H. Yoshida, *Mat. Sci. Eng. A* **182**, 1369-1373 (1994).
38. H. Matsuki, T. Matsuzaki, and A. Suzuki, *Mat. Sci. Eng. A* **182**, 1363-1365 (1994).
39. S. Tsuji and O. Watanuki, *J. Vac. Sci. Tech. B* **12**, 2144-2147 (1994).
40. M. Tejedor, J. A. Garcia, and J. Carrizo, *J. Magn. Magn. Mater.* **118**, 333-338 (1993).
41. H. Kronmüller and H. Grimm, *J. Magn. Magn. Mater.* **6**, 57-60 (1977).
42. P. Kossacki, *J. Magn. Magn. Mater.* **125**, 147-150 (1993).
43. C. D. Graham Jr. and M. R. J. Gibbs, *IEEE Trans. Magn.* **29**, 3457-3459 (1993).
44. P. Szymczak, C. D. Graham, and M. R. J. Gibbs, *IEEE Trans. Magn.* **30**, 4788-4790 (1994).
45. R. H. Smith, G. A. Jones, and D. G. Lord, *IEEE Trans. Magn.* **24**, 1868-1870 (1988).
46. H. Kronmüller, M. Fähnle, M. Domann, H. Grimm, R. Grimm, and B. Groger, *J. Magn. Magn. Mater.* **13**, 53-70 (1979).
47. K. Zaveta, O. V. Nielsen, and K. Jurek, *J. Magn. Magn. Mater.* **117**, 61-68 (1992).
48. K. Kioke, H. Matsuyama, W. J. Tseng, and J. C. M. Li, *Appl. Phys. Lett.* **62**, 2581-2583 (1993).
49. L. J. Heyderman, F. Dick, J. N. Chapman, and M. R. J. Gibbs, *Inst. Phys. Conf. Series* **138**, 325-328 (1993).
50. M. Fähnle, J. Furthmüller, R. Pawellek, and C. Elsässer, in *Proc. 5th Int. Conf. Physics of Magnetic Materials (Madralin, Poland, 1990),* W. Gorzkowski, M. Gutowski, H. K. Lachowicz, and H. H. Szymczak, eds., p. 204-241, World Scientific, London (1991).
51. A. P. Thomas and M. R. J. Gibbs, *J. Magn. Magn. Mater.* **103**, 97-110 (1992).
52. D. Atkinson, P. T. Squire, M. R. J. Gibbs, and J. Yamasaki, *IEEE Trans. Magn.* **29**, 3478-3480 (1993).
53. C. N. J. Wagner, in *Amorphous Metallic Alloys,* F. E. Luborsky, ed., p. 58-73, Butterworths, London (1983).
54. N. Cowlam and G. E. Carr, *J. Phys. F* **15**, 1117-1126 (1985).
55. M. Dahlborg, U. Dahlborg, and C. Svensson Huldt, *Physica A* **201**, 434-440 (1993).
56. L. Fernandez Barquin, J. C. Gomez Sal, J. M. Barandiaran, F. J. Bermejo, and W. S. Howells, *J. Phys. Condens. Matter* **6**, 1645-1652 (1994).
57. Q. A. Pankhurst, J. D. Wicks, R. L. McGreevy, and M. R. J. Gibbs, unpublished, 1995.
58. A. C. Hannon, M. Hagen, R. A. Cowley, H. B. Stanley, and N. Cowlam, *Physica B* **180**, 230-232 (1992).
59. R. A. Cowley, N. Cowlam, and L. D. Cussens, *J. Physique* **49**, 1285-1286 (1988).
60. R. A. Cowley, C. Patterson, N. Cowlam, P. K. Ivison, J. Martinez, and L. D. Cussen, *J. Phys. Condens. Matter* **3**, 9521-9537 (1991).
61. U. Krey, *J. Magn. Magn. Mater.* **6**, 27-37 (1977).
62. H. Kronmüller and J. Ulner, *J. Magn. Magn. Mater.* **6**, 52-56 (1977).
63. J. L. Finney, in *Amorphous Metallic Alloys,* F. E. Luborsky, ed., p. 42-57, Butterworths, London (1983).
64. I. Turek, *J. Magn. Magn. Mater.* **98**, 119-129 (1991).
65. I. Turek and J. Hafner, *Phys. Rev. B* **46**, 247-256 (1992).
66. T. Tomida and T. Egami, *Phys. Rev. B* **48**, 3048-3057 (1993).
67. M. Liebs, K. Hummler, and M. Fähnle, *Phys. Rev. B* **51**, 8664-8667 (1995).
68. R. Lorenz and J. Hafner, *J. Magn. Magn. Mater.* **139**, 209-227 (1995).
69. D. A. Keen and R. L. McGreevy, *J. Phys. Condens. Matter* **3**, 7383-7394 (1991).
70. R. L. McGreevy, *Nucl. Instrum. Meth. A* **354**, 1-16 (1995).
71. J. D. Wicks, *Studies of Disordered Materials,* Oxford University, Ph. D. Thesis, 1993.
72. M. Kopcewicz, *Structural Chem.* **2**, 313-342 (1991).
73. G. Czjzek, J. Fink, F. Gotz, H. Schmidt, J. M. D. Coey, J. P. Rebouillat, and A. Lienard, *Phys. Rev. B* **23**, 2513-2530 (1981).
74. M. Kopcewicz, H. G. Wagner, and U. Gonser, *Solid State Commun.* **48**, 531-533 (1983).
75. M. Kopcewicz, H. G. Wagner, and U. Gonser, *J. Phys. F* **16**, 929-935 (1986).
76. M. Kopcewicz, B. Idzikowski, and A. Wrzeciono, *J. Magn. Magn. Mater.* **125**, 290-296 (1993).
77. B. Window, *J. Phys. E* **4**, 401-402 (1971).

78. J. Hesse and A. Rübartsch, *J. Phys. E* **7**, 526-532 (1974).
79. P. Mangin, G. Marchal, M. Piecuch, and C. Janot, *J. Phys. E* **9**, 1101-1105 (1976).
80. G. Le Caër and J. M. Dubois, *J. Phys. E* **12**, 1083-1090 (1979).
81. A. S. Schaafsma, *Phys. Rev. B* **23**, 4784-4787 (1981).
82. C. L. Chien, *Phys. Rev. B* **23**, 4788-4790 (1981).
83. G. Le Caër and J. M. Dubois, *Physica Stat. Sol. A* **64**, 275-282 (1981).
84. C. Wivel and S. Mørup, *J. Phys. E* **14**, 605-610 (1981).
85. I. Vincze, *Nucl. Instrum. Methods* **199**, 247-262 (1982).
86. M. E. Lines and M. Eibschütz, *Solid State Commun.* **45**, 435-439 (1983).
87 F. Varret, J. M. Grenèche, and J. Teillet, in *Trends in Mössbauer Spectroscopy: Proceedings of 2nd Seeheim Workshop on Mössbauer Spectroscopy, May 1983,* P. Gütlich and G. M. Kalvius, eds., pp. 105-116, University of Mainz, Mainz (1983).
88. G. Le Caër, J. M. Dubois, H. Fischer, U. Gonser, and H. G. Wagner, *Nucl. Instrum. Meth. B* **233**, 25-33 (1984).
89. M. Eibschütz, M. E. Lines, H. S. Chen, and T. Masumoto, *J. Phys. F* **14**, 505-520 (1984).
90. R. E. Vandenberghe, D. Gryffroy, and E. De Grave, *Nucl. Instrum. Meth. B* **26**, 603-609 (1987).
91. M. Miglierini, *Nucl. Instrum. Meth. B* **36**, 475-484 (1989).
92. R. E. Vandenberghe, E. De Grave, and P. M. A. de Bakker, *Hyp. Interact.* **83**, 29-49 (1994).
93 L. X. Dou, R. J. W. Hodgson, and D. G. Rancourt, *Nucl. Instrum. Meth. B* **100**, 511-518 (1995).
94. J. M. Grenèche and F. Varret, *J. Physique Lett.* **43**, L233-L237 (1982).
95. I. Vincze, *Solid State Commun.* **25**, 689-693 (1978).
96. Q. A. Pankhurst, J. Z. Jiang, S. Betteridge, M. R. J. Gibbs, and G. A. Gehring, *J. Phys. Condens. Matter* **7**, 9571-9593 (1995).
97. J. Y. Ping and D. G. Rancourt, *Hyp. Interact.* **71**, 1433-1436 (1992).
98. S. Margulies and J. R. Ehrman, *Nucl. Instrum. Methods* **12**, 131-137 (1961).
99. R. Hollatz, E. Gerdau, and R. Rüffer, *Hyp. Interact.* **28**, 807-810 (1986).
100. M. C. D. Ure and P. A. Flinn, in *Mössbauer Effect Methodology,* I. J. Gruverman, ed., p. 245-262, Plenum Press, New York (1971).
101. D. A. Shirley, M. Kaplan, and P. Axel, *Phys. Rev. B* **123**, 816-830 (1961).
102. J. D. Bernal, *Proc. Roy. Soc. London A* **280**, 299 (1964).
103. P. H. Gaskell, *Nature* **289**, 474-476 (1981).
104. G. Czjzek, *Phys. Rev. B* **25**, 4908-4910 (1982).
105. R. A. Brand, G. Le Caër, and J. M. Dubois, *J. Phys. Condens. Matter* **2**, 6413-6431 (1990).
106. G. Le Caër and R. A. Brand, *Hyp. Interact.* **71**, 1507-1514 (1992).
107. M. Henry, F. Varret, J. Teillet, G. Ferey, O. Massenet, and J. M. D. Coey, *J. Physique Colloq. C1* **41**, 279-280 (1980).
108. J. M. Dubois and G. Le Caër, *Acta Metallurgica* **32**, 2101-2114 (1984).
109. R. J. Pollard, Z. S. Wronski, and A. H. Morrish, *Phys. Rev. B* **29**, 4864-4868 (1984).
110. G. L. Whittle, P. Wells, S. J. Campbell, A. Calka, and A. M. Stewart, in *Rapidly Quenched Metals,* S. Steeb and H. Warlimont, eds., p. 545-548, Elsevier, Amsterdam (1985).
111. P. Raj and A. Sathyamoorthy, *J. Phys. F* **18**, 113-121 (1988).
112. S. H. Ge, G. L. Chen, M. X. Mao, D. S. Xue, C. X. Li, F. S. Li, Y. D. Zhang, W. A. Hines, and J. I. Budnick, *J. Magn. Magn. Mater.* **129**, 207-212 (1994).
113. N. Randrianantoandro, J. M. Grenèche, and F. Varret, *J. Magn. Magn. Mater.* **117**, 93-101 (1992).
114. Y. Yoshizawa, S. Oguma, and K. Yamauchi, *J. Appl. Phys.* **64**, 6044-6046 (1988).
115. G. Rixecker, P. Schaaf, and U. Gonser, *J. Phys. Condens. Matter* **4**, 10295-10310 (1992).
116. M. Miglierini, J. Lipka, and J. Sitek, *Hyp. Interact.* **94**, 2193-2197 (1994).
117. T. Komatsu, *J. Mat. Sci.* **28**, 6295-6302 (1993).
118. B. Arcondo, J. Moya, F. Audebert, and H. Sirkin, *Hyp. Interact.* **83**, 293-297 (1994).
119. R. C. O'Handley, in *Amorphous Metallic Alloys,* F. E. Luborsky, ed., p. 257-282, Butterworths, London (1983).
120. P. Panissod, J. Durand, and J. I. Budnick, *Nucl. Instrum. Meth.* **199**, 99-114 (1982).
121. K. Handrich, *Physica Stat. Sol. B* **32**, K55 (1969).

122. C. L. Chien, *Phys. Rev. B* **18**, 1003-1015 (1978).
123. A. K. Bhatnagar, B. B. Prasad, and R. Jagannathan, *Phys. Rev. B* **29**, 4896-4903 (1984).
124. T. Kaneyoshi and I. Tamura, *Physica Stat. Sol. B* **123**, 525-531 (1984).
125. J. M. Grenèche and F. Varret, *J. Phys. C* **15**, 5333-5344 (1982).
126. Q. A. Pankhurst, M. R. J. Gibbs, and A. P. Thomas, *J. Magn. Magn. Mater.* **104**, 111-112 (1992).
127. Q. A. Pankhurst and M. R. J. Gibbs, *J. Phys. Condens. Matter* **5**, 3275-3288 (1993).
128. H. D. Pfannes and H. Fischer, *Appl. Phys.* **13**, 317-325 (1977).
129. M. Melamud, L. J. Swartzendruber, L. H. Bennett, J. Cullen, and M. Wun-Fogle, *J. Appl. Phys.* **61**, 3644-3646 (1987).
130. J. Z. Jiang, Q. A. Pankhurst, and M. R. J. Gibbs, *Hyp. Interact.* **94**, 2137-2143 (1994).
131. D. Atkinson, P. T. Squire, M. R. J. Gibbs, S. Atalay, and D. G. Lord, *J. Appl. Phys.* **73**, 3411-3417 (1993).
132. D. Atkinson, M. R. J. Gibbs, P. T. Squire, and Q. A. Pankhurst, *J. Magn. Magn. Mater.* **131**, 19-28 (1994).
133. D. Atkinson and P. T. Squire, *IEEE Trans. Magn.* **30**, 4782-4784 (1994).
134. G. Le Gal and F. Varret, *J. Magn. Magn. Mater.* **111**, 115-122 (1992).
135. Q. A. Pankhurst, unpublished results, 1993.
136. R. A. Cowley, N. Cowlam, P. K. Ivison, and J. Martinez, *J. Magn. Magn. Mater.* **104**, 159-160 (1992).
137. F. Bucholtz, K. P. Koo, A. Dandridge, and G. H. Sigel Jr., *J. Magn. Magn. Mater.* **54-57**, 1607-1608 (1986).
138. S. J. Harker and R. J. Pollard, *J. Phys. Condens. Matter* **1**, 8269-8273 (1989).
139. Q. A. Pankhurst, S. Betteridge, J. Z. Jiang, and M. R. J. Gibbs, *IEEE Trans. Magn.* **30**, 4809-4811 (1994).
140. S. Betteridge, *The Application of Polarised Mössbauer Spectroscopy to the Study of Moment Orientations in Magnetic Materials,* University of Liverpool, Ph. D. Thesis, 1995.
141. S. S. Hanna, J. Heberle, C. Littlejohn, G. J. Perlow, R. S. Preston, and D. H. Vincent, *Phys. Rev. Lett.* **4**, 177-180 (1960).
142. U. Gonser, M. Ghafari, H. G. Wagner, and H. Fischer, *J. Physique* **C2**, 126-127 (1979).
143. U. Gonser and H. Fischer, in *Topics in Current Physics, Vol. 25: Mössbauer Spectroscopy II,* U. Gonser, ed., p. 99-137 Springer-Verlag, Berlin (1981).
144. S. Shtrikman, *Solid State Commun.* **5**, 701-703 (1967).
145. F. Varret, P. Imbert, G. Jehanno, and R. Saint-James, *Physica Stat. Sol. A* **27**, K99-K101 (1975).
146. J. Jäschke, J. Pollman, H. D. Rüter, and E. Gerdau, *Il Nuovo Cimento*, in press (1995).
147. Q. A. Pankhurst, unpublished results, 1995.

4

Diffusion Studies by Mössbauer Spectroscopy

G. Vogl

1. INTRODUCTION

1. 1. The Principal Ideas

The idea of using Mössbauer spectroscopy to study diffusion is nearly as old as Mössbauer spectroscopy itself. Very soon after Mössbauer's discovery, in three different papers[1-3] it was pointed out that the new method offered the chance to study diffusion in a "microscopic" way. Singwi and Sjölander[1] wrote the relation between S(**Q**, ω), i.e., the recoil free resonance absorption cross-section, for the sample as a Mössbauer absorber, or the emission probability, for the sample as a Mössbauer source, and the self-correlation function, $G_s(\mathbf{r}, t)$, the latter describing the probability of finding an atom, which was at the origin at zero time, displaced by a vector, **r**, at time t, as

$$S(\mathbf{Q}, \omega) \propto \int \exp[i(\mathbf{Qr} - \omega t) - \pi\Gamma_0|t|/h]G_s(\mathbf{r}, t)d\mathbf{r}dt, \qquad (1.\ 1)$$

where $h\omega/2\pi$ is the energy difference between the source and absorber, and Γ_0 is the natural linewidth. In Equation 1. 1, we have omitted all numerical and geometrical factors not relevant for diffusion and, for Mössbauer absorption, the nuclear absorption cross section for the pertinent γ-ray radiation. The reader should note that in this chapter, the author has elected to use the "scattering terminology," i.e., **Q** instead of **k** for the wave vector, and S(**Q**, ω) instead of $\sigma_a(\mathbf{Q}, \omega)$, for the absorption cross section, in order to obtain formulations directly comparable to those of quasielastic neutron scattering. We think that this advantage outweighs the inconvenience of a terminology differing slightly from the usual one in papers on Mössbauer spectroscopy. Equation 1. 1 indicates that the shape of the absorption cross section, as a function of the energy difference, $h\omega/2\pi$, and the direction of the γ-wave vector, **Q**, relative to the crystal orientation, will depend on the displacement in space and time of the diffusing atom. The probability of recoil-free nuclear γ-ray emission follows essentially the same law. After a Fourier transform in space and time, one arrives at the absorption cross section which is of Lorentzian shape.

G. Vogl • Institut für Festkörperphysik der Universität Wien, Strudlhofgasse 4, A-1090 Wien, Austria.

Mössbauer Spectroscopy Applied to Magnetism and Materials Science, Volume 2, edited by Gary J. Long and Fernande Grandjean. Plenum Press, New York, 1996.

Singwi and Sjölander[1] discussed two extreme cases for Equation 1. 1, first, diffusion via discrete rapid jumps followed by residence at a lattice site for a finite time and, second, continuous diffusion. In most of this chapter we shall deal with the first case, i.e., jump diffusion in crystalline lattices, for which we can neglect the time during which the atom jumps. But in Section 4 we shall briefly look at the other extreme case, for example, diffusion in glasses.

The self-correlation function can be separated into a part arising from lattice vibrations, with a characteristic time scale of picoseconds, and a part arising from diffusion, with a characteristic time scale of nano to microseconds. Because of these very different time scales their effects on a Mössbauer spectrum are completely decoupled. Because the resolution of Mössbauer spectroscopy is determined by the natural linewidth, and therefore by the lifetime of the Mössbauer excited state, which is of the order of 100 ns for iron-57, lattice vibrations and other motions faster than a tenth of a nanosecond, give rise to lines of a width too large for even very fast Doppler drives. Therefore, they influence the Debye-Waller factor, i.e., they lead to a decrease in line intensity, and only motions slower than nanoseconds can usually be resolved through a line broadening. In this case, it would be more correct to speak of the Lamb-Mössbauer factor, but again, for compatibility with other methods, such as quasielastic neutron scattering, we use the scattering terminology.

Just because fast motions enter only into the Debye-Waller factor, does not mean that information about such fast motion is totally inaccessible. In the absence of a measurable line broadening, the temperature and angle dependence of the Debye-Waller factor, f, can offer, and in some cases has provided, the first information about such motion. For polycrystalline materials,

$$f = \exp[-\langle(\mathbf{Q}\mathbf{x})^2\rangle] \qquad (1.\ 2)$$

from which information about the mean-squared displacement, $\langle \mathbf{x}^2 \rangle$, of a motion, or a whole ensemble of motions, can be derived. In Section 4 such cases will be discussed. In contrast, motions on a time scale larger than a few microseconds do not cause sufficient line broadening to be observed, at least with iron-57 Mössbauer spectroscopy. Thus, as with many other physical methods, we have to "live in a limited time window." In terms of the diffusivity, D, the window corresponds to $10^{-14} < D < 10^{-10}\,m^2/s$.

Figure 1 shows how to understand the diffusion induced line broadening in a simple picture of γ-ray emission, a picture which is borrowed from classical electrodynamics. The γ-ray wave train, in the case of a *static atom*, is emitted from a lattice site with the linewidth, Γ_0, which is related to the reciprocal of the nuclear life time, τ_N. In contrast, in case of an *atom jumping* from lattice site to lattice site, the "emitter," i.e., the source atom changes its local position during emission. Therefore the γ-ray wave train is "cut into pieces," which "interfere" at the receiver, i.e., the absorber atom. Because the "fragmented wave trains" are,

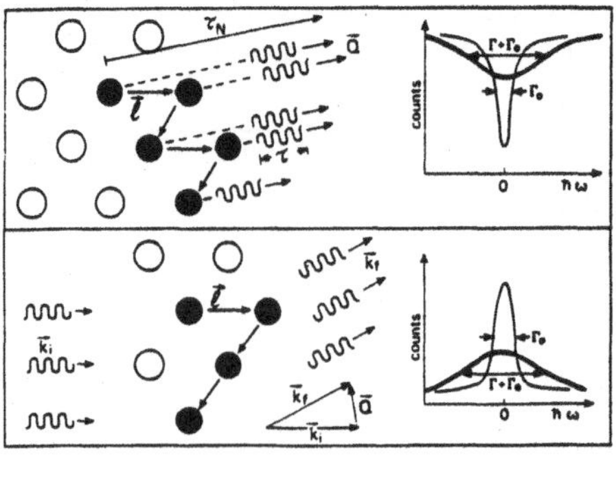

Figure 1. The interference of fragments of the wave train of the incoherent radiation emitted from or scattered at a single atom when that atom visits different sites of a crystal lattice with a jumping motion. In the upper figure a γ-quantum is emitted from a single atom, and the picture for absorption would be analogous. The lower figure shows, for comparison, the scattering of a neutron.

in general, phase-shifted with respect to each other, the phase shift depending on the angle between the γ-ray wave vector, **Q**, and the jump vector, **l**, a smearing of the frequency detected by the absorber will occur, i.e., a broadening, Γ, of the Mössbauer line. In this picture the broadening will be larger the stronger the fragmentation of the wave train, i.e., the shorter the residence time, and will further depend on the fragments' relative phase shifts, i.e., on the relative directions of **Q** and **l**. The picture for absorption is analogous. The process is *incoherent*, i.e., neither is the interference of the γ-ray quanta emitted or absorbed from different atoms taken into account, nor is the interference of neutrons scattered from different atoms.

We repeat here the vital point of diffusion studies with Mössbauer spectroscopy. Through their ability to measure the Fourier transform of the self-correlation function of a single atom, Mössbauer spectroscopy and quasielastic neutron scattering, see Section 5, are unequalled in offering a "microscopic" view of diffusion dynamics. In contrast, the conventional method for measuring diffusion, whose general idea is to determine the diffusion profile of tracer atoms, yields only the net result of a large number of jumps and, therefore, is a macroscopic method. Because both Mössbauer spectroscopy and quasielastic neutron scattering measure the self-correlation function in *time* and *space*, $G_s(\mathbf{r}, t)$, from single crystal studies, both the jump *frequency* and the jump *vector* can be obtained. Thus, these methods are superior to other nuclear methods, such as NMR or perturbed angular correlation studies, which yield only $G_s(t)$ and, therefore, only the jump frequency.

1. 2. Experimental Conditions and Limitations

There are only a very few isotopes that are suitable for Mössbauer studies of diffusion. Probably 90 percent of all studies use iron-57 and exploit its positive features. These features include, first, the availability of strong sources at moderate prices. Typically, for high-temperature studies of alloys, 50 to 100 mCi sources are recommended, however, the latter currently cost more than US$10,000. Second, the iron-57 Debye-Waller factors are reasonably large even at high temperatures, i.e., even in metals up to about 1000°C, and thus measurements are possible in a few days or less. Third, iron is an important element and is a component in many materials of considerable interest, even from the practical point of view, e.g., intermetallic iron alloys which currently face their first applications in turbines.

To the author's knowledge, diffusional line broadening with other Mössbauer isotopes has not been measured to date. However, with other isotopes one can obtain indirect information on diffusion through the temperature dependence of the Debye-Waller factor.[4-6] Even with iron-57, the measuring times required to see reliable line broadening in an intermetallic alloy at high temperatures can be long. Studies in sucrose or a protein at 200 K demand measuring times of one to several days, even with the best instrumentation available. Therefore Mössbauer studies of diffusion are time-consuming and involving a graduate student in such a project should be carefully considered in view of her or his maximum permitted time for thesis work.

Measurements on single crystals, when available, provide maximum information. The preparation of the crystals, either with or without iron-57 enrichment, often demands months and enters into a research paper as just one sentence. These preparations are, to date, an art rather than a science, and particularly the art of growing crystals of a sufficiently large diameter, often as much as one square centimeter, in order to obtain reasonably high counting rates often causes problems, even though the required thickness of an absorber is only about 30 μm, i.e., much less than the crystal thickness of 3 cm, which is optimum for quasielastic neutron scattering. Orientation of the crystal in the furnace is the next difficult task and needs an in-furnace goniometer with, if possible, computer driven circles for automatically changing the crystal orientation.

Sufficiently high diffusivities are needed to see diffusional line broadening, because the time window opens only at a D of ca. 10^{-14} m^2/s. Therefore, for metallic systems rather high temperatures are mandatory, a demand which requires a special type of furnace. It has to have windows which are transparent to low-energy γ-ray radiation, it has to have a small geometry, in order to use an optimum part of the radiation emitted by the source, and it has to allow orientation of a single crystalline specimen. All this poses a number of technical problems which, however, have been solved as may be seen from the drawings in references 7 and 8. Information on such furnaces can be provided by the author.

Even though diffusion studies are a constant struggle for increased count rates, in order to observe reliable broadening of an often weak line, it is advisable to use sufficiently thin absobers in order to remain within the thin-absorber approximation for the Mössbauer absorption. Practically all high-temperature measurements on metallic materials reported herein have met this condition, mostly automatically because of the low Debye-Waller factors. Fitting the spectra with the transmission integral for a non-infinitely thin absorber has never shown significant differences in the results. In an investigation of glassy substances, however, it is sometimes advisable to adapt the sample thickness to the measuring temperature, which requires, e.g., that one measure iron diffusion in sucrose with a thinner sample at 100 K than at 240 K.

1. 3. Chapter Outline

This chapter will not go into detail for most of the experiments conducted up to 1990, for which earlier reviews[9] are available. This is particularly the case for diffusion in simple lattices, see Section 2. We shall, however, discuss earlier results if we judge them necessary for a general comparison, which is, e.g., the case for diffusion in a restricted region, see Section 4.

According to Equation 1. 1, the Mössbauer absorption cross section or emission probability is the Fourier transform in space and time of the correlation function, $G_s(\mathbf{r}, t)$. In sections of this paper we shall deal with this function and, in each special diffusion case, indicate how it derives from the appropriate rate equation. The chapter first reports on diffusion in simple Bravais lattices, mostly metals, in Section 2. In Section 3 we discuss diffusion in more complex non-Bravais lattices, and in Section 4, diffusion in a restricted region or a "cage." Section 5 compares the successes, the possibilities, and the limitations of Mössbauer spectroscopy in diffusion studies with those of quasielastic neutron scattering. We end the chapter with an outlook on possible new developments.

This is surely not the first review of diffusion studies with Mössbauer spectroscopy. In the following we refer to those reviews which have appeared in the 1990s. A shorter review[9] from 1990 contains most of the earlier work, and we shall not repeat all the references which have already been reported, but rather refer the interested reader to the review. Reviews by B. Sepiol[10] in 1993 and G. Vogl[11] in 1994 have concentrated on specific studies, namely diffusion in intermetallic alloys and fast metal diffusion, respectively.

2. THE SIMPLEST CASE

Whereas the Singwi-Sjölander formula applies quite generally to diffusion processes in liquids or solids, Chudley and Elliott[2] considered the particular case of diffusion jumps in an empty lattice. They wrote a rate equation for jump diffusion in an *empty* Bravais lattice as

$$\frac{\partial P(\mathbf{r}, t)}{\partial t} = \frac{1}{N\tau} \sum_k [P(\mathbf{r} + \mathbf{l}_k, t) - P(\mathbf{r}, t)] , \tag{2. 1}$$

where $P(\mathbf{r}, t)$ is the probability of finding a diffusing atom at $\mathbf{r}$, which can jump to N sites with index k over the vector distance, $\mathbf{l}_k$, from $\mathbf{r}$, in the simplest case to a nearest neighbor site. The self-correlation function, $G_s(\mathbf{r}, t)$, is the solution of this equation. Inserting $G_s(\mathbf{r}, t)$

into Equation 1. 1, yields a Lorentzian dependence on ω, the width of which is called the diffusional line broadening, Γ, of the Mössbauer line. For jumps to nearest-neighbor sites one obtains only

$$\Gamma = \frac{h}{2\pi\tau}\left[1 - (1/N)\sum_k \exp(i\mathbf{Q}\mathbf{l}_k)\right]. \tag{2. 2}$$

The half width at half-maximum line broadening, Γ, is proportional to the reciprocal of the atom's residence time, τ, at one lattice site before jumping to any nearest neighbor site where $1/\tau$ is the jump frequency, ν, and is further determined by the sum over a type of structure factor, $\exp(i\mathbf{Q}\mathbf{l}_k)$. It is evident from what has been said above, that maximum information on the diffusion event will be obtained from studies on single crystals, and hence, we shall concentrate on these. For studies on polycrystals, a proper averaging procedure has to be used. For a comparison with results from tracer diffusion studies, and with Section 4, and in particular with the equations for the Brownian oscillator, the reader should note that the diffusion constant is given by the Einstein-Smoluchowski equation, $D = l^2/6\tau$, where we have neglected the influence of correlation between jumps, and l is the average jump path length.

The first successful diffusion studies with Mössbauer spectroscopy were performed by Knauer and Mullen,[12] who studied the diffusion of iron-57 impurity atoms in the cubic lattices of gold and copper. They used polycrystals and demonstrated that problems in comparing the diffusional line broadening with tracer results occur when one tries to compare the results in a straightforward manor. Rather one has to consider correlation factors, and some caution in the polycrystal averaging is necessary. In a study of self-diffusion of iron, full agreement with tracer results was achieved.[7] Full details of the diffusion mechanism are obtainable only when single crystals are used. The key experiments were performed by Mantl et al.[13] and Steinmetz et al.[8] for iron-57 diffusion in aluminum and copper single crystals. Fig. 2 shows a typical result of these investigations.

The anisotropy in diffusional line broadening is usually compared with simulations based on different jump models. Best agreement with theory, see Equation 2. 2, is found for jumps into vacancies on the nearest-neighbor sites. The dashed line in Fig. 2 is the corresponding theoretical curve for self-diffusion. The jump frequency and jump mechanism agree fully with expectations from tracer diffusion experiments. However, interesting details of the jump process, not accessible from a macroscopic method, such as tracer diffusion, are revealed by the Mössbauer studies, as for example, the complex correlated motion of the iron atom and vacancy, which requires at least five jump frequencies for its description. The full line in Fig. 2 demonstrates that the so-called five-frequency model indeed provides an even better fit. The interested reader should consult the original papers or the review by Vogl[9] for more details.

3. DIFFUSION IN MORE COMPLEX LATTICES

In the past decade scientists have been daring enough to start Mössbauer studies of diffusion in more complex systems than the simple metals or diluted alloys, the systems which provided the optimum initial experiments for establishing Mössbauer spectroscopy as an accepted method by the diffusion research community. A few new techniques have been introduced and new theories have been developed and extended, sometimes supported by older theories developed for other techniques, such as for quasielastic neutron scattering.[14]

The growing complexity of the investigated systems necessitated refinement of the theory. The original Singwi-Sjölander-Chudley-Elliott-Krivoglaz theory, describing jumps in Bravais lattices, had been extended to the non-Bravais case by Rowe et al.[15] for quasi-

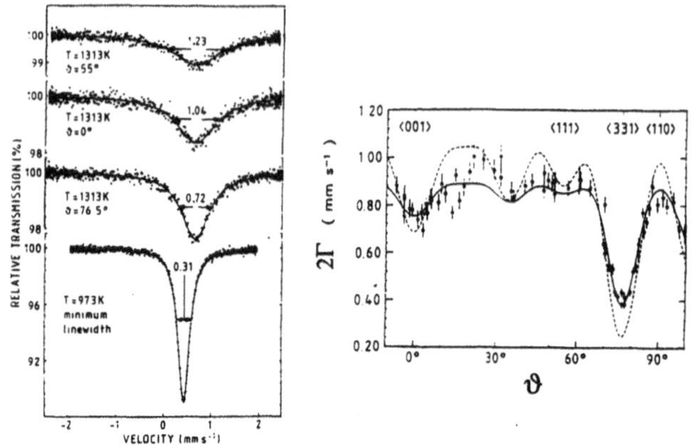

Figure 2. The Mössbauer spectra of iron-57 in a copper single crystal, left.[8] The line width is different for different directions of observation, i.e., for different values of $\exp(i\mathbf{Q}\mathbf{l}_k)$ summed over k. The diffusional line broadening, 2Γ, of the iron-57 Mössbauer spectrum as a function of the observation direction in copper at 1313 K, right.[8] In this figure ϑ denotes the angle between the [001] crystal axis and the observation direction in a plane slightly tilted from the (110) crystal plane. For an interpretation of the full and dashed lines see the text.

elastic neutron scattering, and further developed by several groups.[16] Randl et al.[17] have recently further extended the theory to Mössbauer spectroscopy. We refer the reader to their paper which contains mathematical details too involved for this chapter.

3. 1. A Few Comments on the Theory

Non-Bravais lattices, lattices with more than one atom per crystal unit cell, can be described as consisting of several interpenetrating sublattices. The reason for the greater complexity of diffusion in non-Bravais lattices is evident, because there are now different possible jump paths with different jump frequencies.

In the following we consider a system made up of atoms that jump between sites in such a non-Bravais lattice. Again, as in the simple case, we assume that between two jumps the atoms vibrate about these sites for a time interval, the "residence time," τ, which is very large in comparison with the actual jump time. It should be noted that in general the residence time will be different for the different sublattices. For the sake of simplicity we shall assume that only nearest-neighbor jumps are possible, but an extension to longer jumps is straightforward. Because the sites do not form a Bravais lattice, the m inequivalent sites per primitive unit cell have to be labelled and the set of jump vectors for each site will, in general, connect inequivalent sites. Therefore, we have to write the jump vectors with more indices than was the case in Section 2. Let $\mathbf{l}_{ij}^k$ be the vector from the site of local symmetry, i, to the k-th site of local symmetry, j, and c_j be the probability of occupation of the j-th sublattice site. Further, let each site of the i-th sublattice be surrounded by n_{ij} sites of the j-th sublattice and τ_{ij} be the residence time on a site on the i-th sublattice before jumping to a site on the j-th sublattice. Then the rate equations are

$$\frac{\partial P_i(\mathbf{r}, t)}{\partial t} = \sum_{j,k} \left[\frac{P_j\left(\mathbf{r} + \mathbf{l}_{ij}^k, t\right)}{n_{ji}\,\tau_{ji}} - \frac{P_i(\mathbf{r}, t)}{n_{ij}\,\tau_{ij}} \right]. \qquad (3.\ 1)$$

The self-correlation function $G_s(\mathbf{r}, t)$ is now a weighted sum of the solutions of these rate equations, see Equation 5 of reference 17. Therefore, the absorption cross section is made up of m Lorentzian functions rather than just one such function. The half widths of the Lorentzians are $M_p(\mathbf{Q})$, which are the eigenvalues of a matrix **A**, the weights, w_p, of which are given by

$$w_p = \left|\sum_i \sqrt{c_i}\, b_i^p\right|^2, \tag{3.2}$$

where b_i^p is the i-th component of the p-th eigenvector of matrix **A**, belonging to the eigenvalue M_p. It should be noted that if **A** is non-Hermitian it has first to be made so by a similarity transformation. Because **A** describes the jumps between various sites, it is referred to as the "jump matrix." The elements of matrix **A** are

$$A_{ij} = \frac{1}{n_{ji}\,\tau_{ji}} \sum_k \exp(-i\mathbf{Q}\mathbf{l}_{ij}^k) - \delta_{ij} \sum_j \frac{1}{\tau_{ij}}. \tag{3.3}$$

3. 2. Iron Diffusion in Ordered Intermetallic Alloys

3. 2. 1. The Fe-Si Alloy System Close to Fe_3Si

Fe_3Si is an ordered intermetallic alloy and the prototype of the DO_3 structure. It has a cubic superstructure consisting of four sublattices, labeled α_1, α_2, β, and γ, as is shown in Fig. 3. In the fully ordered structure iron atoms occupy the α_1, α_2, and γ sublattices and silicon atoms occupy the β sublattice. From the Zeeman splitting of the room temperature Mössbauer spectra of Fe_3Si, it was concluded,[18] based on the well established work of Stearns and others,[19] that two-thirds of the iron atoms are on sites with four nearest-neighbor iron atoms, i.e., on the α sublattices, and one-third of the iron atoms are on sites with eight nearest neighbor iron atoms, i.e., on the γ sublattice. This implies that order is practically perfect. For non-stoichiometric alloys a distinct component of iron atoms on the α sites, with five iron nearest neighbors, appears, and even a small component attributable to iron atoms with six iron nearest neighbors can be distinguished. The fractions of these additional components agree with what is expected from statistics.

According to Equation 3. 3, for jumps between three sites, **A** is a 3×3 matrix,

$$A = \begin{pmatrix} -2 & E & E^* \\ E^* & -1 & 0 \\ E & 0 & -1 \end{pmatrix}, \tag{3.4}$$

where E is a function of the structure of the "jump lattice." For the Fe_3Si lattice,

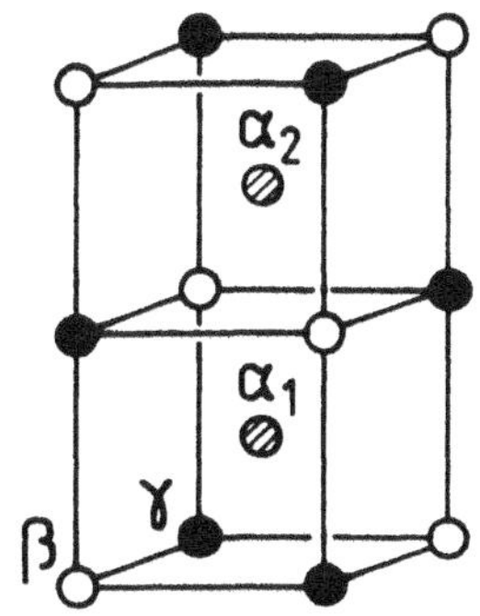

Figure 3. The DO_3 structure of Fe_3Si. In the stoichiometric alloy iron atoms occupy the α_1, α_2, and γ sublattices and silicon atoms occupy the β sublattice.

$$E = \cos(Q_x a)\cos(Q_y a)\cos(Q_z a) + i\sin(Q_x a)\sin(Q_y a)\sin(Q_z a),$$

where Q_x, Q_y, and Q_z are the components of the γ-wavevector, **Q**, which is 73 nm^{-1}, relative to the crystal axes for a lattice parameter, a, which is 0.571 nm for Fe_3Si at 720°C. The zero values in the matrix indicate that there are no jumps between the α_1 and the α_2 sites, which are further apart than nearest neighbors.

For Fe_3Si the Mössbauer absorption probability is then

$$S(\mathbf{Q},\omega) = \sum_p w_p(\mathbf{Q}) \frac{\Gamma_p(\mathbf{Q})}{(\Gamma_p(\mathbf{Q}))^2 + \omega^2}, \qquad (3.5)$$

where this function represents a sum of three Lorentzians, each with a half width at half maximum of $\Gamma_p = \nu M_p$, where ν is the jump frequency from a site on the α_1 or α_2 sublattice into a vacancy on a distinct nearest-neighbor site of the γ-sublattice, and vice versa, and with a weight,

$$w_p = (1/3)\left|\sum_i b_i^p\right|^2,$$

where M_p, with p = 1, 2, and 3, are the eigenvalues and b_i^p, with p = 1, 2, and 3, and i = 1, 2, and 3, are the components of the p-th eigenvector of the jump matrix.

Figure 4 shows the predictions of these calculations. It is evident that, for Mössbauer γ-ray radiation parallel to certain single crystal directions, one Lorentzian line dominates the Mössbauer spectrum. In the [111] direction it is the narrowest of the three Lorentzians, whereas, for the [113] direction it is the broadest. Thus, measurements should be performed parallel to these directions. Indeed, a comparison with the experimental results shown in Fig. 5 confirms these systematic trends.

A striking result of these studies is the unusually high diffusivity calculated from the line broadening with the Einstein-Smoluchowski equation, a result which has in the meantime been confirmed by tracer studies.[20]

Quite recently Sepiol and Vogl[21] have studied how the spectrum changes when the alloy concentration deviates from the exact Fe_3Si stoichiometry. Surprisingly, they find that deviations of as little as one to two atomic percent lead to dramatic changes, changes which the authors attribute to a change in the diffusion mechanism.

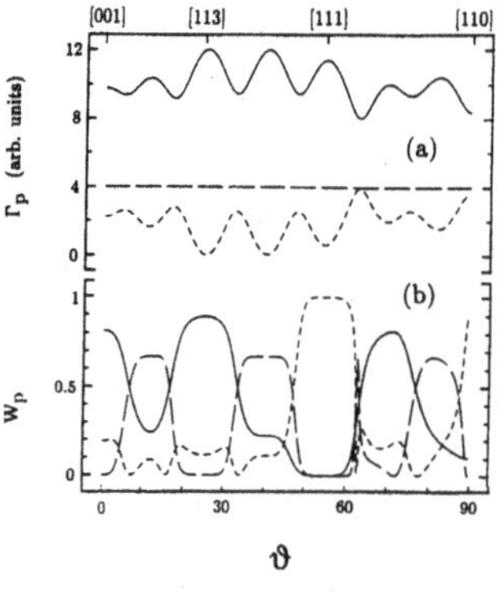

Figure 4. Model predictions for the linewidths, Γ_p, a, and the weights, w_p, b, of the Mössbauer lines due to diffusion on the three iron sublattices, α_1, α_2, and γ, of Fe_3Si, as a function of crystal orientation.

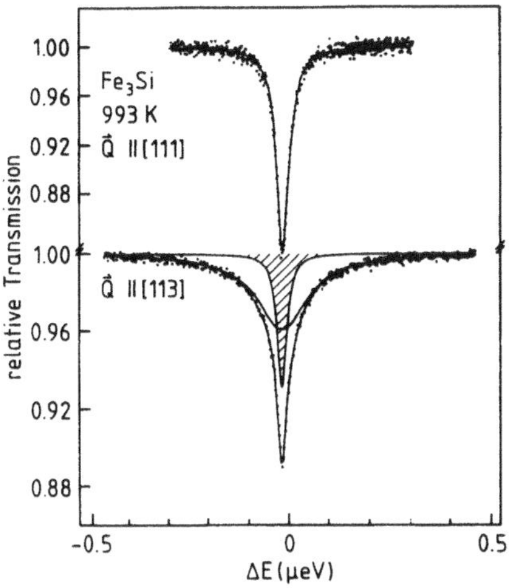

Figure 5. The iron-57 Mössbauer spectra of a Fe_3Si single crystal, measured at 720 K with transmission in the [111] direction, top, and in the [113] direction, bottom. In the [111] direction the spectrum can be well represented by just one Lorentzian, whereas in the [113] direction it clearly consists of at least two lines, a narrow and a broad line. Their fractional intensities are 20 and 80 percent, in agreement with the weights predicted by the model.

3. 2. 2. The Fe-Al Alloy System

The Fe-Al alloy system is another example for which the unique possibilities of Mössbauer spectroscopy have been successfully used to determine the diffusion mechanism in the attractive and important field of ordered intermetallic alloys. A good part of the work has just recently been published.[22,23] Figure 6 shows the iron rich part of the iron-aluminum phase diagram with the temperature/concentration regions of the ordered CsCl-type structure and the disordered bcc structure, referred to as the B2 and A2 structures, respectively, by metallurgists. For temperatures high enough that diffusional Mössbauer line broadening is visible, ca. 1000°C, the low-temperature DO_3 structure is of no interest. Above the line separating B2 from A2 the disordered bcc structure dominates. Of course, both structures are perturbed close to the coexistence line, and it is very instructive to follow the change in the spectrum with a change in alloy composition and to deduce from it the change in the diffusion mechanism.

Figure 7 compares the theoretical models with the measured diffusional line broadening in Fe-Al alloys with different iron concentrations, in order to draw conclusions on the diffusion mechanism. For iron concentrations around 50 and 75 atomic percent, the simple theory of Section 2 suffices, there is just one diffusionally broadened line, and obviously the iron atoms jump effectively between the sites of a Bravais lattice. The diffusion mechanism, however, is different in the two alloys, see the bottom of Fig. 7.

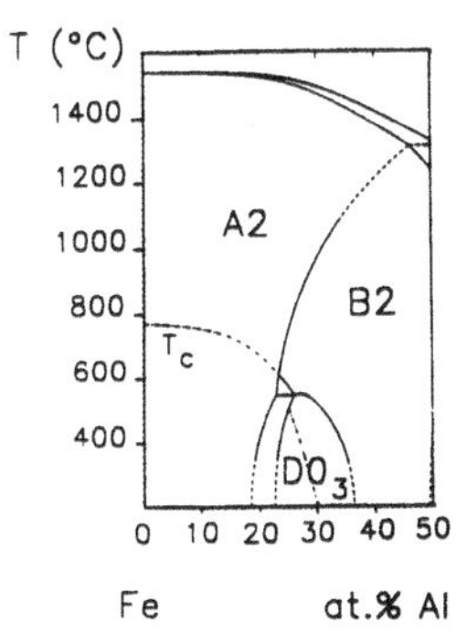

Figure 6. The iron rich portion of the Fe-Al phase diagram.[24]

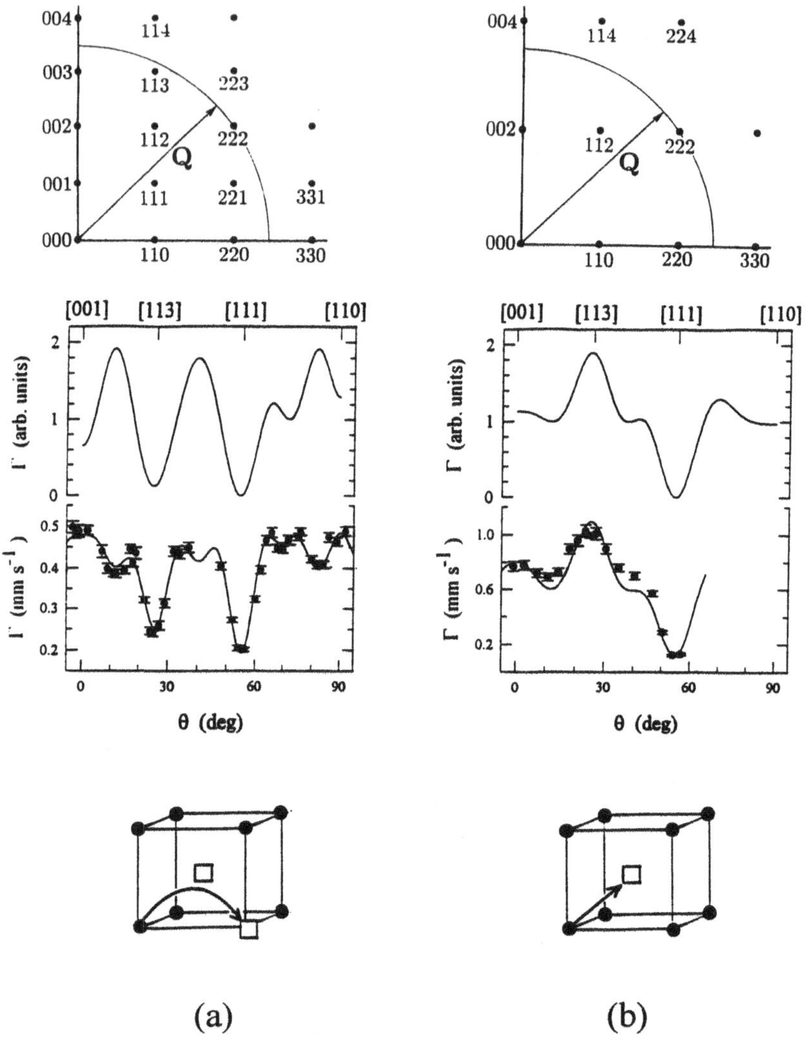

Figure 7. Top, the ($1\bar{1}0$) plane of the reciprocal "jump lattices" for an Fe-Al alloy with a distance of a = 0.297 nm between neighboring atoms along [100]. (a) [100] jumps, i.e., jumps on a simple cubic lattice, (b) [$\frac{1}{2}\frac{1}{2}\frac{1}{2}$] jumps, i.e., jumps onto the nearest-neighbor sites of the other sublattice. Center, upper plot, the expected angular dependence of the diffusional line broadening in the ($1\bar{1}0$) plane as given by Equation 2. 2, (a) and (b) have the same meaning as above. Center, lower plot, the experimental results for the angular dependence of the diffusional broadening, Γ, of the Mössbauer absorption line in the ($1\bar{1}0$) plane. ϑ is the angle from [001] along a great circle containing [001], [111], and [110]. The Fe-Al alloys have different iron concentrations of (a) $Fe_{50}Al_{50}$, ref. 22, and (b) $Fe_{72}Al_{28}$, ref. 25. Bottom, the Fe-Al structure in which ● represents the iron atoms and □ the vacancies. The iron jump mechanisms are deduced from a comparison of the upper and lower curves shown in the center plots for (a) a double jump with a very short residence on an antistructure site on the aluminum sublattice (there is not perfect agreement because the real situation is somewhat more complex, see ref. 22 for details) and (b) a jump to a nearest-neighbor site in the disordered structure.

For iron concentrations between 50 and 72 atomic percent the simple theory of Section 2 fails. Here again the full multiline theory for non-Bravais lattices has to be applied. Fig. 8 shows Mössbauer spectra measured along the [113] crystal direction for three different alloys. It is evident that the spectra change dramatically. At the comparable temperatures of 1020 and 1065°C, the spectra change from a narrow single line for the practically stoichiometric $Fe_{50}Al_{50}$ alloy, through a two line spectrum for $Fe_{66}Al_{34}$, to a strongly diffusionally broadened spectrum for $Fe_{72}Al_{28}$. These changes mirror the changes in the iron diffusion mechanism. Thus, we can conclude that, when changing the iron concentration, its diffusion mechanism changes from jumps nearly directly to a next nearest neighbor site on the iron sublattice, for $Fe_{50}Al_{50}$, via intermediate residence on the antistructure site on the aluminum sublattice, for $Fe_{66}Al_{34}$, to jumps between the completely equivalent sites in the disordered bcc structure of $Fe_{66}Al_{34}$. Notice that, in agreement with Fig. 7, the Mössbauer spectra, measured along the [111] direction, should be narrow in all cases, as is indeed found experimentally. This mirrors the fact that (222) is a reciprocal lattice point of the jump lattice for any jump model, see the top of Fig. 7.

4. LOCALIZED DIFFUSION IN CRYSTALLINE SOLIDS, GLASSES, AND BIOPOLYMERS

A particularly interesting phenomenon is that of localized diffusion of iron atoms in a cage. This type of diffusion is also referred to as bounded diffusion or diffusion in a restricted region. We shall travel through this field by visiting simultaneously crystalline and glassy substances and biopolymers, visits which aim, last, but not least, at showing the parallels between diffusion in simple lattices, on one hand, and more complex motional processes in proteins, on the other hand.

The first to consider the changes in lineshape due to the hopping of Mössbauer atoms in a quasibounded spatial volume were Krivoglaz and Repetskii,[26] who studied theoretically the Mössbauer spectrum of an iron atom bound to and jumping into and around a vacancy. They predicted two differently broadened Mössbauer lines and, even today, it is attractive to study their early paper, a difficult but exciting task. Their predictions were made without any definite feedback from experiments, which is why quasiclairvoyance was mixed with rather unrealistic claims.

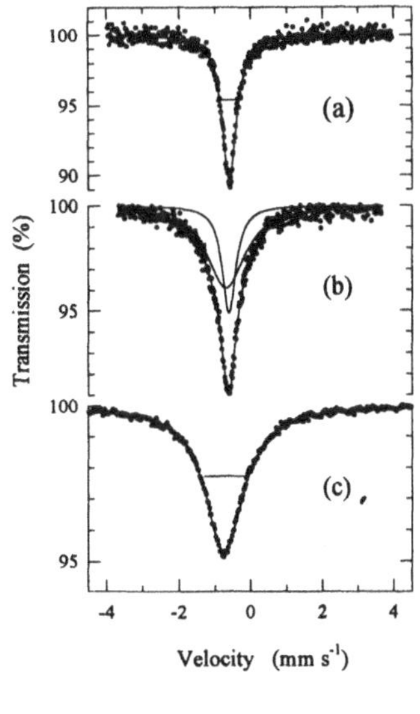

Figure 8. The iron-57 Mössbauer spectra, measured along the [113] direction, for $Fe_{50}Al_{50}$ at 1065°C, top, $Fe_{66}Al_{34}$ at 1020°C, center, and $Fe_{72}Al_{28}$ at 1020°C, bottom.[22, 23, 25]

Diffusion in a restricted region occurs, e.g., when the Mössbauer atom performs jumps between energetically equivalent sites caused by the diffusion of interstitials. The main consequence of this motion is the appearance of both a narrow unbroadened, and a broadened component in the Mössbauer spectrum. Indeed the first Mössbauer experiments to see localized diffusion were of this type. Vogl et al.[27-29] found that iron-57 atoms having trapped irradiation-induced interstitials in metals were pushed to interstitial sites. A drastic and sudden decrease in the Mössbauer line intensity appeared over a narrow temperature range with increasing temperature. This phenomenon was explained by a rapid diffusion which, surprisingly, kept the iron atom in a restricted region which the authors called a "cage." It was most interesting that, after a dramatic intensity decrease, a narrow Mössbauer line remained and decreased further in intensity with a normal Debye-like behavior. The theoretical explanation given by the authors[28,29] is that, in the narrow temperature interval over which the dramatic intensity decrease occurs, in reality very broad lines, with linewidths, Γ_p, in addition to the natural linewidth, appear. The number of broad lines equals the number of inequivalent jump distances in the cage. The broad lines disappear into the background just above a local "phase transition." The full argument runs as follows, in case of restricted diffusion, the rate equation corresponding to Equation 2. 1, is

$$\frac{\partial P(m,n,t)}{\partial t} = \frac{1}{N\tau} \sum_{m'} [P(m',n,t) - P(m,n,t)] \ , \tag{4. 1}$$

where P(m, n, t) is the conditional probability of finding an atom at cage site m, at time t, if it had been at cage site n, at time zero, and τ is the average residence time of an atom before jumping to any of its nearest-neighbor sites in the cage. This approach also assumes that no jumps other than to near-neighbor sites occur. The solutions of Equation 4. 1 are weighted sums of exponential functions, see Equation 9 of reference 29, with multiples of $\nu t = t/\tau$ in the exponents. Together they make up the self-correlation function. The Mössbauer absorption cross section is again a sum of Lorentzian functions of energy $h\omega/2\pi$ with the half widths at half maximum, Γ_p, which are multiples of the jump frequency, ν,

$$S(\mathbf{Q}, \omega) \propto \sum_p f_p(\mathbf{Q}) \frac{\Gamma_p}{(\Gamma_p(\mathbf{Q}))^2 + \omega^2}, \tag{4. 2}$$

where $f_p(\mathbf{Q})$ is the intensity of these lines. Because cage jumps over energy barriers must be thermally activated, the jump frequency, ν, depends on temperature with a simple Arrhenius like behavior,

$$\nu = \nu_0 \exp(-E_{cage}/k_B T), \tag{4. 3}$$

where ν_0 is an attempt frequency, E_{cage} is the activation energy, and k_B is the Boltzmann constant.

Up to this point, there is much analogy with the Mössbauer spectra for diffusion in non-Bravais lattices, see Section 3, but there is also a completely new aspect. The intensity for i = 0, i.e., the ratio of the intensity of the narrow line, corresponding to higher temperatures, at which the broad lines have disappeared into the background, to the total intensity of the Mössbauer spectrum, often called the "Mössbauer fraction," is

$$f_0 = \left| \frac{1}{N} \sum_k \exp(i\mathbf{Q}\mathbf{l}_k) \right|^2 \tag{4. 4}$$

and the narrow line does not disappear even when diffusion becomes rapid. The intensities, $f_p(\mathbf{Q})$, of the broad lines are more complex functions of the scalar product, $\mathbf{Q}\,\mathbf{l}_k$.[28] What remains when the broad lines have disappeared into the background is the narrow line. In the experiments on dilute iron-57 in aluminum single crystals, the intensity showed an angle dependence which permitted the determination of the geometrical structure of the cage through the use of Equation 4. 4.

An exciting development, which in the end was found to have a number of physical aspects in common with diffusion in a cage, began in 1965 from a completely different direction. Gonser and Grant[30] found that the area under the Mössbauer spectrum of iron in hemoglobin showed deviations from a Debye-like behavior at temperatures above 200 K. In 1981 Mayo et al.[31] reproduced this observation and found an additional broad line. The biophysical explanation in the early days was that the protein molecule, originally hemoglobin, and then deoxymyoglobin, oscillated between two conformational substates and that the iron atom followed this motion, thus producing two Lorentzian lines, a narrow, unbroadened line, and a broadened line. Clearly, there is an analogy with localized diffusion in a cage.[27]

For investigations of Mössbauer atom motion in biopolymers, and later on in glasses, it was soon believed that a description of the motion by a succession of abrupt jumps, between a very limited number of sites, was not quite an appropriate view of the non-crystalline state of these materials. Shaitan and Rubin,[32] Parak et al.,[33] and Bauminger et al.,[34] therefore proposed a continuous model of motion in the cage rather than a model of discrete jumps. This led to a description of continuous "bounded" or "cage" diffusion in terms of Brownian motion of an overdamped harmonic oscillator, the Uhlenbeck-Ornstein model.[35] Both Parak et al.[33] and Nowik et al.[36] were able to explain local motion of iron atoms in proteins through the Brownian oscillator.

In the case of a one-dimensional Brownian oscillator it is known that P is described by the Fokker-Planck or Smoluchowski equation,

$$\frac{\partial P}{\partial t} = D\frac{\partial^2 P}{\partial x^2} - \frac{1}{\beta}\frac{\partial (KP)}{\partial x}, \qquad (4.\ 1a)$$

which, for three dimensions becomes

$$\frac{\partial P}{\partial t} = D\Delta P - \frac{1}{\beta}\frac{\partial (\mathbf{K}P)}{\partial \mathbf{r}}, \qquad (4.\ 1b)$$

where β is a friction or damping coefficient and $\mathbf{K}$ is the force acting on the particle, the Mössbauer atom. It becomes clear after some thought, particularly when taking into account the Einstein-Smoluchowski equation, see Section 2, that Equations 4. 1a and b correspond to the rate Equations 2. 1, 3. 1, and 4. 1, for the case of discrete jumps.

Nowik et al.[36] compared the two scenarios, jump diffusion in a cage and harmonically bounded continuous diffusion, and found that descriptions in terms of, either jump diffusion between *many* sites, or continuous diffusion, were equivalent. Thus, there are both similarities and differences between cage diffusion between discrete sites and continuous cage diffusion. First, rate Equation 4. 1 for discrete jumps contains no damping term but does include local restriction by choice of the allowed sites. Second, whereas the rate equations for discrete jumps describe the motion in terms of a jump frequency, ν, the reciprocal of the residence time, and the jump vectors, Equations 4. 1 a and b, for the Brownian oscillator, describe the motion via the diffusion constant, D, and the friction, β. Third, in the case of jump diffusion, $G_s(\mathbf{r}, t)$ are just sums of exponentials, in multiples of νt, given by the solutions of Equation 4. 1. For the Brownian oscillator, $G_s(\mathbf{r}, t)$ is a complex function of exponentials in multiples of the ratio of the harmonic and the damping force constants. Fourth, whereas, in the case of jump diffusion the Mössbauer spectrum is just a sum of Lorentzians, Equation 4. 2, for the Brownian oscillator it is a more complex curve which can, however, be expanded in an infinite series of Lorentzians, see Equation 5 of Nowik et al.[37] Fifth, the Arrhenius type temperature dependence, Equation 4. 3, for the case of discrete jumps has to be replaced by the more complex Kramers equation, or an even more complex equation.[38] Sixth, intensities of the narrow line are comparable. Whereas, for discrete cage jumps the intensity is given by Equation 4. 4, for the Brownian oscillator it is given by

$$A_0 = \frac{1}{Z^2}\left|\int \exp(i\mathbf{Q}\mathbf{r}(T))\, d\mathbf{r}\right|^2, \qquad (4.\ 4a)$$

where

$$Z = \int d\mathbf{r} \exp(-U(\mathbf{r})/k_BT),$$

and U(**r**) is the diffusion potential.[39] It is evident that Equation 4. 4a corresponds to Equation 4. 4 for the intensity of the narrow line, for the case of discrete jumps in the cage.

From the late 1980's on, glass physics began to find increasing interest among Mössbauer experimentalists. Glass forming liquids possess two different relaxation modes, the α-process, which is of a slower diffusional character, and the β-process, which is faster and rather like the local motion of small subentities, such as atom clusters, in the glass. A recent review[40] covers the historical development of Mössbauer studies in the field of glass science, together with recent results. Nienhaus et al.[41] have studied the motion of divalent iron-57 ions in a water-glycerol mixture and could successfully describe the temperature dependent broadening of the spectra with a jump model with a Cole-Davidson distribution of relaxation times instead of a single relaxation time. This is the diffusional motion of the α-process which, however, can not account for the full spectral temperature dependence. An anomalously strong decay of the Debye-Waller factor above about 200 K is attributed to a very fast, high frequency, local motion, i.e., the β-process. The corresponding line broadening is thus so large, probably as a result of hyperfine parameter distributions, that it can not be resolved in the experimental spectra.

Very similar phenomena were found for biomolecules. In proteins, the Mössbauer fraction, f_0, of the narrow line follows the expected temperature dependence only up to about 200 K at which temperature, it decreases dramatically, see the references given in reference 40, and a broad quasielastic line appears. The decrease in the Mössbauer fraction and the broad line can be explained as a result of the rugged energy landscape in proteins and in glasses.[42] This striking similarity suggests that experiments on glasses may pave the way to a better understanding of the Mössbauer results on the motion in biopolymers, studies which have already been underway for ten years.

We finally mention that quite recently, nanosized particles in a polymer network were investigated, a system in which the "cage properties" can be manipulated.[43] By varying the cross-linkage of the polymer chains or, alternatively, the viscosities of the solvating liquids, the freedom of the $Fe(OH)_3$ particles can be varied, in other words, the cage size can be modified. The system thus can pass from free diffusion, if the size of the cage exceeds the wavelength of the Mössbauer radiation, via Brownian diffusion, for a somewhat smaller cage, which yields a spectrum with a narrow line plus a range of broadened lines, down to diffusion in a very narrow cage, of say one tenth of an Ångstrom, which yields a spectrum with only a narrow single broadened line. Fig. 9 compares different cages in crystalline and glassy substances.

5. COMPARISON WITH THE POTENTIAL OF QUASIELASTIC NEUTRON SCATTERING

The original theory of quasielastic line broadening was not designed for Mössbauer spectroscopy, but rather was for quasielastic neutron scattering. In 1957 Brockhouse[47]

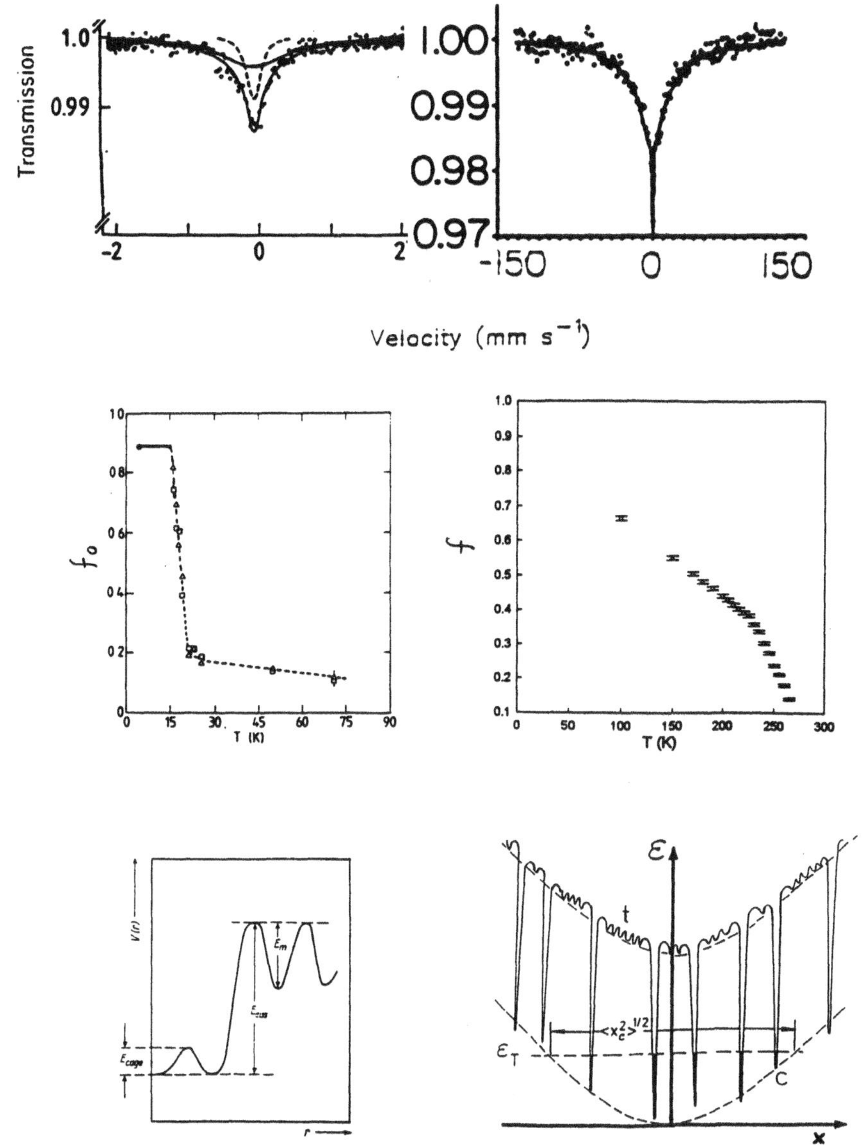

Figure 9. A comparison of the localized diffusion in crystalline, left, and non-crystalline materials, right. Top, the iron-57 Mössbauer spectra with a narrow and broad lines from interstitial iron-57 in an aluminum crystal at 17.2 K measured in the [100] direction, left, and $Fe(OH)_3$ particles made with iron-57 in a two percent cross linked polymer network at 243 K, right. Center, the temperature dependence of the intensity of the narrow iron-57 interstitial Mössbauer absorption line in an aluminum crystal, left, and iron-57 in sucrose, right. Bottom, the one-dimensional models for potential cages for interstitial iron-57 in an aluminum crystal, left, and for Brownian oscillation of the iron-57 motion in a protein, right, where C designates the conformational substates, substates with energy less than ε_T are occupied at thermal equilibrium. Friction is symbolized by wiggles in the transition state, t.[28, 29, 43-46]

defined "quasielastic neutron scattering" as neutron scattering with very small energy changes, changes which he had found to be caused by diffusion in water. The idea behind this and many further quasielastic neutron scattering experiments was to exploit the high incoherent neutron scattering cross section of hydrogen in order to determine the self correlation function of the hydrogen motion. The transfer of the diffusion theory to the Mössbauer effect is most evident in the work of Singwi and Sjölander. In their first paper[48] they had developed a theory of quasielastic neutron scattering and were able to apply that theory to the Mössbauer effect[1] very soon after its discovery. Figure 1b demonstrates the principle of quasielastic neutron scattering, and a comparison of Fig. 1a and b reveals the strong similarity of both methods. Chudley and Elliott's theory[2] was also designed for quasielastic neutron scattering.

Thus, diffusion Mössbauer spectroscopy's "big sister" is doubtlessly incoherent quasielastic neutron scattering and quite a few "tricks" for describing multi-Lorentzian lines have been learned from her. Whether one may regard Mössbauer spectroscopy of today as more successful than quasielastic neutron scattering is a matter of perspective. For concentrated systems and, in particular, for hydrogen motion in biological substances, quasielastic neutron scattering is probably unbeatable if the diffusivity is high enough. The latter point is relevant, because an important drawback of quasielastic neutron scattering, as compared to quasielastic Mössbauer spectroscopy, is the one hundred times weaker energy resolution of quasielastic neutron scattering. However, this may change in the near future with the advent of new spin echo instruments at the Institut Laue-Langevin in Grenoble and other places. A doubtless advantage of quasielastic neutron scattering is the broad spectrum of nuclei accessible for incoherent neutron scattering, whereas quasielastic Mössbauer spectroscopy rests on iron-57, and to a lesser extent on tin-119 and a very few other isotopes. With quasielastic neutron scattering simultaneous information about line broadening in several directions and for *several Q values* can be obtained, which makes measurements about two orders of magnitude faster than for quasielastic Mössbauer measurements. The permanent heavy overload at the very few instruments for quasielastic neutron scattering, and the necessity of using samples with volumes of some several cubic centimeters, in order to exploit the full neutron flux, however, often is a serious practical drawback, which overcompensates for its advantages.

A serious drawback of quasielastic neutron scattering, as compared to Mössbauer spectroscopy, may be the current lack of high-resolution spectrometers, that can reach scattering lengths of **Q** higher than 20 nm^{-1}, even though this situation is improving. The sensitivity to details of a diffusion jump is therefore lower for quasielastic neutron scattering than for Mössbauer spectrocopy. This can be appreciated by remembering what enters into line broadening, see for example Equation 2. 2, or the intensity of the narrow line, see Equation 4. 4, which contains the scalar product, **Ql**, whose variation, and thus information content is, of course, larger the larger the "**Q**-window."

We demonstrate some of the advantages and drawbacks of quasielastic neutron scattering, as compared to Mössbauer spectroscopy, with just one suitably chosen example. In Fig. 10 we show a typical result of quasielastic neutron scattering obtained for nickel diffusion in Ni_3Sb, an intermetallic alloy, which has the Fe_3Si structure and much in common with Fe_3Si, as discussed in Section 3. Two spectra observed at different **Q** values are shown in order to stress the advantage of quasielastic neutron scattering in varying **Q**. In contrast, the poorer energy resolution of quasielastic neutron scattering, as compared to "quasielastic" Mössbauer spctroscopy, is clearly apparent. A recent report[49] compares quasielastic Mössbauer studies on Fe_3Si and neutron studies on Ni_3Sb.

Of course, the possibility of studying the hyperfine interactions and their relaxation is practically missing in quasielastic neutron scattering, a feature which may, in the future, be a serious drawback, especially if the only slightly exploited capacity of diffusion studies, through hyperfine relaxation studies with Mössbauer spectroscopy, become important.

We have recently compared the potential of Mössbauer spectroscopy and quasielastic

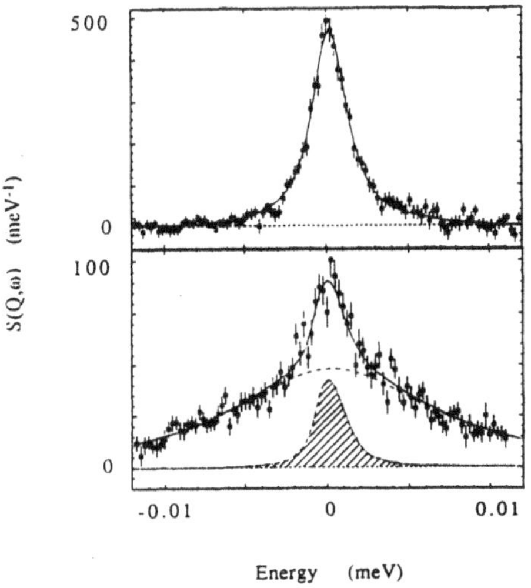

Figure 10. Quasielastic neutron scattering spectra for a $Ni_{72.5}Sb_{27.5}$ intermetallic alloy crystal with $|\mathbf{Q}| = 4.1\ nm^{-1}$, top, and $|\mathbf{Q}| = 18.2\ nm^{-1}$, bottom. The crystal orientation relative to the neutron beam is explained in the original papers.[50,51]

neutron scattering for scientists working in the latter field.[52] An older review[53] compares the potentials and limitations of various nuclear methods in diffusion studies, including not only Mössbauer spectroscopy and quasielastic neutron scattering, but also NMR and perturbed angular correlation studies. We judge that comparison as still describing the state of the art and recommend it to the interested reader.

6. OUTLOOK

A possible future development in Mössbauer studies of diffusion may come through Rayleigh scattering of Mössbauer radiation. Rayleigh scattering would, of course, liberate the field from the ties to iron-57 and would permit the study of diffusion in any material. Until now this method has suffered from much too low an intensity for most applications in diffusion. How much more progress will come from Mössbauer spectroscopy with synchrotron radiation remains to be seen. Diffusion could be studied directly as a function of time thus avoiding one of the Fourier transformations in Equation 1. 1. Working in the *time domain* may be regarded as a more direct way of looking at a process which proceeds with *time* than studying its Fourier transform into the energy domain. Besides there would be other advantages, such as better angular resolution, resulting from the tiny angular divergence of the synchrotron beam, and possibly also faster measurements. The problems facing daring research scientist are considerable.

Actually, much sooner than expected and while we were completing this chapter, the first successful investigation of diffusion in a crystalline solid with synchrotron radiation has been reported.[54] The well investigated diffusion of iron-57 between the three iron sublattices of the intermetallic alloy, Fe_3Si, has provided an opportunity to compare well established, conventional Mössbauer spectra, obtained in the energy domain, with the decay in the time domain of the forward scattered intensity of synchrotron radiation. Fig. 11 shows the first result of an experiment performed in October 1995. Clearly visible are the two decays, in the time domain, decays which correspond to the two eigenvalues of the jump matrix, see Equation 3. 4, which dominate an experiment with a single crystal oriented such that the 14.4 keV γ-ray is parallel with the [113] crystal direction. For sufficiently thin samples the forward scattered synchrotron radiation intensity decays with two characteristic decay rates,

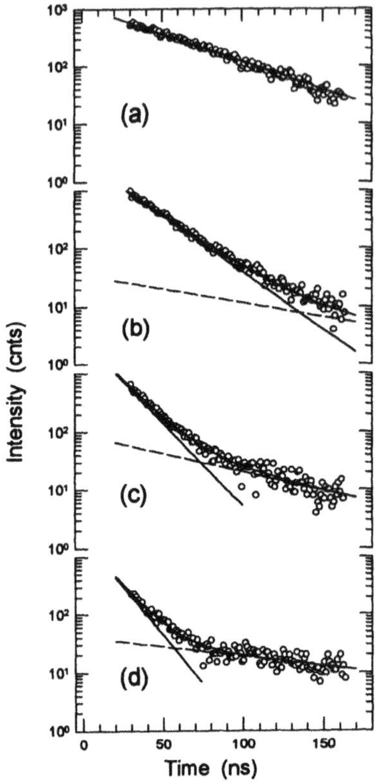

Figure 11. The nuclear forward scattering of synchrotron radiation at a sample made from the same Fe_3Si crystal which yielded the Mössbauer spectra shown in Fig. 5. Transmission was in the [113] direction and the sample was at 876 K, a, 916 K, b, 959 K, c, and 1013 K, d. Above 916 K a two-step exponential decay is clearly visible. The two straight lines indicate the asymptotic behavior of the decay at short and long times.

corresponding to the two differently broadened lines shown in Fig. 5. A first theory of diffusion, though not yet in crystalline material, may be found in reference 55.

ACKNOWLEDGEMENT. I thank B.Sepiol for many enlightening discussions. This work was supported by the Austrian Fonds zur Förderung der wissenschaftlichen Forschung, project No. S5601.

REFERENCES

1. K. S. Singwi and A. Sjölander, *Phys. Rev.* **120**, 1093 (1960).
2. T. Chudley and R. J. Elliott, *Proc. Phys. Soc.* **77**, 353 (1961).
3. M. A. Krivoglaz, *Zh. Eksp. Teor. Fiz.* **40**, 1812 (1961); *Sov. Phys.-JETP* **13**, 1273 (1961).
4. G. Wortmann, M. Tiedtke, E. A. Bychkov, and Y. S. Grushko, *Hyp. Interact.* **56**, 1495 (1990).
5. R. Brüning, D. H. Ryan, and J. O. Ström-Olsen, *Hyp. Interact.* **55**, 917 (1990).
6. H. Shechter, E. A. Stern, Y. Yacoby, R. Brener, and Z. Zhang, *Phys. Rev. Lett.* **63**, 1400 (1989).
7. A. Heiming, K.-H. Steinmetz, G. Vogl, and Y. Yoshida, *J. Phys. F: Metal Physics* **18**, 1491 (1988).
8. K. H. Steinmetz, G. Vogl, W. Petry, and K. Schroeder, *Phys. Rev. B* **34**, 107 (1986).
9. G. Vogl, *Hyp. Interact.* **53**, 197 (1990).
10. B. Sepiol, *Physica Scripta* **T49**, 378 (1993).

11. G. Vogl, *Neutron Scattering from Hydrogen in Materials,* A. Furrer, ed., p. 240, World Scientific, Singapore (1994).
12. R. C. Knauer and J. G. Mullen, *Phys. Rev.* **174**, 711 (1978).
13. S. Mantl, W. Petry, K. Schroeder, and G. Vogl, *Phys. Rev. B* **27**, 5313 (1983).
14. M. Bée, *Quasielastic Neutron Scattering*, Hilger, Bristol (1988).
15. J. M. Rowe, K. Sköld, H. E. Flotow, and J. J. Rush, *J. Phys. Chem. Solids* **32**, 41 (1971).
16. R. Kutner and I. Sosnowska, *J. Phys. Chem. Solids* **38**, 741 (1977); K. W. Kehr, D. Richter, and R. H. Swendsen, *J. Phys. F* **8**, 433 (1978); I. S. Andersen, A. Heidemann, J. E. Bonnet, D. K. Ross, S. K. Wilson, and M. McKergow, *J. Less-Common Met.* **101**, 405 (1984); J. W. Haus and K. W. Kehr, *Phys. Rep.* **150**, 263 (1987); D. Richter, R. Hempelmann, and C. Schönfeld, *J. Less-Common Met.* **172-174**, 595 (1991).
17. O. G. Randl, B. Sepiol, G. Vogl, and R. Feldwisch, *Phys. Rev. B* **49**, 8768 (1994).
18. B. Sepiol and G. Vogl, *Phys. Rev. Lett.* **71**, 731 (1993).
19. M. B. Stearns, *Phys. Rev.* **129**, 1136 (1963); G. Rixecker, P. Schaaf, and U. Gonser, *Physica Stat. Sol. (a)* **139**, 309 (1993).
20. A. Gude, Ph. D. Thesis, Universität Münster (1995).
21. B. Sepiol and G. Vogl, *Hyp. Interact.* **59**, 149 (1995).
22. G. Vogl and B. Sepiol, *Acta Metall. Mater.* **42**, 3175 (1994).
23. R. Feldwisch, B. Sepiol, and G. Vogl, *Acta Metall. Mater.* **43**, 2033 (1995).
24. O. Kubaschewski, *Iron Binary Phase Diagrams*, Springer, Berlin (1982).
25. R. Feldwisch, G. Ploner, B. Sepiol, and G. Vogl, to be published.
26. M. A. Krivoglaz and S. P. Repetskii, *Fiz. Met. Metalloved* **32**, 899 (1971); *Phys. Met. Metallogr. (USSR)* **32**, 1 (1971).
27. G. Vogl, W. Mansel, and P. H. Dederichs, *Phys. Rev. Lett.* **36**, 1497 (1976).
28. W. Petry, G. Vogl, and W. Mansel, *Phys. Rev. Lett.* **45**, 1862 (1980).
29. W. Petry, G. Vogl, and W. Mansel, *Z. Phys. B* **46**, 319 (1982).
30. U. Gonser and R. W. Grant, *Biophys. J.* **5**, 823 (1965).
31. K. H. Mayo, F. Parak, and R. L. Mössbauer, *Phys. Lett.* **82A**, 468 (1981).
32. K. V. Shaitan and A. B. Rubin, *Biophysics* **25**, 809 (1980).
33. F. Parak, E. W. Knapp, and D. Kucheida, *J. Mol. Biology* **161**, 177 (1982).
34. E. R. Bauminger, S. G. Cohen, I. Nowik, S. Ofer, and J. Yariv, *Proc. Natl. Acad. Sci. USA* **80**, 736 (1983).
35. G. E. Uhlenbeck and L. S. Ornstein, *Phys. Rev.* **36**, 823 (1930).
36. I. Nowik, S. G. Cohen, E. R. Bauminger, and S. Ofer, *Phys. Rev. Lett.* **50**, 1528 (1983).
37. I. Nowik, E. R. Bauminger, S. G. Cohen, and S. Ofer, *Phys. Rev. A* **31**, 2291 (1985).
38. H. Frauenfelder, *Hyp. Interact.* **90**, 21 (1994).
39. A. M. Afanasev and V. E. Sedov, *Physica Stat. Sol. (b)* **31**, 299 (1985).
40. G. U. Nienhaus and F. Parak, *Hyp. Interact.*, **90**, 243 (1994).
41. G. U. Nienhaus, H. Frauenfelder, and F. Parak, *Phys. Rev. B* **43**, 3345 (1991).
42. H. Frauenfelder, *Hyp. Interact.*, **90**, 21 (1994).
43. A. S. Plachinda, V. E. Sedov, V. I. Khromov, I. P. Suzdalev, V. I. Goldanskii, G. U. Nienhaus, and F. Parak, *Phys. Rev. B* **45**, 7716 (1992).
44. H. Lichtenegger, *Diplomarbeit*, Universität Wien (1995).
45. G. Vogl, *Physica Stat. Sol. (b)* **144**, 259 (1987).
46. F. Parak and K. Achterhold, private communication.
47. N. Brockhouse, *Suppl. Nuovo Cimento* **9**, 45 (1958); *Phys. Rev. Lett.* **2**, 287 (1959).
48. K. S. Singwi and A. Sjölander, *Phys. Rev.* **119**, 863 (1960).
49. G. Vogl, O. G. Randl, and B. Sepiol, *Ann. Rep.*, Institut Laue-Langevin 1994, Grenoble, p. 67.
50. B. Sepiol, O. G. Randl, C. Karner, A. Heiming, and G. Vogl, *J. Phys.: Condensed Matter* **6**, L43 (1994).
51. G.Vogl,, M. Kaisermayr, and O. G. Randl, submitted to *J. Phys.: Cond. Matter.*
52. G.Vogl, *Physica B: Condensed Matter,* in print.
53. W. Petry and G. Vogl, *Materials Science Forum* **15-18**, 323 (1987).
54. B. Sepiol, A. Meyer, G. Vogl, R. Rüffer, A. I. Chumakov, and A. Q. R. Baron, submitted to *Phys. Rev. Lett.*
55. G. V. Smirnov and V. G. Kohn, *Phys. Rev. B* **52**, 3356 (1995).

5

Analytical Methods for Mössbauer Spectral Analysis of Complex Materials

D. G. Rancourt

1. INTRODUCTION

This chapter discusses the various aspects of Mössbauer spectral analysis and data treatment that are particularly relevant to complex materials. Complex materials are materials whose Mössbauer spectra cannot be described, or resolved, in terms of small numbers of subspectral contributions that are attributable to specific crystallographic sites or discrete local environments. Such subspectral contributions are termed "elemental contributions" and, even in the presence of dynamic effects, are characterized by the natural linewidth, Γ, of the Mössbauer transition and a site or environment-specific set of hyperfine parameters, that may include microscopic dwell times.

Complex materials have spectra consisting of superpositions of large numbers of unresolved elemental contributions and may have chemical complexity, such as in the layer silicates where, despite high crystallinity, the octahedral sites can accommodate a large array of different cations; or structural complexity, such as in amorphous alloys, quasicrystals, mechanically alloyed nanocrystals, intercalation compounds, nanophase composites, multilayers, microporous materials, such as pillared clays, microcrystals, and super-paramagnetic particles; or magnetic complexity, such as when the spin structure, i.e., the spatial distribution of thermally averaged local moments, does not simply map one to one onto the chemical or structural disorder; or any combination of the above. In all cases, the Mössbauer spectra are understood in terms of continuous or quasicontinuous distributions of hyperfine parameters, that may include microscopic times which characterize fluctuations in the local environments.

Some hyperfine parameters, such as the isomer shift, are mainly dependent on the ionic structure, i.e., valence state, others, such as the electric field gradient interactions, are mainly caused by the local chemical and local crystallographic structures, and still others, such as the hyperfine field, are mainly caused by the local magnetism, that is, by the orientations, magnitudes, and fluctuations of the local thermal average atomic magnetic moments.

Unfortunately, many different local environments can cause identical sets of hyperfine parameters or different sets of hyperfine parameters corresponding to identical elemental con-

D. G. Rancourt • Department of Physics, University of Ottawa, Ottawa, Ontario, K1N 6N5, Canada

Mössbauer Spectroscopy Applied to Magnetism and Materials Science, Volume 2, edited by Gary J. Long and Fernande Grandjean. Plenum Press, New York, 1996.

tributions. This is a fundamental limitation of Mössbauer spectroscopy and is part of the "non-uniqueness problem" that we discuss below. As a consequence, the distributions of hyperfine parameters do not directly give distributions of local distortion or local chemical environments but do represent the most information about such local environments that one can hope to extract from the spectra.

The main topics that are discussed in this chapter are related to the problems of absorber thickness spectral distortions, the non-uniqueness of fits to measured spectra, dynamic lineshape effects, and the extraction of hyperfine parameter distributions. In many cases, significant recent advances are described. In other cases, known effects are discussed in the context of complex materials. An effective analytic method for deconvoluting the transmission integral is described that makes full thickness corrections routine and removes the necessity of ever fitting with the transmission integral. Much emphasis is put on hyperfine parameter distribution methods, both fitting and direct extraction methods. A Voigt-based fitting method is described which uses analytic lineshape expressions. Direct extraction methods are surveyed, culminating in Bayesian theory methods which offer error propagation analysis for the distribution, straightforward generalizations to both any elemental lineshape, multi-dimensional distributions in multi-phase systems, and full parameter space mapping.

Because of the lack of space, the reader is encouraged to go to the original articles for details and worked out examples. Many of the analysis routines described in this chapter are available on a non-profit basis. For more information contact the author by e-mail at mosmod@physics.uottawa.ca.

2. ANALYTICAL TREATMENT OF ABSORBER THICKNESS

There are two characteristic absorber thicknesses, the ideal thickness, t_{ideal}, defined to give the largest signal to noise ratio in a given time, and the thin-absorber thickness, t_{thin}, defined to be just small enough to reduce thickness effects to some tolerable, user chosen, amount. The problem arises because often $t_{thin} << t_{ideal}$. We have discussed this in detail and have shown how t_{thin} and t_{ideal} can be easily and reliably calculated, including realistic cases with intrinsically broad absorption lines.[1,2]

In thin single-line source absorption experiments with non-polarizing uniform absorbers the problem is that the absorber resonant cross-section, $\sigma_a'(v)$, is not measured directly. Instead, the measured spectrum, N(v), is given by the well known transmission integral,

$$N(v) = BG - \eta_M f_s \frac{2}{\pi\Gamma_0} \int_{-\infty}^{+\infty} d\psi \frac{\Gamma_0^2/4}{(\psi+v)^2 + \Gamma_0^2/4} \left(e^{-\sigma_a'(\psi)} - 1\right), \qquad (1)$$

where BG is the measured background level, η_M is the part of BG that is from both recoilless and nonrecoilless Mössbauer γ-rays, f_s is the recoilless fraction of the source, and Γ_0 is the natural linewidth of the Mössbauer transition, which is 0.097 mm/s for the 14.4 keV transition in iron-57.

In order to treat thickness effects correctly, therefore, one must either fit with the transmission integral or deconvolute out the absorber cross section from the transmission integral. We have developed a straightforward analytic way of doing the latter, see Equation 15 of reference 3. This means that $\sigma_a'(v)$ can be uniquely determined from the measured spectrum and then fit directly. Alternatively, $\sigma_a'(v)$ can be used to simulate a thin-limit spectrum with Gaussian noise as,

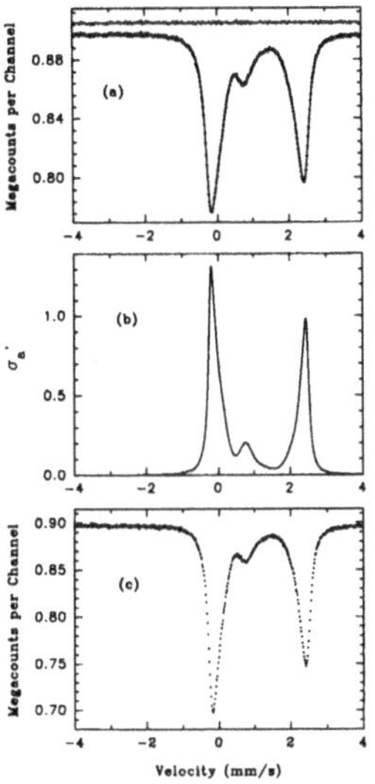

Figure 1. Thickness correction for the room temperature spectrum of a biotite mica sample BS200.[4] The uniform thickness random orientation powder absorber had 18.0 mg-Fe/cm^2. A mean absorber recoilless fraction, f, of 0.7 was assumed. The raw folded spectrum with the ideal Voigt line fit and the difference spectrum, a. The extracted dimensionless intrinsic absorber resonance cross-section, b. The simulated thin-limit spectrum with simulated Gaussian noise, using σ_a' in Equation 2, c.

$$N_{th}(v) = BG - \eta_M f_s \frac{2}{\pi\Gamma_0} \int_{-\infty}^{+\infty} \frac{\Gamma_0^2/4}{(\psi+v)^2 + \Gamma_0^2/4} \sigma_a'(\psi)\, d\psi + n_G(v) \qquad (2)$$

such that $N_{th}(v)$ is fit instead of N(v), using the usual chi squared criterion.

Our method is illustrated in Fig. 1, where the raw folded spectral data, the corresponding extracted absorber cross section, and the corresponding simulated thin-limit spectrum for a thick random orientation absorber of biotite mica are shown. Whereas, in this application, the relative spectral areas and hyperfine parameters do not change dramatically upon thickness correction, the extracted iron(II) quadrupole splitting distribution does.[4]

The method works irrespective of the particular physics which causes the absorber cross section and gives the thickness corrected lineshape in an extraction that is performed once, prior to any physical model fitting. This has a significant advantage in problems where distributions of hyperfine parameters are present because the true Mössbauer linewidth, Γ, can and should be fixed at the known value rather than selected in an ad hoc way to model thickness broadening. The common practice of using a width that corresponds to that of a known single-site absorber, having the same average thickness as the absorber of interest, is incorrect because thickness effects depend on the intrinsic spectral shape, i.e., on the particular absorber cross section.

The only difficulty in applying the method, as always when thickness corrections are performed, is that the parameter, η_M, see Equation 1, must be known. One would be ill advised to attempt fitting with the transmission integral by letting η_M vary as a free parameter because of obvious trade-off problems. One should either measure η_M or obtain it from the normalization condition for the cross section.[3] By definition, the measurement of η_M, should be performed with the absorber in place and can be achieved by pulse height analysis using a large velocity range to ensure being off resonance. In practice, we have found that, within the error that the average absorber recoilless fraction can often be estimated, the normalization condition, Equation 16 of reference 3 and Equation 4 of reference 1, gives satisfactory results, e.g., see reference 4. This is a considerable advantage, given the experimental errors and difficulties in measuring η_M. In any case, either method gives a significant improvement over simply neglecting thickness effects.

Note that, having different site-specific recoilless fractions in a given material or having different phase-specific recoilless fractions in a multi-phase sample does not invalidate the deconvolution method, as long as the conditions of applicability of the transmission integral are satisfied. The generalization to different recoilless fractions is given in reference 1.

Finally, having adopted the above method as a standard procedure, one has every advantage in using the ideal absorber thickness for every measurement, instead of attempting to use 'sufficiently thin' absorbers. Generally, the extra amount of absorber material also helps one to make a uniform absorber, in order to satisfy that condition of applicability of the transmission integral.

3. SOLUTION DOMAINS IN MÖSSBAUER PARAMETER SPACE

One of the major difficulties in Mössbauer spectroscopy is that a given fitting model applied to a given spectrum does not, in general, give a unique solution. This can occur by different mechanisms and is true even in the simplest case of a single-site material giving an elemental pattern consisting of Lorentzian lines. We refer to these difficulties as the non-uniqueness problem.

The causes of the non-uniqueness problem can be divided into three categories that can act alone or concomitantly depending on the circumstances. First, true degeneracy occurs when different sets of hyperfine parameters correspond to exactly the same calculated Mössbauer spectrum. Second, approximate degeneracy occurs when different sets of hyperfine parameters correspond to different calculated Mössbauer spectra that cannot be resolved in the measured spectral data, using a statistical criterion such as chi squared. Third, site trade-off occurs when the hyperfine parameters, and/or sub-spectral areas, of different sub-spectral contributions to the total spectrum trade-off due to line overlap. That is, different sets of fitting parameters for a given, either site-specific or phase-specific, sub-spectral contribution give equally good fits to the overall measured spectrum because the fitting parameters of the other sub-spectral contribution(s) adjust accordingly.

The simplest example of true degeneracy is with a random orientation absorber of single-site non-magnetic material where the quadrupole splitting is given by

$$QS = \frac{1}{2} e^2qQ\left(1 + \frac{1}{3}\eta^2\right)^{1/2}, \tag{3}$$

such that all combinations of the hyperfine parameters, q and η, that satisfy Equation 3 give exactly the same Mössbauer spectrum. This is resolved by using the quadrupole splitting as a single fitting parameter.

The situation is much more complex with a single-site magnetic material exhibiting a sextet pattern. Here both true degeneracy and approximate degeneracy are significant problems. In the perturbation limit, where $e^2qQ/2 \ll g^*\mu_N H$, the quadrupole shift is given by

$$\varepsilon = \frac{1}{4} e^2qQ\left(1 - \frac{1}{2}\left(3 - \eta\cos 2\phi\right)\sin^2\theta\right). \tag{4}$$

This suggests that, even when the full Hamiltonian solution is used, approximate degeneracy will occur when θ, ϕ, η, and q satisfy a condition similar to Equation 4. This is indeed the case outside the perturbation limit.[5]

The problems of true degeneracy and approximate degeneracy in the spectra of single-site materials are significantly more severe with oriented single-crystal or non-random absorber measurements, where the γ-ray orientation, (θ_γ, ϕ_γ), with respect to the electric field gradient axes are additional fitting parameters. Whereas true degeneracy is relatively well understood, neither true degeneracy nor approximate degeneracy have been completely and satisfactorily mapped out for iron-57. Too many spectroscopists simply hope that such problems are not present in their particular application and that parameters resulting from a fit give at least representative values.

With complex materials, both true degeneracy and approximate degeneracy give rise to an added level of non-uniqueness because different local environments in the same material can correspond to sets of hyperfine parameters that are related by either true degeneracy or approximate degeneracy. In other words, a given value of the hyperfine parameter in an extracted distribution, in general, corresponds to an entire family of local structural or chemical environments. For example, many different small structural distortions of the common octahedral coordination environment give the same quadrupole splitting value.

Site trade-off problems are even more notorious and are potentially different for every application. With overlapping spectral contributions the non-uniqueness problem is usually severe. In all cases, one needs to map out the region in the space of all the fitting parameters, that contains all the points corresponding to statistically equivalent fits of the particular measured spectrum. This region is called the solution domain in the Mössbauer parameter space. It represents the "Mössbauer solution," because it contains as much information as can be obtained from the measured spectrum by comparison with the chosen theoretical fitting lineshape model. Usually, the full parameter space does not need to be used in representing the solution domain because some of the parameters, e.g., the center shifts, have almost constant values. Only the parameters that trade-off, i.e., that are dependent, need to be used in visualizing the solution domain.

The boundary of the solution domain in the space of the dependent parameters is determined by a statistical criterion as follows. For example, because, with a correct model and Gaussian spectral noise the value of chi squared has a normal distribution, $P(\chi^2)$, centered on $\chi^2_{av} = \upsilon \equiv M - N$, where M is the number of channels or data points and N is the number of adjustable fitting parameters, with a standard deviation $\sigma_{\chi^2} = \sqrt{2\upsilon}$, we can define the solution domain as the region where $\chi^2 < \chi^2_{av} + 2\sigma_{\chi^2}$, in which case the boundary is given by $\chi^2 = \chi^2_{av} + 2\sigma_{\chi^2} = \upsilon + 2\sqrt{2\upsilon}$. In addition, it might, for illustrative purposes, be preferable to use only a subset of the dependent parameters in showing the solution domain. In this case, only the dependent parameters of primary interest are used, for example, the subspectral areas that give required site populations. The other dependent parameters trade-off with the chosen parameters but are not shown.

The next question is, how does one "reduce" the solution domain? One obvious answer is to collect better statistics, but the ultimate practical approach is to perform other measurements, e.g., x-ray diffraction, chemical analysis, high resolution microscopy, or optical spectroscopy measurements, on the same sample and/or to collect Mössbauer spectra under different experimental conditions, e.g., differing temperature, magnetic field, or pressure, on the same sample. The intersection of solution domains from spectra collected under different conditions can give a significantly reduced solution domain, thereby partially or totally solving the non-uniqueness problem. Simultaneously fitting sets of related spectra, while imposing that certain fitted parameters be the same for all spectra in a set, is a useful approach in this context.

In relating spectra collected under different conditions, or collected for different related samples, one must be particularly careful in selecting the parameters that are expected to be the same. For example, it is usual to assume that the electric field gradient asymmetry parameter, η, is a purely structural property that is the same above and below a magnetic ordering temperature. This is not in general true. Rancourt et al.[5] have shown that magnetic order can induce a non-zero value of η via spin-orbit coupling in an axially symmetric crystallographic environment.

We have recently illustrated the above ideas by applying them to the determination of site populations in annite mica.[6] In this work, spectra collected at 4.2, 77, and 295 K lead to octahedrally coordinated iron(II) and tetrahedrally coordinated iron(III) with populations accurate to ± 0.2 percent of the total iron. Similarly,[7] we have shown that the cis and trans octahedral sites in layer silicates cannot be resolved as previously claimed by many authors.

4. DYNAMIC EFFECTS AND FLUCTUATION LINESHAPES

In this section we make the following main points. First, hyperfine interactions are most often time-dependent and are only effectively static if certain conditions involving the microscopic fluctuation times are met. Second, fluctuations, such as fluctuations in the hyperfine field direction, are isotropic, i.e., paramagnetic only in paramagnetic materials and are anisotropic in ferromagnetic materials, or any time that the various possible microscopic configurations are not equally probable. Third, anisotropic fluctuation lineshapes are very different from isotropic fluctuation lineshapes and have different ranges of effectively static behavior. Fourth, even ferromagnetic systems that have effectively static behavior at low temperatures must exhibit dynamic lineshape effects when the temperature is raised and approaches the Curie point, in a range that can be significantly larger than the critical region.

4. 1. Conditions for Effectively Static Behavior with Isotropic or Anisotropic Fluctuations

Hyperfine interactions occur between the probe nucleus, here iron-57, and the surrounding electrons. Electrons are not static, nor are any of their manifestations in solids, e.g., the local atomic moments, charge distributions, chemical bonds, and electric field gradients. The precise circumstances in which *effectively static* hyperfine parameters can be used have been described elsewhere, see Section 3 of reference 8. For the purpose of this discussion, we consider the analytical lineshape expression first given by Blume and Tjon.[9] Such lineshapes take the time dependence to be discrete and stochastic. Their importance in various systems where hyperfine field fluctuations cause observable effects has been reviewed.[10]

For simplicity we consider the specific case where the asymmetry parameter, η, is zero and H is assumed to jump stochastically between the two directions parallel to the electric field gradient principal axis, $\theta = 0$ and 180°. On average, the field spends a certain microscopic dwell time, $\tau_+ = 1/f_+$, in the $\theta = 0°$ direction and between successive jumps. It then spends an average dwell time, $\tau_- = 1/f_-$, in the $\theta = 180°$ direction before returning to the "plus" direction. We also define a mean jump frequency as $f = (f_+ + f_-)/2$ and half the difference as $\rho = (f_+ - f_-)/2$. In a paramagnetic or superparamagnetic system, $f_+ = f_-$ and $\rho = 0$, whereas in a magnetically ordered ferromagnetic or superferromagnetic, system $f_+ \neq f_-$ and $\rho \neq 0$. Here, the four limiting cases, which lead to effectively static behavior, are, i, the case where f_+, or f_-, is zero and $\rho \neq 0$; ii, the case where $f_+ = f_- = 0$ or $f = \rho = 0$; iii, in the limit when f is large when $\rho = 0$; and iv, in the limit when both f and ρ are large.

Case-i is the true static limit, where τ_+ or $\tau_- = \infty$ and the field spends all of its time in the plus, or minus, electric field gradient direction. The result is that the lineshape is a true Lorentzian sextet corresponding to the value +H , or -H, of the hyperfine field.

Case-ii is also a true static limit where the field is frozen in both possible directions, where both directions have equal a priori probabilities. For a sample containing many probe nuclei, a superposition of two Lorentzian sextets, corresponding to +H and -H, is observed. In our simple example, these two sextets are identical but would not be for arbitrary fluctuation directions.

Case-iii is a fast limit and corresponds to fast paramagnetic fluctuations. Here $\rho = 0$ and it is necessary that the following conditions also be true simultaneously,

$$f >> \Gamma_0,\ f >> \alpha^2/\Gamma_0,\ \text{and}\ f >> v^2/\Gamma_0, \tag{5}$$

where α is the order of the excited state Zeeman splitting and v is the maximum velocity in the range of interest.[8] Then, when $e^2qQ = 0$, the lineshape is a single Lorentzian singlet with a full width at half maximum linewidth of Γ_0 and, when $e^2qQ \neq 0$, it becomes a Lorentzian quadrupole doublet with each line having a linewidth of Γ_0. This is equivalent to the static situation with H = 0 and explains the observed Mössbauer spectra of many paramagnetic materials. As the conditions given by Equation 5 are relaxed, line broadening occurs and neither are the lines Lorentzian, nor are they sums of Lorentzian lines. The last of the conditions given in Equation 5 implies that deviations from the Lorentzian shape can first be seen in the distant spectral line tails.

Finally, case-iv is most relevant to hyperfine field distribution work. It is a fast limit where we require both f and ρ to satisfy the conditions expressed by Equation 5. Only under such conditions do we obtain a true Lorentzian sextet, with all linewidths equal to Γ_0, as the elemental lineshape. It corresponds to an effectively static hyperfine field equal to the true time averaged hyperfine field, $H_{eff} = \langle H \rangle_t = H\rho/f$. The inherent conditions of case-iv are always at least implicitly assumed in any hyperfine field distribution work.

4. 2. Breakdown of Effectively Static Behavior

One might argue that all spin and hyperfine field fluctuations in a magnetic material are always fast in the sense of Equation 5 with iron-57. Nonetheless, even if all frequencies are large, i.e., both f_+ and f_- in our model, near the magnetic ordering temperature no direction can be highly preferred and all $\langle H \rangle_t$ must be small, and have a small ρ. When $\rho = 0$, case-iii will hold, but just below T_c, transitional spectra with non-Lorentzian lineshapes must occur. The temperature range over which such transitional spectra occur is relatively small when the average frequency, f, well satisfies Equation 5, and is relatively large when f only moderately satisfies Equation 5. The latter situation can be recognized by the presence of dynamic temperature dependent broadening of the paramagnetic line occurring above the *independently measured*, i.e., the true T_c.

Many real systems to which hyperfine field distribution analyses have been applied, probably suffer from dynamic effects leading to transitional type elemental sub-spectra over broad temperature intervals about T_c.[10] Examples are fcc Fe-Ni alloys near the Invar concentration.[11-14] It has been argued[10] that this occurs more often than is recognized and that there are good physical reasons for this behavior.

The safest region of applicability of hyperfine field distributions is "as far away from T_c as possible." It is necessary in all applications to look for clear signs of dynamic effects. For example, the critical points in ferromagnetic random and amorphous alloys are as sharp as in any crystalline compound, as is seen in several critical region studies, see, for example, reference 15. Clearly then, the true average moments and average hyperfine fields on all sites must be zero above T_c. Large and temperature dependent spectral broadening above the true T_c must then be taken as a clear indicator of dynamic effects, effects which invalidate any hyperfine field distribution analysis in a broad neighborhood around T_c. Another indicator of dynamic effects is the presence of very long spectral tails which, in a hyperfine field distribution analysis, require field components which are larger than the saturation fields seen at the lowest temperatures in the same absorbers.

Equivalent considerations are also important in quadrupole splitting distribution work where fluctuations in hyperfine fields or in electric field gradients can cause dynamic broadening which must be distinguished from quadrupole splitting distributions with elemental Lorentzian quadrupole doublets. With valence fluctuations, where the ionic state of the probe jumps between two possible values, unequal probabilities or preference for one ionic state is possible and unequal dwell times will again play a role. With fluctuations between different local structural distortions, certain distortions can again be preferred and require longer dwell times.

4. 3. Anisotropic Fluctuation Lineshapes

Whereas paramagnetic relaxation lineshapes are often shown in the literature, anisotropic fluctuation or relaxation lineshapes are rarely shown. As a result, most spectroscopists are not familiar enough with the latter lineshapes to recognize their characteristic shapes. In fact, most spectroscopists rely on their experience with paramagnetic relaxation lineshapes to evaluate the presence or absence of dynamic effects in their spectra. This is a dangerous situation, especially in dealing with complex materials, because the anisotropic fluctuation lineshapes are very different from the paramagnetic ones and are often very similar to the spectra that arise from distributions of effectively static hyperfine parameters.[8, 10-13] As a result, as shown in reference 8, one cannot rule out dynamic effects based on good fits with physical parameters being obtained by a distribution analysis. For these reasons, it is worth showing several simulated anisotropic fluctuation lineshapes and comparing them with their paramagnetic counterparts. This is done in Figs. 2 to 6, which use the simple dynamic lineshape described above and in reference 8, with $g^{*}\mu_N H = 2.24$ mm/s, i.e., $H = 330$ kOe, $e^2qQ = 0$, $\Gamma = 0.2$ mm/s, a zero center shift, and values of $f \equiv (f_+ + f_-)/2$ and $\rho \equiv (f_+ - f_-)/2$ as given on the figures and in the figure captions.

5. HYPERFINE PARAMETER DISTRIBUTIONS

With static hyperfine parameter distributions, an elemental contribution, $E(p_1, ..., p_n;v)$, depends on hyperfine parameters, p_i, where $i = 1, ..., n$, that are distributed according to the joint probability density, $P(p_1, ..., p_n)$, such that the corresponding Mössbauer spectrum is written as

$$M(v) = BG - N(v) = \int ... \int E(p_1, ..., p_n;v)\, P(p_1, ..., p_n)\, dp_1 ... dp_n, \quad (6)$$

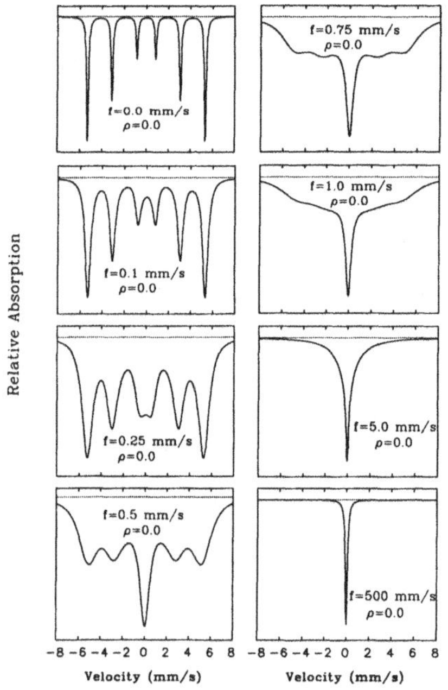

Figure 2. The isotropic $\rho = 0$ fluctuation lineshapes with various values of the average hyperfine field fluctuation frequency, $f \equiv (f_+ + f_-)/2$, as shown. The zero frequency spectrum is the usual static Lorentzian line sextet. At large frequency one obtains the usual zero field Lorentzian line singlet.

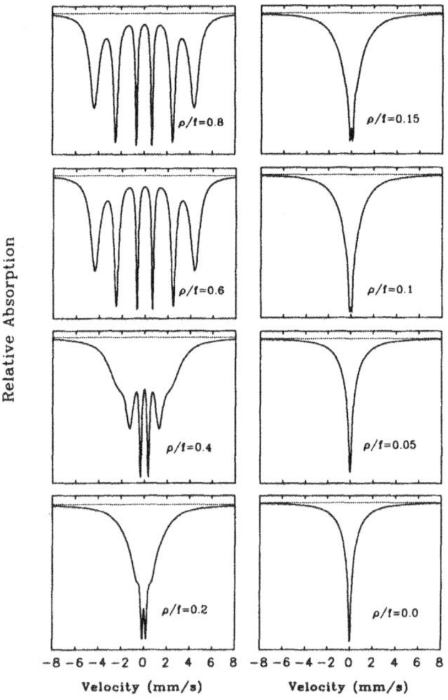

Figure 3. The anisotropic $\rho \neq 0$ fluctuation lineshapes with average hyperfine field fluctuation frequency, $f = 12.5$ mm/s, and various values of $\rho \equiv (f_+ - f_-)/2$, as shown. The $\rho/f = 1.0$ spectrum is not shown because it corresponds exactly to the $f = 0.0$ mm/s spectrum of Fig. 2.

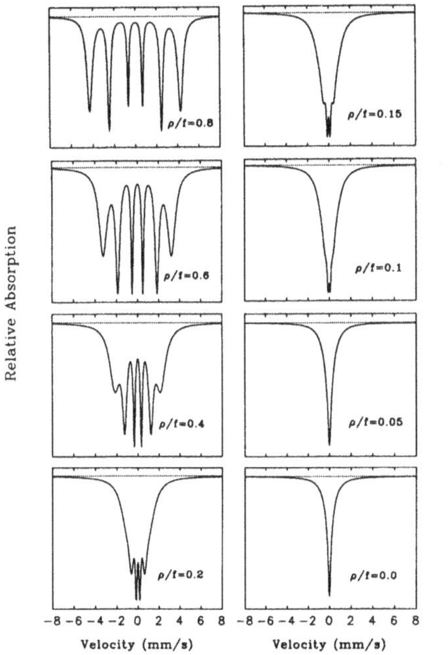

Figure 4. The anisotropic $\rho \neq 0$ fluctuation lineshapes with average hyperfine field fluctuation frequency, f = 25 mm/s, and various values of $\rho \equiv (f_+ - f_-)/2$, as shown. The $\rho/f = 1.0$ spectrum is not shown because it corresponds exactly to the f = 0.0 mm/s spectrum of Fig. 2.

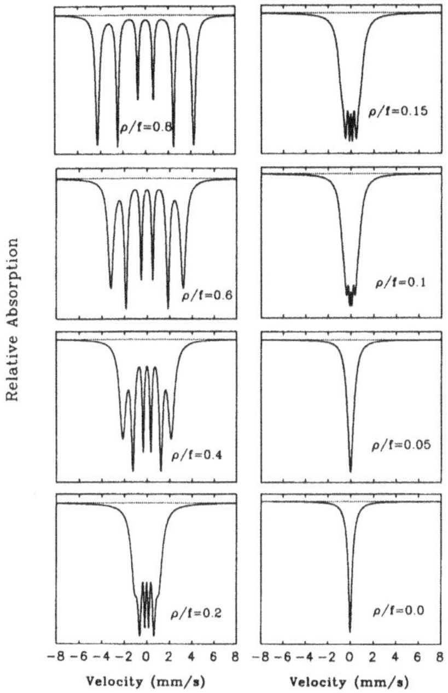

Figure 5. The anisotropic $\rho \neq 0$ fluctuation lineshapes with average hyperfine field fluctuation frequency, f = 50 mm/s, and various values of $\rho \equiv (f_+ - f_-)/2$, as shown. The $\rho/f = 1.0$ spectrum is not shown because it corresponds exactly to the f = 0.0 mm/s spectrum of Fig. 2.

where BG is the background and N(v) is the measured, thin-limit, absorption spectrum. When $E(\vec{p};v)$ is an elemental doublet and only the quadrupole splitting, Δ, is distributed with a probability density, $P(\Delta)$, we talk of quadrupole splitting distributions. The other simple common case is when $E(\vec{p};v)$ is an elemental sextet and only the hyperfine field, H, is distributed. P(H) is then referred to as a hyperfine field distribution.

Cases involving hyperfine parameter distributions give an added dimension of complexity to the non-uniqueness problem because the probability, $P(\vec{p})$, is extremely sensitive to the precise characteristics of the elemental contribution. The latter characteristics are usually not known a priori and include such features as the coupling between the main distributed parameter and other hyperfine parameters. As an example of how extreme the non-uniqueness problem is with distributions, consider that the spectra arising from quadrupole splitting distributions or hyperfine field distributions can always be modeled by distributions of Lorentzian singlet line positions. Even if the exact functional form of $E(\vec{p};v)$ is known, including the values of all parameters that are not distributed, there are significant problems in extracting $P(\vec{p})$ given M(v). In particular, with more than one distributed parameter, $n > 1$, severe non-uniqueness of $P(\vec{p})$ will, in general, persist unless adequate physical constraints are applied. Even with a single distributed parameter and complete knowledge of E(p;v), the extraction of P(p) by inversion is non-trivial because of the presence of spectral noise. In fact, it is an 'ill-defined problem.' We illustrate this in Section 5. 2, where we also describe the effect of an incomplete knowledge of E(p;v). We describe the direct extraction by the combined partial deconvolution method in Section 5. 3, and the most powerful Bayesian

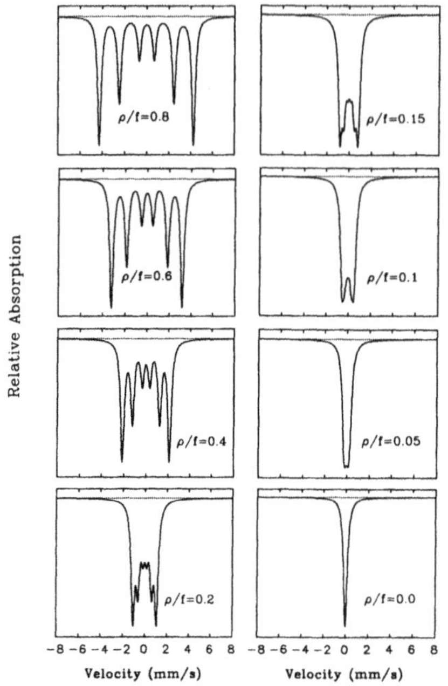

Figure 6. The anisotropic $\rho \neq 0$ fluctuation lineshapes with average hyperfine field fluctuation frequency, $f = \infty$ mm/s, and various values of $\rho \equiv (f_+ - f_-)/2$, as shown. The $\rho/f = 1.0$ spectrum is not shown because it corresponds exactly to the $f = 0.0$ mm/s spectrum of Fig. 2.

direct extraction method in Section 5. 4. But first we describe the Voigt-based fitting method in Section 5. 1. Complete references to older methods are given in reference 16.

A direct extraction method is one that produces a unique, but not necessarily correct, extracted distribution in one step by a mathematical procedure analogous to Fourier transformation, whereas a fitting method uses non-linear least squares minimization to iteratively fit the measured data itself or some suitably treated version of the measured data. Well-defined direct extraction methods have the advantage that they are equivalent to *linear* least squares minimization in that a unique arbitrary shape distribution is extracted for a given known or assumed elemental contribution. The limits of the direct extraction methods are first, the extracted distribution is sensitive, in a non-linear way, to the characteristics of the elemental contribution which are often not completely known and, second, overlapping spectral contributions or "sites" that have overlapping distributions, or distributions of different hyperfine parameters, either cannot be treated at all, e.g., by the combined partial deconvolution-direct extraction method, or can only be treated at a significantly increased, but not prohibitive, cost in computer time, e.g., the Gibbs sampling generalized Bayesian direct extraction method, unless there exists a mapping onto a single site problem.

In contrast, fitting methods involve *non-linear* least squares minimization, and the associated non-uniqueness problem, but they are able to treat characteristics of the elemental contribution as adjustable via additional fitting parameters, and easily treat overlapping spectral contributions or sites. However, with hyperfine field distributions, the Voigt based fitting method, and most other fitting methods, assume the perturbation limit, $e^2qQ/2 << g^*\mu H$. Such assumptions involving the elemental lineshape, such as the perturbation limit and linear-only coupling to non-distributed parameters, are not required for either the combined partial deconvolution-direct extraction or Baysian direct extraction, where the choice of any elemental contribution, including ones that have specified independent distributions of other hyperfine parameters, is possible.

5. 1. Voigt-Based Fitting Method

As described above, different distribution methods have different relative advantages and disadvantages. In our opinion, when it is applicable, the Voigt based fitting method[8] is the most stable and efficient method ever devised. It should be applicable in all cases involving quadrupole splitting distributions or hyperfine field distributions, in the perturbation limit, or coexisting quadrupole splitting distributions and hyperfine field distributions, when all couplings to non-distributed parameters are at most linear and as long as the true distributions are not so pathological that they cannot be represented as sums of Gaussian components. We therefore proceed to describe the Voigt-based distribution fitting method whose central physical assumption is that the distribution can be expressed as a sum of Gaussians with different widths, positions, and relative areas

$$P(p) = \sum_{i=1}^{R} w_i \, G_i(p_{oi}, \sigma_{pi}; p), \tag{7}$$

where

$$G_i(p_{oi}, \sigma_{pi}; p) = \frac{1}{\sigma_{pi}\sqrt{2\pi}} \exp\left[\frac{-(p - p_{oi})^2}{2\sigma_{pi}^2}\right], \tag{8}$$

and

$$\int P(p)dp = \int G_i dp = \sum_{i}^{R} w_i = 1. \tag{9}$$

This assumption is not as limiting as it may seem because the number of terms, R, may be as large as one chooses and the positions, p_{oi}, widths, σ_{pi}, and weights, w_i, are arbitrary. We expect that most real distributions can be expressed in this way. Indeed, in almost all of our applications to both quadrupole splitting distributions and hyperfine field distributions, we find that $R \leq 3$ is sufficient to give statistically acceptable fits with reasonable parameters.

In applying the Voigt based fitting method, one normally starts with $R = 1$ and increases R until acceptable fits are obtained. Increasing R further does not destabilize the solution as the same P(p) is obtained, but is not necessary.[8] The Voigt-based fitting method therefore has a built-in bias towards relatively smooth distributions, i.e., as expressed by a small number of Gaussians, however this is an advantage in that the true distributions probably are smooth compared to the artificial oscillations that are produced by other methods, and the non-uniqueness problem is thereby greatly reduced.

For quadrupole splitting distributions, we assume that the center shift or isomer shift, δ, is linearly coupled to Δ, i.e.,

$$\delta = \delta_0 + \delta_1 \Delta \tag{10}$$

and, with hyperfine field distributions, we assume that both the center shift,

$$\delta = \delta_0 + \delta_1 g^* \mu_N H \tag{11}$$

and the quadrupole shift,

$$\varepsilon = \varepsilon_0 + \varepsilon_1 g^* \mu_N H \tag{12}$$

are linearly coupled to H. It follows that, given Equation 7, the analytical expression for the total Mössbauer lineshape[8] is

$$N(v) = \sum_{i=1}^{R} w_i \sum_{k} V_{ik}\left(A_k + B_k\, p_{oi}, |B_k|\sigma_{pi}, \Gamma, h_k; v\right), \tag{13}$$

where $V_{ik}(\alpha, \beta, \Gamma, h; v)$ is the Voigt function. Here, the sum on k has two terms for quadrupole splitting distributions and six terms for hyperfine field distributions. The parameters A_k and B_k are functions of the coupling parameters, δ_0, δ_1, ε_0, and ε_1, which can be treated as fitting parameters.

A particularly efficient and accurate analytical approximation to the Voigt function is used[3,8,17] and Equation 13 is implemented into a least squares minimization routine. For detailed applications to hyperfine field distributions see references 8 and 11. For applications to quadrupole splitting distributions see references 4 and 18.

5. 2. Limits of Direct Extraction by Matrix Inversion

It is well known that, in principle, single parameter distributions can be obtained from Mössbauer spectra by direct extraction using simple matrix inversion. Here, we investigate the limits of this method in terms of the spectral signal to noise ratio, elemental contribution parameter uncertainties, and matrix dimensions.[19] Further, we specialize the method to the case of hyperfine field distributions and write Equation 6 as

$$M(v) = BG - S(v) = \int F(H; v)P(H)dH, \tag{14}$$

where F(H; v) is an elemental "sextet" in general consisting of eight Lorentzian lines. If we separate the hyperfine field range of interest into N equally spaced field values, H_j, with j = 1, ..., N and use M discrete Doppler velocities, v_i ,with i = 1, ..., M, e.g., the velocity channels in the measured spectrum, then Equation 14 can be written in matrix notation as

$$(M) = [F](P), \tag{15}$$

where (M) is a vector with elements $M(v_i)$, (P) is a vector with elements, $P(H_j)$, and [F] is an M x N matrix with elements, $[F](H_j; v_i) = F(H_j; v_i)dH$, where dH is the chosen field increment or spacing. Then it follows that (P), i.e., the required hyperfine field distribution, can be found by simple matrix inversion, when M = N,

$$(P) = [F]^{-1}(M) \tag{16}$$

where $[F]^{-1}$ is the inverse of [F]. This is the essence of the matrix inversion direct extraction method. The solution given by Equation 16, more generally $(P) = ([F]^T[F])^{-1}[F]^T(M)$, is the well known maximum likelihood or least squares estimator. It arises from Bayesian inference theory when no a priori knowledge of the distribution is assumed.[16]

The first thing that becomes clear in implementing matrix inversion direct extraction is that the matrix [F] is often nearly singular even in double precision, such that an ordinary matrix inversion method, like the Gauss-Jordan elimination method, does not work. That is, even with an infinite signal to noise ratio, and using the known and correct elemental sextet parameters in constructing [F], the extracted hyperfine field distribution has widely oscillating positive and negative values and no relation to the known hyperfine field distribution used in simulated spectral data. This occurs despite the fact that the reconstructed spectrum, using the extracted hyperfine field distribution and Equation 15, is numerically identical to the original, simulated spectrum. We therefore implemented the robust matrix inversion method known as the singular value decomposition method. It allows $M \neq N$ and minimizes the residual $r = |[F](P) - (M)|$ which may be zero.[20] The discussion which follows uses only singular value decomposition methods.

The elemental sextet parameters, Γ, δ_0, δ_1, ε_0, ε_1, and line-1/line-3 and line-2/line-3 intensity ratios, see reference 8, for the simulated spectra are taken to be those obtained when the Voigt based hyperfine field distribution method is used with a single Gaussian component to fit the 200 K spectrum of $Fe_{65}Ni_{35}$.[11] The spectral noise is generated using a random variable, R_v, with range $|R_v| \leq 1$. Uncertainties in the elemental sextet parameters, c, that are used in constructing [F] are generated as $c \rightarrow c + cfR_v$, where f is some fraction. In all cases, the reconstructed spectra, using the extracted hyperfine field distributions and Equation 15, are numerically identical, spectral noise effects notwithstanding, to the simulated spectra; however, the extracted hyperfine field distributions can be and often are significantly different from the known hyperfine field distribution. The latter is taken to be a simple Gaussian P(H) centered on $g^*\mu_N H_0 = 2.07$ mm/s, i.e., for $\langle H \rangle = 30.53$ T, and having a width $g^*\mu_N \sigma_H = 0.1448$ mm/s so as to resemble what is obtained for fcc $Fe_{65}Ni_{35}$, Invar at 200 K, see Table 3 of reference 11. The effect on the extracted hyperfine field distribution of different signal to noise ratios or of different elemental sextet parameter uncertainties, with the same f for all seven elemental sextet parameters in a given extraction, is shown in Fig. 7 for matrix dimensions M = N = 256. One sees that unreasonably large signal to noise ratios and small f values are required for the extraction to be successful. Nonetheless, the method's performance can be improved by using N < M. This is shown for different signal to noise ratios in Fig. 8.

The results shown in Figs. 7 and 8 give a quantitative measure of the limits of the matrix inversion direct extraction method as a hyperfine field distribution method. Although it does not work in most practical situations, this method is conceptually simple and serves to illustrate the mathematical difficulty of the problem. Next we consider how a matrix inversion direct extraction-like method can be successfully implemented via a combined partial deconvolution of the Mössbauer spectrum, see Section 5. 3. After that the very new and most powerful methods based within the framework of Bayesian inference theory, see Section 5.4, are described.

5. 3. Direct Extraction by Combined Partial Deconvolution

Whereas limited success can be obtained with matrix inversion direct extraction by using singular value decomposition for the matrix inversion and N < M, the combined partial deconvolution-direct extraction method[21] is far superior in all respects. It is based on the idea

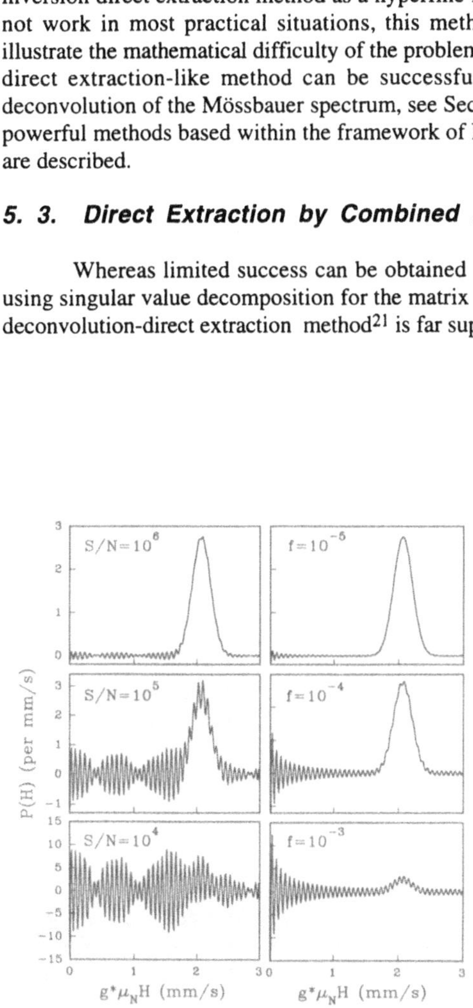

Figure 7. The hyperfine field distributions extracted by the matrix inversion direct extraction method from the simulated spectra with different signal to noise levels and different elemental sextet parameter uncertainties.

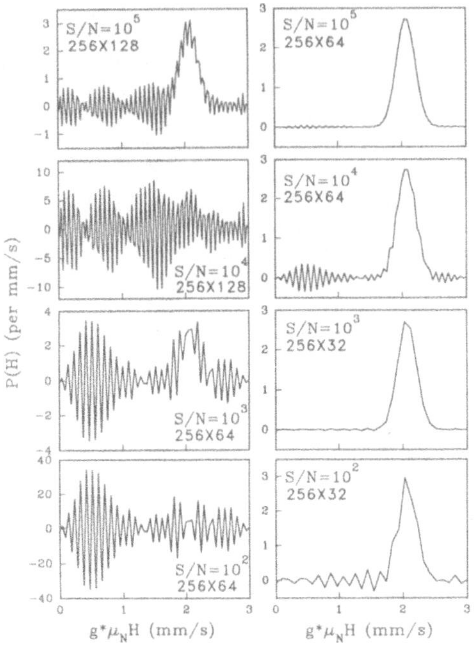

Figure 8. The hyperfine field distributions extracted by the matrix inversion direct extraction method from the simulated spectra with various signal to noise levels and using different size rectangular [F] matrices.

that, when the elemental contribution, E(p; v), is known to be a sum of Lorentzian lines with equal widths, as is the case when experimental artifacts and dynamic lineshape effects are absent, this elemental Lorentzian can be deconvoluted out of the measured spectrum in such a way as to also reliably remove the spectral noise before the inversion is applied to the resulting deconvolution to obtain the desired distribution, P(p).

We write Equation 6 for the single distributed parameter case, with n = 1, as

$$M(v) = BG - N(v) = \int E(p;\, v)P(p)dp. \tag{17}$$

The elemental contribution is written as

$$E(p;v) = \sum_{k} h_k(p)L\left(v_k(p) - v\right), \tag{18}$$

where the sum over k has as many terms as there are Lorentzians in the elemental pattern, i.e., two for quadrupole splitting distributed doublets or eight for hyperfine field distributed generalized sextets, $h_k(p)$ is a p- and k-dependent height, and $L(v_k(p) - v)$ is a Lorentzian centered on $v_k(p)$, i.e.,

$$L\left(v_k(p) - v\right) = \frac{\Gamma^2/4}{\left(v_k(p) - v\right)^2 + \Gamma^2/4}. \tag{19}$$

Here, each $v_k(p)$ is, in general, a complicated function of p because, although p is the only parameter that is independently distributed for n = 1, other, hidden hyperfine parameters of the elemental multiplet may be correlated to p, such that they have associated distributions determined by P(p). For example, the overall center shift of the elemental multiplet may be a single valued function of p, as δ(p), such that, for quadrupole splitting distributions, $v_k(p)$

would be given by $v_k(\Delta) = \delta(\Delta) + (2k - 3)\Delta/2$ with k = 1 and 2. Similarly, with hyperfine field distributions, both the center shift and the quadrupole shift of the elemental sextet can be single valued functions of H, as $\delta(H)$ and $\varepsilon(H)$, respectively. Note that the perturbation limit $e^2qQ/2 << g^*\mu_N H$ is *not* assumed and that the couplings are general and need not be linear, only single valued. It is also straightforward to generalize the combined partial deconvolution-direct extraction method to cases where the hidden parameters have their own known or supposed a priori independent distributions, i.e., that are the same for each value of p in the unknown distribution to be extracted.

Next, M(v) can be written as

$$M(v) = \int T(u)L(u - v)du, \tag{20}$$

where

$$T(u) = \int\left\{\sum_k h_k(p)\delta\left[v_k(p) - u\right]\right\}P(p)dp, \tag{21}$$

as is verified by substitution on using the properties of the Dirac delta function, $\delta[f(x)]$, whose argument is a function of the integration variable. The required partial deconvolution, which is partial because only the elemental Lorentzian, not the elemental multiplet, is deconvoluted out, and noise removal is achieved by fitting the Mössbauer spectrum to an arbitrary number, Q, of Voigt lines,

$$M(v) = \sum_{j=1}^{Q} V_j(v) = \sum_{j=1}^{Q} \int_{-\infty}^{\infty} h_j L(w - v)G_j(w)dw\,, \tag{22}$$

where

$$G_j(w) = \frac{1}{\sigma_j\sqrt{2\pi}} \exp\left[-\left(w_{0j} - w\right)^2 / 2\sigma_j^2\right]. \tag{23}$$

Q must be greater or equal to the number of Voigt lines required to get a statistically ideal fit with reduced chi squared of ~1. On comparing Equations 20 and 22 and using the general properties of Fourier transformations, it follows that

$$G(u) \equiv \sum_{j=1}^{Q} h_j G_j(u) = T(u)\,, \tag{24}$$

where the integral in T(u), Equation 21, can be evaluated to give

$$T(u) = \sum_k \frac{h_k\left(p_k^u\right)}{\left|v_k'\left(p_k^u\right)\right|} P(p_k^u)\,, \tag{25}$$

where $v'_k(p) \equiv \partial v_k(p)/\partial p$ is the partial derivative with respect to p of the position of the k-th Lorentzian line in the elemental doublet and p_k^u is the particular value of p that satisfies the condition $v_k(p) = u$. The central equation of the combined partial deconvolution-direct extraction method is therefore obtained as

$$\sum_k K_k^u P(p_k^u) = G(u)\,, \tag{26}$$

where K_k^u is defined as

$$K_k^u \equiv h_k(p_k^u) / \left|v_k'(p_k^u)\right|,$$

and G(u) is defined in Equation 24. Equation 26 represents an infinite set, one for each value of the continuous variable, u, of stringent conditions that the distribution P(p) must satisfy.

For each value of u, one has a linear equation relating probability densities evaluated at a number of values of p equal to the number of Lorentzians in the elemental multiplet that must be satisfied. With trial P(p) values, a correct P(p) must satisfy all of these conditions. The next task is to devise a practical way in which P(p) can be deduced, given Equation 26.

One takes N equally spaced parameter values p_j, j = 1, ..., N, then, using m-point Lagrange polynomial interpolation, each of the $P(p_k^u)$ values in Equation 26 are uniquely expressed as a linear combination of probability densities, $P(p_j)$, evaluated at m set parameter values. One finds that m = 4 works well.[21] We also use an array of M velocities, u_i, where i = 1, ..., M, that need not be regularly spaced and that, unlike v_is in matrix inversion direct extraction, need not have a direct relation to the measured velocities. This again leads us to matrix inversion where, instead of Equation 16, we now have, for M = N

$$(P) = [A]^{-1}(G) \tag{27}$$

with, in general, the M × N matrix [A] having elements that are sums of products of K_k^u and polynomial interpolation coefficients. (G) is the vector with elements $G(u_i)$, i = 1, ..., M. This completes the description of the combined partial deconvolution-direct extraction method. We obtain (G) by partially deconvoluting the spectrum, thereby concomitantly removing the noise, construct the matrix [A] as prescribed above, find the inverse of [A] by single value decomposition methods, and calculate the unique (P) via Equation 27. An example of an application to the quadrupole splitting distribution in an iron(III) chlorite is given in reference 21. The first applications to hyperfine field distributions are shown in Figs. 9, 10, and 11.

5. 4. *Direct Extraction Using Bayesian Inference Theory*

Here we describe the very new direct extraction method based on Bayesian inference theory, introduced by Dou et al.,[16,22] that we refer to as Bayesian direct extraction. Bayesian inference theory supplies a powerful framework that allows an error propagation analysis of the distribution, the use of any elemental lineshape, multi-dimensional distributions, independent distributions in multi-phase systems, full parameter space exploration for the non-distributed elemental contribution parameters, and variable a priori physical constraints. First we consider its application to hyperfine field distribution analysis.

We rewrite Equation 15 so as to explicitly include the Gaussian spectral noise

$$(M) = [F](P) + (G). \tag{28}$$

We also rescale (M), [F], and (G) by the known standard deviations of the Gaussian noise at each velocity and obtain[16]

$$(m) = [f](P) + (g). \tag{29}$$

According to Bayesian inference theory, the probability of (P) given (m), or the posterior probability, is

$$\boldsymbol{P}(P|m) \sim \boldsymbol{L}(P|m)\,\boldsymbol{P}(P), \tag{30}$$

where $\boldsymbol{L}(P|m)$ is the likelihood of (P) given (m), or the likelihood function and $\boldsymbol{P}(P)$ is the prior probability of (P), i.e., what is known about (P) without knowledge of the data (m).

The likelihood function is equal to the known probability distribution of the noise and is therefore written as

$$\begin{aligned} \boldsymbol{L}(P|m) &= \boldsymbol{P}(P) \\ &\equiv (2\pi)^{-M/2} \exp\{-(g)^T(g)/2\} \\ &= (2\pi)^{-M/2} \exp\{-((m) - [f](P))^T((m) - [f](P))/2\}, \end{aligned} \tag{31}$$

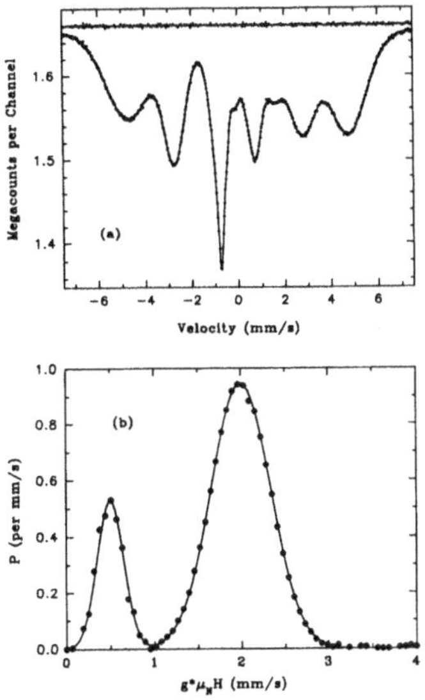

Figure 9. Application of combined partial deconvolution-direct extraction to the case of a hyperfine field distribution consisting of two well separated Gaussian contributions. (a) Simulated spectrum with ideal Voigt line fit and the difference spectrum. (b) Extracted hyperfine field distribution, dots, compared to the known hyperfine field distribution, solid line.

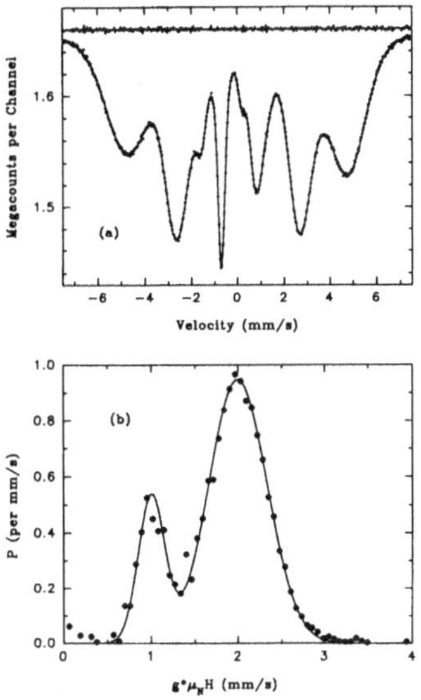

Figure 10. Application of combined partial deconvolution-direct extraction to the case of a hyperfine field distribution consisting of two moderately overlapping Gaussian contributions. (a) Simulated spectrum with ideal Voigt line fit and the difference spectrum. (b) Extracted hyperfine field distribution, dots, compared to the known hyperfine field distribution, solid line.

where M is the number of velocity channels, as defined in Section 5. 2.

As a first application, consider that we have no a priori knowledge of (P), that is, $\boldsymbol{P}$ (P) = constant. It follows that

$$\boldsymbol{P}(\mathrm{P}\,|\mathrm{m}) \sim \exp\{-((\mathrm{m})-[\mathrm{f}](\mathrm{P}))^{\mathrm{T}}((\mathrm{m})-[\mathrm{f}](\mathrm{P}))/2\} \tag{32}$$

and, hence, that the best estimate for (P) which maximizes $\boldsymbol{P}(\mathrm{P}\,|\mathrm{m})$ is

$$(\mathrm{P})_{\mathrm{best}} = ([\mathrm{f}]^{\mathrm{T}}[\mathrm{f}])^{-1}[\mathrm{f}]^{\mathrm{T}}(\mathrm{m}) \tag{33}$$

which, for square [f] matrices, i.e., for M = N, is equivalent to Equation 16. It is the usual least squares estimator, with its inherent problems, see reference 23 and Section 5. 2. The logical next step is to include a priori knowledge of (P), e.g., no negative values, and no discontinuities, etc., and to repeat the calculation. Dou et al.[16] showed that by requiring that the hyperfine field distribution, P(H), have finite second derivatives leads, via an application of the maximum entropy principle, to a prior $\boldsymbol{P}$ (P) that is a Gaussian function of (P) and, consequently, to a posterior probability $\boldsymbol{P}(\mathrm{P}\,|\mathrm{m})$ that is a multivariate normal distribution, centered on a well behaved $(\mathrm{P})_{\mathrm{best}}$ and having known standard deviations for each element $\mathrm{P}(\mathrm{H_j})$, where j = 1, ..., N, of $(\mathrm{P})_{\mathrm{best}}$. This represents an exact analytically well behaved solution to the direct extraction problem, with error analysis of the distribution. Dou et al. have since generalized this approach to the various other cases mentioned above in so far unpublished work.

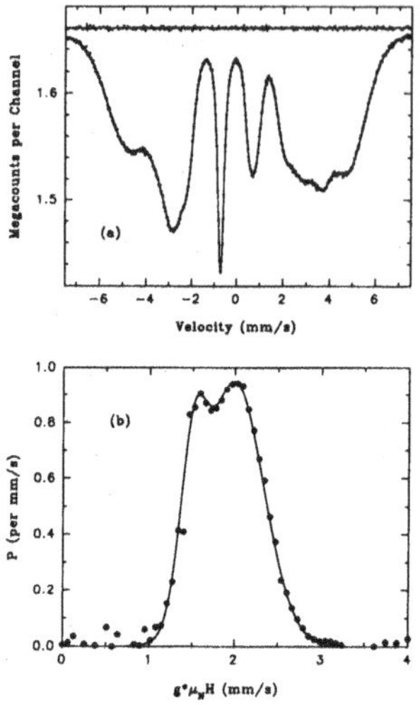

Figure 11. Application of combined partial deconvolution-direct extraction to the case of a hyperfine field distribution consisting of two closely overlapping Gaussian contributions. (a) Simulated spectrum with ideal Voigt line fit and difference spectrum. (b) Extracted hyperfine field distribution, dots, compared to the known hyperfine field distribution, solid line.

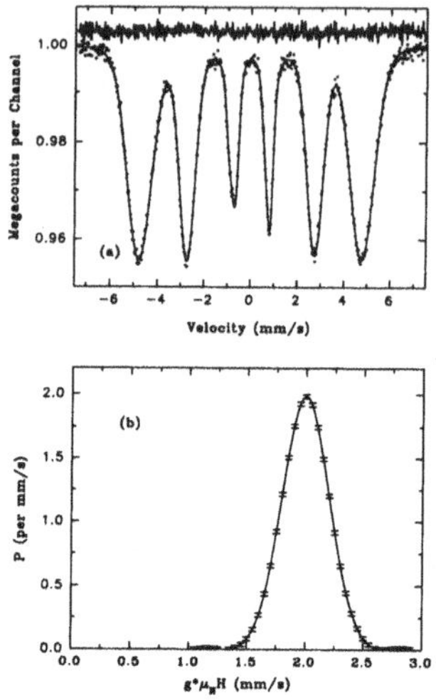

Figure 12. Extracted hyperfine field distribution using Bayesian-direct extraction. The top shows the simulated spectrum, dots, and the Bayesian-direct extraction reconstruction, solid line, based on the extracted hyperfine field distribution. The bottom shows the extracted hyperfine field distribution with error bars. The known hyperfine field distribution is a Gaussian centered at $g^*\mu_N H$ = 2 mm/s, equivalent within error to the extracted distribution.

Examples of applications are given in Figs. 12 to 15 for various hyperfine field distribution shapes. Here, the assumed parameters for both the spectrum simulation and the extraction are zero center shift and e^2qQ = 1.2 mm/s. Otherwise, the heights and positions of the Lorentzians in the elemental sextets are calculated exactly, assuming a random distribution of the hyperfine field directions, where θ and ϕ are averaged for each elemental sextet, at each field value. The known hyperfine field distributions used to simulate the spectra, with simulated Gaussian noise, are full or truncated, $H \geq 0$, Gaussian hyperfine field distributions having standard deviations, $g^*\mu_N\sigma_H$ = 0.2 mm/s, and maximum positions of $g^*\mu_N H_{max}$ = 2, 1, 0.5, and 0.2 mm/s, in Figs. 12 to 15, respectively. In each case, the Bayesian direct extraction method exactly recovers the correct hyperfine field distribution, within errors that are non-zero because of the spectral noise.

6. CONCLUSIONS

The main conclusions relating to the topics covered in this chapter are as follows. Thickness effects can be confidently removed analytically, see Section 2. This should become standard practice, just as folding was adopted to produce flat backgrounds. In our laboratory we routinely generate a thin-limit spectrum, corrected for all thickness effects, for each new spectrum and find that the hyperfine parameters, subspectral areas, and hyperfine parameter distributions are significantly changed on going to the true thin-limit. The main limitation of the thickness correction method is that it applies only to non-polarizing absorbers such as ran-

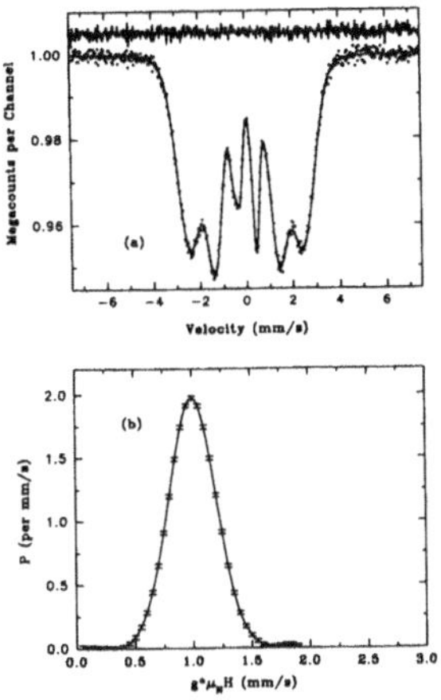

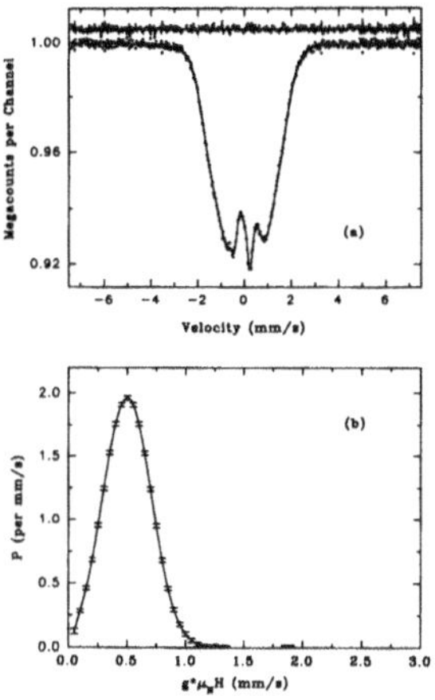

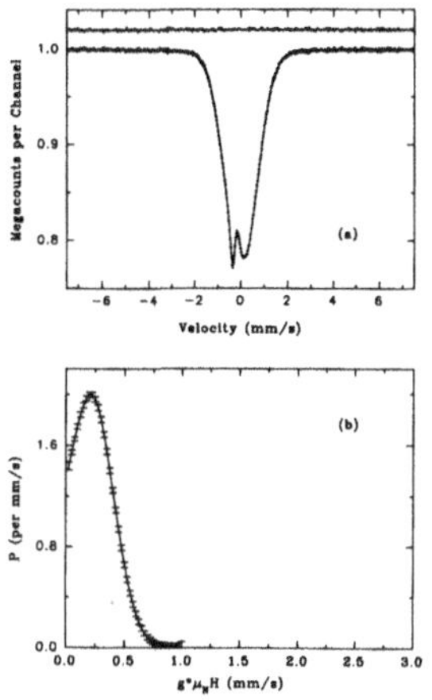

Figure 13, left. Extracted hyperfine field distribution using Bayesian-direct extraction. The top frame shows the simulated spectrum, dots, and the Bayesian-direct extraction reconstruction, solid line, based on the extracted hyperfine field distribution. The bottom frame shows the extracted hyperfine field distribution with error bars. The known hyperfine field distribution is a Gaussian centered at $g^*\mu_N H$ = 1 mm/s, equivalent within error to the extracted distribution.

Figure 14, middle. Extracted hyperfine field distribution using Bayesian direct extraction. The top frame shows the simulated spectrum, dots, and the Bayesian direct extraction reconstruction, solid line, based on the extracted hyperfine field distribution. The bottom frame shows the extracted hyperfine field distribution with error bars. The known hyperfine field distribution is a truncated Gaussian centered at $g^*\mu_N H$ = 0.5 mm/s, equivalent within error to the extracted distribution.

Figure15, right. Extracted hyperfine field distribution using Bayesian direct extraction. The top frame shows the simulated spectrum, dots, and the Bayesian direct extraction reconstruction, solid line, based on the extracted hyperfine field distribution. The bottom frame shows the extracted hyperfine field distribution with error bars. The known hyperfine field distribution is a truncated Gaussian centered at $g^*\mu_N H$ = 0.2 mm/s, equivalent within error to the extracted distribution.

dom orientation powders. General and effective methods for treating polarizing absorbers that have distributions of hyperfine parameters have not yet been developed.

The Mössbauer "solution domain" is a useful conceptual aid in coming to grips with the often severe non-uniqueness problem, see Section 3. Too many spectroscopists underestimate the magnitude of the non-uniqueness problem, do not perform error calculations, and simply hope that parameters resulting from particular fits give at least representative values. The topologies and extensions of the iron-57 solution domains, for even the simplest cases, have not been systematically studied or mapped.

With dynamic effects, see Section 4, anisotropic fluctuation lineshapes are very different from the usual isotropic fluctuation lineshapes and probably occur in many physical situations where their presence has not been recognized.

With hyperfine parameter distributions, see Section 5, within its limited domain of applicability, Voigt based fitting is the most efficient and effective method ever devised. When complete knowledge of the elemental contribution E(p; v) is available and the latter is made up of equal width Lorentzian lines, combined partial deconvolution-direct extraction is an effective and reliable method. Alternatively, when knowledge of the elemental contribution is incomplete, combined partial deconvolution-direct extraction can be used to extract the unique distribution P(p) corresponding to an *assumed* E(p; v) made up of Lorentzian lines. Overall however, although it is very new, Bayesian direct extraction is the most versatile and powerful general approach to the hyperfine parameter distribution problem. It supplies error propagation to the distribution for the first time in Mössbauer spectroscopy and is straightforwardly generalized to any physical situation.

ACKNOWLEDGMENTS. The author thanks J. Y. Ping and Lixin Dou for many helpful discussions and for allowing him to use our unpublished data for Figs. 9 - 11 and Figs. 12-15, respectively. The author also thanks the Natural Sciences and Engineering Research Council of Canada for financial support.

REFERENCES

1. D. G. Rancourt, A. M. McDonald, A. E. Lalonde, and J. Y. Ping, *American Mineralogist* **78**, 1-7 (1993) .
2. J. Y. Ping and D. G. Rancourt, *Hyp. Interact.* **71**, 1433-1436 (1992).
3. D. G. Rancourt, *Nucl. Instrum. Methods Phys. Res. B* **44**, 199-210 (1989).
4. D. G. Rancourt, J. Y. Ping, and R. G. Berman, *Phys. Chem. Minerals* **21**, 258-267 (1994).
5. D. G. Rancourt, J. M. Daniels, and H.-Y. Lam, *Can. J. Phys.* **63**, 1540-1547 (1985).
6. D. G. Rancourt, I. A. D. Christie, M. Royer, H. Kodama, J.-L. Robert, A. E. Lalonde, and E. Murad, *American Mineralogist* **79**, 51-63 (1994).
7. D. G. Rancourt, *Phys. Chem. Minerals*, **21**, 250-257 (1994).
8. D. G. Rancourt and J. Y. Ping, *Nucl. Instrum. Methods Phys. Res. B* **58**, 85-97 (1991).
9. M. Blume and J. A. Tjon, *Phys. Rev.* **165**, 446-461 (1968).
10. D. G. Rancourt, *Hyp. Interact.* **40**, 183-194 (1988).
11. J. Y. Ping, D. G. Rancourt, and R. A. Dunlap, *J. Magn. Magn. Mater.* **103**, 285-313 (1992).
12. D. G. Rancourt, H. H. A. Smit, and R. C. Thiel, *J. Magn. Magn. Mater.* **66**, 121-152 (1987).
13. D. G. Rancourt, S. R. Julian, and J. M. Daniels, *J. Magn. Magn. Mater.* **51**, 83-88 (1985).
14. F. Van der Woude and A. J. Dekker, *Physica Stat. Solidi* **9**, 775-786 (1965).
15. P. Hargraves and R. A. Dunlap, *J. Phys. F* **18**, 553-561 (1988).
16. L. Dou, R. J. W. Hogson, and D. G. Rancourt, *Nucl. Instrum. Methods Phys. Res. B* **100**, 511-518 (1995).
17. J. Puerta and P. Martin, *Applied Optics* **20**, 3923-3928 (1981).
18. J. Y. Ping, D. G. Rancourt, and Z. M. Stadnik, *Hyp. Interact.* **69**, 493-496 (1991).
19. J. Y. Ping and D. G. Rancourt, *Hyp. Interact.* **92**, 1209-1212 (1994).
20. W. H. Press, B. P. Flannery, S. A. Tenkolsky, and W. T. Vetterling, *Numerical Recipes*, Cambridge University Press, Cambridge (1986).
21. J. Y. Ping and D. G. Rancourt, *Hyp. Interact.* **92**, 1203-1207 (1994).
22. L. Dou, R. J. W. Hogson, and D. G. Rancourt, Proceedings of the Int. Conf. Appl. Mössbauer Effect 1995, in press.
23. G. Talenti, *Inverse Problems. Lecture Notes in Mathematics*, Springer, Berlin (1986).

6

Quasicrystalline Materials

Zbigniew M. Stadnik

1. INTRODUCTION

The discovery of an icosahedral Al-Mn alloy by Shechtman et al.[1] extended the dichotomous division of solids into either crystalline or amorphous by introducing the notion of quasiperiodic crystals, or quasicrystals. This new form of matter possesses long-range *quasiperiodic* order and long-range orientational order associated with the classically forbidden fivefold (icosahedral), eightfold (octagonal), tenfold (decagonal), and twelvefold (dodecagonal) symmetry axes.[2-4] A central problem in condensed matter physics is to determine whether quasiperiodicity leads to physical properties which are significantly different from those of crystalline and amorphous materials. The majority of known quasicrystals are either icosahedral (*i*-) or decagonal (*d*-) alloys. A few known octagonal and dodecagonal alloys cannot be produced in sufficient quantities to allow studies of their physical properties.

The first few years of studies of quasicrystals revealed that their physical properties are similar to either their corresponding crystalline or amorphous (*a*-) counterparts.[5] For example, the values of the electrical resistivity, ρ, and its temperature dependence for *i*-$Al_{86}Cr_8Fe_6$ are very similar to those for an amorphous alloy $Al_{87}Ce_{6.3}Fe_{6.7}$, see Fig. 1. It was only later realized that the first quasicrystals, which were thermodynamically *metastable*, possessed significant structural disorder, as manifested in the broadening of the x-ray- and/or electron-diffraction lines. They also contained non-negligible amounts of second phases. The poor quality of these samples hampered the detection of those properties which could be intrinsic to quasiperiodicity. They also led to confusion, especially in the area of the magnetic properties of quasicrystals, where some "unusual" magnetic properties were claimed to have been observed.[8] Some of these properties were later shown[9,10] to result from the presence of magnetic second phases in the icosahedral alloys studied. Therefore the discovery at the end of the 1980's of thermodynamically *stable* quasicrystals was of great importance because these stable, single-phase quasicrystals possess a high degree of structural perfection[11] comparable

Zbigniew M. Stadnik • Department of Physics, University of Ottawa, Ottawa, Ontario K1N 6N5, Canada

Mössbauer Spectroscopy Applied to Magnetism and Materials Science, Volume 2, edited by Gary J. Long and Fernande Grandjean. Plenum Press, New York, 1996.

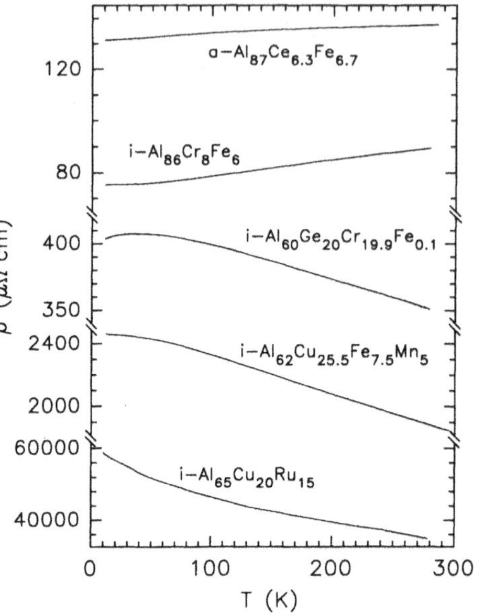

Figure 1. The temperature dependence of the electrical resistivity of some icosahedral and amorphous alloys.[6,7]

to that found in the best periodic alloys.

This chapter will provide an up-to-date review of some of the most unusual properties of quasicrystals and the role Mössbauer spectroscopy played in elucidating a better understanding of these complex and fascinating alloys.

2. STRUCTURAL MODELS OF QUASICRYSTALS

In spite of an enormous theoretical and experimental effort, the atomic structure of quasicrystals is still unknown, although some fundamental structural principles are now understood.[2-4,12] There are three principal models of the atomic structure of quasicrystals. The perfect quasicrystal model is based upon the concept of the Penrose tiling. Within this model, an icosahedral lattice, for example, is described in terms of a three-dimensional cut through a six-dimensional periodic lattice. Thus a high dimensional crystallography has to be employed to derive structural information from the diffraction data. The second model, the so-called random-tiling model, introduces some randomness into the quasilattice by replacing the strict matching rules in the ideal quasicrystal model with weaker ones. And finally, the glass model is based upon a random packing of icosahedral clusters. A more detail description of these models is beyond the scope of this chapter and therefore a reader should consult the literature.[2-4,12]

Quasicrystals, just as amorphous alloys, lack long-range periodic order. Therefore, diffraction methods upon which the bulk of our knowledge on the structure of quasicrystals is based, must be supplemented by local probes, such as Mössbauer spectroscopy, extended x-ray-absorption fine structure, nuclear quadrupole resonance, or nuclear magnetic resonance, if the true atomic structure of quasicrystals is to be understood. This is because some theoretical structural models may produce diffraction patterns which are indistinguishable within the finite experimental resolution, whereas they may produce different sets of hyperfine parameters and thus can be unambiguously differentiated by the local probes. The contribution of Mössbauer spectroscopy to the structural studies of quasicrystals is discussed in Section 4.

3. UNUSUAL PROPERTIES OF QUASICRYSTALS

3. 1. Brief Survey of Properties of Quasicrystals

The new stable icosahedral alloys with a high degree of structural perfection exhibit some unexpected behavior. First, their most salient feature, which is not expected for alloys formed of normal metallic elements, is the very high value of ρ, or the very low value of the electrical conductivity, σ.[13-17] These ρ or σ values are several orders of magnitude larger or smaller than those of the constituent metals or of the amorphous alloys, and are comparable to those of doped semiconductors. This is illustrated in Fig. 1, where the ρ values in i-$Al_{62}Cu_{25.5}Fe_{7.5}Mn_5$ and i-$Al_{65}Cu_{20}Ru_{15}$ are respectively an order and two orders of magnitude larger than those in i-$Al_{86}Cr_8Fe_6$ and a-$Al_{87}Ce_{6.3}Fe_{6.7}$. Such high ρ values correspond to σ values which are comparable to or smaller than the Mott's "minimum metallic conductivity" of 200 $\Omega^{-1}cm^{-1}$ for the metal-insulator transition.[18] It was even suggested[17] that i-Al-Pd-Re alloys are insulators at low temperatures. Second, the temperature coefficient of ρ of these new icosahedral alloys is generally negative, which is inconsistent with the expected behavior for metals. A very small increase of ρ with decreasing temperature is also observed for some metastable icosahedral alloys, such as i-$Al_{60}Ge_{20}Cr_{19.9}Fe_{0.1}$, see Fig. 1. Third, the ρ values are extremely sensitive to the sample composition,[16,17,19] which is reminiscent of doping effects in semiconductors. Fourth, the ρ values of these stable icosahedral alloys increase as their structural quality improves as occcurs upon annealing which removes the defects,[5,17] in contrast to the behavior of typical metals. Other unexpected anomalies in the transport properties of icosahedral alloys involve a very low electronic contribution to the specific heat, large and strongly temperature-dependent Hall coefficients and thermoelectric power, and a non-Drude-like optical conductivity.[11] From a magnetic point of view, the stable icosahedral alloys of high structural quality are unusual in that they are diamagnetic[20,21] in spite of containing a significant concentration of transition-metal atoms.

Quasicrystals of decagonal symmetry combine two structural characteristics; they are quasiperiodic with tenfold symmetry in planes which are stacked with translational periodicity along the tenfold axis perpendicular to them. As expected, their physical properties are anisotropic. For example, ρ has metallic characteristics along the periodic direction and exhibits a nonmetallic behavior, similar to that observed in icosahedral alloys, in the quasiperiodic plane.[22]

3. 2. Models of the Electronic Structure of Quasicrystals

A fundamental question in the physics of quasicrystals is the origin of the unusually low values of σ. The first suggested interpretation, which still prevails, is based upon a Hume-Rothery mechanism,[3-5,11,13] which implies the existence of a pseudogap in the electronic density of states in the vicinity of the Fermi level, E_F. Initially, the Hume-Rothery mechanism was invoked, both from the structural[23] and theoretical[24] side, in relation to the problem of the stability of quasicrystals. It was later linked[25] to the small σ values through the Einstein equation,[18] $\sigma = e^2DN(E_F)$, in which D is the electron diffusion coefficient, i.e., the diffusivity, and $N(E_F)$ designates the density of states at E_F. The correlation between the low σ and the low $N(E_F)$ was based upon the experimental observation of the very small value of the electronic contribution to the specific heat, γ,[5,25] which is directly proportional to $N(E_F)$. Support for the Hume-Rothery mechanism in icosahedral alloys also comes from the results of the NMR[26] and optical conductivity[27] experiments which were interpreted in terms of low $N(E_F)$ or the existence of a pseudogap in the density of states around E_F. The reduced $N(E_F)$ is also consistent with the observed diamagnetism.[20,21] Furthermore, the notion of a structure-induced pseudogap in the density of states around E_F results not only from theory based on the nearly-free-electron approximation[24,28] but is also supported by the electronic structure calculations for the lowest-order crystalline approximants of icosahedral alloys,[29-31] including the approximants containing transition metals atoms.[30,31] Approximants are complex crystalline phases close in composition to quasicrystals and with known structures; their local atomic configurations are close to those expected in the corresponding quasicrystals. Apart from its seeming simplicity, expressed in the relation $Q = 2k_F$,[23] where Q is the magnitude of the reciprocal lattice vector corresponding to the strongest x-ray reflection and k_F is the radius of the Fermi sphere, the Hume-Rothery mechanism is so appealing because it can explain qualitatively why the stable icosahedral alloys have both the lowest values of σ and γ. It can also be used to rationalize qualitatively why stable icosahedral alloys exist only in a rather narrow composition range and why a small composition change can shift the E_F away from the density of states minimum.

The experimental studies mentioned above provide *indirect* information on the density of states at one particular energy, E_F. To determine the electronic structure of quasicrystals, and in particular the possible existence of a pseudogap in the density of states around E_F, spectroscopic techniques which probe density of states *directly*[11,22] must be used. Unfortunately, due to insufficient energy resolution available, these spectroscopic experiments could not unambiguously verify whether the minimum in the density of states around E_F does exist.

Some of the recent low-temperature, high energy-resolution ultraviolet photoemission spectroscopy results[32,33] obtained at a photon energies of 21.2 eV, from He I, and 40.8 eV, from He II, are presented here. They convincingly solve the problem of the minimum of the density of states at E_F.

Figure 2a shows a theoretical He II valence band predicted for an *i*-Al-Cu-Fe alloy.[32] The theoretical calculations[30] of the density of states were performed for a 1/1 approximant of an *i*-Al-Cu-Fe alloy. A clear depression of intensity, which is proportional to the total density of states, towards E_F is clearly seen, with the predicted density of states minimum being slightly above E_F. Additionally, the theory[29-31] predicts the presence of many narrow spikes in the density of states, see Fig. 2a. This spikiness is believed to be a specific property which distinguishes quasicrystals from amorphous and crystalline alloys. However, the predicted pseudogap is not such a property because it is also present in some amorphous and crystalline alloys. To make a meaningful comparison of the theoretical band in Fig. 2a with the measured one, the former has to be multiplied by the Fermi-Dirac distribution function corresponding to the temperature of an experiment, and appropriately broadened to account for the lifetime broadening effects inherent to the ultraviolet photoemission spectroscopy technique and for the finite instrumental resolution.[11,22] Such a broadened band in the vicinity of E_F is shown in Fig. 2b.

Figure 3a shows the valence band of a well-characterized single-phase *i*-$Al_{65}Cu_{20}Fe_{15}$. Two broad features at the binding energies of about -1 and -4 eV are due respectively to states of the iron 3d and Cu 3d character,[34] in good agreement with the theoretical predictions, see

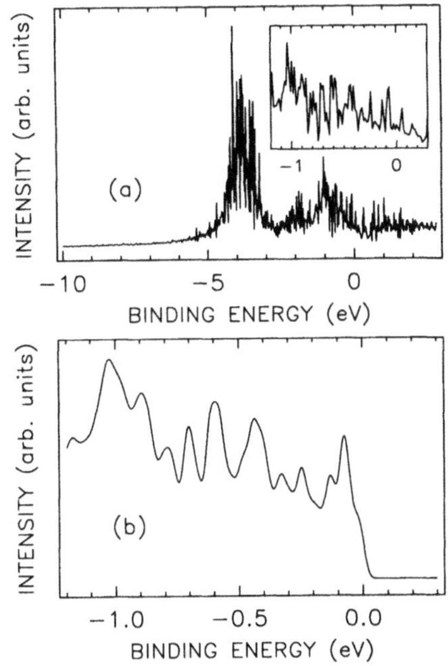

Figure 2. (a) The theoretical 40.8 eV ultraviolet photoemission spectroscopy valence band of the 1/1 approximant of an *i*-Al-Cu-Fe alloy obtained by summing the theoretical partial aluminum, copper, and iron density of states associated with different angular momenta, from ref. 30, weighted by the corresponding photoionization cross sections and by the composition and the number of electrons, as described in reference 22. The inset shows a part of the band around E_F. The binding energy is referred to E_F. (b) Part of the theoretical band close to E_F from (a) which was multiplied by the Fermi-Dirac distribution function at 14 K and convoluted with a Lorentzian, $\Gamma_L^0 = 0.02$ eV^{-1}, to account for the lifetime broadening, see ref. 22, and with a Gaussian with full width at half maximum of 31 meV, to account for the instrumental broadening.

Fig. 2a and reference 30. A sharp intensity decrease towards E_F and an indication of the presence of a Fermi edge can also be noticed, see Fig. 3a. These two characteristics are clearly seen in the valence band measured in the vicinity of E_F, Fig. 3b. Three observations are apparent from Fig. 3b. First, there is a strong intensity decrease towards E_F, which can be interpreted as being due to the presence of a pseudogap centered in the vicinity of E_F. A more quantitative interpretation of this observation is given below. Second, a clear Fermi edge is observed, which constitutes the spectroscopic proof that the system is metallic. A similar earlier study[35] of *i*-$Al_{65}Cu_{21}Fe_{14}$ carried out at room temperature and with the resolution of 233 meV concluded that there is no well-defined Fermi edge in the *i*-Al-Cu-Fe alloy. Third, there is no evidence for the existence of spikes which should be observed clearly for the energy resolution used in the experiment, see Fig. 2b.

One can reproduce the observed structure of the valence band close to E_F, see Fig. 3b, by using a simple model suggested in reference 35. It can be noticed in Fig. 3b that the intensity increases towards E_F for binding energies between -1.2 and about -0.9 eV. The model[35] thus assumes that without a pseudogap, the density of states at 0 K, which can be called the normal density of states, would increase linearly up to E_F, see Fig. 4a. It is then assumed[35] that the normal density of states has a dip in a Lorentzian shape centered at E_F, with the half-width, G_L, and the dip depth, C, relative to the normal density of states. One can obtain a relatively good fit, as shown by the solid line in Fig. 4b, for $G_L = 0.36(1)$ eV and $C = 0.70(1)$. It can be thus concluded that, in agreement with an earlier analysis,[35] the observed intensity depression close to E_F in the valence band of *i*-$Al_{65}Cu_{20}Fe_{15}$ can be accounted for by a pseudogap of the

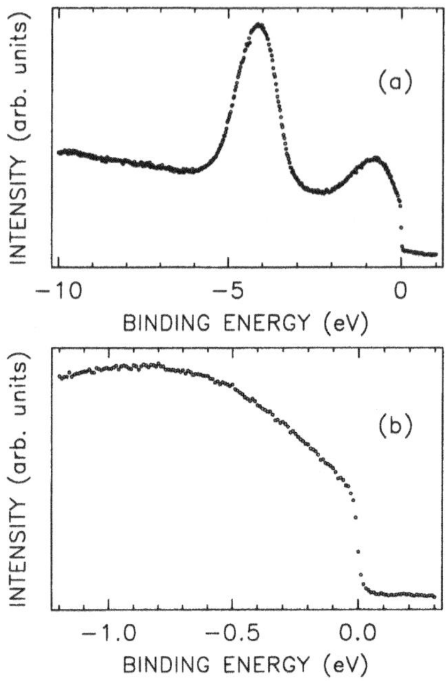

Figure 3 (a). The valence band of i-$Al_{65}Cu_{20}Fe_{15}$ measured at 40.8 eV and 14 K with the energy resolution of 42 meV, see reference 32. The data were not corrected for the background due to the inelastic electron scattering. (b) The valence band of i-$Al_{65}Cu_{20}Fe_{15}$ in the vicinity of E_F measured at 40.8 eV and 14 K with the energy resolution of 31 meV, see reference 32.

half-width of 0.36(1) eV and that the density of states at E_F is reduced by 70(1) percent with respect to the normal density of states at E_F.

The predicted spikiness could not be observed even with an energy resolution of 7.5 meV, see Fig. 5. Such a resolution represents the ultimate experimental conditions now achievable in a photoemission spectroscopy measurement of solids.[36] The temperature evolution of the leading edge of the valence band of i-$Al_{65}Cu_{20}Fe_{15}$ follows the temperature evolution of the Fermi-Dirac distribution function, see Fig. 5b, which additionally confirms the metallic character of the i-$Al_{65}Cu_{20}Fe_{15}$ alloy.

Valence bands similar to those shown in Figs. 3 and 5 were also obtained[32] for other stable icosahedral and dodecagonal alloys, except for the i-Al-Pd-Re system[33] in which no well-defined Fermi edge was observed, as is shown in Fig. 6, which also shows that there is a very low intensity at E_F in this system. This proves that the i-Al-Pd-Re system is nonmetallic.

In conclusion, the ultra-high-resolution ultraviolet photoemission spectroscopy experiments[32,33] provide a direct experimental proof that there is a wide pseudogap in the vicinity of E_F which must be partially responsible for the observed high values of ρ. All known quasicrystals, except for the i-Al-Pd-Re system, are metallic as evidenced by a clearly developed Fermi edge. The lack of the Fermi edge in the i-Al-Pd-Re system is consistent with its insulator behavior at low temperatures.[17] The possible reasons for the absence of the predicted spikiness of the density of states are discussed in Section 5.

There are experimental arguments which indicate that the Hume-Rothery mechanism alone is insufficient to explain the high values of ρ or the low values of σ. The dramatic changes of σ, of up to two orders of magnitude, with composition[13,15-17,19,37,38] and structural

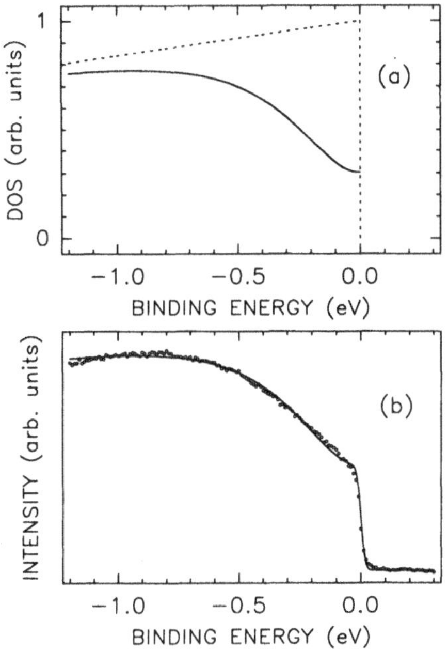

Figure 4. (a) The model[35] of density of states at 0 K which is used to fit the valence band in Fig. 3b. The broken line, which was obtained from a linear fit of the band in Fig. 3b between the binding energies -1.2 and -0.9 eV, represents the normal density of states, without the presence of a pseudogap. The solid line designates the dip which must be subtracted from the normal density of states in order to fit the band in Fig. 3b. (b) The valence band from Fig. 3b, circles, fitted, solid line, to the model density of states shown in (a).

quality[5,13,17,39] have been observed for *i*-samples whose values of γ differ only by up to about 10 percent. For example, the values of σ at 2 K and γ for *i*-$Al_{69}Pd_{19}Re_{12}$ are respectively 1000 $\Omega^{-1}cm^{-1}$ and 0.28 mJ/mol.K^2, see reference 38. The corresponding values for *i*-$Al_{67}Pd_{23}Re_{10}$ are 100 $\Omega^{-1}cm^{-1}$ and 0.25 mJ/mol.K^2, see reference 38. As both σ in the Einstein equation and γ are proportional to $N(E_F)$, the above experimental fact indicates that the small values of σ cannot be associated only with the minimum in the density of states at E_F. This is supported by the observation of the lack of the expected correlation between σ and the electron per atom ratio, e/a, in the *i*-Al-Cu-Fe system.[39] No correlation was also observed between the Hall coefficient and e/a in this system.[39] Additionally, the Einstein equation assumes that σ is proportional to the electron scattering time, which implies that σ should increase by removing the defects. The opposite is observed for icosahedral alloys.[5,16,17,19,40,41] This proportionality also predicts that σ should decrease with increasing temperature, which is contrary to the observed steep increase of σ with temperature at temperatures well above the Debye temperature.[13,17,19] Since σ in the Einstein equation is proportional to the product $DN(E_F)$, the unusually small values of σ may also originate from the reduced D.

The arguments presented above suggest that a simple interpretation based on the Einstein equation is not applicable to icosahedral alloys and alternative approaches are required to explain the unusual properties of these alloys. These new approaches are based on the concepts of tunnelling, localization, and critical states. One of the first of such approaches, which is very qualitative and which claims to give a good explanation of the electronic properties of icosahedral alloys, is based on an internal structural model.[42] It assumes the presence of the conductive icosahedral blocks which are enveloped by an insulating layered-

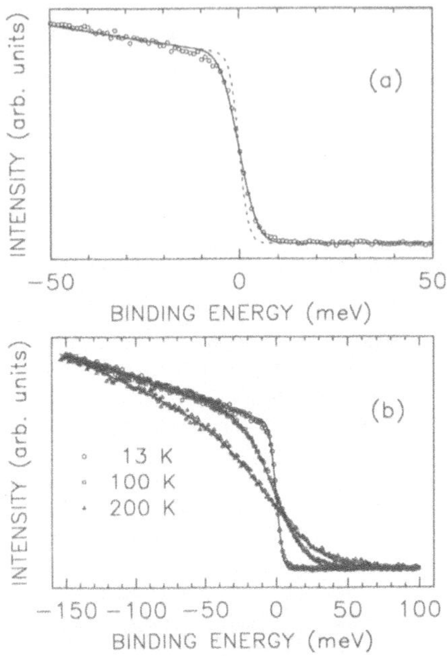

Figure 5. (a) The near-E_F spectrum of i-$Al_{65}Cu_{20}Fe_{15}$ measured at 21.2 eV and 13 K. The solid line is the fit to a linearly decreasing intensity multiplied by the Fermi-Dirac distribution function at 13 K, the broken curve, and convoluted with a Gaussian whose linewidth is 7.5 meV. Note that the step between the data points is 1 meV, see reference 32. (b) The temperature-dependent ultraviolet photoemission spectra of i-$Al_{65}Cu_{20}Fe_{15}$ near E_F measured at 21 eV. The solid lines are the fits to a linearly decreasing intensity multiplied by the Fermi-Dirac distribution function at the corresponding temperature and convoluted with a Gaussian whose linewidth is 7.5 meV, see reference 32.

structure network. For this structural model the electrical conduction should occur via tunnelling. Such tunnelling should lead to deviations from Ohm's law; however, the law was obeyed perfectly in the i-Al-Cu-Fe film for bias voltages varying by seven orders of magnitude.[43] Another approach invokes the concept of localization.[18] Inherent to this concept is the notion of disorder which causes the electrons to become localized, i.e., spatially confined to some region of an icosahedral alloy. Conduction then occurs via thermal activation between the localized states, or hopping. Consequently, the very small values of σ can occur for the finite values of the density of states at E_F. The use of the localization approach to explain the electronic transport properties of icosahedral alloys was suggested by Kimura et al.[44] and was later invoked and elaborated in the following works.[17,45] The third approach is based upon the idea of the new type of electronic states, the so-called critical states, which are intermediate between the extended electronic states in metals and the localized electronic states in insulators. Such states were shown theoretically to exist in one- and two-dimensional quasiperiodic systems.[46] Although it is not clear whether the critical states can occur in realistic three-dimensional quasiperiodic systems, they may have profound consequences on the electronic transport properties of quasicrystals.[13,47,48] It can be concluded that there is no generally accepted explanation of the unusual transport properties of quasicrystals.

A relevant question related to the unusual properties of quasicrystals discussed above is to what extent is the quasiperiodicity essential for their occurrence. The observations[13,26] of similar unusual transport behavior in crystalline approximants indicate that it is the local atomic order, rather than the quasiperiodicity, which determines the properties of quasicrystals.

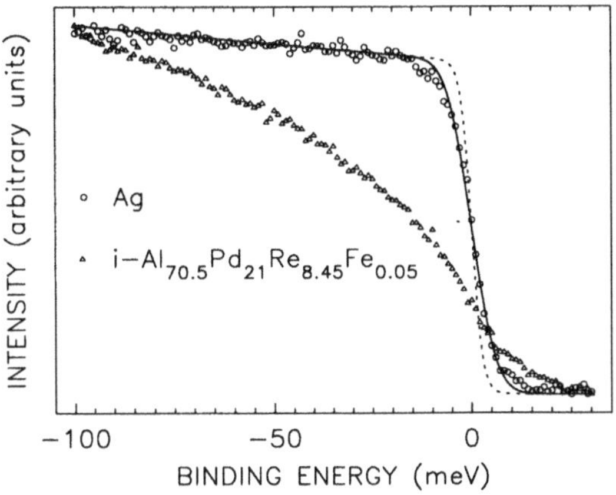

Figure 6. The near-E_F spectra of *i*-$Al_{70.5}Pd_{21}Re_{8.45}Fe_{0.05}$ and of silver evaporated onto *i*-$Al_{70.5}Pd_{21}Re_{8.45}Fe_{0.05}$ measured at 21.2 eV and 15 K, see reference 33. The resolution of the experiment was determined from the fit, solid line, of the silver spectrum to a linearly decreasing intensity multiplied by the Fermi-Dirac distribution function at 15 K, broken curve, and convoluted with a Gaussian whose linewidth was 10 meV.

4. HYPERFINE INTERACTIONS IN QUASICRYSTALS

Mössbauer spectroscopy is a local probe which can be used in studies of quasicrystals to obtain, in principle, two types of information. First, local atomic structure can be studied through the electric quadrupole interactions. The hyperfine parameters derived from the Mössbauer spectra must be compared with those calculated for a particular structural model if the full potential of Mössbauer spectroscopy is to be employed. Such calculations, however, are not trivial, especially for structurally complex quasicrystals. The first such calculations performed for a *d*-Al-Co-Cu alloy[50] will be compared below with the experimental Mössbauer data. Examples of Mössbauer spectra of representative quasicrystals and their analysis will be presented. Second, magnetic properties of quasicrystals can be investigated through the magnetic dipole interaction. Examples of such investigations will be discussed.

4. 1. Nonmagnetic Quasicrystals

Most quasicrystals are not magnetically ordered, but rather they are either paramagnetic or diamagnetic down to liquid helium temperatures.[20] Their Mössbauer spectra thus exhibit the combination of the electric monopole and quadrupole hyperfine interactions.[49] Mössbauer spectra of most nonmagnetic quasicrystals consist of two very broad and structureless lines, with typical values for the full linewidth of the component lines in the range 0.35 to 0.45 mm/s, values which should be compared with the typical value of 0.23 mm/s for the full linewidth of the two inner lines of the Zeeman sextet of a thin α-iron calibration foil. However, as will be discussed below, Mössbauer quadrupole spectra of some quasicrystals exhibit clear structural features superimposed on the two broad lines, which can be, in principle, directly related to possible structural models. The presence of broad component lines shows that iron atoms are not in a single or in a few discrete local environments. The analysis of such Mössbauer spectra has to be then based on the distribution of the hyperfine interaction parameters.[49]

Mössbauer spectra of quasicrystals can be analyzed at two levels of sophistication. If only average values of the isomer shift, δ, and the quadrupole splitting, Δ, are of interest, then a simple fit with, for example, an asymmetric Lorentzian doublet is sufficient. The values of $\bar{\delta}$ and $\bar{\Delta}$ determined in this way have a relative error of less than a few percent. Increasing the number of fitted components, which may not have any physical justification, significantly improves the fit and the resulting $\bar{\delta}$ and $\bar{\Delta}$ values calculated from the fitted parameters of these components are determined more precisely. One gets a rather limited insight into the structural properties of quasicrystals only from the average values, $\bar{\delta}$ and $\bar{\Delta}$, because these values do not change significantly in metallic systems of different structures.

At another level of spectral analysis, one attempts to derive the distribution of quadrupole splittings, $P(\Delta)$, which is closest to the true one. It is the derived $P(\Delta)$ which is of great importance because it is directly related to the local atomic structure. However, deriving this distribution is not a simple task and there are many pitfalls which, if not avoided, lead to perfect fits with the $P(\Delta)$'s which are far from the true ones. A review of various methods of deriving $P(\Delta)$ and of various problems associated with this derivation is given in references 51 and 52 and in Chapter 5 of this book. There are generally two main approaches used to extract $P(\Delta)$. First, either a specific shape of $P(\Delta)$ is assumed, for example, a Gaussian shape or, second, no a priori assumption about the shape of $P(\Delta)$ is made, usually in conjunction with the constrained version[53] of the Hesse-Rübartsch method[54] or the Rancourt-Ping method.[52] The disadvantage of the first approach is that it presupposes the shape of $P(\Delta)$ which may differ considerably from the real $P(\Delta)$. The disadvantage of the Le Caër-Dubois version[53] of the second approach is the possibility of unphysical oscillations in $P(\Delta)$ for inappropriately chosen parameters which must be fixed in the fit. These oscillations do not occur in the Rancourt-Ping method.[52] Usually, a linear relation, $\delta = \delta_0 + \delta_1\Delta$, is assumed,[52,53] where δ_0 and δ_1 are the fitted parameters.

We used both the Le Caër-Dubois and Rancourt-Ping methods in the analysis of Mössbauer spectra of various quasicrystals.[55,56] The $P(\Delta)$'s, which are presented below, turned out to be essentially independent of the method used to extract them from the Mössbauer spectra.[55,56] This indicates that they are close to the true $P(\Delta)$'s. In many cases, the derived $P(\Delta)$ is of, or very close to, a Gaussian shape. One should note that the discussions[51-54] of various methods of deriving $P(\Delta)$ assume an ideal situation of a single-phase sample. In reality, most samples contain a few percent of a second phase/phases which may not be detected in a typical x-ray diffraction measurement. The presence of such a second phase can be detected[55] in Mössbauer spectra of many quasicrystals by carefully analyzing the residuals, i.e., the difference between the Mössbauer spectrum and the fit to a particular $P(\Delta)$. Accepting the presence of clear structural features in the residuals, and thus a value of χ^2 much larger than one, may distort[55] the derived $P(\Delta)$.

All Mössbauer absorbers used in our studies[55,56] were thin and homogeneous. They were prepared by mixing a pulverized alloy with a petroleum jelly, which resulted in a random orientation of the sample particles. The low ratio of the absorber radius to the source-absorber distance assured that the cosine smearing effect was negligible. In all quasicrystals discussed below with the iron concentration smaller than 3 atomic percent, the iron metal used to prepare them was enriched to 95.9 percent in the iron-57 isotope. The velocity scale of all Mössbauer spectra presented here is relative to α-iron at room temperature.

4. 1. 1. *Icosahedral Alloys*

A typical Mössbauer spectrum of a metastable *i*-$Al_{60}Ge_{20}Cr_{19.9}Fe_{0.1}$, as is shown in Figure 7, can be fitted relatively well with a structureless Gaussian-like P(Δ), see Fig. 7c. A close inspection of the residuals in Fig. 7a indicates the presence of some structure due to an unaccounted impurity component. To a first approximation, this component can be taken in the form of a symmetric Lorentzian doublet. Its inclusion in the fit, as is shown in Fig. 7b, leads to a lowering of the χ^2 from 1.30 to 1.12 and to the disappearance of the residuals structure. Although for this particular icosahedral alloy the inclusion of the impurity doublet does not lead to a significant change of P(Δ), see Fig. 7c, it does for other quasicrystals.[55]

Both the Mössbauer spectra of metastable icosahedral alloys and the resulting P(Δ)'s, as illustrated in Fig. 7, are very similar to those characteristic for amorphous alloys.[58] It was thus initially thought that metastable icosahedral alloys are highly disordered and that Mössbauer spectroscopy will not be able to reveal any structural differences between these alloys and the corresponding amorphous counterparts. That this is not the case is illustrated in Fig. 8 in which the Mössbauer spectra and the resulting P(Δ)'s are compared for the $Al_{75}Cu_{14.85}Fe_{0.15}V_{10}$ alloy which can be produced both in an amorphous and icosahedral state.[59] The Mössbauer spectrum of *a*-$Al_{75}Cu_{14.85}Fe_{0.15}V_{10}$, see Fig. 8a, and the resulting P(Δ), Fig. 8(d), are very similar to those characteristic for the *i*-$Al_{60}Ge_{20}Cr_{19.9}Fe_{0.1}$ alloy, as is shown in Fig. 7. However, the Mössbauer spectra of *a*- and *i*-$Al_{75}Cu_{14.85}Fe_{0.15}V_{10}$ and the resulting P(Δ)'s, see Fig. 8, differ significantly, as may be noted by the presence of some structure at about -0.2 mm/s in the spectrum of *i*-$Al_{75}Cu_{14.85}Fe_{0.15}V_{10}$. An attempt to fit the spectrum of *i*-$Al_{75}Cu_{14.85}Fe_{0.15}V_{10}$ with one P(Δ) component and an impurity component, Fig.

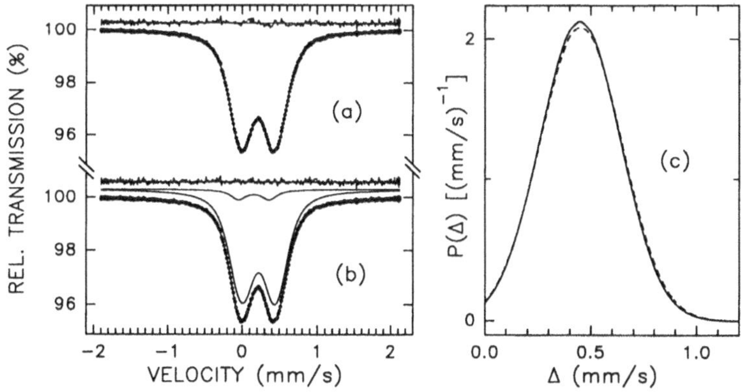

Figure 7. The room temperature Mössbauer spectrum of *i*-$Al_{60}Ge_{20}Cr_{19.9}Fe_{0.1}$ fitted with (a) one P(Δ) component, the solid line in (c), and with (b) the P(Δ), the broken line in (c) plus impurity components. The component spectra are also shown in (b). The residuals, multiplied by a factor of three, are shown above each spectrum.

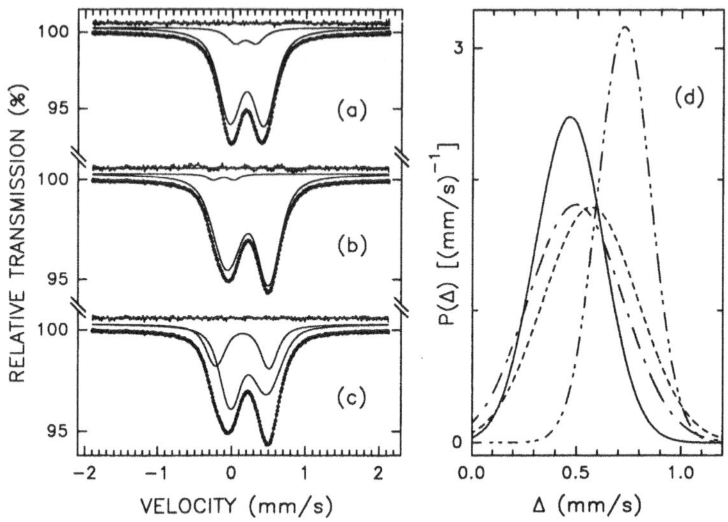

Figure 8. The room temperature Mössbauer spectra of (a) a-$Al_{75}Cu_{14.85}Fe_{0.15}V_{10}$ fitted with the P(Δ) and impurity components, and of i-$Al_{75}Cu_{14.85}Fe_{0.15}V_{10}$ fitted with the P(Δ) and impurity components (b), and with two P(Δ) components (c). The P(Δ)'s corresponding to fits in (a), the solid line, (b), the broken line, and (c), the dash-dot and dash-double-dot lines, are shown in (d). The residuals, multiplied by a factor of three, are shown above each spectrum.

8(b), does not produce the structureless residuals, $\chi^2 = 1.46$. Such residuals can be obtained only from a fit with two P(Δ)'s, see Fig. 8c, which results in $\chi^2 = 1.04$. These two P(Δ)'s are indicative that there must be two different classes of transition metal environments in i-$Al_{75}Cu_{14.85}Fe_{0.15}V_{10}$. Contrary to what one would expect, the $\overline{\Delta}$ value for i-$Al_{75}Cu_{14.85}Fe_{0.15}V_{10}$ is significantly larger than that for a-$Al_{75}Cu_{14.85}Fe_{0.15}V_{10}$, at 0.575(1) and 0.468(1) mm/s, respectively.

Another example showing the distinct differences in Mössbauer parameters corresponding to three different structures involves the Al-Cu-Fe system. For a particular composition in this system, one can produce samples in the amorphous, icosahedral, or cubic phases.[55,60] The $\overline{\Delta}$ values determined from the Mössbauer spectra[55] of the a-, i-, and c-$Al_{64}Cu_{24}Fe_{12}$, as are shown in Fig. 9, are respectively 0.535(1), 0.396(1), and 0.419(1) mm/s. The corresponding $\overline{\delta}$ values are 0.239(1), 0.230(1), and 0.247(2) mm/s. Note also that the P(Δ) of i-$Al_{64}Cu_{24}Fe_{12}$, which is a stable icosahedral alloy, is significantly narrower than that of metastable amorphous and cubic phases, see Fig. 9d, and than those P(Δ)'s of other metastable i-alloys, see Figs. 7c and 8d. This difference in width of P(Δ) can be related, at least partially, to the fact that the method by which they are produced results in the metastable samples containing various defects, reflected in broadening of the x-ray diffraction Bragg peaks,[55] whereas the stable i-$Al_{64}Cu_{24}Fe_{12}$ has no such defects, as is reflected in narrow Bragg

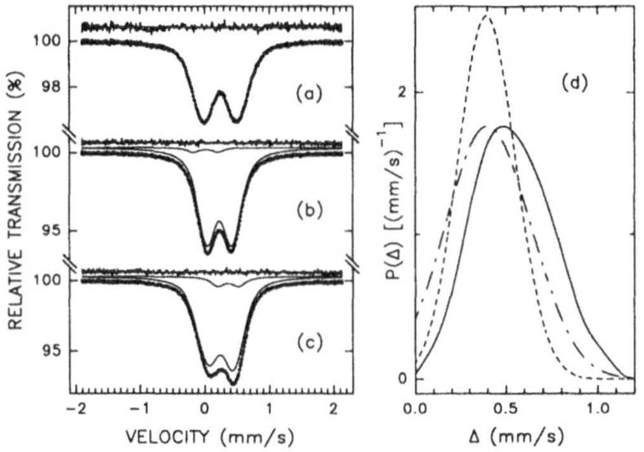

Figure 9. The room-temperature Mössbauer spectra of $Al_{64}Cu_{24}Fe_{12}$ in (a) the amorphous state fitted with one P(Δ), component, the solid line in (d), (b) the icosahedral state fitted with one P(Δ), the broken line in (d), and impurity components, and (c) the crystalline state fitted with one P(Δ), the dash-dot line in (d), and impurity components. The component spectra are also shown in (b) and (c). The residuals, multiplied by a factor of three, are shown above each spectrum.

peaks.[55] Nevertheless, the very fact that in the stable, structurally "perfect" *i*-$Al_{64}Cu_{24}Fe_{12}$ the P(Δ) has a finite and non-negligible width shows that there must be some type of structural and/or chemical disorder present even in the stable, structurally perfect *i*-alloys. The nature and role of this disorder will be discussed in Section 5.

Other stable *i*-alloys of high structural quality also exhibit the presence of P(Δ) in their Mössbauer spectra, as is shown in Fig. 10,[55,56] which supports the idea of the existence of some sort of topological and/or chemical disorder in these alloys. The values of the full width at half maximum of the P(Δ)'s corresponding to *i*-$Al_{64}Cu_{24}Fe_{12}$, Fig. 9, *i*-$Al_{65}Cu_{20}Fe_{7.5}Ru_{7.5}$, Fig. 10, and *i*-$Al_{70.5}Pd_{21}Re_{8.45}Fe_{0.05}$, Fig. 10, are 0.366(1), 0.451(1), and 0.358(2) mm/s, respectively. One notices that these values are very similar for the ternary stable alloys. The larger value for the quaternary *i*-$Al_{65}Cu_{20}Fe_{7.5}Ru_{7.5}$ alloy is mainly due to its increased chemical disorder. Also the $\bar{\Delta}$ values for these three *i*-alloys are quite similar, with values of 0.396(1), 0.410(1), and 0.404(2) mm/s, respectively. Furthermore, these alloys have a similar, Gaussian-like shape of P(Δ). These arguments suggest that the structure of *i*-Al-Cu-TM, where TM is Fe, Ru, and Os, see reference 55, and *i*-Al-Pd-Re, see reference 56, must be similar. Unfortunately, there are at present no theoretical calculations which would predict the P(Δ) for possible structural models of these alloys.

The Mössbauer spectra of *i*-alloys discussed above had essentially no structure, except for the spectrum of *i*-$Al_{75}Cu_{14.85}Fe_{0.15}V_{10}$ shown in Fig. 8, and consequently the corresponding P(Δ)'s had single-peaked, Gaussian-like shapes. Such a shape was also observed for other *i*-alloys.[61,62] Recent extensive studies[55] of various quasicrystals show that there are quasicrystalline systems with structured Mössbauer spectra, and therefore, with structured shapes of P(Δ)'s. This should be very helpful in placing stringent constraints on various structural models. A few examples of such Mössbauer spectra are presented below.

The Mössbauer spectrum of metastable *i*-$Al_{70}Pd_{20}Fe_{10}$ alloy[55,63] exhibits a very strong asymmetry, as is shown in Fig. 11. Its fit with one P(Δ) is clearly unsatisfactory as evidenced by the value of χ^2=1.30 and the structure present in the residuals. However, this fit already indicates that the shape of P(Δ) has a bimodal character, see Fig. 11c. The fit with one P(Δ) and an impurity component reduces χ^2 to 0.96, but the relative area under the impurity compo-

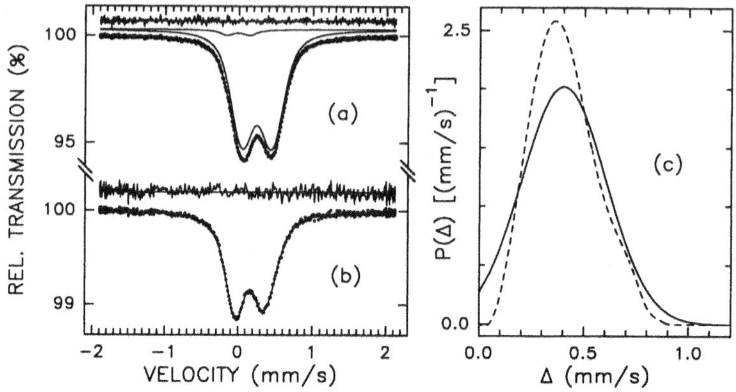

Figure 10. The room-temperature Mössbauer spectra of (a) *i*-$Al_{65}Cu_{20}Fe_{7.5}Ru_{7.5}$ fitted with one P(Δ) component, the solid line in (c), and impurity components and (b) *i*-$Al_{70.5}Pd_{21}Re_{8.45}Fe_{0.05}$ fitted with one P(Δ) component, the broken line in (c). The residuals, multiplied by a factor of three, are shown above each spectrum.

nent is 29 percent, which is much above the impurity concentration of a few percent expected from the x-ray diffraction spectrum.[55,63] An excellent fit with $\chi^2 = 0.93$ is obtained with two P(Δ) components, as is shown in Fig. 11b. It can be thus concluded there are two distinct classes of iron environments in this alloy, and any structural model has to take this fact into account.

The Mössbauer spectra[55] of stable *i*-$Al_{70}Pd_{20}Mn_{9.97}Fe_{0.03}$ and *i*-$Al_{70}Pd_{20}Cr_5Fe_5$ show distinct asymmetry and some structure. They cannot be fitted with one P(Δ), but satisfactory fits are obtained assuming the presence of two P(Δ)'s, see Fig. 12.

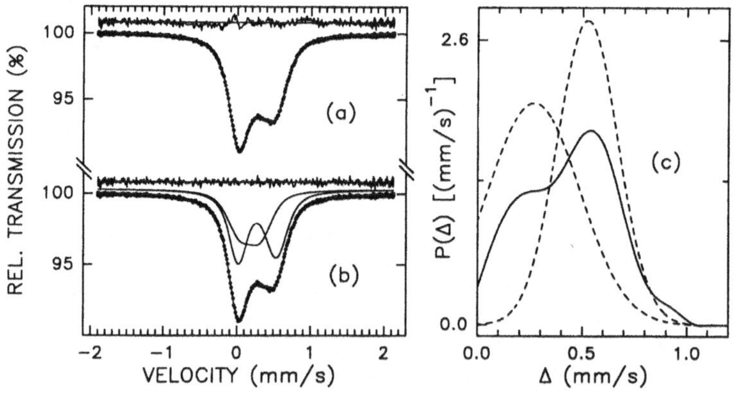

Figure 11. The room-temperature Mössbauer spectrum of *i*-$Al_{70}Pd_{20}Fe_{10}$ (a) fitted with one P(Δ) component, the solid line in (c) and (b) fitted with the two P(Δ) components, the broken lines in (c). The component spectra are also shown in (b). The residuals, multiplied by a factor of three, are shown above each spectrum.

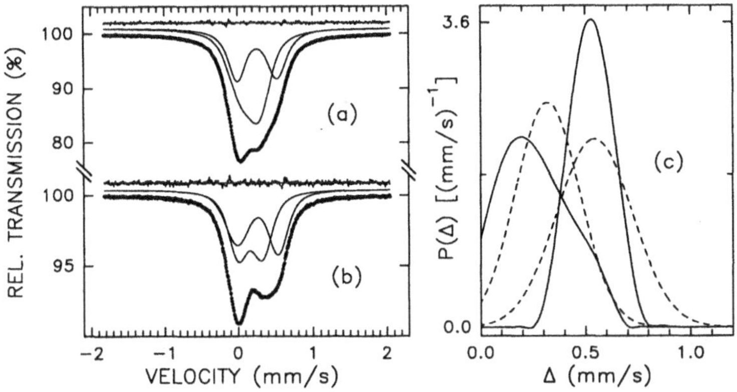

Figure 12. The room-temperature Mössbauer spectra (a) of i-$Al_{70}Pd_{20}Mn_{9.97}Fe_{0.03}$ fitted with two P(Δ) components, the solid lines in (c), and (b) of i-$Al_{70}Pd_{20}Cr_5Fe_5$ fitted with two P(Δ) components, the broken lines in (c). The component spectra are also shown. The residuals, multiplied by a factor of three, are shown above each spectrum.

4. 1. 2. Decagonal Alloys

Mössbauer spectra of decagonal alloys[55,64,65] generally possess more structure than those of i-alloys. For some d-systems a single-peaked, Gaussian-like P(Δ) is observed. This is illustrated in Fig. 13 for the stable[63,66] d-$Al_{75}Pd_{15}Fe_{10}$ alloy. As expected, a slightly narrower P(Δ) is obtained, see Fig. 13(c), for an annealed sample as compared to the P(Δ) of an as-cast sample.[66]

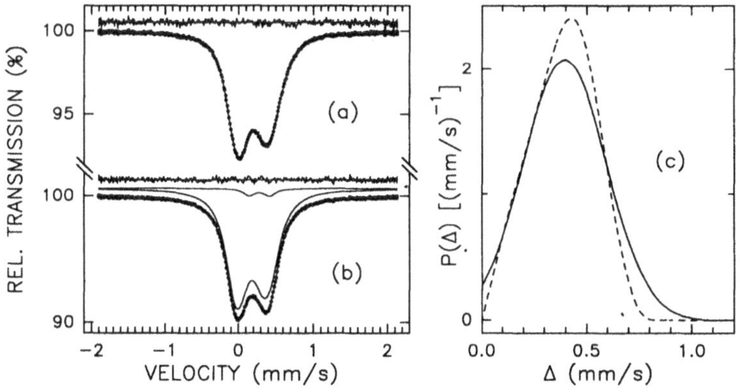

Figure 13. The room-temperature Mössbauer spectra of d-$Al_{75}Pd_{15}Fe_{10}$ (a) as cast and fitted with one P(Δ) component, the solid line in (c), and (b) after annealing, fitted with the P(Δ), the broken line in (c), and impurity components. The component spectra are also shown in (b). The residuals, multiplied by a factor of three, are shown above each spectrum.

The Mössbauer spectrum[55] of the metastable d-$Al_{70}Ni_{15}Fe_{15}$ alloy[67] cannot be fitted successfully with one P(Δ) component, as is clearly evidenced by the residual spectrum shown in Fig. 14a. Nevertheless, the derived P(Δ) is of a bimodal-type, see Fig. 14c, which indicates the possibility of two different classes of iron environments. An attempt to fit this spectrum with an additional impurity component[55] improves the fit, but the relative area of this component is 17 percent. This, however, is incompatible with the expected impurity concentration of a few percent from the x-ray diffraction pattern.[55] An excellent fit is obtained with two P(Δ) components, see Fig. 14b, which strongly suggests the existence of two structurally distinct classes of environments around the iron atoms in the d-$Al_{70}Ni_{15}Fe_{15}$ alloy.

The unquestionable evidence for the presence of at least two distinct classes of environments around transition metal atoms in some Al-transition metal d-alloys comes from the Mössbauer spectra of stable $Al_{70}Co_{15}Ni_{14.9}Fe_{0.1}$, see reference 65, and $Al_{65}Co_{15}Cu_{19.9}Fe_{0.1}$, see references 64 and 65, d-alloys. A visual inspection of the spectrum[55,65] of d-$Al_{70}Co_{15}Ni_{14.9}Fe_{0.1}$, see Fig. 15a, shows its clear asymmetry and the structure of the left absorption line. It is thus obvious that one needs at least two components, see Figs. 15b and c, to account for these asymmetry-structure features.

An even clearer structure in the left absorption line can be seen in the Mössbauer spectrum[55,64,65] of d-$Al_{65}Co_{15}Cu_{19.9}Fe_{0.1}$ which is shown in Fig. 16. There is a striking similarity of the shapes of P(Δ)'s in both alloys, compare Figs. 15c and 16c; the low-Δ P(Δ) is single-peaked, whereas the high-Δ has a bimodal character. Also the $\bar{\delta}$ and $\bar{\Delta}$ values are quite similar. This strongly suggests that the quasicrystalline structures must be very similar in both alloys.

It should be stressed here that, because of the so-called non-uniqueness problem, see Chapter 5 of this book, the P(Δ)'s derived here from the zero-field Mössbauer spectra of quasicrystals may not be unique. Within the boundary of the solution domain in the Mössbauer parameter space, other, slightly different P(Δ)'s also fit the spectra well.[55,65] The solution domain, and thus the possible P(Δ)'s, can be significantly narrowed by performing in-field Mössbauer spectroscopy. What seems to be certain, however, is the presence of two distinct classes of iron environments in many quasicrystalline systems.

There are several different structural models which attempt to describe the structure of d-Al-Co-Cu alloy.[12] One of them is the model suggested by Burkov.[68] Kramer et al.[50] calculated

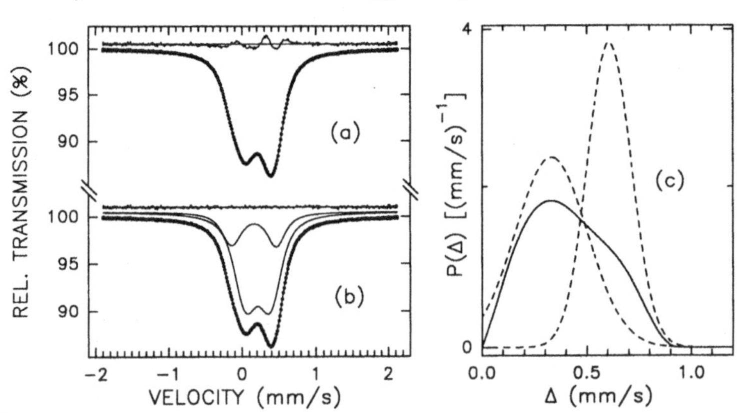

Figure 14. The room-temperature Mössbauer spectrum of d-$Al_{70}Ni_{15}Fe_{15}$ fitted with (a) one P(Δ) component, the solid line in (c), and (b) two P(Δ) components, the broken lines in (c). The P(Δ) component spectra are also shown in (b). The residuals, multiplied by a factor of three, are shown above each spectrum.

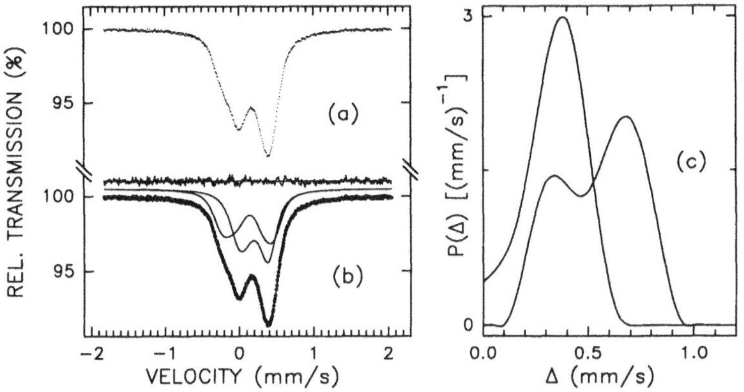

Figure 15. The room-temperature Mössbauer spectrum of d-$Al_{70}Co_{15}Ni_{14.9}Fe_{0.1}$, (a) the unfitted spectrum, (b) the spectrum fitted with two P(Δ) components, the solid lines in (c). The residuals, multiplied by a factor of three, are also shown in (b).

for this model, which corresponds to the composition $Al_{62}Co_{19}Cu_{19}$, the Mössbauer hyperfine parameters and the resulting Mössbauer spectrum. This theoretical spectrum is compared in Fig. 17 with the fit from Fig. 16b. It is clear from this comparison that theory and experiment do not agree. There are two major reasons for this, either the Burkov model[68] is incorrect or the calculations, due to their inherent limitations,[50] require some refinement. It should be noted that similar calculations carried out for c-$Al_{13}Fe_4$ resulted in a modest agreement between the predicted and measured Mössbauer spectra.[50] Recent calculations[69] for the composition $Al_{61.8}Co_{17.7}Cu_{20.5}$, which are based on a modified Burkov model in which copper atoms

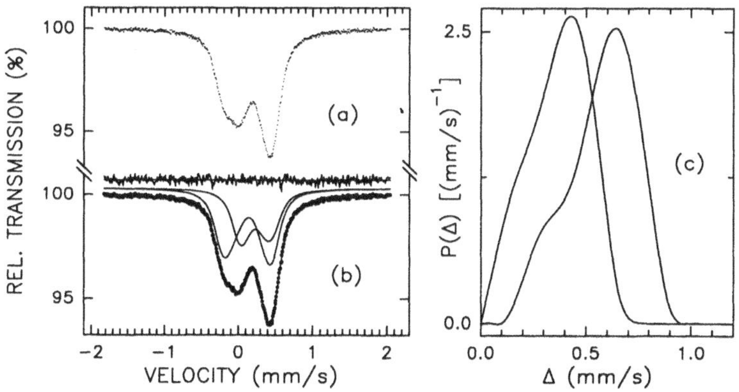

Figure 16. The room-temperature Mössbauer spectrum of d-$Al_{65}Co_{15}Cu_{19.9}Fe_{0.1}$. (a) the unfitted spectrum, (b) the spectrum fitted with two P(Δ) components, the solid lines in (c). The residuals, multiplied by a factor of three, are also shown in (b).

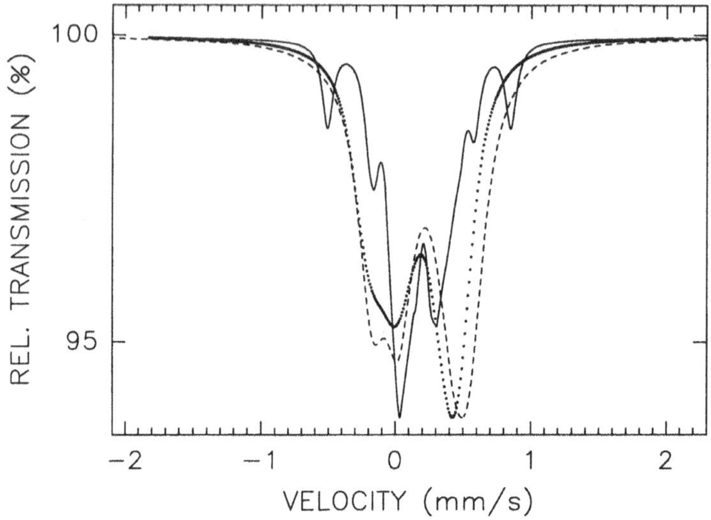

Figure 17. A comparison between the fit from Fig. 16(b), the dots, and the theoretical Mössbauer spectra calculated for the Burkov model,[50] the solid line, and for the modified Burkov model,[69] the broken line. The theoretical spectra were shifted horizontally so that their average isomer shift coincides with that of the fit.

occupy some of the cobalt positions, lead to a theoretical spectrum which is in a much better agreement with the experimental spectrum, see Fig. 17.

The experimental observation of two distinct P(Δ)'s, as shown in Figs. 15 and 16, implies the existence of two basic building blocks, or clusters, in the structure of the *d*-Al-Co-Ni and *d*-Al-Co-Cu alloys. This observation would thus favor structural models which are based on two such blocks. Recently, a new structural model of *d*-Al-Co-Cu has been proposed.[70] It assumes that there are two primary atom arrangements, or clusters, formed from an icosahedron by removing the atoms on two opposing vertices to form pentagonal polyhedra. The large pentagonal polyhedron cluster consists of ten aluminum atoms surrounding a cobalt or copper atom at the centre. The model does not differentiate between the cobalt and copper atoms whose atomic radii differ by only about two percent. In the smaller cluster, five cobalt or copper atoms replace the aluminum atoms in one pentagon, and an aluminum atom is located in the interior. These clusters are then properly connected to form aggregates, which in turn can be arranged to form a crystalline or quasicrystalline structure.[70] The geometry of these model clusters implies that one could expect qualitatively two different classes of transition metal sites, a conclusion that agrees with the experimental observation, see Fig. 16. Obviously, calculations of the Mössbauer hyperfine parameters for these two-cluster model are required to convincingly determine its compatibility with the Mössbauer results.

4. 1. 3. In-field Mössbauer Spectroscopy of Quasicrystals

Due to the nature of the 3/2 → 1/2 Mössbauer transition in iron-57, zero-field spectra discussed above provide only a distribution of the absolute values of

$$\Delta = 1/2\, eV_{zz}Q(1 + \eta^2/3)^{1/2}.$$

However, more stringent constraints can be imposed on possible theoretical models of the local atomic structure of quasicrystals if both the distributions of V_{zz} and η are known, as was successfully demonstrated for amorphous alloys.[71] In principle, such distributions can be extracted from in-field Mössbauer spectra.

A review[49] of in-field Mössbauer spectroscopy results for quasicrystals indicates that this method had a rather limited impact on elucidating structural properties of quasicrystals. This is because all studied quasicrystals, except for one case of *i*-$Al_{65}Cu_{20}Fe_{15}$, see reference 62, were metastable and thus possessed a significant amount of strain induced by the method of production; this resulted in a broadening or washing out of possible fine structural features in P(Δ). The stable *i*-Al-Cu-Fe alloys exhibit a single-peaked, Gaussian-like P(Δ), see Figs. 9b and d, and the moderate external magnetic fields applied[62] do not produce enough structure in the measured Mössbauer spectra. The main result of the in-field experiments[49] is the observation of a preponderance of the negative values of V_{zz} in *i*-alloys and of the positive values of V_{zz} in *d*-alloys. It is not clear, however, how this result relates to the atomic structure of the studied quasicrystals.

The in-field Mössbauer spectroscopy, with applied magnetic fields larger than about 7 T, has a potential for making a significant impact in structural studies of high-quality stable quasicrystals which exhibit a clear structure in their P(Δ)'s. The Mössbauer results, however, must be related to theoretical calculations of the electric field gradient tensor for various possible structural models. The latter is a challenging endeavour, but the first attempt has already been completed.[50,69]

4. 1. 4. Two Classes of Manganese Sites

The notion of two separate *classes* of manganese sites, not to be confused with the erroneous concept[49] of the two distinct manganese sites, in paramagnetic Al-Mn-based *i*-alloys, distinguished by the presence or absence of a localized magnetic moment, was suggested on the basis of various experimental results.[20,72] Perhaps the most convincing experimental proof came from the in-field Mössbauer measurements.[73,74] The magnetic susceptibility measurements[73,74] on the paramagnetic *i*-Al-Mn-Si alloys, in which a fraction of manganese atoms was substituted by iron atoms, showed that the effective magnetic moment per transition metal did not change with substitution. This is contrary to what is expected if all manganese atoms carry a magnetic moment. The in-field Mössbauer spectra showed, see Fig. 18, that the measured hyperfine magnetic field turned out to be the same as the applied magnetic field.[73,74] This is possible only for iron atoms with no local magnetic moment. Thus iron atoms substitute for only one class of manganese atoms - the class with zero magnetic moment.

The origin of these two classes of manganese sites seems to be related to the volume occupied by manganese atoms.[72,74] For large volumes manganese atoms carry no magnetic moment, whereas for small volumes manganese atoms possess a magnetic moment. Model electronic structure calculations carried out for an $Al_{80}Mn_{20}$ quasicrystal[75] support the concept of two classes of manganese sites. They predict the existence of manganese sites with a

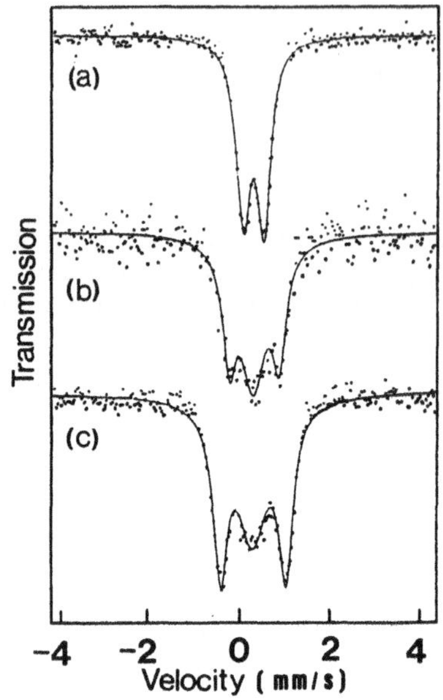

Figure 18. The Mössbauer spectra of *i*-$Al_{72}(Mn_{0.99}Fe_{0.01})_{22}Si_6$ measured at 4.2 K in external magnetic fields of (a) 0.0 T, (b) 3.0 T, and (c) 4.5 T. Figure obtained from reference 73.

quasigap in the density of states at the Fermi level and no magnetic moment, and of manganese sites with a peak in their density of states and a magnetic moment. As is discussed below, these two classes of magnetic sites are also present in magnetically ordered icosahedral alloys.

4. 2. Magnetically Ordered Icosahedral Alloys

The majority of known quasicrystals are not magnetically ordered. They exhibit paramagnetic, diamagnetic, or spin-glass properties.[20] There are reports[8] on the observation in metastable Al-Fe-Ce, Al-Mn-Si, Al-Ge-Mn, and Al-Ge-Cu-Mn icosahedral alloys of an unusual ferromagnetic behavior characterized by a very small value of magnetization and relatively high value of the Curie temperature. Unfortunately, the samples in these alloys cannot be produced in a single-phase form. Therefore, one has to carry out careful investigations with several techniques to separate the magnetic contributions from the possible magnetically-ordered second phases and from the icosahedral phase. As is shown below for the *i*-Al-Ge-Cu-Mn system, Mössbauer spectroscopy, combined with x-ray diffraction and magnetization measurements, proved very useful in separating magnetic contributions, which are intrinsic to the icosahedral phase, from those resulting from the impurities.

A thorough analysis of x-ray diffraction patterns[9,10] clearly indicates the presence in *i*-Al-Ge-Cu-Mn alloys of a ferromagnetic impurity, AlGeMn, whose Curie temperature of 519 K is very close to the values reported for these alloys.[9,10] Furthermore, the temperature dependence of the magnetization of the *i*-Al-Ge-Cu-Mn samples is very similar to that of the AlGeMn

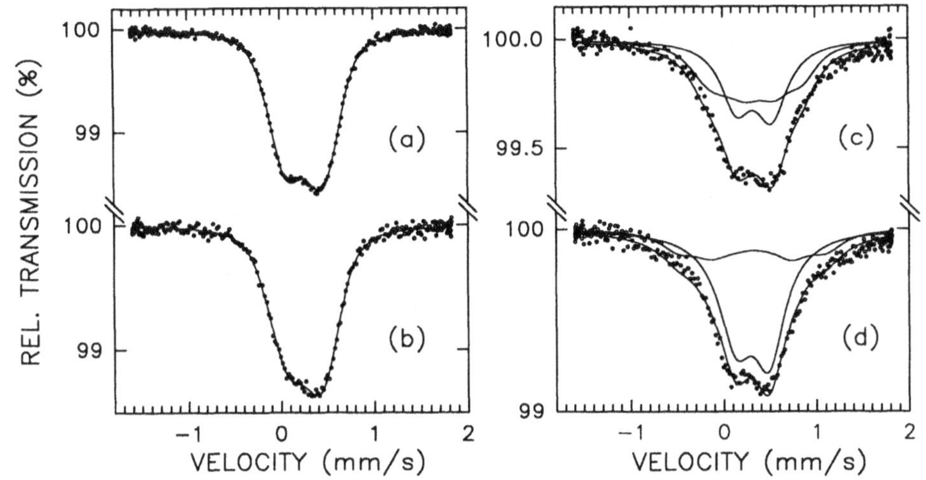

Figure 19. The Mössbauer spectra of i-$Al_{40}Ge_{25}Cu_{10-x}Fe_xMn_{25}$ obtained at room temperature for samples with x values of 0.06, (a), and 3, (b), fitted with one Gaussian P(Δ) component, the solid lines, and at 4.2 K for samples with x values of 0.06, (c), and 3, (d), fitted with the paramagnetic and magnetic components, as described in the text. Figure obtained from reference 76.

ferromagnet, except for a temperature region below about 30 K, where the magnetization suddenly increases nonlinearly.[10] In addition, the room-temperature Mössbauer spectra[9,10,76] of i-$Al_{40}Ge_{25}Cu_{10-x}Fe_xMn_{25}$ are clearly paramagnetic doublets, see Fig. 19a and b.

The 4.2 K Mössbauer spectra of i-$Al_{40}Ge_{25}Cu_{10-x}Fe_xMn_{25}$ cannot be fitted successfully with one magnetic component, as is explained in detail elsewhere.[9] A successful fit requires magnetic and nonmagnetic components,[9,76] see Figs. 19c and d. This means that there are also two classes of magnetic transition metal atoms in magnetically ordered icosahedral alloys and that iron atoms enter both of them. Furthermore, as is evidenced by the areas of the magnetic and nonmagnetic components, see Figs. 19c and d, iron atoms enter randomly both classes of transition metal sites for very small iron concentrations. However, for larger concentrations, iron atoms seem to enter preferentially the nonmagnetic class of transition metal sites.

It was concluded from the combination of the magnetization and the 4.2 K and room-temperature Mössbauer spectral studies[9,76] that the transition from a paramagnetic to a magnetic state, which is intrinsic to the i-Al-Ge-Cu-Mn system, must occur below 100 K. The temperature of this transition was determined from the analysis of the Mössbauer spectra measured at different temperatures.[10] An additional broadening of the spectra below 30 K is reflected in the divergence of the electric quadrupole interaction parameters,[10] as is shown in Fig. 20. It can be thus concluded that, in spite of the presence of the ferromagnetic AlGeMn impurity in the i-Al-Ge-Cu-Mn system, a careful analysis of Mössbauer spectra, combined with other experimental techniques,[9,10,76] made it possible to determine the intrinsic ordering temperature of about 30 K in this system and the value of iron magnetic moment of about 0.2 to 0.3 μ_B for the iron atoms entering the magnetic transition metal sites.

Recently, new magnetically-ordered, metastable, single-phase icosahedral alloys in the Al-B-Pd-Mn were discovered,[77] which are characterized by high values of ca. 500 K for their ordering temperatures[78] and of the magnetization, for example, 18.9 emu/g for i-$Al_{64}B_6Pd_{15}Mn_{15}$ at 4.2 K in a field of 2 T.[78] NMR studies[78] showed that the manganese atoms

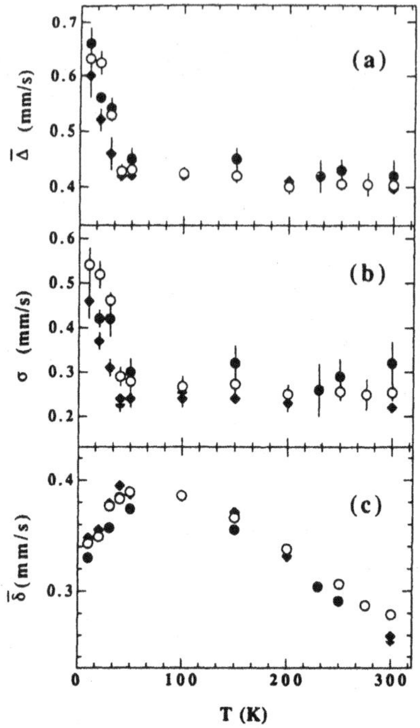

Figure 20. The temperature dependence of $\bar{\Delta}$ (a), the standard deviation σ of P(Δ) (b), and $\bar{\delta}$ (c) derived from fitting the Mössbauer spectra of *i*-$Al_{40}Ge_{25}Cu_{10-x}Fe_xMn_{25}$ [(♦) for x=0.1, (•) x=0.3, and (○) x=3, with one P(Δ) component. Figure obtained from reference 10.

with and without a magnetic moment coexist in these alloys. The Mössbauer spectra[79] of *i*-$Al_{64}B_6Pd_{15}Mn_{14.925}Fe_{0.075}$ at room-temperature and at 5.0 K, see Figs. 21a and b, show no evidence of ordering of the iron atoms. They can be fitted with two P(Δ)'s, as is shown in Fig. 21c. This confirms the NMR result[78] concerning the existence of the nonmagnetic class of

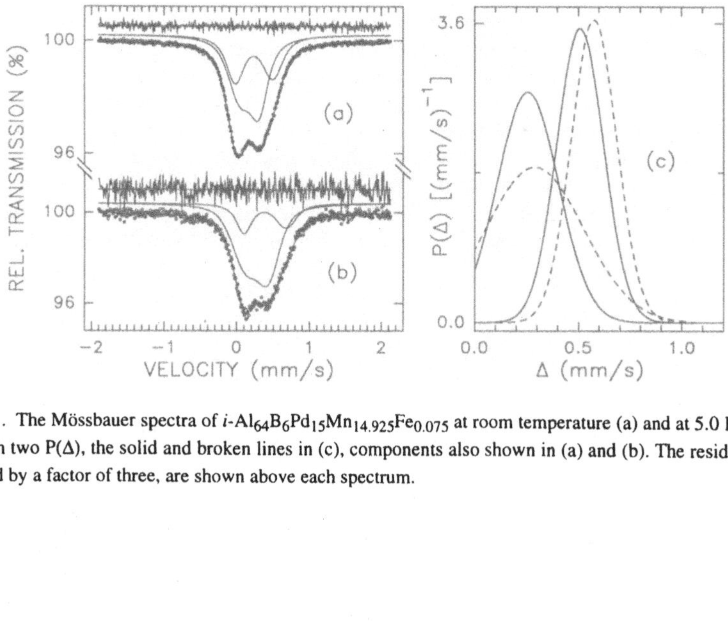

Figure 21. The Mössbauer spectra of *i*-$Al_{64}B_6Pd_{15}Mn_{14.925}Fe_{0.075}$ at room temperature (a) and at 5.0 K (b) fitted with two P(Δ), the solid and broken lines in (c), components also shown in (a) and (b). The residuals, multiplied by a factor of three, are shown above each spectrum.

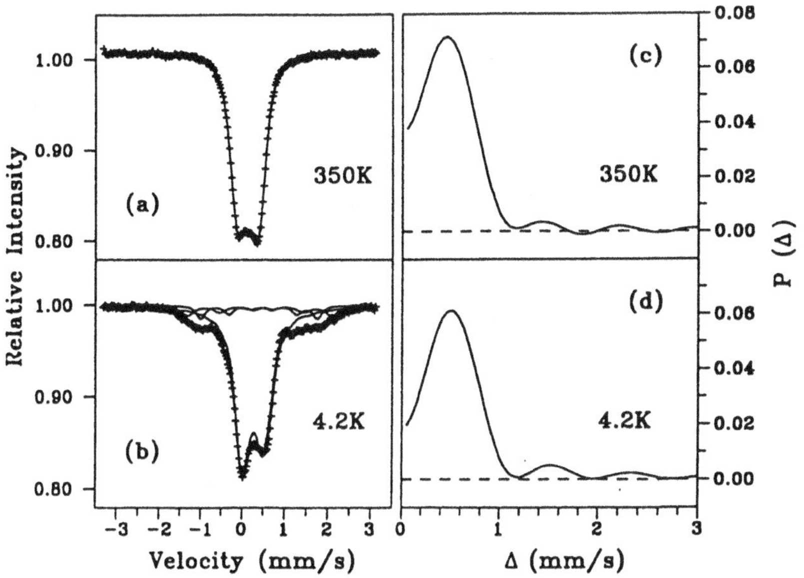

Figure 22. The Mössbauer spectra of *i*-$Al_{62.5}B_{7.5}Pd_{15}Fe_{15}$ measured at 350 K (a) fitted with one P(Δ), (c), and at 4.2 K (b) fitted with the P(Δ), (d), and the Zeeman components. The component spectra are indicated in (b). Figure obtained from reference 80.

manganese sites and indicates that there are two distinct environments around nonmagnetic Mn(Fe) atoms in this system.

It was recently discovered that completely replacing manganese atoms by iron atoms in the *i*-Al-B-Pd-Mn system retains both the icosahedral symmetry and the magnetism.[80] The Mössbauer spectra, shown in Fig. 22, of an icosahedral alloy of this system with an ordering temperature of 305 K provide unquestionable evidence for the existence of the magnetic and nonmagnetic classes of iron sites.

5. QUASICRYSTALS AND DISORDER

As mentioned in Section 1, before the discovery of stable quasicrystals, the available metastable samples contained significant amount of disorder and second phases which hindered the detection of properties which might be inherent to quasiperiodicity and led to significant confusion[8] in many areas of the physics of these alloys. Many stable quasicrystals are labelled "perfect," or "phason-free" because, from a diffraction point of view, their Bragg peaks are extremely sharp and their widths scale appropriately with the components of Q.[2-4,81] Yet, there is experimental evidence, which is summarized below, which indicates that these perfect quasicrystals contain some sort of disorder, which must then influence their physical properties.

The Mössbauer results reviewed above clearly indicate the presence of P(Δ) also in stable phason-free quasicrystals. Similar distributions are observed with other local probes, such as NMR and nuclear quadrupole resonance spectroscopy.[82] Such distributions of the hyperfine parameters can only occur if there is a chemical and/or topological disorder in the investigated samples.

The diffuse scattering is often observed in x-ray and electron diffraction patterns of high-quality quasicrystals.[83] Such scattering indicates that some disorder is present in the diffracting structure. A recent study on the propagation of acoustic shear waves in a single-grain *i*-Al-Pd-Mn alloy shows the similarity of the acoustic properties of this alloy to those of amorphous metals.[84] The success of quantum interference theories,[3-6,14] which were originally developed for disordered conductors, in accounting for the temperature and field dependencies of the electrical conductivity and magnetoresistance of many stable quasicrystals also indicates the importance of disorder. The presence of disorder may also explain why the spikiness of the density of states predicted by theory, see Fig. 2, is not observed even in the ultra-high resolution ultraviolet photoemission spectroscopy valence bands, see Figs. 3 to 6. The presence of two magnetic classes of transition metal sites may also be related to disorder.[61]

There seem to be at least two qualitative reasons for the presence of disorder in high-quality stable quasicrystals. First, the concept of quasiperiodicity, by its very nature, stipulates that no two atoms in a quasilattice have exactly the same environment, which can be viewed as a sort of topological disorder. This must be in part responsible for the observed distribution of the hyperfine parameters studied with local probes. It may also cause the quasicrystals to become electronically disordered alloys, and hence, account for the success of the quantum interference theories. Second, there are at present about two dozen stable quasicrystals, see references 11, 12, and 22, all of which are ternary alloys. Clearly, one expects some chemical disorder to be present in such ternaries and such disorder is expected to lead to a smoothing out of the sharp characteristics of their physical properties, for example, of the spikiness in their density of states near the Fermi level, and/or their broadening, for example, compare the full width at half maximum of the P(Δ) for *i*-$Al_{64}Cu_{24}Fe_{12}$ with that for *i*-$Al_{65}Cu_{20}Fe_{7.5}Ru_{7.5}$, see Figs. 9d and 10c. Thus it seems that disorder has to be included in any attempts to explain the unusual physical properties of quasicrystals.

6. CONCLUDING REMARKS

Two major themes have been discussed in this chapter. The first theme is related to the electronic structure of quasicrystals because it is the knowledge and understanding of this structure which eventually will enable us to fully understand the unusual properties of these complex materials. The second theme is associated with the contribution of the recent Mössbauer spectral studies to elucidating some aspects of the structure and magnetism of quasicrystals.

Although the nature of the surprising transport properties is still unsolved, it seems that they may result from the combination of the low density of states at the Fermi level, localization effects, and the presence of the critical states. The quasiperiodicity by itself does not seem to be essential for the occurrence of these effects as they are also observed in structurally complex crystalline approximants.

After the first few years of confusion associated with poor quality samples and the problematic interpretation of their spectra, Mössbauer spectroscopy has emerged as an important experimental technique in studies of structural and magnetic properties of quasicrystals. It was instrumental, among other things, in establishing the occurrence of two classes of transition metal sites in many quasicrystals. It also provides, when used critically, reliable information on the quadrupole interaction distributions. However, this information by itself is not sufficient to answer the questions related to the structure of quasicrystals. It must be compared with the theoretically predicted P(Δ)'s for possible structural models. Such a process is still in its infancy.

Magnetically ordered quasicrystals require in-field Mössbauer measurements to determine their type of ordering. Such measurements are also needed for diamagnetic and paramagnetic quasicrystals in order to impose further constraints on the derived hyperfine parameters from zero-field spectra, which is essential for using these parameters to discriminate between possible structural models. It is the belief of the author that Mössbauer spectroscopy is especially useful when used in combination with other experimental techniques, especially in the area of magnetism.

All Mössbauer spectral studies performed so far have been done with the iron-57 isotope. However, new quasicrystals with zinc and the rare-earth elements are now available. Mössbauer spectroscopy on other nuclei would undoubtedly provide valuable data on the local properties of these new quasicrystals.

ACKNOWLEDGEMENTS. This work was supported by the Natural Sciences and Engineering Research Council of Canada. The author is grateful to Professors A.-P. Tsai, A. Inoue, K. Edagawa, and S. Takeuchi for providing high-quality quasicrystalline samples and to Professor Y. Baer in whose laboratory the ultra-high-resolution photoemission experiments were carried out.

REFERENCES

1. D. Shechtman, I. Blech, D. Gratis, and J. W. Cahn, *Phys. Rev. Lett.* **53**, 1951 (1984).
2. A. I. Goldman and K. F. Kelton, *Rev. Mod. Phys.* **65**, 213 (1993); K. F. Kelton, *Int. Mat. Rev.* **38**, 105 (1993).
3. C. Janot, *Quasicrystals: A Primer*, Oxford University Press, Oxford (1992).
4. D. P. DiVincenzo and D. P. Steinhardt, eds., *Quasicrystals, The State of the Art*, World Scientific, Singapore (1991); A. I. Goldman and M. Widom, *Ann. Rev. Phys. Chem.* **42**, 685 (1991).
5. S. J. Poon, *Adv. Phys.* **41**, 303 (1992), and references therein.
6. Z. M. Stadnik and M. Akbari-Moghanjoughi, unpublished results.
7. Z. M. Stadnik, G. W. Zhang, M. Akbari-Moghanjoughi, A. P. Tsai, and A. Inoue, *Int. J. Mod. Phys.*, in press.
8. R. C. O'Handley, R. A. Dunlap, and M. E. McHenry, in *Handbook of Magnetic Materials*, K. H. J. Buschow, ed., Vol. 6, p. 453, Elsevier, Amsterdam (1991), and references therein.
9. Z. M. Stadnik and G. Stroink, *Phys. Rev. B* **43**, 894 (1991); **44**, 4255 (1991).
10. S. Nasu, M. Miglierini, and T. Kuwano, *Phys. Rev. B* **45**, 12778 (1992); M. Miglierini and S. Nasu, *J. Phys. Soc. Japan* **60**, 2135 (1991).
11. Z. M. Stadnik, G. W. Zhang, A. P. Tsai, and A. Inoue, *Phys. Rev. B* **51**, 4023 (1995), and references therein.
12. W. Steurer, *Mater. Sci. Forum* **150-151**, 15 (1994).
13. D. Mayou, C. Berger, F. Cyrot-Lackmann, T. Klein, and P. Lanco, *Phys. Rev. Lett.* **70**, 3915 (1993); F. S. Pierce, P. A. Bancel, B. D. Biggs, Q. Guo, and S. J. Poon, *Phys. Rev. B* **47**, 5670 (1993).
14. M. A. Chernikov, A. Bernasconi, C. Beeli, and H. R. Ott, *Europhys. Lett.* **21**, 767 (1993).
15. B. D. Biggs, S. J. Poon, and N. R. Munirathnam, *Phys. Rev. Lett.* **65**, 2700 (1990).
16. Y. Honda, K. Edagawa, A. Yoshioka, T. Hashimoto, and S. Takeuchi, *Japan J. Appl. Phys. A* **9**, 4929 (1994).

17. F. S. Pierce, S. J. Poon, and Q. Quo, *Science* **261**, 737 (1993); F. S. Pierce, Q. Quo, and S. J. Poon, *Phys. Rev. Lett.* **73**, 2220 (1994).
18. N. Mott, *Conduction in Non-Crystalline Materials*, Clarendon, Oxford (1993).
19. C. Berger, T. Grenet, P. Lindqvist, P. Lanco, J. C. Grieco, G. Fourcaudot, and F. Cyrot-Lackmann, *Solid State Commun.* **87**, 977 (1993).
20. S. Matsuo, T. Ishimasa, H. Nakano, and Y. Fukano, *J. Phys. F* **18**, L175 (1988); Z. M. Stadnik, G. Stroink, H. Ma, and G. Williams, *Phys. Rev. B* **39**, 9797 (1989), and references therein.
21. T. Klein, C. Berger, D. Mayou, and F. Cyrot-Lackmann, *Phys. Rev. Lett.* **66**, 2907 (1991); P. Lanco, T. Klein, C. Berger, F. Cyrot-Lackmann, G. Fourcaudot, and A. Sulpice, *Europhys. Lett.* **18**, 227 (1992); P. Lanco, C. Berger, F. Cyrot-Lackmann, and A. Sulpice, *J. Non-Cryst. Solids* **153-154**, 325 (1993).
22. Z. M. Stadnik, G. W. Zhang, A. P. Tsai, and A. Inoue, *Phys. Rev. B* **51**, 11358 (1995), and references therein.
23. P. A. Bancel and P. A. Heiney, *Phys. Rev. B* **33**, 7917 (1986); A. P. Tsai, A. Inoue, and T. Masumoto, *Mater. Trans., Japan Inst. Met.* **30**, 464 (1989); *ibid.* **30**, 666 (1989); A. Inoue, A. P. Tsai, and T. Masumoto, in *Quasicrystals*, T. Fujiwara and T. Ogawa, eds., p. 80, Springer-Verlag, Berlin (1990).
24. J. Friedel and F. Dénoyer, *C. R. Acad. Sci. Paris* **305**, 171 (1987); A. P. Smith and N. W. Ashcroft, *Phys. Rev. Lett.* **59**, 1365 (1987); V. G. Vaks, V. V. Kamyshenko, and G. D. Samolyuk, *Phys. Lett. A* **132**, 131 (1988); J. Friedel, *Helv. Phys. Acta* **61**, 538 (1988); *Philos. Mag. B* **65**, 1125 (1992).
25. K. Kimura, H. Iwahashi, T. Hashimoto, S. Takeuchi, U. Mizutani, S. Ohashi, and G. Itoh, *J. Phys. Soc. Japan* **58**, 2472 (1989); J. L. Wagner, B. D. Biggs, and S. J. Poon, *Phys. Rev. Lett.* **65**, 203 (1990).
26. F. Hippert, L. Kandel, Y. Calvayrac, and B. Dubois, *Phys. Rev. Lett.* **69**, 2086 (1992); E. A. Hill, T. C. Chang, Y. Wu, S. J. Poon, F. S. Pierce, and Z. M. Stadnik, *Phys. Rev. B* **49**, 8615 (1994).
27. S. E. Burko, T. Timusk, and N. W. Ashcroft, *J. Phys.: Condens. Matter* **4**, 9447 (1992); L. Degiorgi, M. A. Chernikov, C. Beeli, and H. R. Ott, *Solid State Commun.* **87**, 721 (1993); D. Macko and M. Kaspárková, *Philos. Mag. Lett.* **67**, 307 (1993); X. Wu, C. C. Homes, S. E. Burkov, T. Timusk, F. S. Pierce, S. J. Poon, S. L. Cooper, and M. A. Karlow, *J. Phys.: Condens. Matter* **5**, 5975 (1993).
28. M. A. Fradkin, *J. Phys.: Condens. Matter* **4**, 10497 (1992); A. E. Carlsson, *Phys. Rev. B* **47**, 2515 (1993).
29. M. Windisch, M. Krajcí, and J. Hafner, *J. Phys.: Condens. Matter* **6**, 6977 (1994), and references therein.
30. G. Trambly de Laissardière and T. Fujiwara, *Phys. Rev. B* **50**, 5999 (1994), and references therein.
31. M. Krajcí, M. Windisch, J. Hafner, G. Kresse, and M. Mihalkovic, *Phys. Rev. B* **51**, 17355 (1995).
32. Z. M. Stadnik, D. Purdie, M. Garnier, Y. Baer, A. P. Tsai, and A. Inoue, unpublished results.
33. Z. M. Stadnik, D. Purdie, M. Garnier, Y. Baer, K. Edagawa, and S. Takeuchi, unpublished results.
34. Z. M. Stadnik and G. Stroink, *Phys. Rev. B* **47**, 100 (1993).
35. M. Mori, S. Matsuo, T. Ishimasa, T. Matsuura, K. Kamiya, H. Inokuchi, and T. Matsukawa, *J. Phys.: Condens. Matter* **3**, 767 (1991).
36. M. Grioni, D. Malterre, and Y. Baer, *J. Low Temp. Phys.* **99**, 195 (1995).
37. U. Mizutani, Y. Sakabe, T. Shibuya, K. Kishi, K. Kimura, and S. Takeuchi, *J. Phys. Condens. Matter* **2**, 6169 (1990).
38. T. Takeuchi, Y. Yamada, U. Mizutani, Y. Honda, K. Edagawa, and S. Takeuchi, *Int. J. Mod. Phys.*, in press.
39. P. Lindqvist, C. Berger, T. Klein, P. Lanco, F. Cyrot-Lackmann, and Y. Calvayrac, *Phys. Rev. B* **48**, 630 (1993).
40. K. Wang, C. Scheidt, P. Garoche, and Y. Calvayrac, *J. Phys. (Paris)* **2**, 1553 (1992).

41. S. Takeuchi, *Mater. Sci. Forum* **150-151**, 35 (1994).
42. J. C. Phillips, *Phys. Rev. B* **47**, 7747 (1993); *ibid* **47**, 2522 (1993); *Solid State Commun.* **83**, 379 (1992); J. C. Phillips and K. M. Rabe, *Phys. Rev. Lett.* **66**, 923 (1991).
43. T. Klein and O. G. Symko, *Phys. Rev. Lett.* **73**, 2248 (1994).
44. K. Kimura and S. Takeuchi, in *Quasicrystals, The State of the Art* (Ref. 1), p. 313; K. Kimura, K. Kishi, T. Hashimoto, and S. Takeuchi, in *Quasicrystals*, K. H. Kuo and T. Ninomiya, eds., p. 233, World Scientific, Singapore (1991).
45. D. N. Basov, F. S. Pierce, P. Volkov, S. J. Poon, and T. Timusk, *Phys. Rev. Lett.* **73**, 1865 (1994).
46. H. Tsunetsugu, T. Fujiwara, K. Ueda, and T. Tokihiro, *J. Phys. Soc. Japan* **55**, 1420 (1986); *Phys. Rev. B* **43**, 8879 (1991); M. Kohmoto, B. Sutherland, and Ch. Tang, *ibid.* **35**, 1020 (1987); L. S. Levitov, *Europhys. Lett.* **7**, 343 (1988); K. Niizeki and T. Akamatsu, *J. Phys.: Condens. Matter* **2**, 2759 (1990); H. Tsunetsugu and K. Ueda, *Phys. Rev. B* **43**, 8892 (1991); B. Passaro, C. Sire, and V. G. Benza, *ibid.* **46**, 13751 (1992); C. Sire, B. Passaro, and V. Benza, *J. Non-Cryst. Solids* **153-154**, 420 (1993).
47. S. J. Poon, F. S. Pierce, and Q. Quo, *Phys. Rev. B* **51**, 2777 (1995).
48. N. P. Lalla, R. S. Tiwari, and O. N. Srivastava, *J. Phys.: Condens. Matter* **7**, 2409 (1995).
49. Z. M. Stadnik, *Hyp. Interact.* **90**, 215 (1994), and references therein.
50. P. Kramer, A. Quandt, M. Schlottmann, and T. Schneider, *Phys. Rev. B* **51**, 8815 (1995).
51. S. J. Campbell and F. Aubertin, in *Mössbauer Spectroscopy Applied to Inorganic Chemistry*, G. J. Long and F. Grandjean, eds., Vol. 3, p. 183, Plenum, New York, (1989); G. Le Caër and R. A. Brand, *Hyp. Interact.* **71**, 1507 (1992); R. E. Vandenberghe, E. De Grave, and P. M. A. de Bakker, *ibid.* **83**, 29 (1994).
52. D. G. Rancourt and J. Y. Ping, *Nucl. Instrum. Meth. Phys. Res. B* **58**, 85 (1991).
53. G. Le Caër and J. M. Dubois, *J. Phys. E* **12**, 1083 (1979).
54. J. Hesse and A. Rübartsch, *J. Phys. E* **7**, 526 (1974).
55. Z. M. Stadnik, A. P. Tsai, and A. Inoue, unpublished results.
56. Z. M. Stadnik, K. Edagawa, and S. Takeuchi, unpublished results.
57. A. Inoue, H. M. Kimura, T. Masumoto, A. P. Tsai, and Y. Bizen, *J. Mater. Sci. Lett.* **6**, 771 (1987); H. M. Kimura, A. Inoue, Y. Bizen, T. Masumoto, and H. S. Chen, *Mater. Sci. Eng.* **99**, 449 (1988).
58. C. L. Chien and K. M. Unruh, *Phys. Rev. B* **29**, 207 (1984); M. Maurer, J. M. Friedt, and J. P. Sanchez, *J. Phys. F* **15**, 1449 (1985); Z. M. Stadnik, F. Müller, G. Stroink, and M. Rosenberg, *J. Non-Cryst. Solids* **156-157**, 319 (1993).
59. S. Garçon, P. Sainfort, G. Regazzoni, and J. M. Dubois, *Scripta Met.* **21**, 1493 (1987); A. P. Tsai, A. Inoue, and T. Masumoto, *Japan J. Appl. Phys.* **26**, L1994 (1987).
60. N. Kataoka, A. P. Tsai, A. Inoue, T. Masumoto, and Y. Nakamura, *Japan J. Appl. Phys.* **27**, L1125 (1988); C. L. Chien and M. Lu, *Phys. Rev. B* **45**, 12793 (1992); M. Lu and C. L. Chien, *Hyp. Interact.* **71**, 1525 (1992).
61. R. A. Brand, G. Le Caër, and J. M. Dubois, *J. Phys.: Condens. Matter* **2**, 6413 (1990); *Hyp. Interact.* **55**, 903 (1990); Z. M. Stadnik, G. Stroink, G. Lamarche, and A. Inoue, *J. Phys. Soc. Japan* **60**, 3829 (1991); Z. M. Stadnik and F. Müller, *Philos. Mag. B* **71**, 221 (1995).
62. J. Teillet and B. Bouchet-Fabre, *Hyp. Interact.* **55**, 1077 (1990).
63. A. P. Tsai, A. Inoue, and T. Masumoto, *Mater. Trans., Japan Inst. Met.* **34**, 155 (1993).
64. M. Eibschütz, M. E. Lines, H. S. Chen, and F. A. Thiel, *Phys. Rev. B* **46**, 491 (1992).
65. Z. M. Stadnik, A. P. Tsai, and A. Inoue, in *Aperiodic '94, Proc. Int. Conf. Aperiodic Crystals*, G. Chapuis and W. Paciorek, eds., p. 259, World Scientific, Singapore (1995).
66. A. P. Tsai, A. Inoue, and T. Masumoto, *Philos. Mag. Lett.* **64**, 163 (1991); A. P.

Tsai, T. Masumoto, and A. Yamamoto, *ibid.* **66**, 203 (1992).
67. A. P. Tsai, A. Inoue, and T. Masumoto, *Mater. Trans., Japan Inst. Met.* **30**, 150 (1989); A. P. Tsai, N. Kataoka, A. Inoue, and T. Masumoto, *Japan J. Appl. Phys.* **29**, L1696 (1990).
68. S. E. Burkov, *Phys. Rev. B* **67**, 12325 (1993).
69. H. Teuscher, private communication.
70. S. G. Song and E. R. Ryba, *Philos. Mag. B* **69**, 707 (1994).
71. M. Maurer and J. M. Friedt, *Hyp. Interact.* **27**, 135 (1986), and references therein.
72. M. Eibschütz, M. E. Lines, H. S. Chien, and J. V. Waszczak, *Phys. Rev. B* **38**, 10038 (1988); M. Eibschütz, M. E. Lines, H. S. Chien, J. V. Waszczak, G. P Espinosa, and A. S. Cooper, *ibid.* **41**, 4606 (1990).
73. K. Edagawa, H. Ino, S. Nasu, K. Kimura, S. Takeuchi, T. Shinjo, K. Koga, T. Shimizu, and H. Yasuoka, *J. Phys. Soc. Japan* **56**, 2629 (1987); H. Ino, K. Edagawa, K. Kimura, S. Takeuchi, and S. Nasu, *Mater. Sci. Forum* **22-24**, 437 (1987).
74. M. Eibschütz, M. E. Lines, H. S. Chien, J. V. Waszczak, G. Papaefthymiou, and R. B. Frankel, *Phys. Rev. Lett.* **59**, 2443 (1987); *J. Appl. Phys.* **63**, 4063 (1988).
75. A. E. Carlsson, *Int. J. Mod. Phys. B* **7**, 293 (1993).
76. Z. M. Stadnik and G. Stroink, *Hyp. Interact.* **69**, 643 (1991).
77. Y. Yokoyama, A. Inoue, and T. Masumoto, *Mater. Trans., Japan Inst. Met.* **33**, 1012 (1992); *Mater. Sci. Eng. A* **181-182**, 734 (1994).
78. T. Shinohara, Y. Yokoyama, M. Sato, A. Inoue, and T. Masumoto, *J. Phys. Condens. Matter* **5**, 3673 (1993); *Mater. Sci. Eng. A* **181-182**, 798 (1994).
79. Z. M. Stadnik, G. W. Zhang, Y. Yokoyama, and A. Inoue, unpublished results.
80. C. R. Lin, C. M. Lin, S. T. Lin, and I. S. Lyubutin, *Phys. Lett. A* **196**, 365 (1995).
81. C. A. Guryan, A. I. Goldman, P. W. Stephens, K. Hiraga, A. P. Tsai, A. Inoue, and T. Masumoto, *Phys. Rev. Lett.* **62**, 2409 (1989); A. P. Tsai, H. S. Chen, A. Inoue, and T. Masumoto, *Phys. Rev. B* **43**, 8782 (1991).
82. A. Shastri, F. Borsa, D. R. Torgeson, and A. I. Goldman, *Phys. Rev. B* **50**, 4224 (1994); A. Shastri, F. Borsa, D. R. Torgeson, J. E. Shield, and A. I. Goldman, *ibid.* **50**, 15651 (1994).
83. K. Hradil, T. Proffen, F. Frey, K. Eichhorn, and S. Kek, *Philos. Mag. Lett.* **71**, 199 (1995); L. E. Levine and J. L. Libbert, *ibid.* **71**, 213 (1995).
84. N. Vernier, G. Bellessa, B. Perrin, A. Zarembovitch, and M. De Boissieu, *Europhys. Lett.* **22**, 187 (1993).

7

The Structure of Industrial Glasses Probed by Mössbauer Spectroscopy

C. E. Johnson, K. F. E. Williams, and J. A. Johnson

1. INTRODUCTION

Glassy materials, whether man-made or naturally occurring, have been recognised for centuries though attempts to understand their nature are relatively recent.[1] They are produced by rapidly cooling the material after melting and are non-crystalline. Only certain materials may readily form glasses. Unlike crystalline materials, where there is a periodic structure which may be determined by diffraction techniques, there is no direct way of determining the arrangement of atoms in glasses. Their diffraction patterns consist of diffuse rings from which the radial distribution function of atoms may be deduced. Indeed unlike crystals, which have a unique three-dimensional structure which depends on their composition, temperature, and pressure, a glass does not have a unique structure but it depends upon the conditions under which it was formed, principally the rate at which it was cooled through the glass transition point. The free energy of the material is higher in the glassy, disordered, state than in the crystalline, ordered, state. Hence thermodynamically the crystal is the stable equilibrium state. However, a glass has strong directional covalent bonds which must be broken in order to be converted to a crystal which is more ionic, hence the time taken for a glass to reach its equilibrium crystalline state may be infinite for all practical purposes.

Figure 1 shows a comparison of the structures, in two dimensions, of a crystalline and glassy form of a M_2O_3 oxide and a glass with a sodium modifier. Notice that there is short-range order in the MO_3 units in both the glass and the crystal, but long-range order exists only in the crystal. The glass transition is illustrated in Fig. 2 which shows the volume plotted against temperature. For a crystal, solidification is a first-order change with a sharp discontinuity in the volume, whereas, when forming a glass, the material may be supercooled below the melting point and the glass transition is a second-order change with a discontinuity in the slope of the curve but without a discontinuity in the volume. The glass transition temperature may be determined by differential thermal analysis measurements.

C. E. Johnson and K. F. E. Williams • Department of Physics, University of Liverpool, Liverpool L69 3BX, UK. *J. A. Johnson* • School of the Built Environment, Liverpool John Moores University, Liverpool, UK.

Mössbauer Spectroscopy Applied to Magnetism and Materials Science, Volume 2, edited by Gary J. Long and Fernande Grandjean. Plenum Press, New York, 1996.

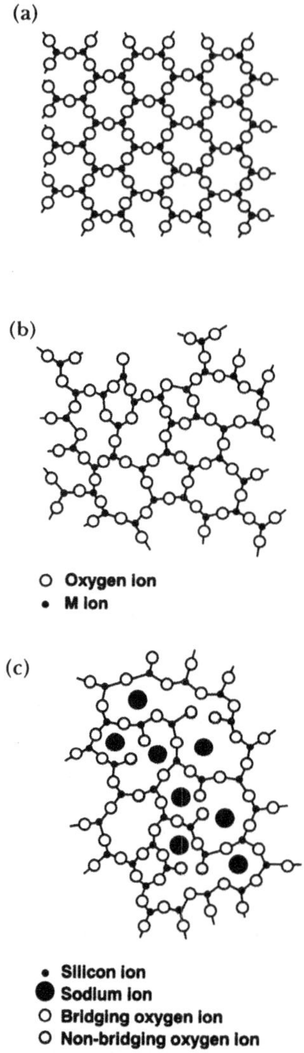

Figure 1, left. A schematic of the two-dimensional structure of M_2O_3 in a crystalline, upper, and a non-crystalline form, centre, and the positions of network modifier ions and their associated non-bridging oxygens, lower.

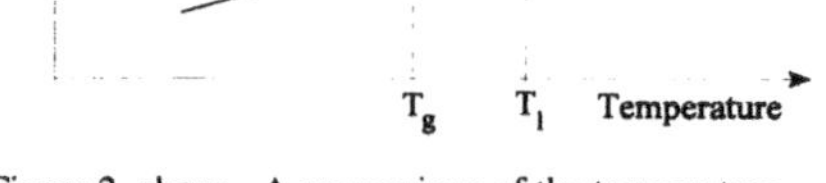

Figure 2, above. A comparison of the temperature variation of the volume of a glass and crystal. T_l is the melting point and T_g is the glass transition point.

The oxides which form glasses, i.e., the "network formers," include SiO_2, B_2O_3, GeO_2, P_2O_5, and As_2O_3, and are characterised by strong covalent bonding. Some other oxides, e.g., Al_2O_3, are not network formers by themselves but nevertheless they do not necessarily weaken the bonding in the network and are known as "intermediates." Finally the strongly ionic oxides, e.g., Na_2O, K_2O, and CaO, do not form glasses and weaken the bonding when incorporated in a glass network; they are known as "network modifiers."

The industrial uses of glasses are enormous and widespread, the most obvious being the transparent glass used for windows, bottles, and decorative articles. Flat glass is a soda-lime-silicate glass and has a large volume of industrial production. High quality glass is needed for optical instruments. Glass used for making bottles is of a lesser quality and has an SnO_2 film on the surface for protective purposes. Borosilicate or pyrex glasses have a low expansivity and are used to make fireproof ovenware for cooking. But there are many non-transparent glasses which have industrial applications in "high tech" devices because of their special electronic properties, e.g., phosphate glasses, which are semiconductors. Also there are glass ceramics - partially crystallised materials - which can be tailored to have the mechanical or thermal properties required for commercial applications. In all of these materials a knowledge of the structure is important for development and advances in technology, and the Mössbauer effect has played a part in providing such knowledge. None of the major constituents of glass is a Mössbauer isotope, but tin is an important ingredient in commercial surface coatings and glazes, and diffuses into the surface in the float glass process, and iron is added to make tinted glasses so there have been many studies using iron-57 and tin-119.

The glasses which will be discussed in this chapter are silica based glasses, which are transparent to visible light. Silica, SiO_2, in its crystalline state contains silicon tetrahedrally coordinated to oxygen. In vitreous silica the SiO_4 tetrahedra are still present and each oxygen bridges two such tetrahedra, but neighbouring tetrahedra are no longer oriented in an ordered way with respect to each other, see Fig. 1b for the analogous two-dimensional structure.

Pure SiO_2 has a high melting point, T_m, of 2003 K and glass transition temperature, T_g, of 1453 K. In flat glass manufacture several other oxides, mainly network modifiers, are added to the SiO_2 in order to reduce these temperatures and to improve the properties of the resulting glass. Usually Na_2O is added to lower the melting point and then CaO to increase the durability. The effect of adding small amounts of Na_2O is that the Na^+ ions fit into the interstitial sites with approximate octahedral coordination in the network and, to keep the charge neutral, results in two non-bridging oxygen ions per sodium atom thus breaking the network structure, see Fig. 1b. If too many sodium ions are added then crystallisation will occur. Another additive is Al_2O_3 which is an intermediate, i.e., it may either substitute for silicon with tetrahedral coordination, or may enter interstitially with octahedral coordination.

It is not possible to predict in all cases whether the added atoms will behave as network formers, i.e., substitutional, tetrahedrally coordinated, covalently bonded atoms, or as modifiers, i.e., interstitial, octahedrally coordinated, ionically bonded atoms, or as intermediates. Because it is very important for glass manufacturers as well as research scientists to have this information, spectroscopic techniques, such as Mössbauer, nuclear magnetic resonance, and nuclear quadrupole resonance spectroscopy, make a valuable contribution.

2. MÖSSBAUER SPECTROSCOPY

The hyperfine interactions which affect a Mössbauer spectrum have several features which can provide information on the state of the atoms. The isomer shift and, to a lesser extent the quadrupole splitting, can determine the ionisation state and the degree of covalency. The probablility of the Mössbauer effect, the *f*-factor, determines the strength of the interatomic forces. All of this information can be used to infer the coordination number of the Mössbauer atom. Because the Mössbauer nuclide is a nuclear probe and sees the properties of the atomic wave function through the hyperfine interactions, it can usually distinguish different oxidation states of the same element more easily than atomic spectroscopies.

A very large number of measurements have now been made of the Mössbauer spectra of glasses and an extensive literature now exists on the subject. The results have been summarised and references given in review articles by Kurkjian,[2] Coey,[3] Müller-Warmuth and Eckert,[4] Tomandl,[5] Nishida,[6] and others. We now consider the main features of Mössbauer spectra and how they may be used in the study of glasses.

2. 1. Linewidth

The lines in the Mössbauer spectra of glasses are broad, as expected for a disordered material, in which the atoms are in different environments, each with slightly different isomer shifts and quadrupole splittings. This has been noted since the first observation of a Mössbauer spectrum of iron-57 in an amorphous material, fused quartz, by Pollack *et al.*[7] This measurement also showed that the Mössbauer effect is not restricted to crystals, although the theory of the effect is based on phonons in a crystalline lattice.

2. 2. The Centre Shift

The position of the centroid of a Mössbauer spectrum depends upon the temperature through the second-order Doppler shift, and on the chemical state of the Mössbauer nucleus via the isomer shift, which is a measure of the s-electron density at the nucleus.

2. 2. 1. The Second Order Doppler Shift

The second order Doppler shift, which arises from the motion of the Mössbauer nucleus, is given, according to the Debye model, by

$$\delta^{SOD} = -\frac{9k\Theta_D}{16\,Mc}\left[1 + 8\left(\frac{T}{\Theta_D}\right)^4 \int_0^{\Theta_D/T} \frac{x^3\,dx}{e^x - 1}\right],$$

where Θ_D is the Debye temperature and T is the temperature. At high temperatures, i.e., for $T > \Theta_D$, the δ^{SOD} is independent of Θ_D and is linear with temperature according to

$$d\delta^{SOD}/dT = -3k/2Mc.$$

Any difference between this and $d\delta/dT$ must arise from the temperature dependence of the isomer shift, $d\delta^I/dT$. This could arise from changes in volume or from the excitation of higher electronic, or chemical, states and has been observed in binary SnO-SiO_2 glasses.[8] At low temperatures, i.e., for $T < \Theta_D$, the data could be used in principle to estimate Θ_D, but not as accurately as from measurements of the *f*-factor, see Section 2. 3.

2. 2. 2. The Isomer Shift

The isomer shift gives a measure of the s-electron density at the Mössbauer nuclei and is given by,

$$\delta^I = (2/3)\pi Ze^2\Delta\langle R^2\rangle\Delta\rho_s(0),$$

where $\Delta\langle R^2\rangle$ is the change in the mean square nuclear radius in the Mössbauer transition, $\Delta\rho_s(0)$ is the difference in electron densities at the Mössbauer nuclei in the source and absorber, and Ze is the nuclear charge. Hence the isomer shift is sensitive to the covalency of the Mössbauer atom and may be used to assign coordination number of the atom. Thus it is an important parameter in determining the local structure in a glass and has to be calibrated for each Mössbauer isotope. Although the effects of covalency and coordination number on the isomer shift are well known to Mössbauer spectroscopists, it seems worth outlining here the main features of the isomer shift for the two most frequently used isotopes, iron-57 and tin-119.

For iron-57 the nucleus has a smaller radius in the excited state than in the ground state, i.e., $\Delta\langle R^2\rangle$ is negative. Thus an increase in shift corresponds to a decrease in s-electron density. The d-electrons screen the nucleus from the s-electrons so that an increase in shift can also be an indication of an increase of the number of d-electrons on the atom.[9] Ionic Fe^{3+} and Fe^{2+} have the configurations $3d^5$ and $3d^6$, respectively, and hence Fe^{2+} has a larger shift than Fe^{3+}. The presence of covalent bonding, or delocalisation of the d-electrons, for both Fe^{3+} and Fe^{2+}, is indicated by a reduction in isomer shift. The isomer shift is also smaller for tetrahedral than for octahedral coordination. These conclusions are illustrated in Table 1, which gives some isomer shift data, relative to metallic iron at room temperature, for several compounds of iron.

The behaviour of tin-119 differs from that of iron-57 in two respects. First, $<\Delta R^2>$ is positive, which is the situation for most nuclei and, second, the covalent bonding involves s- or p-electrons. Because $\Delta<R^2>$ is positive, δ^I increases with $\rho_s(0)$. Table 2 lists the results of isomer shift measurements, relative to $CaSnO_3$ at room temperature, on some inorganic compounds of tin. Ionic Sn^{2+} and Sn^{4+} have the $5s^2$ and $5s^0$ configurations, respectively, so that Sn^{2+} has the more positive shift. The effect of covalency has been discussed by both Flinn[10] and by Parish.[11]

Covalent Sn^{4+} has electrons in the 5s shell and the isomer shift increases with increasing covalency. Also, four coordinate Sn^{4+} has a more positive shift than six coordinate Sn^{4+}, as can be seen from Table 2. The stannic tetrahalides have the anions tetrahedrally coordinated to the tin and the isomer shift is more positive than for the corresponding octahedrally coordinated ion.

For Sn^{2+} the situation is more complex because of the non-bonding pair of 5s electrons and the tendency in the solid state to polymerise. However, the effect of covalency is to reduce the number of s-electrons, i.e., to decrease the shift, as is illustrated in Table 2, where it is seen that SnI_2 has a more negative shift than $SnCl_2$. The isomer shift for tetrahedrally coordinated Sn^{2+} is more negative than that for octahedrally coordinated Sn^{2+}. Thus these effects go in the opposite direction to those in Sn^{4+}.

2. 3. Quadrupole Splitting

Both iron-57 and tin-119 have $3/2 \rightarrow 1/2$ γ-ray transitions, and hence a quadrupole doublet is observed in a non-magnetic Mössbauer spectrum. This is a measure of the asymmetry of the electronic charge distribution on the atom. It is zero for cubic symmetry and is small, typically 0 to 1 mms^{-1} for Fe^{3+} and Sn^{4+}, which have an almost spherical charge distribution. It is generally large, typically 2 to 3 mms^{-1} for Fe^{2+} and Sn^{2+}, which have asymmetric electron wave functions.

Table 1. Effect of Covalency and Coordination Number on the Iron-57 Isomer Shift

Compound	Symmetry	Oxidation State	δ,[a] mms^{-1}
$FeCl_4^-$	tetrahedral	iron(III)	0.30
$FeBr_4^-$	tetrahedral	iron(III)	0.36
Fe_2O_3	octahedral	iron(III)	0.50
$FeCl_3$	octahedral	iron(III)	0.53
$FeBr_3$	octahedral	iron(III)	0.55
$FeBaSi_4O_{19}$	tetrahedral	iron(II)	0.87
$FeCl_4^{2-}$	tetrahedral	iron(II)	1.01
$FeBr_4^{2-}$	tetrahedral	iron(II)	1.02
$FeCl_2$	octahedral	iron(II)	1.20
$FeBr_2$	octahedral	iron(II)	1.22
$FeSiF_6{\cdot}6H_2O$	octahedral	iron(II)	1.42

[a]Values obtained at 77 K relative to iron metal at room temperature.

Table 2. Effect of Covalency and Coordination Number on the Tin-119 Isomer Shift

Compound	Symmetry	Oxidation State	δ,[a] mms^{-1}
$SnCl_6^{2-}$	octahedral	tin(IV)	0.51
$SnBr_6^{2-}$	octahedral	tin(IV)	0.84
SnI_6^{2-}	octahedral	tin(IV)	1.24
$SnCl_4$	tetrahedral	tin(IV)	0.80
$SnBr_4$	tetrahedral	tin(IV)	1.10
SnI_4	tetrahedral	tin(IV)	1.45
$SnCl_2$	tetrahedral	tin(II)	4.15
$SnBr_2$	tetrahedral	tin(II)	4.06
SnI_2	tetrahedral	tin(II)	3.98

[a]Values obtained at 77 K relative to $CaSnO_3$ at room temperature.

Although the quadrupole splitting does not give an unambiguous indication of covalency or coordination number, it can provide a useful confirmation of the chemical state of an atom deduced from the isomer shift. In the case of Sn^{2+} it would be zero in the absence of covalency, so its value increases with the number of p-electrons in the chemical bond. Because the number of s-electrons decreases with covalency, there would be a negative correlation between the magnitudes of the isomer shift and quadrupole splitting.[9]

2. 4. The f-factor

The area of a Mössbauer spectrum is governed by the fraction of the γ-rays which are emitted or absorbed without energy loss, the so-called Debye-Waller factor or *f*-factor. It is a measure of the strength of the atomic bonding, and it is usual to analyse the data in terms of either the Einstein model of a solid, which assumes that all the atoms vibrate with the same frequency, or the Debye model, which assumes a simplified frequency distribution that is parabolic up to a maximum cut-off frequency. Neither model is strictly applicable, but they are both convenient for comparing the bonding strength in different materials.

According to the Debye model, f is given by,

$$f = \exp\left\{-\frac{3\,E_R}{2\,k\Theta_D}\left[1 + 4\left(\frac{T}{\Theta_D}\right)^2 \int_0^{\Theta_D/T} \frac{x\,dx}{e^x - 1}\right]\right\},$$

where E_R is the recoil energy of the nuclide. In the high temperature, $T > \Theta_D$, limit, ln f is proportional to temperature according to

$$\ln f = -\frac{6\,E_R T}{k\Theta_D^2}.$$

Hence a plot of ln f against temperature yields a straight line at high temperatures with a slope of $-\,6E_R/k\Theta_D^2$ and may be used to determine Θ_D.

A measurement of Θ_D may be used to assign the coordination number of the atom in a glass because, in tetrahedral coordination, the network formers, the atoms which replace silicon in strongly covalent bonds, have a high Θ_D. In contrast, the octahedrally coordinated atoms, the network modifiers, are interstitial and hence are less strongly bonded, and thus have a low Θ_D.

A further potential source for structural information lies in the anisotropy of the *f*-factor, or the Goldanskii-Karyagin effect.[12] This shows up as a difference in area of the lines of the

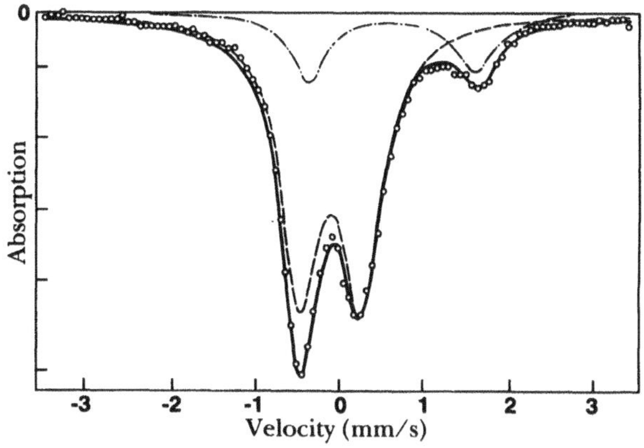

Figure 3. The iron-57 Mössbauer spectrum of an alkali-silica-iron glass showing the presence of both Fe^{2+} and Fe^{3+}. Figure obtained from reference 13.

quadrupole split spectrum, but the effect is generally small and it is not easy to establish its presence unambiguously, especially in glasses.

3. *IRON IN SILICATE GLASSES*

Most glass contains Fe_2O_3, which is present as an impurity in the silica used in glass fabrication. It sometimes produces a green coloration, but generally it is present in such low concentrations that it does not affect the glass properties adversely. It is frequently added in larger amounts in order to colour the glass. Iron in glass usually occurs as a mixture of Fe^{2+} and Fe^{3+} ions which are coordinated to oxygen. A large number of Mössbauer studies have been made on iron in silicate glasses. Fig. 3 shows the spectrum of an alkali-silica-iron glass.[13] The measurements are made mainly to determine the proportion in each oxidation state and the coordination number. Both of these may be fairly readily determined from the isomer shift, see Table 1, and confirmation of the coordination number and the role of the iron atoms in the glass, from the values of Θ_D, which may be estimated from measurements of the *f*-factor.

Despite the large number of measurements the structure and role of iron in glass are still not fully understood. Isomer shift data of Frischat and Tomandl[13] suggest that, in low concentrations, the Fe^{3+} ions in alkali silicate glasses are four-coordinate and so appear to substitute for silicon, whereas Fe^{2+} is six-coordinate and occupies interstitial sites. For higher Fe^{3+} concentrations the number of Na^+ network modifiers is not sufficient to hold the Fe^{3+} in tetrahedral coordination and an increasing amount goes into octahedral coordination. These conclusions seem to be confirmed by optical and electron spin resonance spectroscopy.[14] However, there is no general rule, and in other glasses, e.g., phosphate glasses, iron always has octahedral coordination.[6]

Nishida[6] has obtained structural information on a wide variety of glasses, such as borate, phosphate, vanadate, tellurate, and germanate, as well as silicate glasses, by using small amounts of iron-57 or tin-119 as a probe for determining the coordination number of the glass former. He suggests that, when $\Theta_D > 280$ K for Fe^{3+} or Sn^{4+}, these ions are network formers, whereas if $\Theta_D < 270$ K they are network modifiers.

At low temperatures glasses which contain very small amounts of Fe^{3+} may show magnetic hyperfine splitting arising from slow electron spin relaxation.[13]

4. TIN IN FLOAT GLASS

4. 1. The Float Glass Process

Nearly all the flat glass manufactured today is made by the float glass process. In this process a ribbon of molten glass at 1100°C is formed by passing over a bath of molten tin in a controlled, reducing atmosphere, as is shown schematically in Fig. 4. A continuous ribbon is produced which is cooled to 600°C on the tin bath. In this way the glass is produced with the same high degree of flatness as is found in the surface of the liquid tin. The temperature is carefully controlled so that the irregularities disappear and the surfaces of the glass become flat and parallel under the influence of the forces of gravity and surface tension. Tin metal is the most suitable support material to use owing to its appropriate density, low vapour pressure, small interaction with the glass, and its availability and relatively low cost. Prior to the invention of this process by Pilkington,[15] high quality flat glass could only be produced by grinding plate glass, a very expensive operation.

Float glass is an alkali-lime-silicate glass with a typical composition, in mole percent of SiO_2 (73%), Na_2O (11%), CaO (8%), MgO (6%), Al_2O_3 (0.5%), K_2O (0.3%) with smaller amounts of other elements including 0.05 percent iron.

Although the interaction between the tin and glass is small, tin does diffuse into the first few microns of the lower surface of the glass where it is in contact with the molten metal. A much smaller amount of tin is also found in the upper surface, presumably having been carried there through the atmosphere. It is important to know the chemical state of the tin and the concentration profile in order to be able to attempt to minimise the undesirable properties which it confers on the glass. One such property, known as "bloom," is a hazy appearance which arises from the presence of both Sn^{2+} and Sn^{4+} in the surface layer. This is important in applications where the glass has to be subsequently processed, e.g., toughened by thermal or chemical treatment, because the Sn^{2+} is then oxidised to Sn^{4+}, and the surface becomes wrinkled as a result of the different sizes of the ions, impairing the optical quality of the glass. Tempering is carried out as a toughening process and also to produce curved sheets as, for example, for car windshields. The glass is heated at 600°C or above and is then quenched placing the surface under compressive stress and the internal region under tension. This toughens the glass because it is strong under compression. It is, therefore, important for the glass to contain as small amount of tin as possible. To check if the tin concentration is too high for thermal treatment to be successful a bloom test is performed. A sample is heated at the toughening temperature for a few minutes and then the surface is viewed to check for any signs of microscopic wrinkling.

The float glass process has been in use for over 30 years. During this time considerable improvements have been made to the process by extensive monitoring which has led to increased understanding of the process. The detailed mechanism by which tin diffuses into the bottom surface is not known. Also the immediate surface structure and properties have not been determined. It is known that, if there is oxygen contamination in the molten tin, then

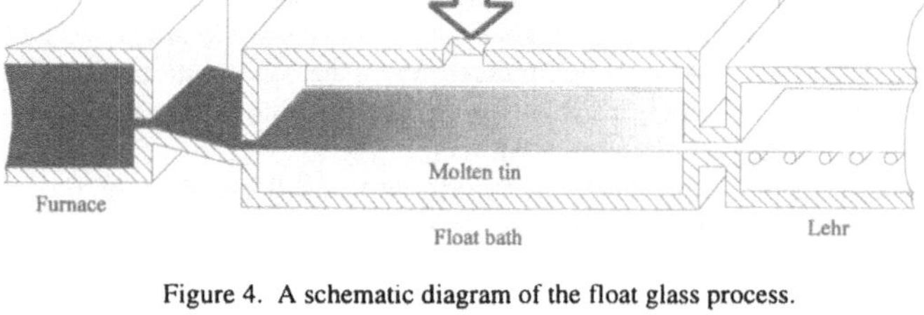

Figure 4. A schematic diagram of the float glass process.

there is increased tin up-take at the glass-metal interface. A fuller understanding of these aspects of the float glass process could lead to reduced tin losses and to reduced product loss during thermal toughening.

Because there is a reducing atmosphere in the float glass process, it is presumed that the tin enters the glass as Sn^{2+}, but it could be quickly oxidised as it diffuses into the bulk of the glass. Analytical methods, such as x-ray photoelectron spectroscopy and extended x-ray absorption fine structure studies, for examining the surface of float glass, have given contradictory results on the oxidation state of the tin present. It is fortunate, therefore, that tin-119 is one of the easier Mössbauer isotopes to work with, but there is considerable difficulty in applying it because of the small amount of tin in the surface layer. An early Mössbauer measurement[16] on powdered surface samples suggested that both Sn^{2+} and Sn^{4+} were present, but only Sn^{4+} was detected in the x-ray photoelectron spectrum of the same sample. This illustrates the point made earlier that the hyperfine interactions of different oxidation states usually show greater differences than the electronic properties.

4. 2. Mössbauer Spectral Results

4. 2. 1. Tin Oxides

In glass the tin is coordinated to oxygen, so it is useful to compare the Mössbauer spectral results with that of the oxides. Crystalline SnO_2 has the rutile structure, with tin in octahedral coordination, and has an isomer shift of -0.02 mms^{-1} and a quadrupole splitting of 0.50 mms^{-1}. The stable, black, form of crystalline SnO has a tetragonal structure with four-coordinate tin forming SnO_4 pyramids and having an isomer shift of 2.65 mms^{-1} and a quadrupole splitting of 1.36 mms^{-1} at 300 K. Collins *et al.* [17] have made measurements on the amorphous oxides and found at 300 K higher shifts and quadrupole splittings of 0.13 and 0.70 mms^{-1} for amorphous SnO_2 and 2.87 and 1.71 mms^{-1} for amorphous SnO.

4. 2. 2. Tin in Alkali Silicate Glasses

Mössbauer studies of tin in alkali-silicate glasses have been made by many workers. Mitrofanov and Sidorov[18] compared silicate and borate glasses and investigated the effect of different alkali modifiers. They concluded that the tin had six-fold coordination. Dannheim *et al.*[19] found evidence from their isomer shift measurements that, at low tin concentrations, both Sn^{2+} and Sn^{4+} were four-coordinate, but as the concentrations were increased there was a transition to six-coordination. Bartenev *et al.*[20] compared the temperature dependence of the *f*-factor in alkali-free and sodium silicate glasses containing tin, and found that the effect of the presence of the sodium was to increase the value of Θ_D for both Sn^{2+} and Sn^{4+}. Their spectra showed an anisotropy in the *f*-factor in the alkali-free glasses which disappeared when sodium was present. They deduced that Sn^{2+} had a tetrahedral oxygen environment in alkali-free glass which became six-coordinate in the presence of sodium modifiers, whereas Sn^{4+} was six-coordinate in all the glasses.

4. 2. 3. Tin in Float Glass

The first successful measurement of the tin-119 Mössbauer spectrum of the surface of a commercial float glass specimen was made by Principi *et al*.[21] who used conversion electron Mössbauer spectroscopy, which is especially sensitive to the surface layer. The resulting spectra are shown in Figs. 5 and 6. The two oxidation states were clearly distinguished, Sn^{2+} being the major constituent. For both these states the chemical shift and quadrupole splitting were different from that of the corresponding amorphous oxides, which in turn were higher than for the crystalline oxides, as is shown in Table 3. The exact proportion of each oxidation state present could not be determined from a single room temperature measurement because

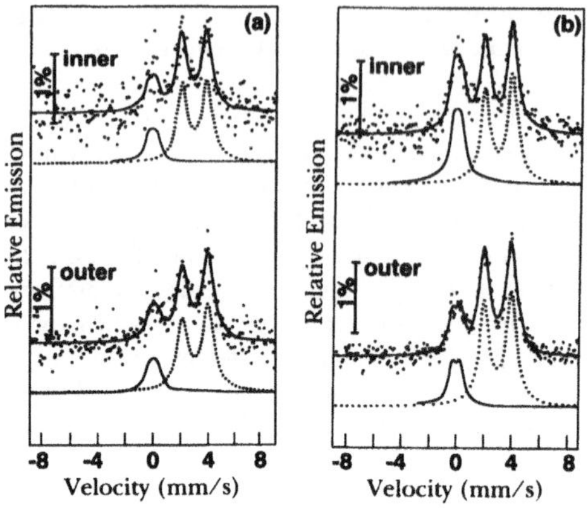

Figure 5. The room temperature conversion electron Mössbauer spectra of tin-119 in float glass taken from different parts of the plant. Reprinted from reference 21, with kind permission from Elsevier Science-NL, Sara Burgerhartstraat 25, 1055 KV Amsterdam, The Netherlands

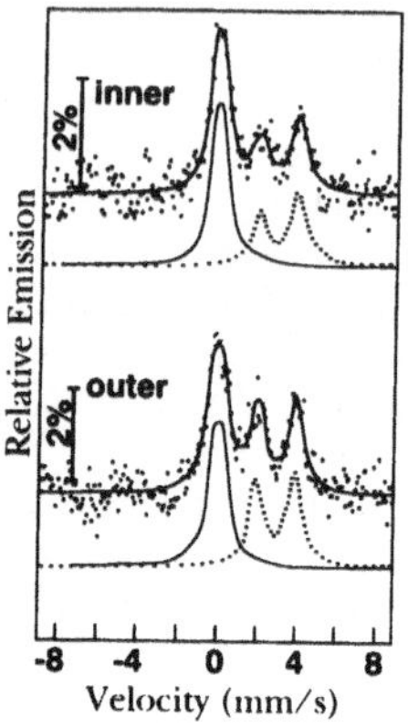

Figure 6. The room temperature conversion electron Mössbauer spectrum of tin-119 in float glass after heating at 730°C. Reprinted from reference 21 with permission.

the *f*-factors were quite different, that of Sn^{4+} being the greater. A comparison was made of samples taken from different positions in the ribbon and significant differences were found in the Sn^{2+}/Sn^{4+} ratio, differences which were ascribed to the variations in oxygen concentration across the ribbon. In another experiment a sample was heated for 10 minutes at 730°C, the conditions that produce bloom, and the expected decrease in the Sn^{2+}/Sn^{4+} ratio was observed, see Figs. 5 and 6.

Measurements on float glass, which has been artificially doped with tin in order to simulate the behaviour of diffused tin, were made by Holland *et al.*[8] and by Johnson *et al.*[22] The float glass was remelted with stannous oxalate under the appropriate conditions to try to maintain the tin in the Sn^{2+} state. It was found that up to 7 mol percent of SnO could be added without crystallisation taking place. It was therefore easy to obtain good Mössbauer spectra, as is shown in Fig. 7. In these samples most of the tin was in the Sn^{4+} state despite the reducing conditions used, although some Sn^{2+} was present. The isomer shifts and quadrupole splittings of both ionisation states were close to those at the surface of untreated float glass. There was a slight dependence of these parameters on tin content, both of them increasing as the amount of tin was increased. The isomer shift was very sensitive to the presence of other elements, and

Table 3. The Isomer Shift[a] of Tin-119 in Oxides and Glasses

Materials	Tin(II)		Tin(IV)		Ref.
	77 K	300 K	77 K	300 K	
SnO crystalline	2.70	2.65			
SnO amorphous	-	2.87			17
SnO_2 crystalline			0.02	-0.02	
SnO_2 amorphous				0.13	17
Float glass	2.84		-0.20		22
Tin-doped float glass	2.87		-0.20		22
SnO_2 coating			0.04	-0.04	23

[a]Values in mms^{-1} relative to $CaSnO_3$ at room temperature.

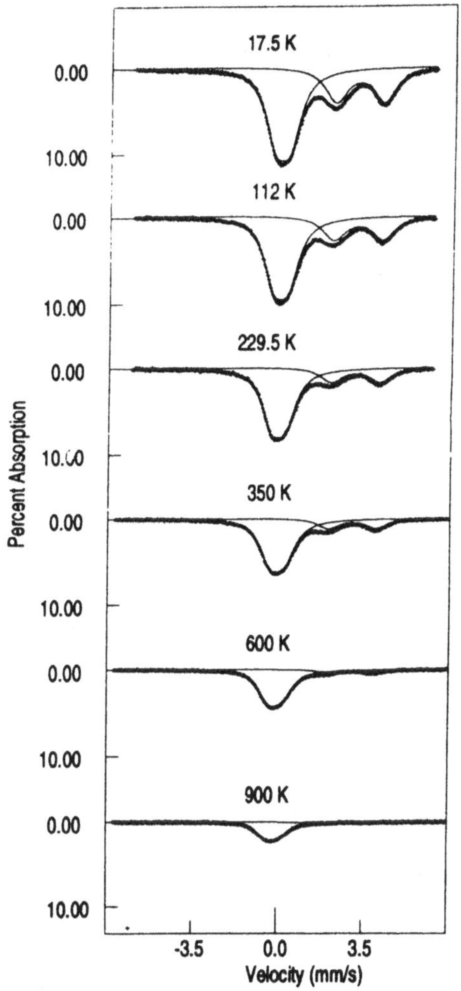

Figure 7. The tin-119 Mössbauer spectra of tin-doped float glass as a function of temperature. Figure obtained from reference 22.

increased when the concentration of aluminium was increased, and increased still further, when all the added elements, principally sodium and calcium, were omitted. Because of the ease of obtaining spectra it was possible to carry out measurements at several temperatures so that the *f*-factors could be determined. Fig. 7 shows spectra measured between 14 and 900 K for the sample containing the most tin, i.e., 6.2 mol percent SnO. Fig. 8 shows the variation of the *f*-factor for each oxidation state as a function of temperature. It is clearly seen that the *f*-factor of Sn^{2+} is smaller than that of Sn^{4+}, and at 900 K it is practically zero, whereas the Sn^{4+} spectrum can still be observed. In terms of the Debye model the value of Θ_D was found to be 319 K for Sn^{4+} and 200 K for Sn^{2+}.

Johnson *et al.*[22] have also obtained transmission Mössbauer spectra for float glass surfaces by using samples comprising of a stack of several discs, each ground down to about 100 microns thickness. In this way the amount of tin in the sample was increased compared with that at a single surface, and a stronger Mössbauer signal was obtained. A further improvement was made by grinding several discs into a powder and packing it into a perspex

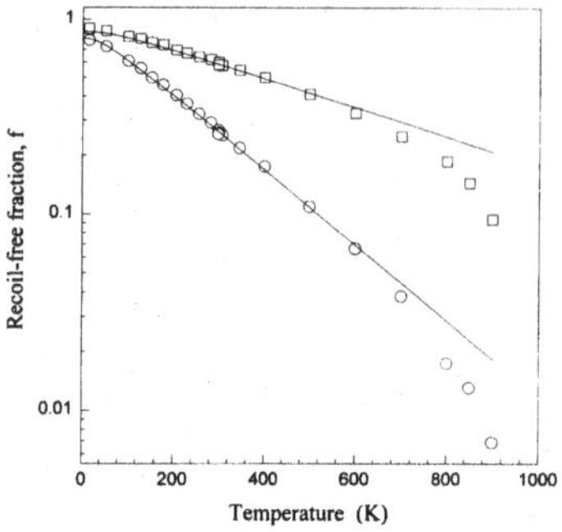

Figure 8. The f-values of tin-119 in tin-doped float glass as a function of temperature for Sn^{2+}, O, and Sn^{4+},□, Figure obtained from reference 22.

sample holder, thereby making possible measurements over a range of temperatures. The resulting spectrum, see Fig. 9, showed that most of the tin was in the Sn^{2+} state, in agreement with the work of Principi *et al.*,[21] and gave parameters which also agree. From the temperature variation of the area under the spectrum the Debye temperatures in float glass were found to be 260 K for Sn^{4+} and 185 K for Sn^{2+}. Thus the doped float glass had significantly higher values of Θ_D, presumably the effect of the extra remelting heat treatment.

Williams[23] has used Mössbauer spectroscopy to compare glass from different float plants and found that the relative amount of stannic ions varied between 16 and 43 percent. Measurements on glass from different regions of the same ribbon showed that there was less Sn^{4+} at the edges than in the centre, i.e., where there is a greater concentration of oxygen in the tin. Measurements were also made after heat treatment and indicated that oxidation and crystallisation had occurred depending on the time, temperature, and atmosphere used. Also reported were measurements of the depth profile of tin in both oxidation states obtained by using samples polished to remove 1.5, 3.5, and 7.5 μm from the surface. The results showed that the majority of tin near the surface is Sn^{2+}, more than half of it occurring in the first 3 μm, and over 80 percent of the Sn^{4+} is deeper than 3 μm.

4. 3. The Nature of Tin in Float Glass

Tables 3 and 4 summarise the isomer shifts, quadrupole splittings, and Debye temperatures of tin in glasses. As already mentioned, tin in float glass gave the same isomer shift and quadrupole splitting as remelted float glass doped with tin, so the chemical state of the tin atoms, both Sn^{2+} and Sn^{4+}, was the same on the surface as in the bulk of the glass. However, the Debye temperature was larger for the remelted samples indicating stronger binding, perhaps due to a tighter packing of the SiO_4 tetrahedra, which in turn would impinge on the tin atoms.

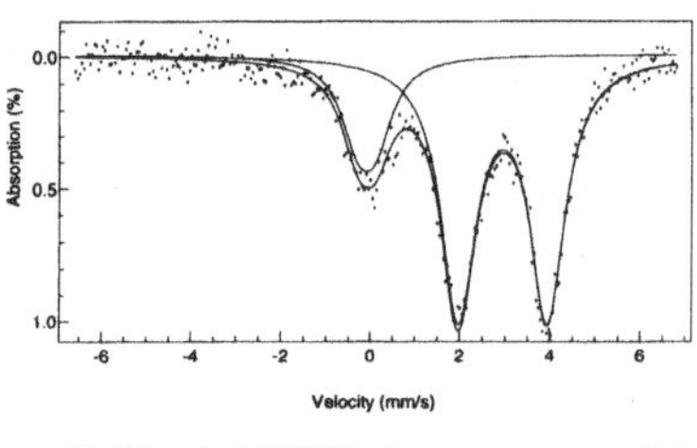

Figure 9. The tin-119 Mössbauer spectrum of tin in float glass measured at 77 K. Figure obtained from reference 22.

Table 4. Quadrupole Splitting and Debye Temperatures of Tin-119 in Oxides and Glasses

	Quadrupole Splitting, mm/s		Debye Temperature, K	
Material	Tin(II)	Tin(IV)	Tin(II)	Tin(IV)
SnO_x crystalline	1.36	0.50	204	306
SnO_x amorphous	1.71	0.70	-	-
Float glass	1.96	0.44	185	260
Tin doped float glass	1.96	0.47	200	319

The isomer shift data suggest that the tin-oxygen bonds in float glass are more ionic than those in crystalline SnO_2 and SnO. The lower values of Θ_D observed are consistent with this and agree with the findings of Collins *et al.*[17] for the amorphous oxides. For Sn^{2+} this could be an indication of a higher coordination number, compared with the four coordination in SnO, suggesting that Sn^{2+} is a network modifier, which is not unlikely. The results for Sn^{4+} in float glass are not as easy to interpret. Sn^{4+} has a strong preference for octahedral coordination, though there is evidence that it can enter the glass network as an intermediate, with tetrahedral coordination and strong covalent bonds.

5. TIN IN GLASS COATINGS

A very important area of industrial glass production is the coatings which are deposited on the glass surface. These are in the forefront of environmental protection measures and are used to reduce the heat losses from buildings. Low emissivity glass, e.g., the Pilkington 'K' glass, is produced by coating float glass with tin oxide, SnO_2. This has the property of reflecting a high proportion of infrared heat radiation while remaining transparent to visible radiation. The coating is applied by chemical vapour deposition during manufacture. The properties of these coatings are thought to relate to their thickness, orientation, and the structure and environment of the tin atoms. Hence measurement of the tin environment, as a function of the depth of the coating, should help to characterise and evaluate the process parameters required for the desired properties.

The Mössbauer spectrum of an SnO_2 coating[23] on float glass measured on a sample containing several surface layers is shown in Fig. 10. The thickness of each coating was about 100 μ. The isomer shift and quadrupole splitting values are given in Tables 3 and 4, and show that the SnO_2 is crystalline, and this conclusion is supported by the observation of a narrow linewidth. The unequal areas of the two lines show that there is some texture in the coating.

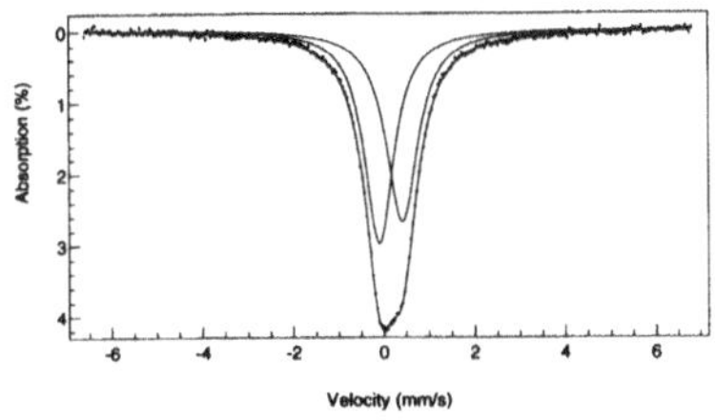

Figure 10. The tin-119 Mössbauer spectrum of tin in a glass coating. Figure obtained from reference 23.

6. CONCLUSIONS

Mössbauer spectroscopy has provided a valuable probe of the structure and bonding of atoms in glass. Its great strength is that it gives separately resolved spectra for each oxidation state of an atom. Hence it can be used as a method of chemical analysis to determine the amount of each oxidation state present in a glass sample, and also it can provide data on covalency, coordination number, and binding strength.

One of many interesting results is that measurements on tin, which has diffused into the surface of float glass during manufacture, show that both Sn^{2+} and Sn^{4+} are less strongly bound than in the bulk material. Possibly this is a surface effect, or a result of the tin not having diffused to an equilibrium state at the low temperature and during the short time that the glass is in contact with the tin bath.

A large number of measurements have been made and much has already been learned, but there is a great deal which is not understood, and much work remains to be done. Further careful work promises to yield important data which will be of both scientific and technological interest.

REFERENCES

1. For a review of the properties and theory of glass behaviour see S. R. Elliott, *Physics of Amorphous Materials*, Longman, London (1984).
2. C. R. Kurkjian, *J. Non-Cryst. Solids* **3,** 157 (1970).
3. J. M. D. Coey, *J. Phys. (Paris)* **35,** C6-89 (1974).
4. W. Müller-Warmuth and H. Eckert, *Phys. Rep.* **88,** 91 (1982).
5. G. Tomandl in *Glass: Science and Technology*, D. R. Uhlmann and N. J. Kreidl, eds., Vol. 4B, Academic Press, New York (1990).
6. T. Nishida, *J. Non-Cryst. Solids* **177,** 257 (1994).
7. H. Pollack, M de Costa, and S. Amelinckx, in *Mössbauer Effect,* D. M. J. Compton and A. H. Schoen, eds., p. 298, Wiley, New York (1962).
8. D. Holland, M. M. Karim, C. E. Johnson, K. F. E. Williams, and J. A. Johnson, *Fundamentals of Glass Science and Technology,* Supp. to Revista della Stazione Sperimentale del Vitro, Vol. XXIII, p. 223 (1993).
9. L. R. Walker, G. K. Wertheim, and V. Jaccarino, *Phys. Rev. Lett.* **6,** 98 (1961).
10. P. A. Flinn, in *Mössbauer Isomer Shifts*, G. K. Shenoy and F. E. Wagner, eds., North-Holland, Amsterdam (1978).
11. R. V. Parish, *Prog. Inorg. Chem.* **15**, 101 (1972).
12. V. I. Gol'danskii, G. M. Gorodinskii, S. V. Karyagin, L. A. Korytko, L. M. Kryzhanskii, E. F. Markorov, I. P. Suzdalev, and V.V. Khrapov, *Dokl. Acad. Nauk. SSSR* **147,** 127 (1962); *Sov. Phys. Dokl.* **147**, 766 (1963).
13. G. H. Frischat and G. Tomandl, *Glastechn. Ber.* **42**, 182 (1969); **44**, 173 (1971).
14. A. Montenero, M. Frigger, D. C. Giori, N. Belkhiria, and L. D. Pye, *J. Non-Cryst. Solids* **84**, 45 (1986).
15. L. A. B. Pilkington, *Proc. Roy. Soc.* **A314,** 1 (1969).
16. J. S. Sieger, *J. Non-Cryst. Solids* **19,** 213 (1975).
17. G. S. Collins, T. Kachnowski, N. Benczer-Koller, and M. Pasternak, *Phys. Rev. B* **19**, 1369 (1979).
18. K. P. Mitrofanov and T. A. Sidorov, *Soviet Physics-Solid State* **9**, 693 (1967).
19. H. Dannheim, H. J. Oel, and G. Tomandl, *Glastechn. Ber.* **49**, 170 (1976).
20. G. M. Bartenev, G. M. Magomedov, A. D. Tsyganov, and Yu. V. Zelenev, *Sov. J. Glass Phys. Chem.* **2,** 114 (1976).
21. G. Principi, A. Maddalena, A. Gupta, F. Geotti-Bianchini, S. Hreglich, and M. Verità, *Nucl. Instr. Meth. B* **76,** 215 (1993).
22. J. A. Johnson, C. E. Johnson, K. F. E. Williams, D. Holland, and M. M. Karim, *Hyp. Interact.* **95,** 41(1995).
23. K. F. E. Williams, Ph.D. Thesis, University of Liverpool (1995).

8

High Pressure Mössbauer Spectroscopy: The Second Generation

Moshe P. Pasternak and R. Dean Taylor

1. INTRODUCTION

Pressure is one of the most effective methods available to solid-state scientists to alter properties of matter. Since its implementation by Bridgman early this century, application of high pressure research in matter has provided substantial information on the properties of matter, in all aspects. Today, as a result of the development of static high pressure devices based on the diamond-anvil cell, experimenters can reach pressures in the megabar region generating energy densities in matter of the order of keV/nm^3. With such energy densities insulators with gap energies in the eV regions become metals, new structural and electronic phases become stable, and new aspects of magnetism may be revealed. The diamond anvil cell thus has become a very powerful ultra-high-pressure device, helping scientists discover new states of matter. Some of the modern diamond anvil cells generating pressures into the Mbar region can fit into the palm of the hand and allow a variety of sophisticated measurements to be performed on materials even though samples are of microscopic dimensions. The principles underlying the diamond anvil cell, its pressure calibration, its applications as of 1983, and some potential uses, are extensively described in the review paper by Jayaraman.[1]

The goal of this chapter is to introduce the Mössbauer spectroscopist to the fascinating field of high pressure studies using the diamond anvil cell and to lead them into the second generation of high pressure Mössbauer spectroscopy research. As will be shown, the contributions of Mössbauer spectroscopy to research into the state of matter at very-high-pressures are significant, because this technique is unique in its measurement of solid properties on the atomic scale and nicely complements x-ray, Raman, and infrared spectroscopy in diamond anvil cells. Mössbauer spectroscopy is the only viable probe today for magnetic studies at pressures above 100 kbar. Due to the relatively large absorption cross sections, most Mössbauer isotopes have adequate sensitivity to probe the sub-milligram sized samples required in diamond anvil cells.

M. P. Pasternak • School of Physics and Astronomy, Tel Aviv University, Ramat Aviv, Tel Aviv 69978, Israel. *R. D. Taylor* • Los Alamos National Laboratory, P. O. Box 1663 MS-K764, Los Alamos, NM 87545, USA.

Mössbauer Spectroscopy Applied to Magnetism and Materials Science, Volume 2, edited by Gary J. Long and Fernande Grandjean. Plenum Press, New York, 1996.

In this chapter we assume the reader is familiar with the fundamentals of Mössbauer spectroscopy and with hyperfine interaction terminology, but not necessarily with diamond anvil cell methods. After an introduction in Section 1, Section 2 is devoted to the methodology of the diamond anvil cell, emphasizing features relevant to Mössbauer spectroscopy. Section 3 presents numerous applications, and finally in Section 4 we discuss the pressure limitations using radioactive sources, applications of synchrotron radiation, and speculate on possible future scientific applications.

1. 1. The First Generation

About three years after the discovery of the Mössbauer effect,[2] Hanks published a short paper,[3] suggesting an application of Mössbauer spectroscopy to the high pressure state of matter. The paper was entitled, *Pressure Dependence of the Mössbauer Effect.* Hanks apparently used a wrong formula to calculate the recoil-free fraction, f, for cesium, but stated, "we find that a pressure of 5000 atm will bring f to an (observable) value of 4×10^{-3}, i.e., will cause sufficient change in the crystal binding so as to change f by two orders of magnitude." Six months later Pound, Benedek, and Drever published[4] results of the first high pressure experiment with Mössbauer spectroscopy. Citing these authors, "The source, about 2 mCi of cobalt-57 diffused into iron, was enclosed in a beryllium-copper pressure bomb equipped with a one-half-inch thick beryllium window. The bomb was cemented to the ferroelectric transducer and connected to a Bridgman press by a stainless steel tube." The pressure bomb used for this experiment is shown in Fig. 1. The highest pressure achieved was 3000 kg/cm^2 or 3 kbar.

The pioneering work in high pressure Mössbauer spectroscopy of *the first generation* and the vast majority of published works, starting in the early 1960's and into the mid-1970's, may be attributed to Drickamer and his collaborators, who also published several excellent reviews.[5,6] Although their studies were primarily conducted with the iron-57 Mössbauer isotope either as sources or absorbers, their success, experience, and consequent progress in the high pressure methodology encouraged many other Mössbauer spectroscopy groups to initiate high pressure studies with a variety of isotopes. Effects of pressure on the electronic configuration of matter were studied with iron-57, tin-119, tellurium-125, europium-153, dysprosium-161, ytterbium-170, tantalum-181, and gold-197. An excellent review on Mössbauer studies at high pressure to 1975 has been published by Holzapfel.[7]

In the *first generation* of high pressure Mössbauer spectroscopy, the cells used were limited to pressures of ca. 20 GPa; however, the maximum pressure commonly was ca. 10 GPa.[8] This pressure is adequate for observing phase transitions in many of the covalent and some of the metallic materials.[8] Due to the bulky construction of the cells, experiments at cryogenic temperatures and in external magnetic fields were not an easy task. For observation of structural and electronic phase transitions in oxides, minerals, and molecular crystals, pressures up to and above 100 GPa are required. Such ultimate hydrostatic pressures can be achieved only with diamond-anvil cells.

1. 2. P = F/A

Pressure is obtained by applying a force on an certain area. One Newton applied on a square meter creates a pressure of one Pascal, 1 Pa = 1 N/m^2, which is the SI pressure unit. A common unit for expressing high pressure is the kilobar, 1 kbar = 10^3 bar = 10^9 dynes/cm^2. In high pressure studies the gigapascal, 1 GPa = 10^9 Pa, is a more appropriate unit, 1 GPa = 10 kbar and 1 Mbar = 10^6 bar = 100 GPa. The pressure at the center of the earth is estimated to be 300 GPa. Using the ocean depths as a generator of *hydrostatic* pressure, one may achieve pressures of at most 0.1 GPa at a depth of 10,000 m! At this pressure the force exerted on 1 cm^2 corresponds to 1 metric ton.

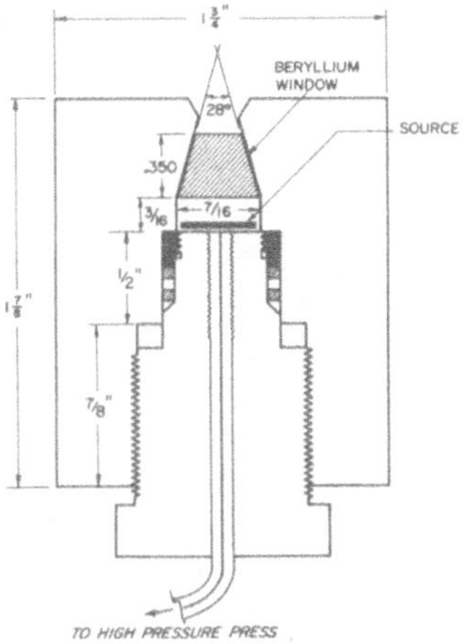

Figure 1. A cross sectional view of the pressure cell, i.e., pressure bomb, used for the first Mössbauer spectroscopy experiment under pressure.[4]

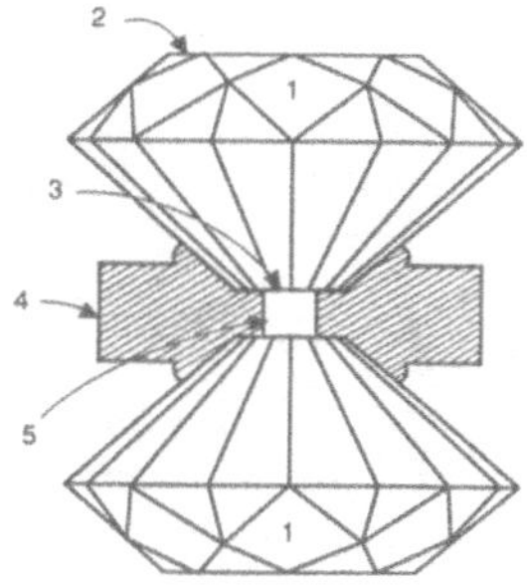

Figure 2. The heart of a diamond anvil cell, see text.[24]

It is obvious that the practical generation of *very* high pressures cannot rely on an ever larger external force, but rather on a reduced size of sample. Thus, to create hydrostatic pressures in the 10 to 100 GPa range in the laboratory one must induce uniaxial forces on areas as small as practical. A one ton force exerted upon a 0.3×0.3 mm^2 area will generate pressures approximating 1 Mbar! To convey such forces to the samples one needs extremely hard materials that will preserve their elastic properties even at this pressure. High quality crystalline diamond anvils, the hardest substance known, are quite transparent to x-rays and light and are the only present material for such anvils. Anvils made of 0.20 to 0.35 carat diamonds are adequate, and their prices are affordable. Finally, to transform the uniaxial pressure into a hydrostatic pressure one needs to contain the samples within hard metal gaskets. The main factors that contributed to the significant advances in high pressure research in recent years are attributed to the development of diamond anvils, proper gasketing materials, hydrostatic transmitting media, progress in powerful miniature sources of electromagnetic radiation at many wavelengths, and the availability of sensitive spectrometers.

1. 3. Principles of the Diamond Anvil Cell

The basic configuration of all diamond anvil cells is shown in Fig. 2. The diamonds *(1)* modified to have flat culets *(3)* are pressed into the sample, which is confined in the sample chamber *(5)* formed in a preindented metal gasket *(4)*. A closure force is applied to the tables *(2) via* appropriate backing plate and closure assemblies. Numerous styles of diamond anvil cells have been developed, and many of them carry the names of the laboratories, e.g., the *NBA cell*[9] or of the developers, e.g., the *Mao-Bell cell,*[10] the *Holzapfel cell,*[1] or the *Merrill-Bassett cell.*[11] Diamond anvil cells can be classified into two main types, **opposed-plate** cells and **piston-cylinder** cells. Whereas cells of the first class are relatively simple, inexpensive, easy to handle, and suitable for pressures up to ca. 30 GPa, with say 600-μm culets,[12] cells of the second class are more robust, more difficult to align, more precise in their closure, and more expensive, but they are suitable for pressures up to and above 100 GPa when using culets sizes of about 150 μm or smaller. Cells typical of the opposed-plates class are shown in Figs. 3 and 4, and one of the piston cylinder class is shown in Fig. 5. The triangular Merrill-Bassett cell, designed for x-ray diffraction but also used for Mössbauer

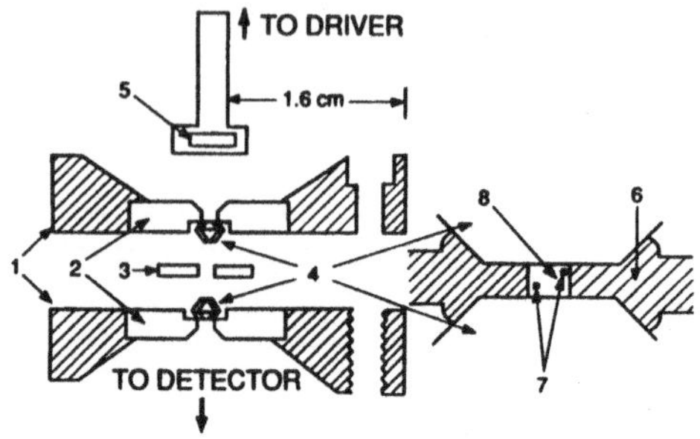

Figure 3. Cross section of the triangular Merrill-Bassett cell.[24]

spectroscopy, is the simplest working extension of the basic configuration and is shown in cross section in Fig. 3, along with a schematic representation of the Mössbauer source *(5)*, absorber *(8)*, and detector. The sample *(8)*, ruby chips *(7)*, and hydrostatic medium are placed in a hole drilled in the preindented gasket material *(6)* placed between the flat parallel faces, or culets, of two opposed anvils *(4)* and are subjected to pressure when squeezed by the three screws acting on the backing plates *(2)* via the cell assembly plates *(1)*.

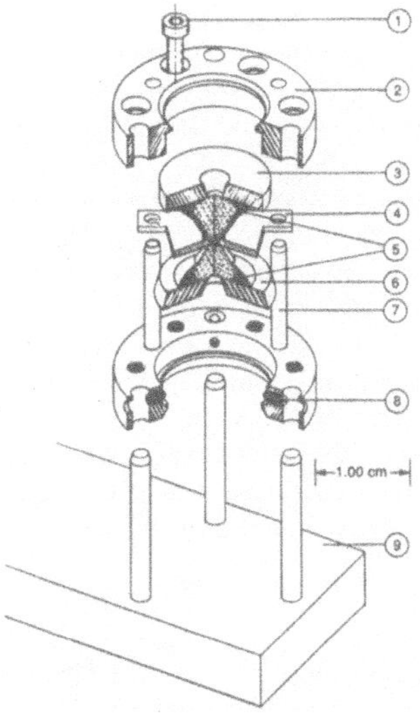

Figure 4. A miniature anvil cell of class 1 built at Tel Aviv University. Six Allen screws (1) inserted in the upper ring (2) apply a firm, uniform force on the upper backing plate (3), into which an anvil (5) is epoxied. This backing plate is fixed into the upper ring. The lower backing plate (6) can be laterally adjusted, using three adjusting screws (8), to allow centering of the culets. The gasket (4) is held fixed using the two guiding pins (7). Some cocking of the closure rings is allowed to obtain an initial interference quality parallelism to the culet faces. A holder (9) is particularly useful for pressures above 10 GPa.[26]

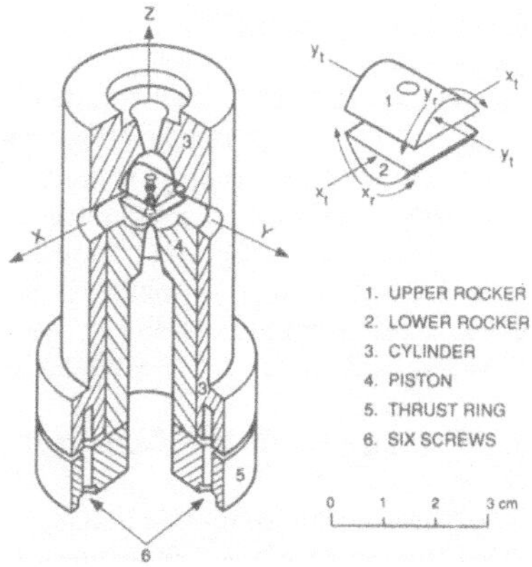

Figure 5. A hardened Inconel piston-cylinder cell made at LANL for Mössbauer spectral studies in the 100 GPa pressure range. The source shown (7) is positioned as close to the anvils as is practical. A "pencil" shaped holder for the ^{57}Co(Rh) 'point' source allows an even closer approach. The anvils are attached to two crossed, opposing hemi-cylindrical WC backing plates (1 and 2) which can be mutually adjusted for anvil positioning and parallelism. Six screws (6) apply force via the cap ring (5) to the close-fitting piston (4) contained in the cylinder portion of the main body (3) of the cell. The gasket, not shown, is attached to the piston.[24]

Diamonds. Diamonds for anvils are usually selected from flawless, gem-quality stock. The *brilliant* cut diamonds are specially faceted with 16 sides to provide a more nearly circular culet face. Anvils are available from commercial sources.[13] Sizes vary from 0.2 to 0.4 carat depending on the particular need. The culets are cut and polished parallel to the table, see Fig. 2, to a very high degree of accuracy. Diamonds with very low birefringence are often selected for pressures beyond 50 GPa. For some optical applications it is desirable to select diamonds having low luminescence. For Mössbauer spectroscopy at pressures up to 40 GPa one typically selects 0.25 to 0.33 carat anvils with culets of 0.6 to 0.7 mm diameter. Beveling of the culet helps to relieve stress gradients and reduce edge fractures.

Gasketing. The use of a gasket for the containment of pressure was first demonstrated by Van Valkenburg.[14] This significant discovery permitted using the diamond anvil cell as a quantitative tool for high pressure research. Besides providing for hydrostatic sample containment, the gasket acts as a supporting ring by extruding around the diamond culets, reducing failures of the anvils due to lateral stress and edge stress. Gaskets are made of hard metals or alloys, such as tempered 301 stainless steel, *Inconel X750, Waspalloy*, as well as from high-Z materials such as $Ta_{90}W_{10}$ and rhenium. The sample cavity is made by drilling a hole of about a half to a third the diameter of the culet flat. Final gasket thickness typically varies between 25 to 80 μm. Fortunately, this is just the ideal thickness range for many natural or partially enriched, Mössbauer absorbers. Thicker gaskets tend to be unstable showing hole drift at the higher pressures.

Manometry. Pressure calibration in diamond anvil cells is performed by the ruby fluorescence method.[15] This method based on the quasi-linear pressure dependence of ruby, i.e., Cr^{3+}-doped Al_2O_3, R-lines at 692.7 and 694.2 nm, simplifies enormously the *in-situ* determination of pressure, and its distribution, in a diamond anvil cell. The ruby R lines are excited by blue or green light from Cd-He or Ar-ion lasers, and the fluorescence wavelengths are measured with a standard monochromator. The pressure dependence of the wavelength shift, $\Delta\lambda$, has been experimentally calibrated using the established equation of states of several elements.[16] The expression suitable for pressures to at least 100 GPa is given as

$$P = 0.03808[(\Delta\lambda/694.2 + 1)^5 - 1], \qquad (1)$$

where P is in GPa and $\Delta\lambda$ in nm. For pressures up to 30 GPa Equation 1 is linear to a very good approximation, and the accepted value for the R-line shift is 0.365 nm GPa^{-1}. The temperature dependence of the shift is known and can be used to establish the dependence of the pressure on temperature for each diamond anvil cell used. It has been shown that for the cells used by the authors the temperature dependence of the pressure on cooling is less than 10 percent. The degree of hydrostaticity in the diamond anvil cell is determined by measuring a number of points, i.e., individual ruby chips, over the sample area.

2. METHODOLOGY OF MÖSSBAUER SPECTROSCOPY WITH DIAMOND ANVIL CELLS

The first reported high-pressure Mössbauer experiment done at 15 GPa with diamond anvils was in 1965 by Herber and Spijkerman.[17] Measurements were carried out in $^{119}SnO_2$ using a conventional *NBA* cell modified for diamond anvils. Presumably no gasket was used, and no description is provided about the pressure calibration or sample confinement. A short report by Huggins et al.[18] on a Mössbauer spectroscopy-diamond anvil cell study appeared in *Carnegie Institution Washington Year Book* for 1974. However, the crucial modern studies that led into the present state of the art of Mössbauer spectroscopy in a diamond anvil cell were effectively initiated in the early 1980's at Los Alamos with Fe(^{57}Co) source experiments[19] to a pressure of 21 GPa and at the University of North Carolina with β-Sn absorber studies[20] to a pressure of 30 GPa. The first International Conference on the Applications of the Mössbauer

Effect at which Mössbauer spectral studies in a diamond anvil cell were presented was in 1985, in Leuven. Papers by Pasternak et al.[21] dealt with studies of $^{129}I_2$ molecular crystals to a pressure of 35 GPa at cryogenic temperatures and by Nasu et al.[22] with ε-Fe and Fe_2O_3 at room temperature to 45 and 76 GPa, respectively. These early publications[23] with iron-57, iodine-129, and europium-151 motivated many of the existing high-pressure Mössbauer spectroscopy laboratories either to switch or to add a diamond anvil cell capability to their facilities.

Certain aspects of the methodology of Mössbauer spectroscopy in a diamond anvil cell have been previously reported by us.[24] In what follows we outline in more detail the needs of a diamond anvil cell laboratory emphasizing the features related to Mössbauer spectroscopy.

2. 1. A Diamond Anvil Cell Mössbauer Spectroscopy Laboratory

Unlike conventional pressure cells, the diamond anvil cell often allows easy access to both low and high temperatures[25] with existing cryostats and furnaces and to high applied magnetic fields using superconducting magnets. In addition to the usual standard Mössbauer laboratory equipment, the following items are needed for Mössbauer spectroscopy in a diamond anvil cell: diamond anvil cells, diamond anvils, ruby chips, gasket materials and pressure transmitting media, hole drilling devices, conventional and inverted microscopes, pressure calibration units, special tools and accessories, and cryostats.

Diamond anvil cells used in many laboratories for absorption Mössbauer spectroscopy are patterned after cells used for x-ray spectroscopy modified to provide a somewhat larger conical aperture in the backing plates and to allow a closer spacing between the source and absorber. A sketch of a cell belonging to *class 1*, built at Tel Aviv University,[26] the TAU cell, is shown in Fig. 4. It was specially designed to be small enough for 25 mm superconducting magnet bores or the top-loading access diameter of many commercial cryostats. Unlike the triangular-shaped Merrill-Bassett cell, see Fig. 3, with its three tightening screws, the miniature TAU cell is circular with six smaller screws that provide the necessary force and rather fine pressure control. The cells have been constructed either from *Maragin* steel or Cu-Be and have achieved pressures up to 50 GPa with 0.3 carat anvils with 16 sided, 0.5 mm culets. The 45° opening in the backing plates that tapers down to a hole of 0.5 to 1.0 mm provides a convenient solid angle for source and detector and an easy close approach for point sources. Due to its simplicity and commercial availability[27] it is now widely used in high pressure laboratories in Europe and Japan. A piston-cylinder cell of *class 2*, constructed at Los Alamos and particularly adapted for Mössbauer spectroscopy, is shown in Fig. 5. It is based on the Mao-Bell cell, but with a simplified six-screw closure scheme reducing its size so it can be used in most Mössbauer cryostats. With this cell, a pressure of 120 GPa has been reached using anvils with 0.2 mm culets.[28] The diamond anvil cell used at Osaka University[22,29] for pressures to 80 GPa is a modified Bassett type cell and is shown in Fig. 6.

Diamond anvils. The technical specifications for the anvils for Mössbauer spectral applications are similar to those required for x-ray diffraction. Typically, gem quality, brilliant-cut diamonds are modified by cutting and polishing the culet tip to provide a small flat parallel to the gem table. Sixteen-sided faceting yields a more nearly circular tip. To attain the highest pressures, a bevel is often added to the flat culet tip. Adequate transmission of the γ-rays through the diamonds may be a problem. In Fig. 7 we present transmission curves for several commonly used Mössbauer transitions. The typical height of a pair of 0.25 to 0.33 carat anvils is 3.5 mm. Thus, for tin-119 or europium-151 Mössbauer spectroscopy the anvils allow 65 percent transmission, whereas for iron-57 only 35 percent of the 14.4-keV γ-rays is transmitted. Reducing the anvil weight to 0.15 carat, thus increasing the transmission of the 14.4 keV to 50 percent, was only partially successful; the highest pressure achieved was only 10 GPa,[30] due in part to the decreased ratio of the table/culet areas. Despite the relatively large absorption of the cobalt-57 radiation by the 0.3 carat diamond anvils, excellent

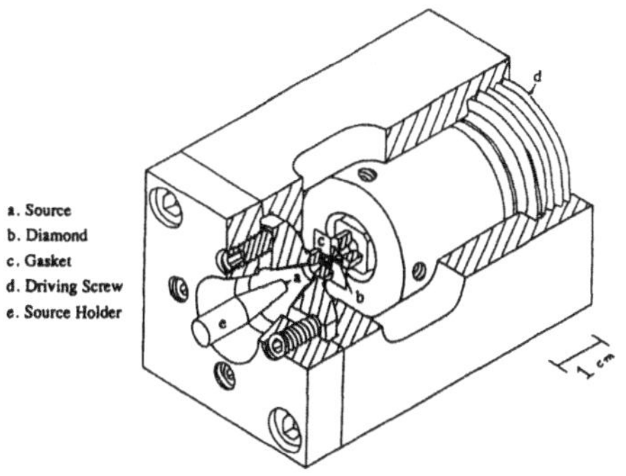

Figure 6. Cross section of a piston cylinder cell used at Osaka University. Force is applied through a single central screw (e) to piston (d) acting on diamond anvils (b) aligned by screws (c) acting on crossed hemi-cylinders (a). The highest pressure reported with this type of cell was 80 GPa.[29]

spectra can be obtained with point sources of 7 mCi, see Section 3. We also note that Mössbauer spectroscopy in a diamond anvil cell for tantalum-181 and other low energy Mössbauer isotopes is impractical due to the severe absorption of even the thinnest usable diamond anvils.

Ruby chips, gasket materials, and pressure transmitting media. Coarse ruby powders or chips are made from commercial quality synthetic ruby rods. Grains usually of the order 5 to 20 μm diameter are spread over the central sample area of both culet surfaces. Prior to final closure of the cell, it is important to verify the presence of visible ruby chips at the sample, especially with opaque absorbers. The radial spread of ruby chips is important for determining pressure distributions.

The metal-gasket plays a dual role in Mössbauer spectroscopy in a diamond anvil cell. It serves to contain the pressure on the sample and to collimate the γ-rays. A suitable non-ferrous metal for gasketing is a $Ta_{90}W_{10}$ commercial alloy. Due to its high Z, it is an excellent gasket/collimator material to pressures of 50 GPa. It may be purchased in forms of rods, sheets, or wires. Work hardening through rolling to ca. 0.25 mm and preindenting is considered important. The latter is carried out in a number of press/release steps to relieve radial stress on the anvils. Other useful high Z gasket materials are rhenium, tungsten, and $Pt_{80}Ir_{20}$. Rhenium is almost universally used for pressures near and above 100 GPa. It is

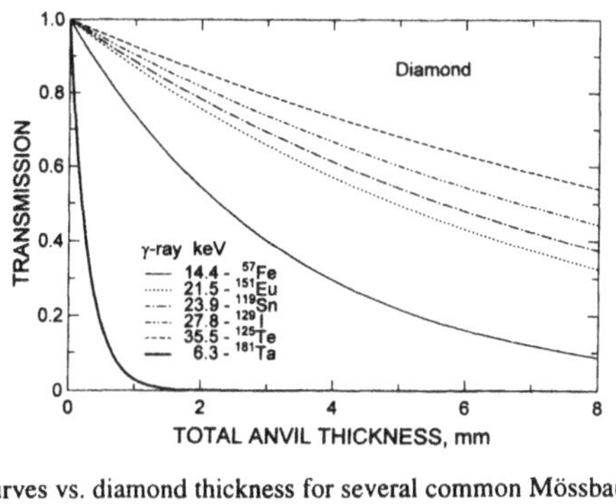

Figure 7. Transmission curves vs. diamond thickness for several common Mössbauer isotopes. The typical on axis thickness of a pair of 0.3 carat diamonds is 3.5 mm.

appreciably harder than $Ta_{90}W_{10}$ and is more difficult to drill. Critical edge considerations for a particular γ-ray may influence the choice of materials for collimation.

Up to 11 GPa a mixture of methanol-ethanol has commonly been used as a satisfactory pressure transmitting medium.[1] For intermediate pressures up to 60 GPa, argon is suitable, whereas for very high pressures in the 100 GPa region, helium or hydrogen is recommended.[1] For argon the best method to fill the sample cavity is to use the cryogenic liquid. The precooled cell filled with the sample material and ruby chips is immersed in a small dewar filled with liquid argon. The tightening screws are loosened to allow the liquid in and then re-tightened. Argon solidifies at ca. 2 GPa at 300 K but has little shear strength. Helium or hydrogen is normally introduced into the sample chamber as a compressed gas.[1]

Hole drilling devices are necessary for making the sample cavities in the indented gaskets. In the *class 1* cells it is convenient to use its alignment pins to locate the gasket sheet reproducibly for all subsequent gasket operations. High-speed steel and carbide or diamond tipped drills, in sizes from 50 to 500 μm, are available from circuit board, machine tool, and jewelry manufacturing suppliers. Miniature drilling machines are available; however, hand drilling the softer gaskets is practical using a pin-vise-type holder and a 16 to 40 × viewer. Holes can also be located and drilled by low-power spark-erosion machines. The latter is recommended for holes under 100 μm and/or for very hard gaskets such as rhenium or tungsten.

Conventional and inverted microscopes. Several types of microscopes are essential in the diamond anvil cell laboratory: (1) a simple *student* microscope with magnification up to 50 for aligning the anvils and for drilling and inspecting the gaskets, (2) a high optical quality metallurgical-type microscope, with several objectives having an extra-long focal length to accommodate the working distance imposed by the diamond anvil cell, and (3) an inverted microscope with a reticule to allow precise interference alignment of the anvils and to follow any changes in the sample with pressure, such as size, shape, color, and opacity.

The pressure calibration unit is vital to a diamond anvil cell facility. It consists of three main parts: a metallurgical-type microscope, a laser for exciting the ruby lines, and an optical spectrometer. In Fig. 8 we show the various components of such a unit, which includes a microscope *(1)* such as the Olympus BH-2, which is comprised of an illuminator *(2)*, *X-Y* stage *(3)* to allow fine positioning of the diamond anvil cell *(4)*, high quality 10 × and 20 × objectives having an extra long working distance of about 10 mm *(5)*, a viewing illuminator *(6)*, and a dichroic mirror/filter assembly *(9)*. Light from a laser *(7)* enters the microscope through *(10)* and is reflected onto the sample by the dichroic filter/mirror but allows the passage of red light to the eye-piece of the microscope *(11)* and via fiber optics *(12)* into the spectrometer *(13)*. The output of a photomultiplier, *(14)* selected for its red sensitive photocathode, is amplified and fed into an analog to digital card of a personal computer which also interactively controls the spectrometer stepper-motor and allows analysis of the wavelength data. A suitable, simple spectrometer is the Oriel 77200 with 0.25 m focal length and 0.2 nm resolution configured with a step controller, analog to digital card, and personal computer operating software. The excitation laser *(7)* can be sharply focused onto individual grains of ruby in the diamond anvil cell. *Ion Laser Technology* produces air-cooled, 25 to 300 mW Ar-ion lasers that are excellent for this purpose and use an externally controlled power supply *(8)*. Green and blue lines can be used simultaneously or individually to excite the red lines of the ruby, but at the highest pressures, where the R-lines tend to weaken, one may give a better signal above possible competing diamond fluorescence.

Tools and accessories, such as tweezers, magnetic and nonmagnetic needles, drills, and vises are needed for delicate operations dealing with microgram amounts of materials and μm dimensions. A hand rolling mill, drill-locating templates, digital calipers, and "point micrometers," are extremely useful in preparing the gaskets. A glove-box with a hygrometer and a microscope for loading of moisture sensitive samples is recommended. Boil-off from a liquid nitrogen storage dewar is a ready supply of dry gas.

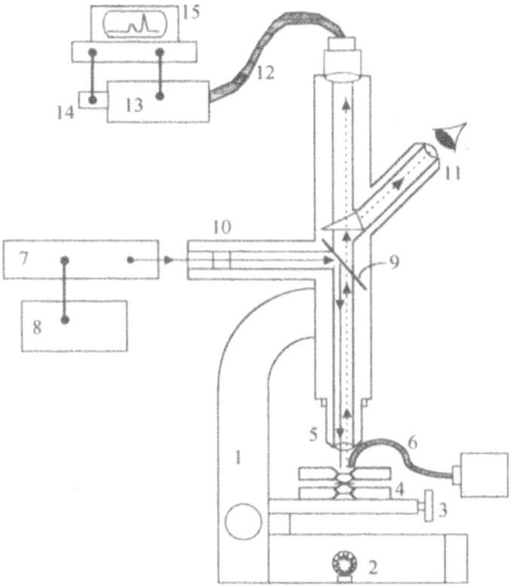

Figure 8. A diamond anvil cell pressure calibration unit is comprised basically of a microscope (1), an excitation laser (7), and spectrometer (13), see text for details.

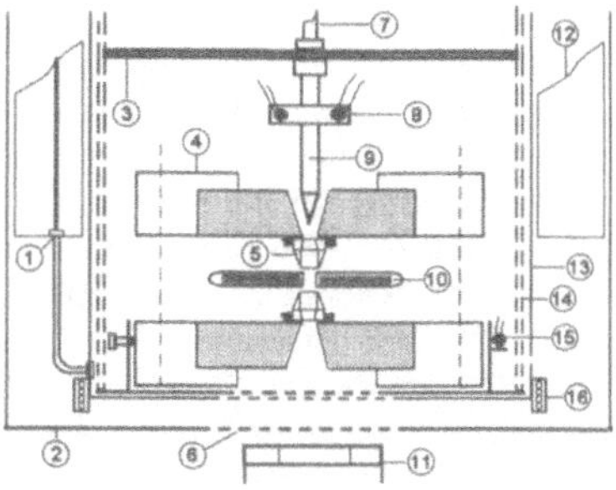

Figure 9. A schematic representation, not to scale, of an apparatus used for iron-57 Mössbauer spectroscopy in a miniature diamond anvil cell in a helium flow cryostat: (1) needle-valve and capillary; (2) vacuum mantle and liquid nitrogen reservoir (not shown); (3) flat centering spring; (4) cell platens with six closure screws; (5) anvils epoxied to the backing-plates; (6) mylar window; (7) stainless steel extension rod; (8) Si-diode thermometer and heater for the source; (9) aluminum source holder with point source at tip; (10) gasket and collimator; (11) detector; (12) liquid helium reservoir; (13) central sample tube; (14) probe assembly; (15) a Si-diode thermometer for the absorber; (16) vaporizer heater.[24]

Cryostats suitable for conventional Mössbauer spectroscopy generally can be adapted to a diamond anvil cell operation. The requirement that the Mössbauer source be quite close to the diamond anvil cell means that the source and absorber are at the same temperature, especially when using a cryogen bath. If the source exhibits a hyperfine splitting at low temperatures, modest heating of the source may be practical where exchange gas coupling is used. A schematic of a flow cryostat with heaters is shown in Fig. 9. Operational details are given in the last article in reference 24.

2. 2. Characteristics of Diamond Anvil Cell Mössbauer Spectroscopy

In this section we discuss particular features of the methodology with regard to the specific problems for various Mössbauer isotopes that have been used in Mössbauer spectroscopy in diamond anvil cell studies. These features are related to: optimization of counters, gasketing and collimation of the γ-rays, sample thickness, geometry of source-absorber-detector, and available sources. Common to all Mössbauer isotopes is the fact that source diameters are considerably larger than the diamond anvil cell absorbers. To obtain the best count rates and signal/noise ratios, source-diamond anvil cell-detector distances need to be optimized. Operating Mössbauer drives at a higher frequency and lower amplitude may help. The interplay of the choice of gasket material and its thickness necessary for hole stability and adequate collimation and of the optimum Mössbauer sample thickness lead to necessary compromises. Full or partial enrichment of the Mössbauer samples may be necessary. Even with enriched isotopes the sample thickness and area for some Mössbauer isotopes, like zinc-67, may exceed the practical limits imposed by the diamond anvil cell and its gasket.

Iron-57. This most important Mössbauer isotope is particularly suited for studies of magnetism at high pressure. The absorption of the 14.4-keV γ-rays by the diamond anvils is a prime experimental issue in iron-57 Mössbauer spectroscopy in a diamond anvil cell, as discussed above. However, commercially available ^{57}Co-Rh *point-sources* with an activity of up to 10 mCi on a 0.25 mm^2 area, or 4000 mCi/cm^2, have been proved viable to pressures above 100 GPa. The *Amersham* source is made by electroplating 40 mCi of carrier-free cobalt-57 onto a 1 × 1 mm, 6 μm thick, rhodium foil and fully diffusing the activity to homogenize and minimize the magnetic ordering temperature. It is then cut into four 0.5 × 0.5 mm pieces each mounted in a customer-supplied aluminum source holder shaped to fit into the diamond anvil cell backing plate, see Fig. 9. Adequate spectra can be obtained with 250 μm diameter, enriched absorbers in hours, instead of days with conventional sources.

Tin-119. Commercial 5 mCi $Ca^{119m}SnO_3$ sources are normally employed in tin-119 Mössbauer spectroscopy in a diamond anvil cell.[31] Due to relatively large source diameter of ca. 5 mm, the optimal source-diamond anvil cell distance is ca. 5 mm. A high-Z gasket such as $Ta_{90}W_{10}$ with a 30 to 40 μm sample cavity thickness is sufficient to collimate the 23.9 keV γ-ray. A 25 μm thick palladium filter selectively reduces the nearby x-rays in the γ-ray path. Standard photomultipliers with a 1 mm thick NaI(Tl) scintillator are commonly used as detectors. Typical count rates are 50 c/s, and high quality spectra for enriched absorbers can be accumulated in 12 to 24 hours.

Tellurium-125. With the 35.6 keV γ-ray energy an additional external collimator is necessary. Recent tellurium-125 Mössbauer spectral studies in a diamond anvil cell by Vulliet et al.[32] used the same external tantalum collimator and gasket as for ytterbium-170.[33] A 2 mm diameter $Mg_3{}^{124}TeO_6$ pellet was neutron activated for two months to produce a ^{125m}Te, $t_{1/2}$ = 60 day, source. Isotope enrichment required is in the 20 percent range, and the source to diamond anvil cell distance is 1 to 2 mm. Absorber cavity dimensions are 250 μm diameter and 50 μm thick. With a germanium detector typical count rates of 50 c/s, and effects of ca. 15 percent were obtained when both source and absorber were at liquid helium temperature yielding a spectrum in about 24 hours.

Iodine-129. Absorbers for Mössbauer spectroscopy in a diamond anvil cell with iodine-129 are prepared in milligram quantities from $Pd^{129}I_2$. This long lived radioisotope is a fission product, and its isotopic abundance in commercial stock, NaI in sulfite solution, is ca. 75 percent. A ^{129m}Te source is made by neutron irradiating about 11 mg of $Mg_3{}^{128}TeO_6$ for about one month in a 10^{14} *n*/cm^2 neutron flux and post-annealing at 500 °C. The source is in a form of an ca. 2 mm diameter pellet located ca. 1 to 2 mm from the backing plate of the diamond anvil cell. Absorbers are loaded in 300 μm diameter × 30 μm thick cavities in $Ta_{90}W_{10}$ gaskets. The collimation for the 27.8 keV γ-ray and close by unresolved x-rays is not as efficient as for tin and europium; typically 30 percent of the photons are transmitted adding to the background. In this respect, a better gasket material would have been $Pd_{80}Ir_{20}$, which has been tested[34] in the case of gadolinium-155 but not with iodine-129. Despite the high background, excellent spectra are obtained in 24 to 32 hours. A typical count rate for Mössbauer spectroscopy in a diamond anvil cell with a one month old source, $t_{1/2}$ = 33 day, is 1000 c/s.

Europium-151. $^{151}SmF_3$ commercial sources of 100 to 150 mCi are used for europium-151 Mössbauer spectroscopy in a diamond anvil cell. Some old samarium-151 made from fission products was actually enriched. As with tin-119, the $Ta_{90}W_{10}$ gasket serves also as a good collimator for the 21.5 keV γ-rays. Typical source diameter is 4 mm, and count rates are 40 to 60 c/s for 300 μm diameter and 30 μm thick absorbers. With natural europium absorbers, good spectra can be obtained[23] within 24 hours at room temperature. Standard photomultipliers with a 1 mm thick NaI(Tl) scintillator or intrinsic germanium are suitable as detectors.

Gadolinium-155. For the 86 keV gadolinium-155 Mössbauer transition, even a high Z gasket material of normal thickness will not stop the γ-rays. Mulder and Thiel,[34] using a TAU cell, added an external gold collimator with a funnel-like shape to a 0.1 mm tungsten sample gasket. The source was a neutron irradiated $SmPd_2$ pellet enriched with samarium-154. The source was 1.2 mm thick and 2.0 mm diameter and was located ca. 3 mm from the absorber. Effects of the order of 0.1 percent are typical of measurements at liquid helium temperatures. Counting times were heroic.

Ytterbium-170. The 84.3 keV γ-ray energy in ytterbium-170 necessitates a collimator solution similar to that for gadolinium-155. The first ytterbium-170 Mössbauer spectral work in a diamond anvil cell was published by Meyer et al.,[33] who used a $Ta_{90}W_{10}$ gasket preindented to 50 μm and an external tantalum collimator with a 99.9 percent effectiveness. The isotope enrichment was 70 percent, and pressure media of argon or methanol-ethanol mixtures were used. The source was made from a 2 mm diameter pellet of $^{170}Tm^{11}B_{12}$ which was irradiated for 5 days in a flux of 10^{14} n/cm^2 and annealed at 1100 °C for 5 hours. Absorbers were typically 250 μm diameter and source-cell distances vary between one and two mm.

3. HIGH PRESSURE MÖSSBAUER SPECTROSCOPY MEASUREMENTS

3. 1. Phase Transitions in Magnetic Materials

Soon after its discovery, the Mössbauer effect was recognized as an excellent tool for studies of magnetism on an atomic scale. A substantial fraction of Mössbauer spectral studies were and are being performed on magnetic materials, and the significance of its contribution to magnetism can hardly be overstated. The combination of magnetic information available from the hyperfine or the transferred hyperfine interactions and the high cross-section for γ-resonance makes Mössbauer spectroscopy in a diamond anvil cell the only available tool for studies of magnetism at pressures above ca. 10 GPa. Furthermore, a rich variety of the materials that are magnetic contain or can be doped with iron-57. Major constituents of the Earth's interior are iron-bearing magnetic insulators, and investigating their magnetic and electronic properties under extreme pressure and temperature conditions is of great importance to the whole geophysical community. These antiferromagnetic insulators are often referred to as Mott insulators. With increased pressure, some undergo a *spin-crossover* transition that precedes either a structural or an insulator-metal phase transition, i.e., the Mott transition.

3. 1. 1. Insulator-metal Transitions of Mott Insulators

Introduction. In recent years compounds involving 3d transition metal elements have been the focus of intensive research in materials science because their electrical and magnetic properties may be substantially altered either by doping or by changing pressure or temperature. Recent examples include high-T_c superconductivity in doped rare-earth copper-oxide materials,[35] unusual "giant" magnetoresistance effects in doped manganese oxide materials,[36] temperature-tuning of an isostructural insulator to metal transition in rare-earth nickelate compounds,[37] and a pressure-induced isostructural insulator to metal transition concurrent with a collapse of antiferromagnetism in transition metal halides,[38] see below. More than half of the binary transition metal compounds are antiferromagnetic insulators under ambient conditions, counter to the conventional electronic band structure expectation that those with partially filled *3d* bands should be metallic. These magnetic insulators have local moments and large optical gaps of 2 to 4 eV, and the energy band gap persists above T_N. These magnetic compounds are generally referred to as Mott insulators.

External pressure is a non-invasive and controlled means of varying the insulating gap in these systems; for example, gap-closure may occur as a result of the bandwidth broadening associated with decreased lattice dimensions. This is in contrast to doping narrow band systems where the complications of electronic and structural disorder may lead to aberrations that do not occur with pressure alone.[39]

Our recent studies of magnetism under extreme conditions have focused on Mössbauer spectral pressure studies of *3d* transition metal compounds that are Mott insulators at ambient pressure. If iron-57 Mössbauer spectroscopy is used for the magnetic studies, then the pressure dependence of the transition metal electronic and spin-state may be investigated directly, given that the link between magnetism and electronic structure can be ascertained. Examples will be given below to illustrate the rich pressure response for a variety of Mott insulators having iron in different initial oxidation states. To help interpret the results of Mössbauer spectral pressure studies, it is necessary to understand the origin of the insulating gap and its relation to the electronic structure of the constituent atoms of the solid at ambient pressure.

Electronic structure of 3d transition metal compounds. It has been recognized that the localization of carriers and corresponding insulating behavior in the Mott insulator is a result of strong on-site Coulomb repulsion, U, which is not mitigated by the restricted range of kinetic energies available to a narrow d-band system. An energy gap occurs in the narrow *3d* band, splitting it into an upper and a lower Hubbard sub-band separated by U, which is greater than the d-band-width, W, that is mainly determined by the crystal structure, i.e., the lattice parameter. This is illustrated in Fig. 10. Doping or external pressure may be used to vary U or W, respectively, whereupon there are often marked changes in physical properties; for example, the compound may undergo an insulator to metal transition because of band-filling/depletion or gap-closure. In view of the origin of the "Coulomb" gap, it is anticipated that a pressure-induced gap-closure, and its associated insulator to metal transition, does not necessarily involve a structural change and should be accompanied by a collapse of the magnetic moment, a process often called the Mott transition. Mössbauer spectroscopy may be used to track the evolution with pressure of magnetism in Mott insulators; iodine-129 Mössbauer spectroscopy was used to detect a Mott transition in NiI_2 for the first time.[40]

It is evident from Fig. 10 that the insulating gap can be of the Mott-Hubbard d-d type or of the so-called charge-transfer p-d type, with the gap designated as Δ. The gap will depend on whether the transition metal-d or anion-p band is the highest occupied band. Other possible gaps include those of transition metal d-transition metal $4s$ or even anion p-transition metal $4s$ character, depending on the relative magnitudes of U and Δ and the relative positions of the $4s$ band-edge. The nature of the insulating gap depends on the crystal structure and the constituent ions that make up the solid; for example, Δ is directly related to the electronegativity of the anion and the Madelung potential.[41] It also determines the mode of carrier delocalization in the metallic state upon closure of the gap. Charge fluctuations involving transition metal ions only, $d^n + d^n \rightarrow d^{n-1} + d^{n+1}$, where d^n is the electronic configuration of the transition metal site, are expected for a Mott-Hubbard intra-band gap closure. Whereas ligand-to-metal charge transfer of the type $d^n \rightarrow d^{n+1}, \underline{L}$, where $\underline{L}$ is a ligand hole, is expected for a charge transfer inter-band gap closure.

It should be noted that the d states of a transition metal ion consist of an orbital multiplet split by crystal-field effects; for example, in cubic symmetry the d states split into two-fold, e_g, and three-fold, t_{2g}, orbitals broadened into bands due to translational symmetry. Therefore, in a more detailed version of Fig. 10, the gap U would be present in both e_g and t_{2g} sub-bands. With increasing formal valence of the metal, Δ systematically decreases and becomes very small or even negative for compounds involving unusually high valences, such as iron(IV). This is a case where there is strong hybridization, i.e., charge transfer, between p and d states resulting in a $d^n \underline{L}$ hybridized valence band that is primarily of anion-p character. The result of strong hybridization is to split-off a $d^n \underline{L}$ discrete state, effectively

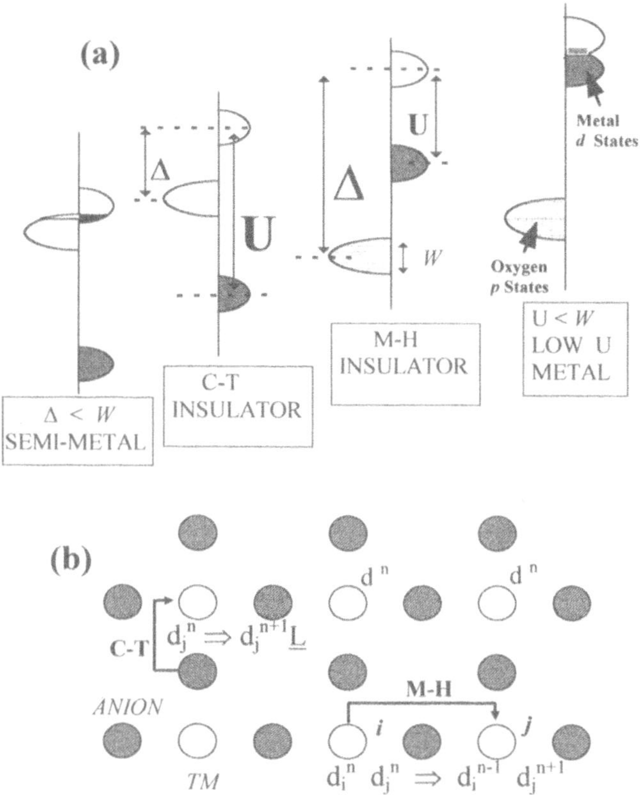

Figure 10. A schematic diagram of the insulating gaps that occur in the band structure of 3*d* transition metal compounds due to the on-site electron-electron Coulomb repulsion, *U*. (a) The highest fully occupied anion *p* band and the 3*d* band are shown on the left and right side of the vertical axis, respectively. The effect of *U* is to split the narrow 3*d* band into an upper and lower subband. Shaded bands represent occupied states, and unoccupied states are left unshaded. The bandwidth, *W*, is assumed to be the same for all three bands for simplicity. The nature of the gap, charge transfer or Mott-Hubbard, depends on the relative positions of the three bands. (b) Representation of the most likely charge-fluctuations that occur in an ionic lattice consisting of transition metal ions and closed-shell anions, depending on the nature of the original gapped state.[54]

introducing a gap into the valence band; carrier delocalization occurs by ligand-to-ligand charge transfer, $d^n\underline{L} + d^n\underline{L} \rightarrow d^n + d^n\,\underline{L}^2$, upon closure of the gap.[42]

The origin and nature of the insulating gaps depicted in Fig. 10 emerged from a theory developed by Zaanen, Sawatzky, and Allen[43] in an attempt to understand the nature of the gaps in transition metal halides, oxides, and sulfides and to interpret photoelectron spectra of these compounds. It is now well-established that the physical properties of transition metal compounds may be drastically modified in the regimes of both the pressure-temperature and the composition-temperature phase diagrams near the insulator to metal transition, where the Zaanen, Sawatzky, and Allen framework is not necessarily applicable but where there is a dearth of alternative theories. Suitable theories should be formulated in terms of parameters involving U, Δ, and the bandwidth, W. Experiments in which these parameters are varied, e.g., pressure experiments, represent a crucial test of theory and thus provide incentives for the development of appropriate alternative theories to understand the effect of perturbations on the gap state in narrow band systems, especially close to an insulator to metal transition.

The simultaneous occurrence of an isostructural insulator-metal transition and of *d*-electron delocalization are inherent features of a Mott transition. Delocalization coincides with a transition from the magnetic to a non-magnetic state. Therefore, to probe experimentally the pressure-induced Mott transition one needs high-pressure capabilities for x-ray diffraction, resistivity, and magnetism. Such properties for pressure-induced Mott transition were first demonstrated in NiI_2 by Pasternak et al.,[40] and similar combined studies in CoI_2 and in FeI_2 followed.[44,45]

Transition Metal-Iodides. In this section we summarize the results reported for the antiferromagnetic-insulators NiI_2, CoI_2, and FeI_2, using iodine-129 and iron-57 Mössbauer spectroscopy in a diamond anvil cell. NiI_2 and FeI_2 crystallize in the CdI_2- $P3m$, and CoI_2 in the $Cd(OH)_2$-$C3m$ structures.[46] These structures consist of alternating layers of weakly interacting I-transition metal-I and can be described as a hexagonal lattice with three NiI_2 or FeI_2 units or one CoI_2 unit per unit cell. At ambient pressure they order antiferromagnetically[47] with T_N = 75, 12, and 9 K for NiI_2, CoI_2, and FeI_2, respectively. Each iodine is located at the top of a trigonal pyramid of three equivalent metal nearest neighbors. Along the c-axis the iodine atom is axially symmetric, C_{3v}. In the spin-ordered state, the divalent transition metal ions located within a sublattice plane, i.e., *ab*-plane, are ferromagnetically ordered, and the hyperfine field, H_{hf}, at the iodine nucleus is proportional to the vector sum of the surrounding divalent transition metal moments, see Fig. 11. Thus, H_{hf} serves as a local probe for the sublattice magnetization and its structure at each temperature and pressure. The magnetic structure at ambient pressure has been determined by neutron diffraction.[48] Ambient pressure iodine-129 Mössbauer studies of the iodides agree with the neutron diffraction magnetic structure of NiI_2, namely a ferromagnetically ordered sublattice with the nickel moments canted at ca. 55° with respect to the *c*-axis.

The samples for Mössbauer spectral studies were prepared in milligram amounts by direct metal-vapor reaction of high purity metals with a stoichiometric amount of elemental $^{129}I_2$, see Section 2. 2. The reaction took place at approximately 700 °C in evacuated quartz tubes. Due to their hygroscopic nature, the samples had to be loaded into the diamond anvil cell under a dry nitrogen atmosphere. For x-ray diffraction and conductivity measurements, samples with natural isotopic abundance were also prepared. X-ray diffraction was carried out at 300 K by the angle-dispersive method using a molybdenum monochromatic radiation source, and the resistance measurements versus pressure and temperature were obtained using a quasi-four point method.

Mössbauer measurements were carried out at temperatures in the range of 4 to 320 K and at pressures in the range of 0 to 32 GPa. From these measurements the pressure dependencies of H_{hf}, T_N, and the isomer shift were deduced. In the paramagnetic state the spectrum was fitted assuming an axially symmetric electric field gradient. At temperatures below T_N, the experimental data were least-squares fitted using the following Hamiltonian to calculate the transition energies and intensities,

$$H = \mu \bullet H_{hf}\,(I_z/I) + \{e^2q_{zz}Q(3\cos^2\theta - 1)/[8I(2I - 1)]\} \bullet \{3I_z^2 - I(I - 1)\}, \quad (2)$$

where μ is the nuclear moment, I_z and I_z^2 are spin operators, H_{hf} is the hyperfine field, $e^2q_{zz}Q$ is the quadrupole coupling, and θ is the angle between H_{hf} and the principal axis of the electric field gradient tensor, eq_{zz}, the latter being along the c-axis. In fitting the spectra at $T < T_N$, $e^2q_{zz}Q(3\cos^2\theta - 1)$ was fixed to be zero.

The onset of the Mott transition is clearly seen in Fig. 12 where we observe the pressure evolution of the hyperfine field for CoI_2. Beyond a critical pressure, P_c, the magnetic state collapses, and the material becomes metallic. In Fig. 13 the temperature variation of the $Ni^{129}I_2$ spectra at 6.7 GPa is shown. From the temperature and pressure dependencies of the

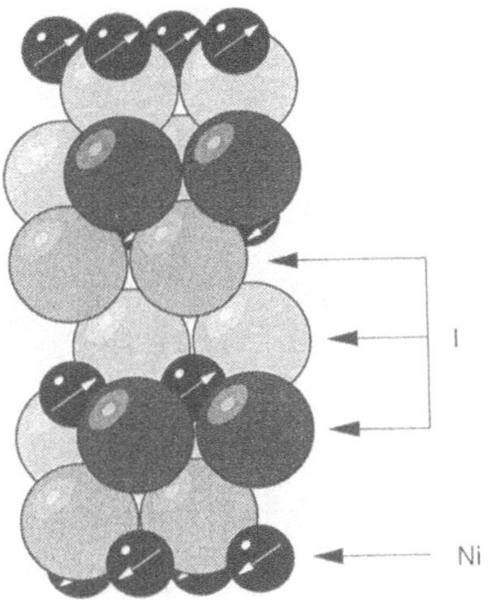

Figure 11. The structure of NiI_2 or FeI_2 at $T<T_N$. The Ni^{2+} moments form a 55° angle with respect to the *c*-axis. Each iodine sees three Ni nearest neighbors, and the transferred hyperfine interaction at the ^{129}I nucleus is a measure of the moments on the Ni. At $T>T_N$ the moments are disordered; for $P>P_c$ there are no moments at any temperature, and the $H_{hf}(I)$ is zero.[38]

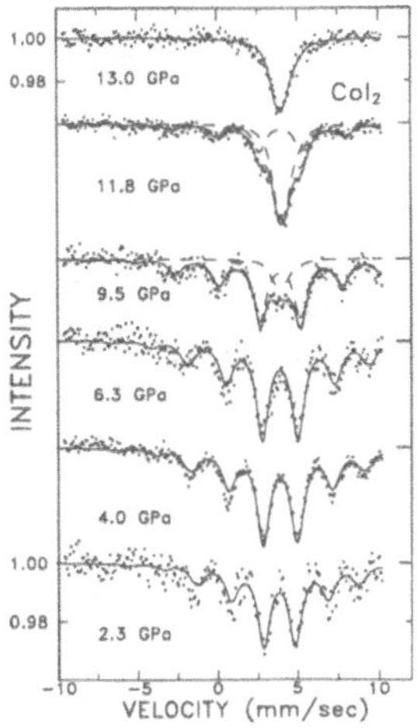

Figure 12. Typical Mössbauer spectra of CoI_2 at 4 K at several pressures. The solid line is a theoretical fit using the Hamiltonian of Eq. 2. Note the coexistence of the magnetic correlated and non-magnetic metallic states at 9.5 and 11.8 GPa.[44,45]

hyperfine field, the pressure dependence of T_N could be derived. A summary of the pressure variation, in terms of volume decrease, of T_N, H_{hf}, and the isomer shift for NiI_2, is shown in Fig. 14. Note that the hyperfine field is zero beyond P_c = 19 GPa. No change in the isomer shift is observed at P_c; this supports the belief that no crystallographic change is involved. Indeed the equation of state, as shown in Fig. 15, shows a smooth decrease in volume, without discontinuity, at and around P_c. Finally, in the pressure dependence of the electrical resistance, shown in Fig. 16, the insulator-metal transition is observed at P_c.

The probe for the magnetic ordering and the pressure-induced Mott transition in the transition metal iodides is the transferred hyperfine field at the iodine-129 nucleus. No direct information is available regarding the presence of paramagnetism in the metallic state. However, with FeI_2[49,50] another Mössbauer probe is available, namely iron-57. Both Mössbauer probes reveal a substantial increase in the T_N with increasing pressure, an increase which is similar to the trend observed in the nickel and cobalt cases. At ca. 20 GPa the breakdown of the correlated state is concluded from the collapse of hyperfine field at the iodine-129 nucleus. However, with iron-57 one clearly observes an onset of a paramagnetic relaxation spectrum at the lowest temperature measured, 4.2 K. The paramagnetic regime, not observed via the transferred hyperfine interaction, persists to 26 GPa with a characteristic hyperfine field at the iron-57 nucleus of ca. 30 T, as compared to a hyperfine field of 12 T in the antiferromagnetic state, see Fig. 17. The onset of the metallic state is detected by measuring the pressure and temperature dependences of the electrical resistance in the 23 to 30 GPa range, see Fig. 18. As can be seen the slope of the temperature dependence of the resistance for pressures greater than 23 GPa is definitely positive, indicative of a metallic state. The pressure-temperature phase diagram of FeI_2 is shown in Fig. 19. This was the first observation of a pressure-induced Mott transition with an intermediate paramagnetic state.

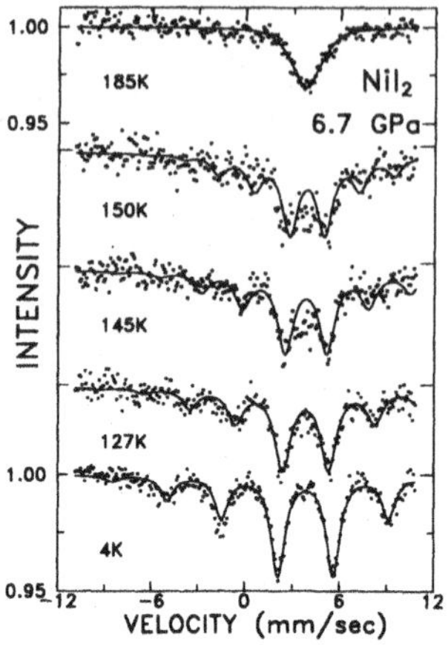

Figure 13, above. $Ni^{129}I_2$ Mössbauer spectra as a function of temperature taken at 6.7 GPa. Note the onset of the ordered state at 185 K. The spectrum shown in (e) is recorded at 26.3 GPa, beyond P_c = 19 GPa. The solid line fits are theoretical spectra based on Eq. 2.[40]

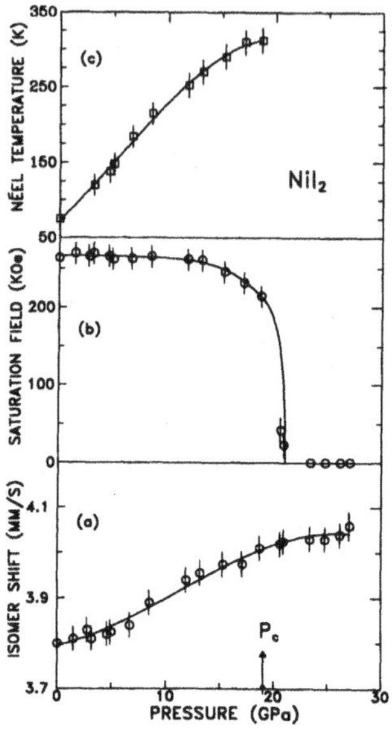

Figure 14, right. The pressure dependence of the isomer shift, hyperfine field and Néel temperature of $Ni^{129}I_2$. Note the abrupt collapse of the hyperfine field at P_c = 19 GPa. Both isomer shift and Néel temperature vary smoothly to the highest pressure. The hyperfine field is practically constant with pressure, an indication of the insensitivity of the *d*-bandwidth to decreasing volume.[40]

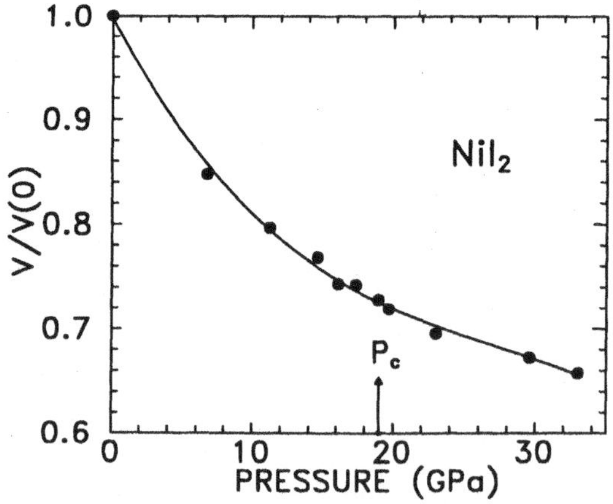

Figure 15. The equation of state of NiI_2 as measured by x-ray diffraction. Note the smooth variation of V(P) around P_c = 19 GPa. The solid curve is a least-squares fit to the Birch-Murnaghan equation of state.[40]

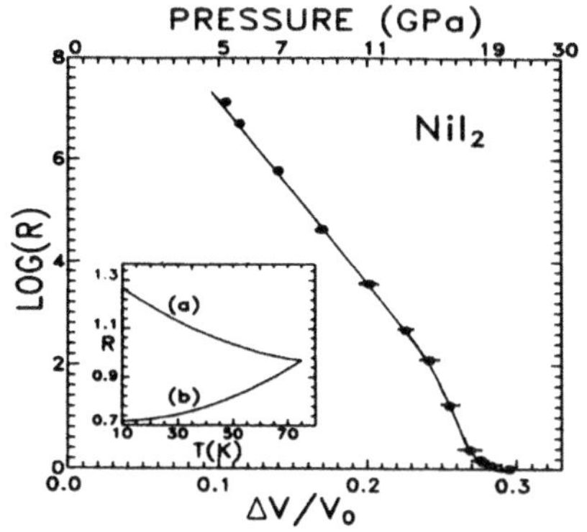

Figure 16. The volume dependence of log R of NiI_2. The resistance drops linearly to $\Delta V/V_0 = 0.25$, then drops rapidly, reaching a plateau near P_c. Inset: R(T) normalized at 77 K. Curve (a) corresponds to the semiconductor antiferromagnetic state at P = 17.5 GPa and curve (b) to the metallic, non-magnetic state at P = 21 GPa.[40]

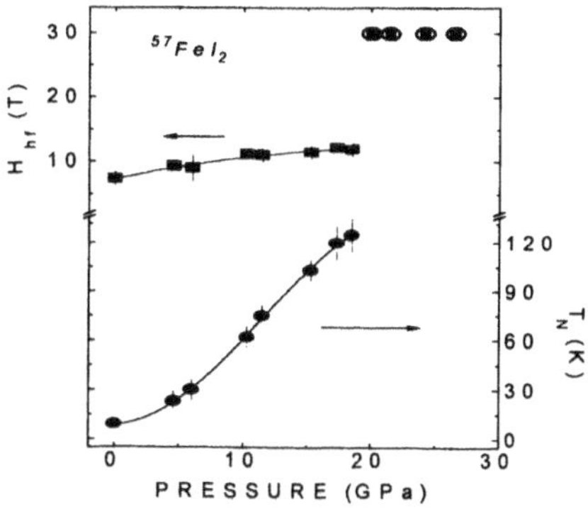

Figure 17. The pressure dependence of the saturation iron-57 hyperfine field and Néel temperature in FeI_2. H_{hf} increases monotonically up to 18 GPa. At $18 < P < 28$ GPa, within a metallic-paramagnetic regime, $H_{hf} = 30$ T. At $P > 28$ GPa the iron ion loses its moment.[49]

3. 1. 2. *Spin-crossover Phase Transitions*

Many of the pressure-induced phase transitions occurring in transition metal-oxide compounds are due to spin-crossovers. These interesting properties can currently be studied in iron compounds due to the progress in precise chemical synthesis of the oxides and in iron-57 Mössbauer spectroscopy in a diamond anvil cell. Some of the most interesting materials are the strontium, calcium, and lanthanum perovskites. The iron-based perovskites have iron

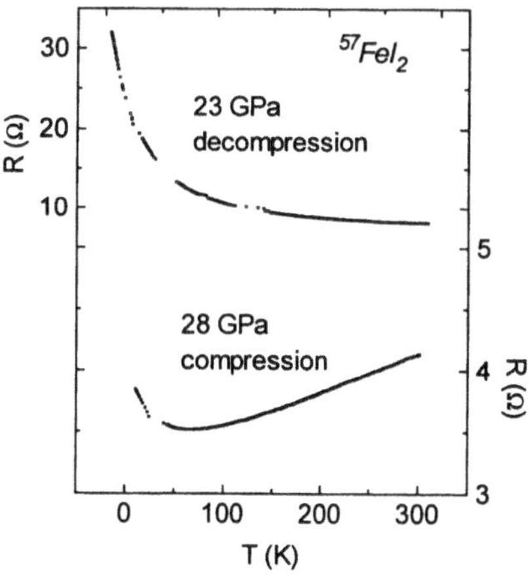

Figure 18. The temperature dependence of the resistance of FeI_2 for various pressures in the vicinity of the pressure-induced Mott transition. The band gap persists in the Mott insulator up to 23 GPa. Between 23 and 28 GPa the material is a metal with a magnetic moment. Above 28 GPa FeI_2 is a non-magnetic metal.

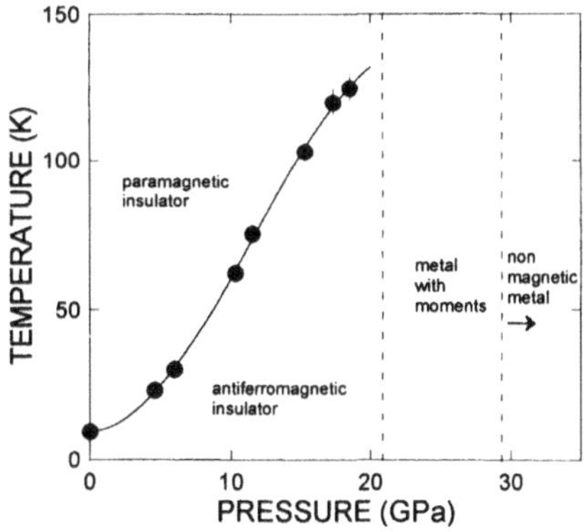

Figure 19. The pressure-temperature phase diagram of FeI_2. The curve separating the paramagnetic and antiferromagnetic states is the $T_N(P)$.[49]

in the unusually high oxidation state of iron(IV) ($3d^4$). These compounds have already been investigated quite extensively by Mössbauer spectroscopy in separate experiments at cryogenic temperatures and at high pressure.[51] The first measurements of this type were by Takano et al.[52] in $CaFeO_3$, see Fig. 20.

Recently, another tetravalent iron-based compound, Sr_2FeO_4, has been the subject of considerable investigation.[29,49,53,54] The compound is an antiferromagnetic semiconductor having a Néel temperature of ca. 60 K. It is an isostructural analogue of the antiferromagnetic insulator, La_2CuO_4, the parent compound of the original strontium-doped high-T_c materials discovered by Bednorz and Müller.[35] Sr_2FeO_4 is tetragonal and has a two dimensional type structure comprised of sheets of FeO_6 octahedra perpendicular to the c-axis with each iron(IV) ion in D_{4h} symmetry. Iron-57 Mössbauer spectroscopy at ambient pressure shows that the single iron site quadrupole doublet at room temperature evolves with decreasing temperature into a complex magnetic spectrum consisting of at least four sites having saturation hyperfine field values in the range 25 to 33 T.[55] This has been attributed to a spiral-spin structure; from neutron diffraction studies on the same sample, there is no evidence of a crystallographic change down to 4 K.

Raman and optical spectroscopy studies at pressures of 20 and 38 GPa have recently been performed on this compound.[56] An oxygen-derived Raman mode disappears at about 5.5 GPa. The optical reflectivity increases substantially in the near-infrared region as the pressure rises above 6 GPa; at pressures greater than 13 GPa it saturates at a value considerably higher than found at ambient pressure. From this behavior it has been speculated that the semiconductor gap narrows under pressure to culminate in a semiconductor-metal transition.

Recent iron-57 Mössbauer, x-ray diffraction, and electrical-resistance pressure studies[54] on Sr_2FeO_4 complement these previous studies. T_N at 9 and 19 GPa was estimated from Mössbauer zero velocity thermal scans or temperature dependent studies of the magnetic spectra appearing below T_N. Because the Sr_2FeO_4 was unenriched, each magnetic spectrum required several days to obtain sufficient statistics for meaningful comparison with ambient pressure results. Mössbauer spectral parameters, determined as a function of pressure, have been plotted in Fig. 21. The isomer shift decreases smoothly in the measured pressure range, consistent with an increase in the density and with preliminary synchrotron x-ray measurements that showed no structural change up to 27 GPa. The quadrupole splitting tends

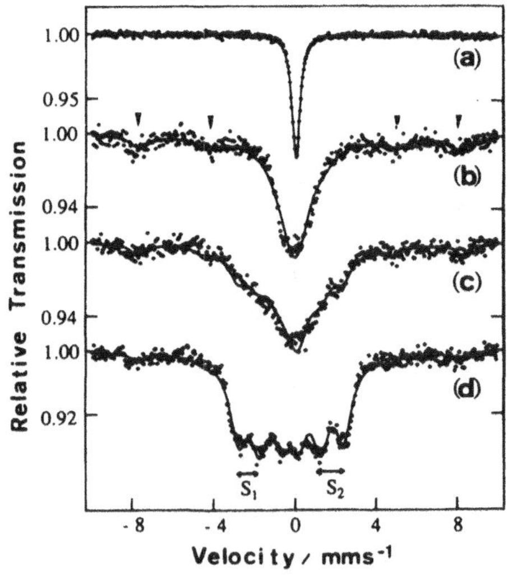

Figure 20. The room temperature iron-57 spectra of $CaFeO_3$ at (a) 0.1 MPa, (b) 21 GPa, (c) 30 GPa, and (d) 41 GPa. The arrows allude to a $Ca_2Fe_2O_5$ magnetic impurity. The magnetic splitting due to the onset of the antiferromagnetic interaction is evident in (c) and (d). The authors' conclusion[52] is that the iron(IV) state is preserved to the highest pressure and no disproportionation is observed.

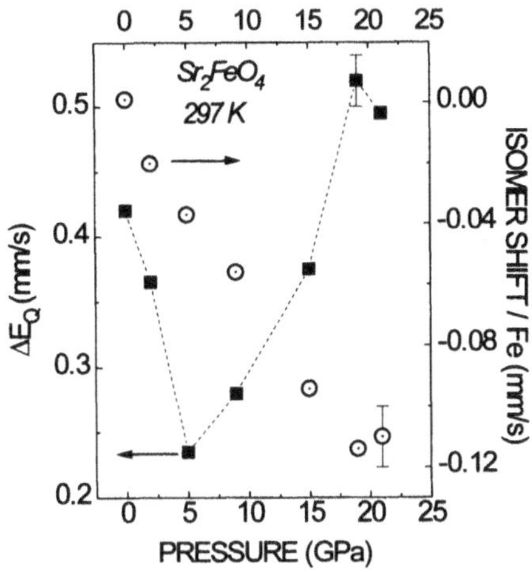

Figure 21. Pressure dependence of the isomer shift and quadrupole splitting of Sr_2FeO_4 at room temperature.[49]

towards a minimum value at ca. 5.5 GPa. Low temperature Mössbauer spectra of Sr_2FeO_4, measured in a diamond anvil cell at 9 GPa, exhibit sharper magnetic features as compared to ambient pressure[54,55] and at 19 GPa become simpler yet. The maximum value of the hyperfine field is estimated to be ca. 25 T, and there is a substantial quadrupole shift. The temperature dependence of the resistance, measured in a diamond anvil cell at pressures to ca. 20 GPa, showed that a semiconductor-metal transition occurs at 17(1) GPa; the metallic behavior is clearly evident at 19 GPa.[54]

The iron local environment is a slightly elongated octahedron under ambient conditions. As the c/a ratio of the tetragonal unit cell is decreased by pressure, the original elongated octahedron condenses into a regular configuration having all iron-oxygen bond lengths almost equal, whereupon the electric field gradient is expected to attain its minimum value. Subsequent added pressure may then produce a new distortion of the octahedra with an increased electric field gradient. An oxygen-derived Raman mode disappears at ca. 5.5 GPa,[56] possibly because it is no longer Raman-active in a tetragonal unit cell having regular FeO_6 octahedral units.

The main conclusions derived from the Mössbauer spectral experiments under pressure are, (1) in the metallic state the compound is still magnetic with a T_N of ca. 130 K, (2) the original magnetic structure, deduced to be spiral-spin-type in the two dimensional structure at ambient pressure, changes considerably under pressure, (3) there is no evidence of charge disproportionation of the type $2\ Fe^{4+} \rightarrow Fe^{3+} + Fe^{5+}$, and (4) the electronic and magnetic properties are similar to those of the cubic perovskite, $SrFeO_3$, found by Nasu[29] to be metallic and magnetic with T_N of ca. 131 K at ambient pressure.

Carrier delocalization in the metallic state of compounds having the transition metal atom in an unusually high oxidation state is supposed to occur by ligand-to-ligand charge transfer,[57] $d^{n+1}\underline{L} + d^{n+1}\underline{L} \rightarrow d^{n+1} + d^{n+1}\underline{L}^2$. This may barely influence the magnetic moment of the iron atom in Sr_2FeO_4 and thus explain the persistence of magnetism even within the metallic state at 19 GPa.

3. 2. Pressure-induced Amorphization

Increasing experimental evidence indicates that the topology of structural disorder induced by hydrostatic pressure is distinctly different from that of classical amorphous glasses produced by rapid cooling from the melt. Pressure-induced amorphization has been observed for over ten years. Recently, with an increasing number of studies employing a variety of experimental methods, and with the implementation of promising theoretical models, it is being realized that the mechanism of pressure induced amorphization may be an entirely new phenomenon in the solid state. Some examples of the unusual properties not encountered in conventional amorphous materials are, (1) the recovery of the crystalline state upon decompression, (2) the memory effect, in which a single crystal undergoing a pressure-induced amorphization recovers its original translational and orientational properties upon decompression, and (3) the retention of the original Néel temperature in the amorphous phase. It is obvious that determining the nature of the disorder on the atomic scale is essential for developing the mechanism of pressure-induced amorphization.

Amorphization is determined by the complete loss of sharp x-ray diffraction patterns and the weakening or loss of optical vibrational frequencies related to the long range translational modes. The usefulness of optical spectroscopy for structural studies of the amorphized state is further impaired when the sample is opaque, say due to metallization. In what follows we summarize the information obtained by Mössbauer spectroscopy, in conjunction with some other spectroscopic methods, on ML_4 molecular crystals, where M is Ge or Sn and L is Br or I and a few oxides containing iron.

3. 2. 1. Germanium and Tin Halide Molecular Crystals

The ML_4 molecular crystals constitute an excellent model for studying pressure-induced amorphization. SnI_4 was one of the first crystals to have been amorphized by pressure. Methods such as Raman,[58] x-ray diffraction,[59] tin-119 and iodine-129 Mössbauer spectroscopy,[60] and conductivity[61] measurements were used to study the mechanism of the amorphization process. Combined x-ray diffraction, Ge-K edge x-ray absorption spectroscopy, and iodine-129 Mössbauer spectroscopy were employed[62] to study the valence and structure, at the crystalline to amorphous phase transition, in the amorphous state of GeI_4. The ML_4 compounds are the only known *molecular crystals* to undergo pressure-induced amorphization. The studies of SnI_4 and GeI_4 were conducted to 30 GPa. Both materials undergo insulator-metal transitions at or near the amorphization pressure.[58,59] In these studies it was unambiguously concluded that amorphization is closely linked to a fundamental change in the intermolecular bonding character. The process of molecular association, i.e., polymerization, takes place with increased density in the cases of GeI_4 and SnI_4 with a reversible crystallization under decompression, near ambient pressure. Despite an increase in density, as measured by the increase in isomer shift at the amorphous phase, the principal Ge-I distances increase.[62] This is a novel discovery attained only by means of atomistic probes such as x-ray absorption spectroscopy and Mössbauer spectroscopy.

***SnI₄* and *GeI₄*.** First indications of crucial changes in the band-structure, suggesting an insulator to metal transition in SnI_4, were reported by Pasternak and Taylor.[60] Once the amorphous/metallic state was formed, a new single line in the tin-119 Mössbauer spectrum was detected that differed dramatically from that of the low-pressure phase, see Fig. 22. The isomer shift value of 1.5 mm/s, characteristic of an insulating crystalline SnI_4 with tin(IV), changed discontinuously to 3.7 mm/s, indicative of a large increase in $\rho_s(0)$ of about one s electron/atom. This was accounted for by the gap-closure, upon metallization, of the *valence*, I^- - $5s^25p^6$ and conduction, Sn^{4+} - $5s^05p^0$ bands.

Iodine-129 Mössbauer spectroscopy proved to be essential for determining the structure of GeI_4 and SnI_4 upon amorphization. At pressures smaller than P_c the single quadrupole-split site, with a negative e^2qQ and $\eta = 0$, is typical of the C_{3v} symmetry of the iodine σ-bonded to the central germanium or tin ion. It is characteristic of the low pressure phase, see Fig. 23.

At pressures greater than P_c, a second spectral component appears, characterized by a positive and reduced value of e^2qQ and $\eta = 0.6$, indicative of the formation of π-bonding.

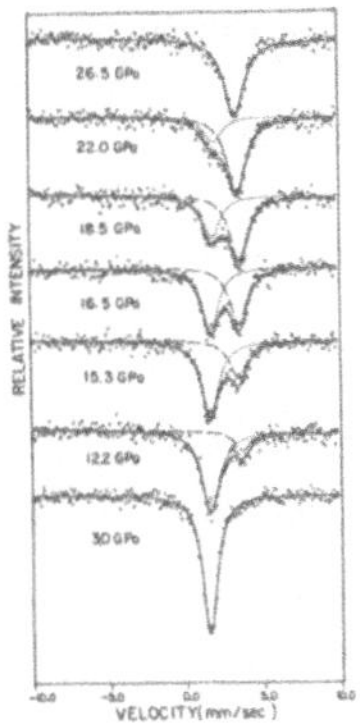

Figure 22, above. The pressure evolution of the high pressure state of tin(IV) due to metallization, i.e., band-gap closure, of SnI_4.[60]

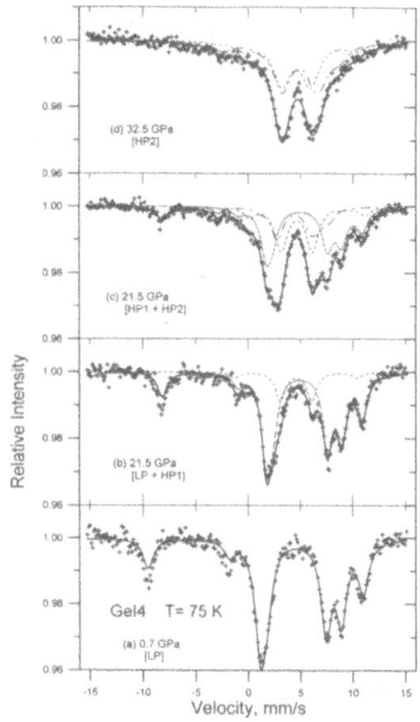

Figure 23, right. Typical $Ge^{129}I_4$ spectra with increasing pressure. The solid line is the convolution of the various subspectra corresponding to different iodine sites, see text.[62]

This high pressure phase, designated as HP1, is due to a strong I–I intermolecular overlap resulting in *molecular association*. In the case of SnI_4, HP1 and the low pressure phase coexist up to 22 GPa, and HP1 dominates the amorphous phase up to 35 GPa, the highest pressure utilized. HP1 is modeled as disordered chains of $(SnI_4)_n$ where two out of four iodines are *intermolecularly* bonded. In GeI_4 the low pressure phase exists to 14 GPa; HP1 coexists with the low pressure phase in the 14 to 20 GPa pressure range, and at 21 GPa a new phase is formed where all four iodines overlap intermolecularly forming a HP2 phase. This three dimensional phase has not been observed in SnI_4 although it may appear at pressures higher than 35 GPa.

Unique information concerning changes in the iodine $5p$ valence-band can be derived from the isomer shift. In general, the isomer shift is proportional to the s-electron density at the nucleus. Chemical bonding of the iodides primarily involves the outer $5p$ valence shell, with the $5s^2 5p^{6-h_p}$ configuration, where h_p is the number of p-holes. The relationship between the isomer shift, IS, and h_p, with respect to a $Mg_3{}^{129m}TeO_6$ source in the $1 > h_p > 0$ range is well established,[63] namely,

$$IS = 1.50h_p + 3.58 \text{ mm/s}. \quad (3)$$

The abrupt increase in isomer shift both in the HP1 and HP2 components is explained as due to a depletion of the p-electrons of about 0.2 electrons per iodine in its valence band. This charge-transfer process, $I_{5p} \xrightarrow{e^-} Ge_{4s4p}$, can explain the metallization of GeI_4 and SnI_4 in the amorphous phase. In this phase the intermolecular van der Waals interaction is replaced by a metallic/covalent bond with increased Ge-I distances. The simultaneous transition to a metallic and amorphous phase is also observed in the analogous SnI_4. The close relationship

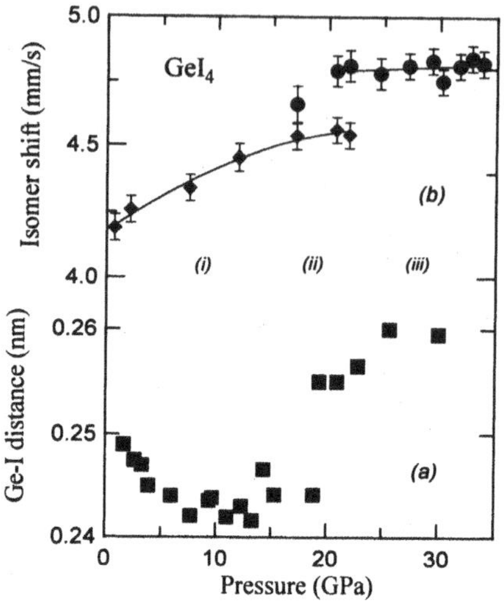

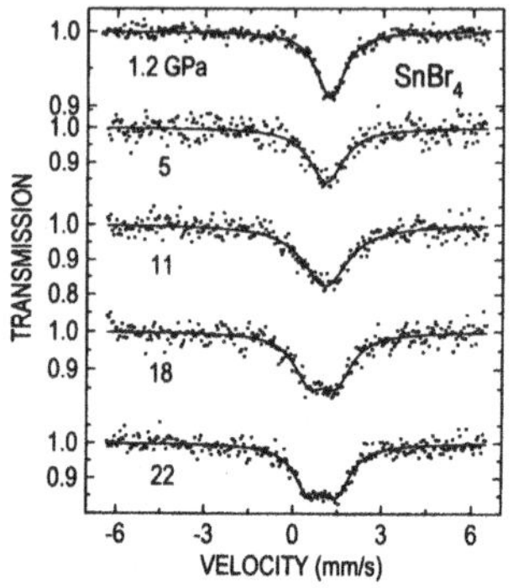

Figure 25, above. Typical $^{119}SnBr_4$ spectra at 80 K for various increasing pressures. The dotted line is the isomer shift for tin(II). Spectra for pressures greater than 5 GPa were fitted with a quadrupole splitting, see text.[65]

Figure 24, left. (a) Ge-I interatomic distances in GeI_4 as a function of increasing pressure. The pressure regions corresponding to the pure molecular-crystal phase, the intermediate range where both crystalline and amorphous phases coexist, and to the pure amorphous phase are indicated as i, ii, and iii. In the ii range, the Ge-I distances are the average derived from the Ge K-edge x-ray absorption spectroscopy data. (b) The ^{129}I isomer shift data. The solid diamond and circle symbols correspond to the low and high pressure phases, respectively. Notice the coinciding decrease of the Ge-I distance with increase in isomer shift in the i region, the coexistence of two values of isomer shift in the intermediate ii phase, and the discontinuous increase in isomer shift in the iii phase. The solid line is a guide to the eye. Values of the isomer shift are with respect to $Mg_3{}^{129m}TeO_6$ at 70 K.[62]

between the isomer shift and the Ge-I distances, as deduced from the x-ray absorption experiments, is described in Fig. 24. The question then arises whether the occurrence of metallization and amorphization in the ML_4 molecular crystals are correlated. Apparently not; recent high pressure optical[64] and tin-119 Mössbauer[65] studies in the analogous $SnBr_4$ definitely show that the amorphous state is reached at around 14 GPa and is not accompanied by metallization.

SnBr₄. In Fig. 25 the pressure evolution of $^{119}SnBr_4$ spectra is presented.[65] Unlike $^{119}SnI_4$, a new single-line spectrum, characterized by a large isomer shift, is not detected. Instead, at a pressure greater than 5 GPa the single line spectrum, characteristic of a symmetric tetrahedron, evolves into a quadrupole-split doublet characteristic of a distorted tetrahedron. The cause for this distortion has been attributed to the formation of a $(SnBr_4)_2$ dimer. Experimental support for this dimerization comes from the pressure dependence of the area under the Mössbauer spectral absorption peak. At a pressure greater than 5 GPa, see Fig. 26, an abrupt increase in area is observed. The Raman studies[66] did not reveal any new modes suggesting hardening of the phonon spectra, thus, it has been concluded that a molecular crystal composed of $(SnBr_4)_2$ dimers would be compatible with such an increase in the effective mass, M, as it appears in the recoil-free fraction, f, expression,

$$f = \exp\left\{\frac{-3E_\gamma^2}{Mc^2k_B\theta_D}\left[\frac{1}{4}+\left(\frac{T}{\theta_D}\right)^2\int_0^{\theta_D/T}\frac{x}{e^x-1}dx\right]\right\}. \tag{4}$$

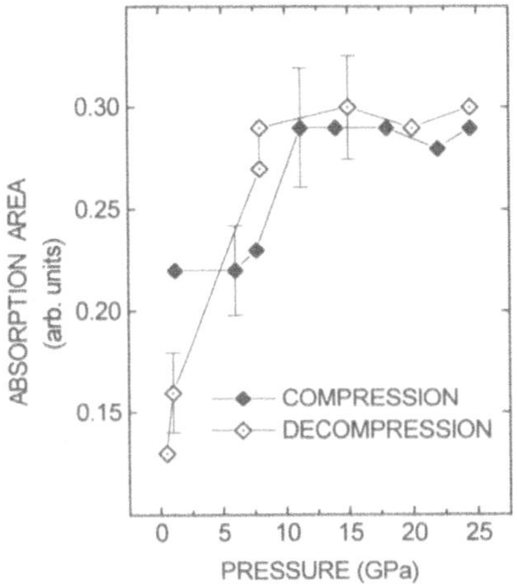

Figure 26. Pressure evolution of the area under the absorption peak. An abrupt increase is observed at pressures greater than 5 GPa and is attributed to the formation of a molecular crystal composed of $(SnBr_4)_2$ dimers.[65]

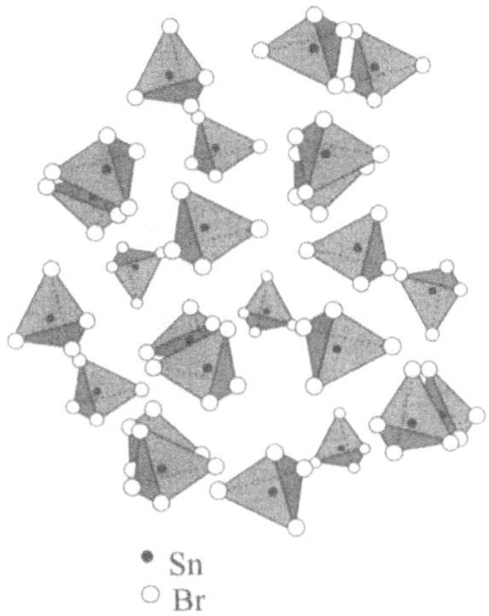

Figure 27. Proposed model of a disordered $SnBr_4$ molecular crystal present at pressures between 5 and 25 GPa. The building blocks are dimers of $(SnBr_4)_2$.[65]

In Fig. 27 a schematic of the proposed model for the disordered molecular crystal is depicted.

3. 2. 2. Iron-based Compounds

Pressure amorphization of an antiferromagnetic oxide has been reported only for *fayalite*, Fe_2SiO_4. Crystalline Fe_2SiO_4 orders antiferromagnetically at 64 K and Mössbauer spectral studies by Hafner et al.[66] reveal two magnetic sites, M1 and M2. A sample enriched in the iron-57 isotope was synthesized by heating the component metal-oxides under a controlled oxygen fugacity and was characterized by x-ray diffraction and Mössbauer spectroscopy. Subsequently it was pressure-induced into a glassy state at a pressure greater than 40 GPa and 300 K.[67] The pressure-induced amorphized material retains its disordered state even after pressure release, facilitating measurements outside the diamond anvil cell with Mössbauer spectroscopy, x-ray diffraction, and susceptibility[68] on the same sample unencumbered by the diamond anvil cell.

Mössbauer spectra measured between 70 and 300 K display a broad line quadrupole split spectrum with a splitting of ca. 3 mm/s. Upon cooling, antiferromagnetic ordering is discernible at 65 K, see Fig. 28, and below T_N a complex spectrum is observed characteristic of a wide distribution of hyperfine fields. There are strong indications that, as in the crystalline state, there are two magnetic sites, M1 and M2, with M2 ordering at a lower temperature of ca. 28 K.

In contrast to pressure-induced amorphization, previous studies of magnetic glasses produced by rapid quenching from the melt have revealed that either there is no magnetic ordering over the observed temperature range[69] or T_N is significantly decreased due to the loss of structural order.[70] Thus, the fact that T_N is unaffected by the loss of long range crystalline order implies that, for one class of sites, M1, the magnetically ordered state is antiferromagnetic rather than frustrated or speromagnetic. This magnetic behavior of disordered matter is quite unique; further studies with $Fe(OH)_2$ and $FePO_4$ are now in progress[71] to further study this phenomenon.

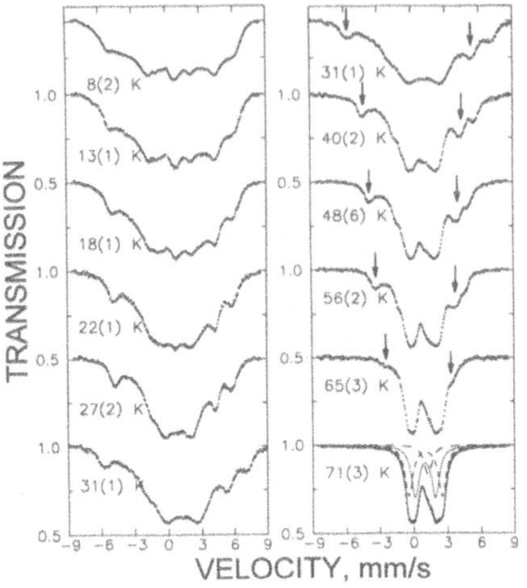

Figure 28, left. Temperature dependence of the Mössbauer spectra of amorphized Fe_2SiO_4. Note the onset of the magnetic interaction (arrows) at 65 K. The onset at T_N = 65 K is attributed to the magnetic site, M1. It is assumed that M2 orders at the lower T_N, see text.[68]

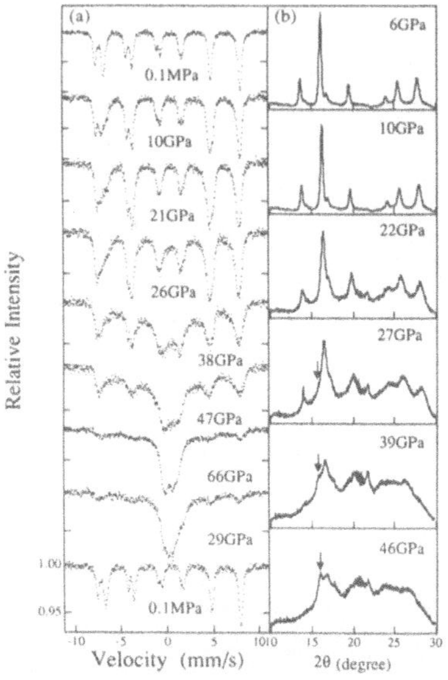

Figure 29, right. Pressure evolution of the iron-57 Mössbauer spectra of Fe_3O_4 at 300 K. Note the onset of the non-magnetic component at pressures greater than 26 GPa observed both in the Mössbauer spectrum, left side, and x-ray diffraction (arrow) lines. The hysteresis with pressure decrease is clearly detected at 29 GPa.[74]

3. 3. Miscellaneous Mössbauer Spectral Studies

3. 3. 1. Fe_2O_3, Fe_3O_4, and $Fe_{1-x}O$

These major iron-oxide compounds have been the subject of several Mössbauer spectral studies in a diamond anvil cell in recent years. The first iron oxide Mössbauer study in a diamond anvil cell was that of *hematite*, Fe_2O_3, by Syono et al.[72] followed by Nasu et al.[22] Experiments were carried out at 300 K using a piston/cylinder Bassett cell. A gradual transition into a non-magnetic or paramagnetic state was observed at 50 GPa. The highest pressure achieved by Nasu et al.[22] was 72 GPa, a pressure at which only a quadrupole split spectrum was detected. The isomer shift of the new phase is lowered considerably suggesting an increase in the s-electron density consistent with the sharp decrease in volume of the new phase as detected by recent x-ray diffraction measurements.[73]

A high pressure Mössbauer spectral study in a diamond anvil cell of magnetite, Fe_3O_4, was carried out by Pasternak et al.[74] to 66 GPa at 300 K. This cubic inverse spinel compound is one of the oldest known magnetic materials and orders ferrimagnetically at 851 K. It was one of the first cases where the diamond anvil cell was employed for high pressure x-ray diffraction studies.[75] Those studies showed that magnetite undergoes a crystallographic phase transition into a monoclinic phase at a pressure of ca. 25 GPa with a possibility that the cations undergo an increase in coordination number. However, the Mössbauer parameters of the high pressure phase were consistent with the retention of the original four and six coordination of the tetrahedral, A, and octahedral, B, sites, respectively. A most interesting and important finding was that the monoclinic phase is non-magnetic, at least for temperatures above 300 K, see Fig. 29. The relative iron population of the A:B sites is preserved at 1:2. In the monoclinic phase, the iron sites are significantly distorted as is evident from the presence of the quadrupole interaction. The loss of the magnetic phase could be a strong indication of a high pressure metallic state associated with a Mott transition, see Section 3. 1. 1. No indications are perceived of an "electron localization," i.e., stabilization of iron(II) and iron(III) ions, at the B-site. The phase transition shows a significant pressure-hysteresis as can be seen in Fig. 29.

Ideal stoichiometric *wüstite*, FeO in the rock salt structure, does not appear to be a stable phase. At ambient pressure it contains a large number of cation vacancies which form clusters in which iron(III) is tetrahedrally coordinated by oxygen vacancies.[76] A representative formula is,

$$Fe^{2+}_{1-3x}\ Fe^{3+}_{2x}\ V_x\, O^{2-}\ , \tag{5}$$

where V stands for an oxygen vacancy. A detailed phase diagram of wüstite at ambient pressure has been reported by Hazen and Jeanloz,[77] and a review of the Mössbauer spectral and structural studies has recently been published by Long and Grandjean.[78] Wüstite is an antiferromagnetic insulator which orders antiferromagnetically at $T_N \sim 195$ K. $Fe_{1-x}O$ occurs in nature with $x > 0.05$, but recently, enriched samples synthesized by McCammon,[79] by quenching from high-pressure and high temperature, resulted in more nearly stoichiometric $Fe_{0.977}O$ as determined by x-ray diffraction. The magnetic hyperfine interaction at ambient pressure and below T_N results in a complex Mössbauer spectrum.[78] Recently, studies by Nasu[29] at 300 K to a pressure of 66 GPa with samples prepared by McCammon from enriched iron-57 have shown that the spectra at $P > 26$ GPa are considerably simpler, see Fig. 30, and support a single resolved magnetic Mössbauer site with negligible quadrupole shift. This leads to the conclusion that the high-pressure phase stabilizes with an ordered cubic structure with a predominant iron(II) valence state. Preliminary studies by Pasternak et al.[28] at temperatures from 4 to 300 K and at pressures up to 110 GPa have shown that a new non-magnetic phase appears above ca. 60 GPa, a phase which has the same isomer shift. Hence, probably no structural changes are involved. Such a transition seems to be the "classical" pressure-induced Mott transition, namely an insulator-metal phase transition.

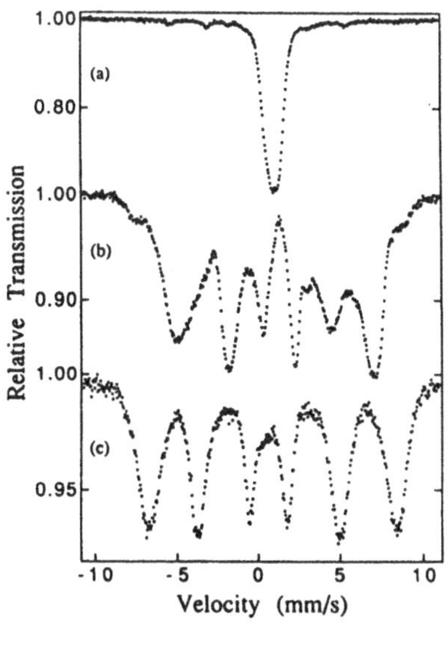

Figure 30. Comparison of $^{57}Fe_{0.977}O$ spectra obtained under various conditions: (a) 0.1 MPa at 300 K, (b) 0.1 MPa at 10 K, and (c) at 66 GPa at 300 K. Note the differences between (b) and (c).

3. 3. 2. Iron Metal

Studies of the high pressure properties of iron are of crucial importance not only for basic understanding of magnetic metals under extreme density conditions at ordinary temperatures but also for modeling of the earth's interior.[80] Under high pressure alone ordinary bcc α-iron transforms into the more closely packed hcp structure of ε-iron. This transformation takes place in the regime of 5 to 20 GPa, pressures often accessible before the advent of the diamond anvil cell. There have been many pressure studies of iron employing a variety of experimental methods ever since the early days of high pressure research.[81]

The iron-57 Mössbauer effect is a natural tool for studies of the α to ε-iron transition properties. Being the only practical technique for magnetism studies in a diamond anvil cell, Mössbauer spectroscopy can vividly contrast the Zeeman-split α-iron and the non magnetic ε-iron.[18,82] Thus, by following the relative area under the absorption spectra of the two phases, one can determine the relative abundance of the two phases. Mössbauer spectral studies in a diamond anvil cell were performed both with a cobalt-57 source[19] diffused into iron and with an enriched iron-57 absorber.[83] Pure ε-iron showed no magnetic ordering down to 0.030 K.[19] Furthermore, ε-iron subjected to applied magnetic fields up to 4.5 T at low temperatures, showed no moment on the iron atom.[82] Typical spectra at various pressures and 300 K are shown in Fig. 31. The α to ε-iron transformation is sluggish with a transition width of 1 to 15 GPa depending on details of how the pressure is effected.[84] Likewise, the transformation back to α-iron upon reducing the pressure is quite sluggish producing a large hysteresis in the ε-iron abundance *vs.* pressure behavior. The transition and its hysteresis for a particular cell and pressuring medium are shown in Fig. 32. This graph distinctly illustrates that the choice of pressure medium may be important in Mössbauer experiments in a diamond anvil cell. We note that the sharper transition widths are associated with the media most likely to provide hydrostaticity.

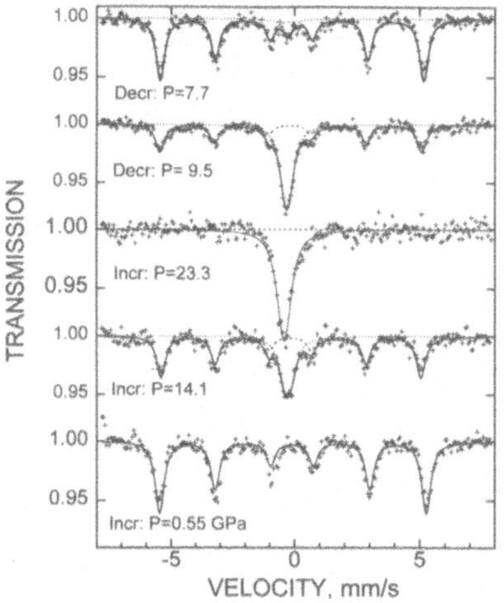

Figure 31. Mössbauer spectra of iron metal enriched to 85 percent in iron-57. The appearance of non-magnetic ε-iron is clearly evident at 14.1 GPa. At 23.3 GPa ε-iron is the only component present. Upon reducing the pressure remnants of this phase are still observed at 7.7 GPa.[83]

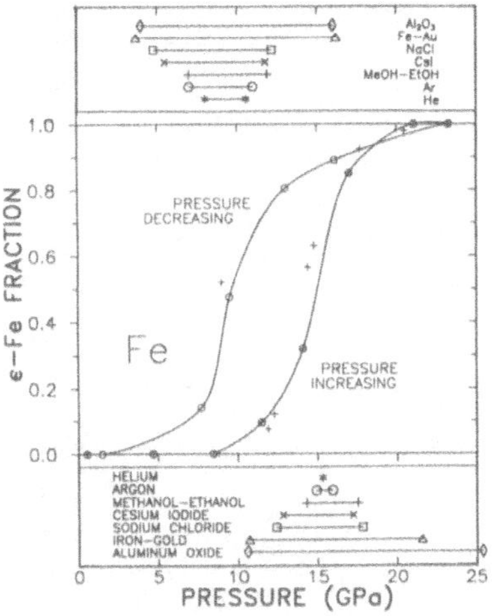

Figure 32. The abundance of ε-iron as a function of increasing and then decreasing pressure. The ⊕, ∅ symbols denote an iron-57 absorber, and the + symbols an $^{57}Co(Fe)$ source experiment. The lower and upper sections of the graph show the influence of different pressure media on the transition-width under compression, lower, and decompression, upper, as measured by several methods.[83]

3. 3. 3. Molecular Iodine, I_2

Since the early days of high pressure research, extensive studies of molecular hydrogen were focused on the elusive $H_2 \rightarrow H + H$ transformation[85] in the solid. Solid monatomic hydrogen has been predicted to be an excellent metal, even a superconductor, and possibly to be pseudo-stable upon release of the pressure. Considering all of the diatomic molecular solids, I_2 was the only case that has showed definite metallic behavior[86] at high pressures and some experimental evidence for molecular dissociation.[87] Because of metallization and opaqueness to visible light, Raman studies[88] to 30 GPa were inconclusive; they did show broad lines near 210 cm^{-1} which were characteristic of a persistence of the I_2 intramolecular stretching mode.

The well resolved iodine-129 Mössbauer spectra of enriched I_2 provides information on the sign of the quadrupole interaction, the asymmetry parameter, and the isomer shift.[89] It is the ideal probe to show, on the atomic scale, the existence of pressure-induced molecular dissociation because the monatomic phase is likely to be a cubic structure. The orientation of the principal axes of the electric field gradient, V_{xx}, V_{yy}, V_{zz}, in I_2 had been well established. V_{zz} lies along the molecular axis of I_2, see Fig. 33a, and its positive sign, eQ being negative, is due to the p_z hole deficiency in the closed $5s^25p^6$ configuration. Due to a non-negligible molecular overlap even at ambient pressure, V_{zz} is not axially symmetric as expected for pure van der Waals forces in a molecular crystal, rather η is 0.18.

To date studies performed in a diamond anvil cell by Pasternak et al.[90] have helped to elucidate various stages of molecular transformations in I_2 under pressure that precede any final dissociation. The absorber used was prepared by the thermal decomposition of $Pd^{129}I_2$, and the vacuum distilled I_2 crystals were loaded in a 40 μm thick, 210 μm diameter chamber drilled in a $Ta_{90}W_{10}$ gasket/collimator, see Section 2. 1. The ^{129m}Te source was prepared by irradiating a 3 mm diameter pellet of Mg_3TeO_6 enriched with tellurium-128 in the Oak Ridge high-flux reactor for about 14 days. Measurements were performed with both source and absorber at 4 K and with pressures up to 30 GPa. The findings can be summarized as follows, see Fig. 34.

The low pressure pure molecular-crystal phase persists up to ca. 15 GPa. In this pressure range the value of $|e^2qQ|$ decreases by about 20 percent, and the asymmetry parameter doubles. At pressures greater than 16 GPa, a new spectrum, HP1, appears characterized by a *positive* e^2qQ with a magnitude considerably lower than that of the low pressure phase, and by an asymmetry parameter of 0.8. These values for the HP1 phase are independent of pressure and are consistent with a zigzag, one dimensional polyiodide chain, see Fig. 33b. At this pressure iodine is conducting and the temperature dependence of its resistance has a positive slope.[87] At pressures greater than 26 GPa, a third component having a similar quadrupole interaction is evident but with an asymmetry parameter of 0.5. This was interpreted as due to the formation of a two-dimensional polyiodide structure in the *bc* plane that is responsible for a further decrease in the resistance and in the slope of its temperature dependence. All three components persisted to the highest pressure measured, a strong indication of the *sluggishness* of the transformation. The main conclusion of this work was to point out that molecular crystals under high pressure can be conducting without the necessity of dissociation. Indeed, experimental optical evidence now exists that hydrogen at the highest pressures, i.e., above 400 GPa, is still molecular but is conducting.

3. 3. 4. EuO and Eu Metal

Among rare-earth systems, the europium compounds and intermetallics have been the focus of high pressure studies particularly in relation to magnetism and intermediate valence phenomena. In this respect, europium-151 and europium-153 Mössbauer spectroscopy has contributed enormously to conventional high pressure studies.[91] The isomer shift and the

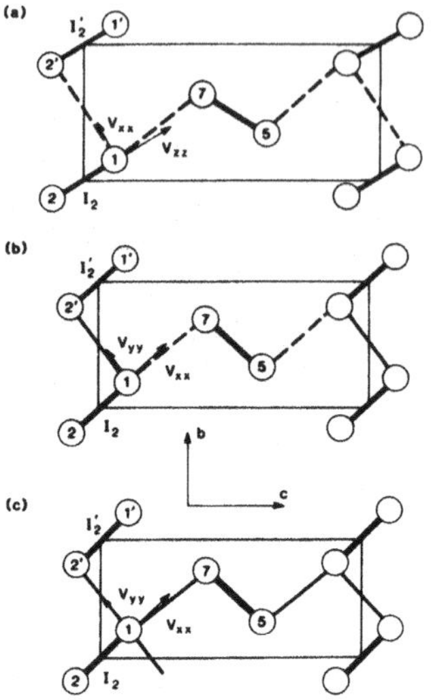

Figure 33. Schematic representation of the stages proposed in the high-pressure structural deformation of iodine. (a) The low-pressure molecular phase. The principal axis of the electric field gradient, V_{zz}, is along the molecular axis, V_{xx} is in the b-c plane, and V_{yy} is perpendicular to the b-c plane. (b) The I_2 - I_2 zigzag chain. In this intermediate HP1 phase, $15 < P < 21$ GPa, V_{zz} is perpendicular to the b-c plane. (c) The 2-dimensional HP2 structure.[90]

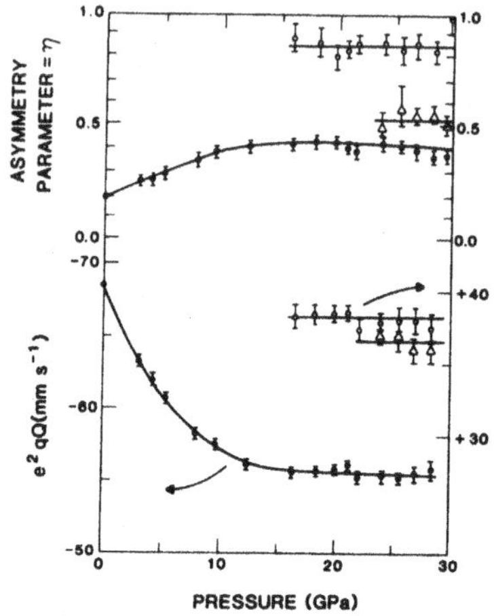

Figure 34. The pressure variation of e^2qQ and the asymmetry parameter, η, of iodine. The symbols, •, Δ, and o, correspond to the parameters of the low pressure, LP, the intermediate high-pressure, HP1, and the high pressure, HP2, phases, respectively.[90]

hyperfine field are excellent probes for investigating the interplay between intermediate valence and magnetism, respectively, particularly in europium(II) compounds. By taking the isomer shift of europium(III) materials to be zero mm/s, that of europium(II) materials to be about – 8 to – 13 mm/s, there is a tremendous sensitivity for studying intermediate valence. Factors other than valence changes may affect the isomer shift and must be considered. An important point connected with magnetism due to the 4*f*-moment system is the question of the electronic ground state for which the electrons are able to fluctuate between the 4*f* orbitals and the conduction band, yielding an intermediate valent system. Because of its simple cubic fcc crystallographic structure and the pure $^8S_{7/2}$ ground state, the ferromagnetic semiconductor EuO is ideal for experimental and theoretical studies related to these phenomena.

The first europium-151 Mössbauer spectral studies in a diamond anvil cell of EuO up to 30 GPa were reported by Taylor and Farrell.[92] Their data for the pressure dependence of the isomer shift showed a linear dependence of the isomer shift on the reduced volume, V/V_0, and indications for a significant change in the slope of the volume dependence of the isomer shift at a pressure of ca. 14 GPa. Later, these and some new EuO data were analyzed by Abd-Elmeguid and Taylor.[93] They concluded that, first, contrary to previous predictions no pronounced europium(II) to europium(III) valence change was observed even at the highest volume decrease, V/V_0, of 20 percent at 31 GPa and, second, the change in slope of the volume dependence of the isomer shift at pressures greater than 14 GPa was due to an appreciable reduction in the energy gap, E_x, a reduction which was compatible with previous high pressure studies of optical reflectivity[94] and electrical resistivity.[95] The reduction in the gap is consistent with the *positive* slope of dΔIS/dT above 14 GPa, indicating a possible thermal admixture of the europium(III) excited state into the europium(II) ground state. Third, the hyperfine field increases monotonically with pressure up to 25 GPa where it levels off to the highest pressure measured, 31 GPa and fourth, this phenomenon and the onset of a maximum in T_C at 25 GPa were due to the competition between two contradictory mechanisms, the ***exchange interaction*** between, and the ***hybridization*** of, the *4f* ground and the *5d, 6s* excited states of the europium(II). This picture is consistent with theoretical predictions by Matlak and Nolting.[96]

Earlier, a similar situation was investigated by Farrell and Taylor[97] in metallic europium. Experiments were carried out with a Merrill-Bassett cell with 1 mm diameter culets and 0.3 mm diameter absorbers. The highest pressure needed was 14 GPa, and typical spectra at 44 K and various pressures are shown in Fig. 35. The conclusion of these studies was that intermediate valence and magnetic ordering coexisted in this pressure range.

3. 3. 5. Tellurium-125, Gadolinium-155, and Ytterbium-170

Three new isotopes have recently been employed for Mössbauer spectral studies in a diamond anvil cell. Whereas gadolinium-155 requires extensive data acquisition time and special strong sources, very high pressure studies with tellurium-125 and ytterbium-170 are possible with rather conventional sources and reasonable data acquisition times. In due time these isotopes, like iron-57, tin-119, iodine-129, and europium-151, will find numerous applications in Mössbauer spectral studies in a diamond anvil cell. We summarize some current results for these less used isotopes.

Elemental tellurium has been the subject of extensive high pressure studies. Its various high pressure structures were determined from a series of x-ray diffraction studies[98] to pressures of 40 GPa. Tellurium metallizes at 4 GPa becoming a superconductor at 7.4 K.

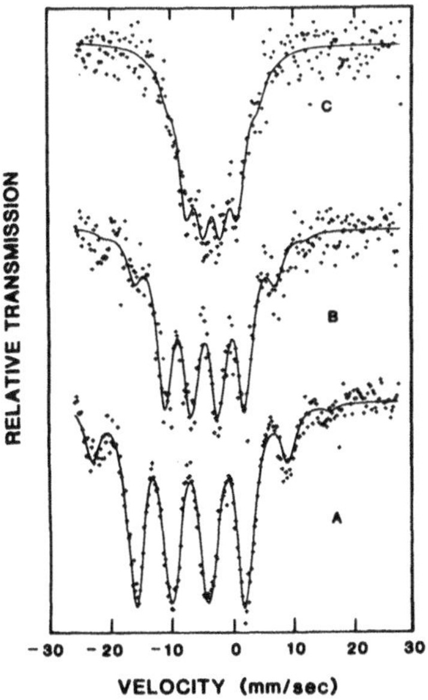

Figure 35. Mössbauer spectra at 44 K of Eu metal at (A) 0, (B) 6.2, and (C) 9.8 GPa. The $^{151}SmF_3$ source was also at 44 K.[97]

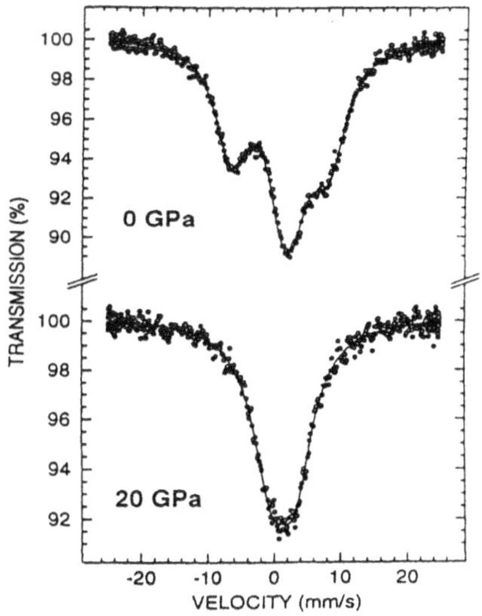

Figure 36. ^{125}Te spectra of $MnTe_2$ at 0 and 20 GPa at 4.2 K. The source is $Mg_3{}^{125m}TeO_6$.[32]

Mössbauer high pressure studies at 10 GPa have been carried out by Ladewig et al.[99] who investigated the pressure behavior of the quadrupole splitting. Mössbauer spectral studies in a diamond anvil cell by Vulliet et al.[32] extended these studies to 20 GPa with increasing, then decreasing pressure. Their results can be summarized as follows. The large quadrupole splitting of 7.60(5) mm/s observed at ambient pressure decreases dramatically with increasing pressure, reaching about 4 mm/s at 20 GPa. The decrease in quadrupole splitting in the monoclinic and the orthorhombic phases at 5 and 7 GPa, respectively, is explained as a result of the enhanced *inter-chain* interaction. According to band-structure calculations, the bonds retain their strong covalent character within the zigzag chains. The continuous decrease in the quadrupole splitting above 10 GPa results from the onset of structures with higher symmetry due to larger coordination numbers. The complete offset of the quadrupole interaction at higher pressures could imply the complete chain dissociation into monatomic tellurium.

$MnTe_2$ is a cubic pyrite antiferromagnetic semiconductor in which the Te_2^{2-} units are situated along the various (111) directions. In the first Mössbauer studies of $Mn^{125}Te_2$ by Pasternak and Spijkervet,[100] from the combined magnetic and quadrupole interactions below T_N of 90.1 K, the spin orientation was deduced to be along the axially symmetric V_{zz} at $\theta = 0°$ or 180°; this is also the direction of the principal axis of Te_2^{2-}. Later neutron work[101] confirmed this finding. The Mössbauer experiments in a diamond anvil cell by Vulliet et al.[32] to 20 GPa yielded the following. Up to 8 GPa the hyperfine field and θ do not change, and e^2qQ remains practically unchanged, see Fig. 36. The magnetic, low pressure component coexists with a non-magnetic and poorly resolved quadrupole doublet with ΔE_Q of ca. 3.5 mm/s. At pressures above 12 GPa, the non-magnetic component is the only contribution to the spectra. No significant changes in the 1.36(5) mm/s isomer shift were observed between the magnetic and non-magnetic phases. Preliminary studies showed a sharp increase in T_N with increasing pressure.

The authors assigned the high pressure, non-magnetic phase to a marcasite-type phase similar to $CoTe_2$ or $FeTe_2$ that results from a high-spin, S = 5/2, to low-spin, S = 1/2, transition or a Mott insulator-metal transition.

Gadolinium metal and intermetallic compounds were studied under pressure by Mulder and Thiel[34] using a TAU-diamond anvil cell. Measurements with gadolinium metal were performed at 4.2 K for several pressures, up to 17 GPa. Under pressure the metal undergoes phase transitions, first at 1.5 GPa from hcp to a Sm-type structure, then at 6.5 GPa to a double hexagonal closed packed, dhcp, structure, and finally at 24 GPa to the completely cubic fcc structure.[102] No distinct features were detected in the gadolinium-155 Mössbauer spectra that could be ascribed to these crystallographic changes. The isomer shift decreases with increasing pressure indicative of the increase in the electron density at the nucleus. A linear volume dependence of the isomer shift was observed from which a value of

$$\lambda_V = dIS/d\ln(V/V_0) = 0.73(8)\ \text{mm/s} \qquad (6)$$

was derived. The absolute value of the hyperfine field decreases with increasing pressure and increases with increasing isomer shift, suggesting that the main factor in the change of hyperfine field is the increased density of the polarized 6*s* electrons at the nucleus. Band structure calculations by Coehoorn[103] could not explain the changes in the V_{zz} for the different structures at high pressure, but explained the trend in the reduction in hyperfine field with increasing pressure.

Mössbauer spectroscopy and x-ray diffraction measurements in a diamond anvil cell were also carried out for the intermetallic compounds, $GdCo_5$, $GdRu_2Si_2$, and $Gd_2Co_{17}N_3$. These compounds are related to the equivalent samarium strong permanent magnets. Mössbauer spectra of the cobalt and ruthenium silicide compounds, at various pressures, are shown in Fig. 37. The spectra were fitted with the full Hamiltonian, yielding results for the electric field gradient, hyperfine field, isomer shift, and θ. The following conclusions were based on the pressure variation of the Mössbauer parameters. Changes in V_{zz} were moderate. Sizable changes in the hyperfine field and isomer shift were observed. In $GdCo_5$, the hyperfine field reaches a minimum at 9.6 GPa, then levels-off to a similar, ambient pressure value of -6.2 T at 18 GPa. In $GdRu_2Si_2$, the hyperfine field remains constant up to 17 GPa and in $Gd_2Co_{17}N_3$ there is a decrease in hyperfine field, which reaches -8 T at 9.8 GPa. The decrease in isomer shift is consistent with the increase in the *s*-density at the nucleus.

Mössbauer experiments in a diamond anvil cell on gadolinium-155 compounds proved to be extremely difficult. Accumulation of each spectrum required weeks as compared with one to two days for the previous described isotopes.

Ytterbium-oxide, Yb_2O_3. High-pressure studies in a diamond anvil cell were carried out by Meyer et al.[33] According to the authors, the main motivation for this work was to test the feasibility of ytterbium-170 Mössbauer spectral studies with a diamond anvil cell. Mössbauer spectral data were combined with synchrotron-radiation x-ray diffraction, carried out in the energy-dispersive mode, to obtain information on phase transitions and the volume dependence of the electric field gradient. The experiments were carried out with the 84.3 keV, $0 \rightarrow 2$ transition in ytterbium-170 enriched to 70 percent. The sample used was the cubic, *C*-type, Yb_2O_3. The quadrupole-split spectra recorded at ambient pressure, *C*-type, and after decompression from 20 GPa, *B*-type, are shown in Fig. 38. Typical counting times are ca. 5 days. The *C*-type spectrum shown in Fig. 38a was fitted with two quadrupole interactions by assuming a 3:1 ratio for their relative abundance. Typical values for the e^2qQ were 11 mm/s for the cubic C_{3i}-symmetry site and 25 mm/s with an asymmetry parameter of 0.1 for the non-cubic C_2-symmetry site. The isomer shift was insensitive to pressure changes. In the *B*-type monoclininc phase, spectra were fitted with three sites with e^2qQ values of 13, 21, and 27 mm/s and η-values varying from 0.1 to 0.4.

The *C* to *B* transition that occurs in Yb_2O_3 on compression up to ca. 14 GPa, as determined from the x-ray diffraction measurements at room temperature, is confirmed by the Mössbauer spectral results at 4.2 K. After decompression to ambient pressure, the *B*-type structure is retained. The authors[33] also modelled the electric field gradient of the *C*-type and its volume dependence and provided an estimate of the valence contribution of the 4*f* electrons to the electric field gradient.

These Mössbauer spectral measurements in a diamond anvil cell were crucial to establish the use of ytterbium-170 Mössbauer spectroscopy for high pressure applications. It should encourage further studies in ytterbium-based heavy fermion materials of current interest[104] with the appeal of viable magnetic studies at pressures greater than 10 GPa.

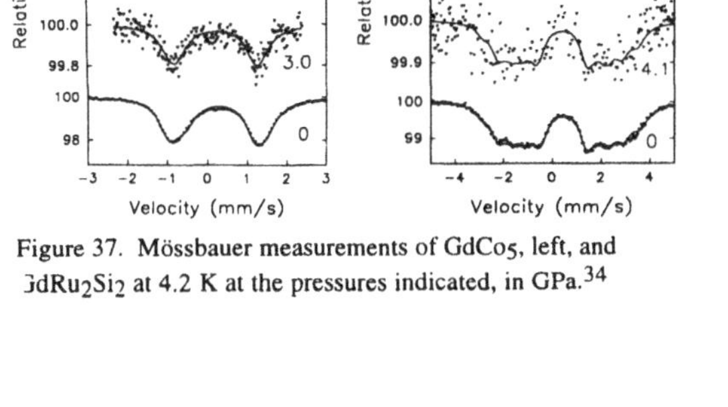

Figure 37. Mössbauer measurements of $GdCo_5$, left, and $GdRu_2Si_2$ at 4.2 K at the pressures indicated, in GPa.[34]

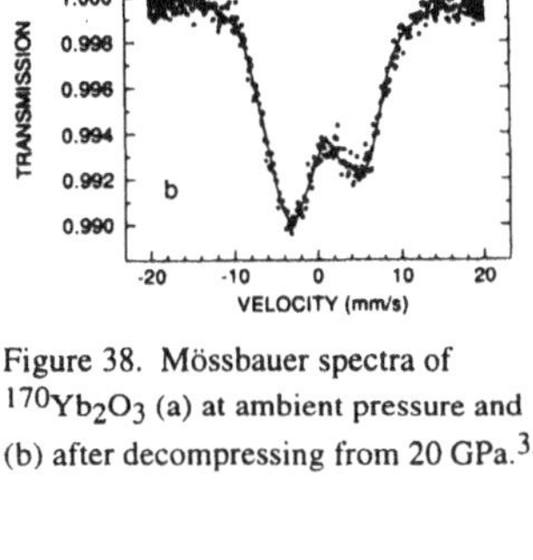

Figure 38. Mössbauer spectra of $^{170}Yb_2O_3$ (a) at ambient pressure and (b) after decompressing from 20 GPa.[33]

4. HIGH PRESSURE MÖSSBAUER SPECTROSCOPY; QUO-VADIS?

4. 1. The Pressure Limits with Radioactive Sources

Undoubtedly, the highest pressure will be achieved with the smallest absorber diameter. To be able to obtain significant statistics in reasonable spectra recording time, the highest specific activity *and* source strength are desired. At present, based on our and other groups experience, we estimate the highest pressure likely to be available for five Mössbauer isotopes given in Table 1.

Table 1. Highest Pressure Available

Isotope	Pressure limit, GPa
Iron-57	≤ 200
Tin-119	≈ 100
Tellurium-125	> 100
Iodine-129	≈ 50
Europium-151	≈ 40

We have assumed a spectra recording time of less than 100 hours and optimization of (1) source geometry and size, (2) absorber thickness, (3) source-absorber-detector geometry and, (4) detector efficiency, see Section 2. 2. 2.

Iron-57. The source used is a commercial Amersham "point source" of ca. 10 mCi deposited in 6 μm thick rhodium foil of 0.5 × 0.5 mm^2. We believe that sources up to 20 mCi could be deposited in the same foil area. However, the source would have to be maintained at temperatures higher than T_{sg}, the magnetic ordering temperature of the Co-Rh alloy formed due to the relatively large concentration of the magnetic impurities. With "old" sources this temperature may increase further, in part, due to the formation of ^{57}Fe-Rh. Culets of ca. 100 μm and absorbers of ca. 50 μm diameter will be required.

Tin-119. The source used currently is a commercial 5 mCi $Ca^{119m}SnO_3$ with a 5 mm diameter. The maximum achievable pressure is ~50 GPa using ca. 500 μm and 250 μm diameter culets and absorbers, respectively. However, if sources of higher specific activity can be produced, either by doubling the irradiation period, say for one year, or in a neutron flux of ca. 10^{15} n cm^{-2} s^{-1} for the same period as currently used, it will be possible to reach pressures close to 100 GPa. This will be achieved by reducing the source diameter, allowing for shorter source-absorber distances without compromising the count-rate.

Tellurium-125. The source strength, made by neutron irradiation of pellets, can be increased up to ten-fold. This will be achieved by using thicker source-pellets and longer irradiation time in the presently available neutron flux. In that case one could easily exceed 100 GPa pressures.

Iodine-129. With the presently available neutron flux used to produce 3 mm source-pellets of $Mg_3{}^{129m}TeO_6$, the optimal conditions for source strength and specific activity have been achieved.[24] However we expect that pressures up to 80 GPa could be a reasonable goal, assuming a neutron flux of 10^{15} n cm^{-2} s^{-1} becomes available.

Europium-151. We do not anticipate improved samarium-151 source strength in the near future. With available sources the pressure limit is estimated at 40 GPa.

4. 2. Applications with Nuclear-Resonance Synchrotron Radiation

The feasibility of *nuclear-resonance synchrotron radiation* to iron-57 Mössbauer spectroscopy in a diamond anvil cell was first demonstrated by Nasu[105] at KEK, Japan, and recently by the group at the European Synchrotron Radiation Facility,[106] Grenoble, France. The naturally collimated, extremely monochromatic, and narrow beam of photons emanating from a port makes it ideal for Mössbauer spectral studies in a diamond anvil cell not only for iron-57 but for a whole series of isotopes. Undoubtedly this technique will provide the most suitable source for the ever smaller samples required for Mössbauer studies in a diamond anvil cell, near and above 100 GPa, once the technical problems are solved. The *third generation* synchrotron facilities in Grenoble, European Synchrotron Radiation Facility, and Argonne, Advanced Photon Laboratory, have installed nuclear-resonance synchrotron radiation beam lines dedicated to Mössbauer studies. General reviews on nuclear-resonance synchrotron radiation theory and applications are available.[107] In what follows we present some results of these pioneering works of nuclear-resonance synchrotron radiation applications to Mössbauer spectral studies in a diamond anvil cell.

Studies of $SrFeO_{2.97}$. Nasu[105] used a Bassett type diamond anvil cell with an iron-57 enriched $SrFeO_{2.97}$ sample, mixed with methanol/ethanol as the pressurizing medium, placed in a 150 μm hole drilled in a *Waspalloy* gasket. The diamond anvil cell was placed in the synchrotron beam of the National Laboratory for High Energy Physics, KEK, facility. The schematic arrangement of the experiment is shown in Fig. 39. The angular flux density of the third harmonic of the undulator radiation at the 14.4 keV was estimated as 1.71×10^{15} photons s^{-1} $mrad^{-2}$ per 0.1% bandwidth resulting in an estimated Mössbauer photon flux of 1.4×10^4 photons cm^{-2} s^{-1}. With this radiation the nuclear-resonance synchrotron radiation scattering in a diamond anvil cell has been observed for the first time. The time-mode spectrum of $SrFeO_3$ at 44 GPa at 300 K is shown in Fig. 40. The quantum-beat modulated pattern was analyzed using the following expression,[108]

$$I(t) \propto e^{-\tau} \frac{\chi}{\tau} J_1^2\left(\sqrt{4\chi\tau}\right)\left\{1 + \sum_{ij} a_{ij} \cos((\omega_i - \omega_j)t)\right\}, \tag{7}$$

where χ is the effective absorber thickness, τ is the normalized lifetime of the excited state, a_{ij} is the amplitude of the quantum modulation related to the Clebsh-Gordan coefficients and the polarization of the photons, $\omega_{i,j}$ denotes the frequencies of the resonance for hyperfine-split levels, and the summation is carried out for all combinations of i and j.

The α-iron to ε-iron phase transition. The α-iron to ε-iron phase transition was chosen as a test-case for the first high pressure Mössbauer spectra recorded at the European Synchrotron Research Facility.[106] The experimental setup for the nuclear-resonance synchrotron radiation experiment is shown in Fig. 41. The experimental results for various pressures of 0, 7, 15, 17, and 34 GPa at 300 K, are depicted in Fig. 42.

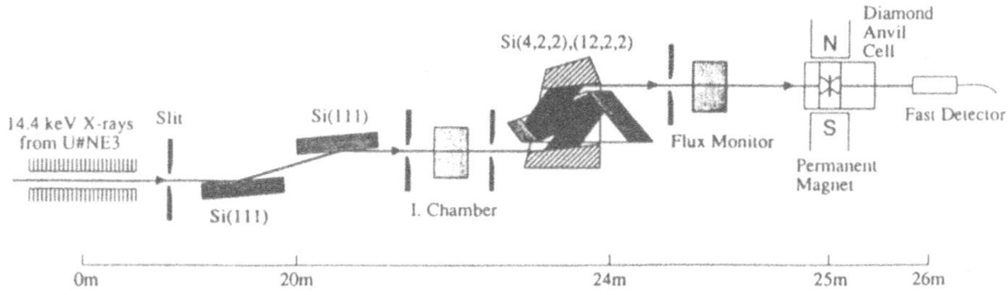

Figure 39. Schematic arrangement of a nuclear forward scattering experiment using a diamond anvil cell at undulator beam line NE3 of the TRISTAN accumulating ring.[105]

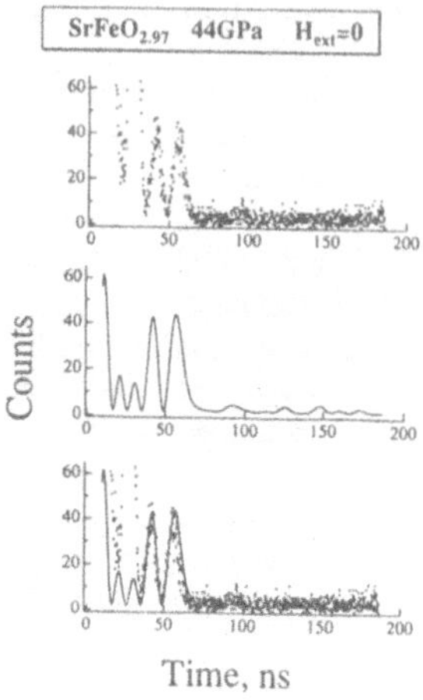

Figure 40. The time-delay spectrum and the simulated results for $SrFeO_{2.97}$ using a magnetic hyperfine parameter of 23 T and a sample thickness of 18 μm.[105]

Attempts to simulate the mixed α/ε phase by weighted coherent superposition of the two phases using expressions similar to Equation 7 were not successful. Partial success was achieved by imposing two incoherent and one coherent contributions. The authors[101] concluded that there are strong indications that the shape of the Mössbauer time-pattern depends not only on the effective thickness and the hyperfine parameters but also on the size and distribution of the magnetic and non-magnetic domains in the intermediate transition regions. This feature, if truly established, could yield novel experiences not available with conventional Mössbauer spectroscopy in a diamond anvil cell.

The conclusion from these two pioneering works is that nuclear-resonance synchrotron radiation studies with the diamond anvil cell are feasible. Furthermore, the possibility of applying large external fields is rather appealing. This feature is rather impractical with a conventional source due to its proximity to the diamond anvil cell.

4. 3. Concluding Remarks

Recent years have witnessed a constant growth in the number of high pressure studies involving Mössbauer spectroscopy in diamond anvil cells. During this period, cells originally designed and used for other spectroscopic methods have been adapted, and new cells more appropriate for Mössbauer spectral studies have been developed. To help overcome the stringent requirement for minute absorber sizes, requirements that are trending to ever smaller amounts in the quest for even higher pressures, some success has been achieved in producing compact, high-areal-density sources. The culmination of such efforts allow diamond anvil cell-based Mössbauer spectroscopy with absorbers at hydrostatic pressures in the 100 GPa regime. The cells are small enough to allow temperature and/or applied magnetic field studies under pressure.

Presently the stimulus for reaching ever higher pressures in Mössbauer spectral work is the desire for studies of metal-iron oxides, which play an extremely important role in earth-sciences, and to provide further studies of insulator-metal transitions in correlated systems. High pressure iron-57 Mössbauer spectroscopy is presently the only adequate tool for investigating the intricate influence of reduced interatomic distances upon the electronic structure and the magnetic state of iron based compounds. Pressure needed for such studies is in the 100 GPa range and beyond.

With the full implementation of nuclear-resonance synchrotron radiation, we expect the trends of high pressure Mössbauer studies to follow closely those of other diamond anvil cell methods, such as optical spectroscopy and x-ray diffraction. One of the most interesting,

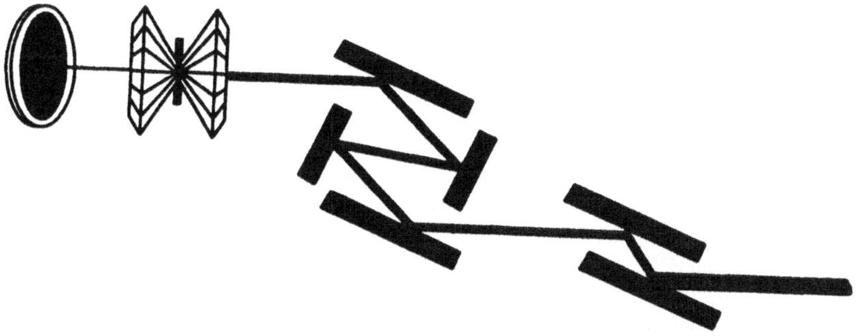

Figure 41. Schematic for the nuclear-resonance synchrotron readiation experiments at the European Synchrotron Research Facility.[106]

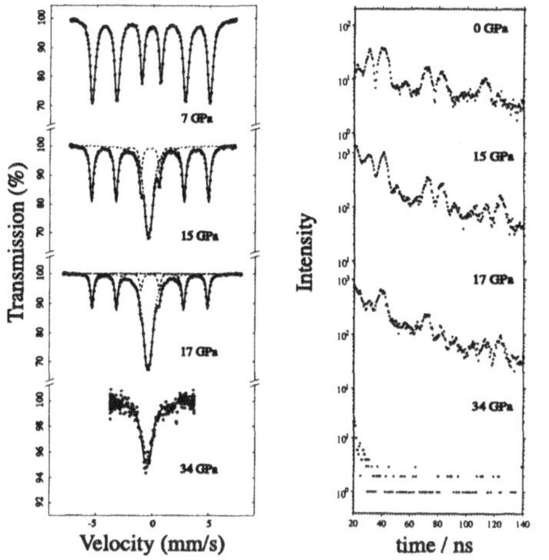

Figure 42. Mössbauer spectra of a 2.6 μm iron foil enriched to 90 percent iron-57 at various pressures demonstrating the α to ε phase transition. The left column corresponds to the conventional Mössbauer source and the right column to the nuclear-resonance synchrotron radiation pattern.[106]

unique, and recently developed subjects will be studies of phonon density of states under pressure using the inelastic channel of the nuclear resonant scattering.[109] The phonon density of states is obtained by monitoring the K-fluorescence radiation of the de-exciting nuclei while the energy of the incident synchrotron radiation is tuned with a meV high-resolution monochromator across the 14.4-keV resonance. It is anticipated that other subjects will be of interest such as *in-situ* studies of chemical reactions under extreme conditions of pressure and temperature and magnetism in metals and insulators.

ACKNOWLEDGMENTS. The authors wish to thank Giovanni Hearne for fruitful discussions and suggestions and Maria Hearne for several drawings. We also acknowledge partial support from the US Department of Energy, the US-Israel Binational Science Foundation, Grant No. 92-00028, and the Israel Science Foundation Grant No. 451/94-1.

REFERENCES

1. A. Jayaraman, *Rev. Modern. Phys.* **55**, 65 (1983).
2. R. L. Mössbauer, *Z. Physik* **151**, 124 (1958); *Naturwissenshaften* **45**, 538 (1958).
3. R. V. Hanks, *Phys. Rev.* **124**, 1319 (1961).

4. R. V. Pound, G. B. Benedek, and R. Drever, *Phys. Rev. Lett.* **7**, 405 (1961).
5. H. G. Drickamer, in *Solid State Physics*, F. Seitz and D. Turnbull, eds., Vol. 17, p. 1, Academic Press, New York (1965); H. G. Drickamer, *Chem. Br.* **9**, 353 (1973); H. G. Drickamer and C. W. Frank, *Ann. Rev. Phys. Chem.* **23**, 39 (1972); H. G. Drickamer, R. W. Vaughan, and G. K. Lewis, Jr., *Comments Solid State Phys.* **1**, 163 (1968).
6. H. Frauenfelder and R. Ingalls, in *Applications of Mössbauer Effect in Chemistry and Solid State Physics*, Technical Report Series, No. 50, p. 37, International Atomic Energy Agency, Vienna (1966).
7. W. B. Holzapfel, CRC Critical Reviews in Solid State Sciences, 89 (1975).
8. M. Abd-Elmeguid and G. Wortmann, private communications.
9. G. J. Piermarini and S. Block, *Rev. Sci. Instrum.* **46**, 973 (1975).
10. H. K. Mao and P. M. Bell, *Carnegie Institution of Washington Year Book* **77**, 824 (1979).
11. L. Merrill and W. A. Bassett, *Rev. Sci. Instrum.* **45**, 290 (1974).
12. The maximum pressure reported with a 300-μm culet anvil was 50 GPa. See: E. Sterer, M. P. Pasternak, and R. D. Taylor in *High-Pressure Science and Technology - 1993*, S. C. Schmidt, J. W. Shaner, G. A. Samara, and M. Ross, eds., p. 1453, American Institute of Physics, New York (1994).
13. Current suppliers of diamond anvils are: Harris Diamond Corp., agents of Drukker Diamond Anvils, 100 Stierli Court, Mount Arlington, NJ 07856 USA, Tel: +1 (201) 770-1420, FAX: +1 (201) 770-1549; Lazare-Kaplan & Sons, Inc., 666 5th Ave., New York, NY 10103; Ramot of Tel Aviv University, 32 H. Levanon St., Tel-Aviv 61392, Israel, Tel: +972-3-5450113, FAX: +972-3-6429865, e-mail: anvils@ccsg.tau.ac.il; D. Drukker & Zn, 12 Sarphatikade, 1001 MC Amsterdam, The Netherlands, Tel: (020)-267321.
14. A. Van Valkenburg, Conférence Internationale Sur les Hautes Pressions, Le Creusot, Saône et Loire, France (1965).
15. R. A. Forman, G. J. Piermarini, J. D. Barnett, and S. Block, *Science* **176**, 284 (1972).
16. H. K. Mao and P. M. Bell, *Science* **200**, 1145 (1978).
17. R. H. Herber and J. Spijkerman, *J. Chem. Phys.* **42**, 4312 (1965).
18. F. E. Huggins, H. K. Mao, and D. Virgo, *Carnegie Institution of Washington Year Book* **74**, 405 (1975).
19. G. Cort, R. D. Taylor, and J. O. Willis, *J. Appl. Phys.* **53**, 2064 (1982).
20. J. N. Farrell, Ph.D. Thesis, University of North Carolina, 1984; L. Chow, P. A. Dean, J. N. Farrell, P. A. Magill, and L. D. Roberts, *Phys. Rev. B* **33**, 3039 (1986).
21. M. Pasternak, J. N. Farrell, and R. D. Taylor, *Hyp. Interact.* **28**, 837 (1985).
22. S. Nasu, K. Kurimoto, S. Nagatomo, S. Endo, and F. E. Fujita, *Hyp. Interact.* **29**, 1583 (1985).
23. R. D. Taylor and J. N. Farrell, *J. Appl. Phys.* **61**, 3669 (1987).
24. R. D. Taylor and M. P. Pasternak, *Hyp. Interact.* **53**, 159 (1990); M. P. Pasternak and R. D. Taylor, *Hyp. Interact.* **62**, 89 (1990); R. D. Taylor and M. P. Pasternak, *High Pressure Res.* **9**, 263 (1992); R. D. Taylor and M. P. Pasternak, *Hyp. Interact.* **72**, 241 (1992); G. R. Hearne, M. P. Pasternak, and R. D. Taylor, *Hyp. Interact.* **92**, 1155 (1994); and G. R. Hearne, M. P. Pasternak, and R. D. Taylor, *Rev. Sci. Instrum.* **65**, 3787 (1994).
25. The high temperature limit depends on the characteristics of the metals from which the cell and gaskets are made. With some Inconel alloys and rhenium, temperatures well above 1000 K can be reached. At high temperatures the diamond anvils should be kept either in high vacuum or in a non-oxidizing atmosphere.
26. E. Sterer, M. P. Pasternak, and R. D. Taylor, *Rev. Sci. Instrum.* **61**, 1117 (1990).

27. The cell can be purchased as a kit from Ramot of Tel Aviv University, 32 H. Levanon St., Tel Aviv 61392, Israel, Tel: +972-3-5450113, FAX: +972-3-6429865, e-mail: anvils@ccsg.tau.ac.il.
28. M. P. Pasternak, R. D. Taylor, R. Jeanloz, X. Li, and C. McCammon, to be published.
29. S. Nasu, *Hyp. Interact.* **90**, 59 (1994).
30. M. P. Pasternak and R. D. Taylor, unpublished.
31. M. Pasternak and R. D. Taylor, *Phys. Rev. B* **37**, 8130 (1988).
32. P. Vulliet, J. P. Sanchez, J. Thomasson, B. Malaman, and R. Welter, in Proceedings of ICAME 95, Rimini, Italy, to be published.
33. C. Meyer, J. P. Sanchez, J. Thomasson, and J. P. Itié, *Phys. Rev. B* **51**, 12187 (1995).
34. F. M. Mulder, Ph.D. Thesis, Rijksuniversiteit of Leiden, 1994; F. M. Mulder and R. C. Thiel, *Europhys. Lett.* **25**, 657 (1994); and F. M. Mulder and R. C. Thiel, *Rev. Sci. Instrum.* **65**, 707 (1994).
35. J. G. Bednorz and K. A. Müller, *Z. Phys. B* **64**, 189 (1986).
36. S. Jin, M. McCormack, and T. H. Tiefel, *J. Appl. Phys.* **76**, 6929 (1994).
37. J. B. Torrance, P. Lacorre, A. I. Nazzal, E. I. Ansaldo, and C. H. Niedemayer, *Phys. Rev. B* **45**, 8209 (1992).
38. M. P. Pasternak, R. D. Taylor, and R. Jeanloz, *J. Appl. Phys.* **70**, 5956 (1991); M. P. Pasternak, R. D. Taylor, and R. Jeanloz, in *Frontiers of High-Pressure Research*, H. D. Hochheimer and R. D. Etters, eds., p. 227, Plenum Press, New York (1991).
39. S. A. Carter, T. F. Rosenbaum, M. Lu, H. M. Jaeger, P. Metcalf, J. M. Honig, and J. Spalek, *Phys. Rev. B* **49**, 7898 (1994).
40. M. P. Pasternak, R. D. Taylor, A. Chen, C. Meade, L. M. Falicov, A. Giesekus, R. Jeanloz, and P. Yu, *Phys. Rev. Lett.* **65**, 790 (1990).
41. J. B. Torrance, P. Lacorre, C. Asavaroengchai, and R. Metzger, *J. Solid State Chem.* **90**, 168 (1991).
42. T. Mizokawa, H. Namatame, A. Fujimori, K. Akeyama, H. Kondoh, H. Kuroda, and N. Kosugi, *Phys. Rev. Lett.* **67**, 1638 (1991).
43. J. Zaanen, G. A. Sawatzky, and J. W. Allen, *Phys. Rev. Lett.* **55**, 418 (1985).
44. M. P. Pasternak, G. Hearne, E. Sterer, R. D. Taylor, and R. Jeanloz, in *AIP Conference Proceeding No. 319*, S. C. Schmidt, J. W. Shaner, G. A. Samara, and M. Ross, eds., p. 335, AIP Press, New York (1994); G. Hearne, E. Sterer, M. P. Pasternak, and R. D. Taylor, *ibid*, p. 1461.
45. M. P. Pasternak and R. D. Taylor, *Physica C* **209**, 113 (1993).
46. R. W. G. Wyckoff, *Crystal Structures*, Vol. I, Interscience, New York (1963).
47. L. G. Van Uitert, H. J. Williams, R. D. Sherwood, and J. J. Rubin, *J. Appl. Phys.* **36**, 1029 (1965).
48. S. R. Kuindersma, J. P. Sanchez, and C. Haas, *Physica B* **111**, 231 (1981).
49. G. R. Hearne, M. P. Pasternak, and R. D. Taylor, *Il Nuovo Cimento*, in press.
50. G. Hearne, M. P. Pasternak, G. Rozenberg, E. Sterer, and R. D. Taylor, to be published.
51. S. E. Dann, M. T. Weller, D. B. Currie, M. F. Thomas, and A. D. Al-Rawwas, *J. Mater. Chem.* **3**, 1231 (1993).
52. M. Takano, S. Nasu, T. Abe, K. Yamamoto, S. Endo, Y. Takeda, and J. B. Goodenough, *Phys. Rev. Lett.* **67**, 3267 (1991).
53. S. Nasu, T. Abe, K. Yamamoto, S. Endo, M. Takano, and H. Takeda, *Hyp. Interact.* **70**, 1063 (1992).
54. G. R. Hearne, M. P. Pasternak, G. Kh. Rozenberg, A. P. Milner, and R. D. Taylor, in Proceedings of ICAME 95, Rimini, Italy, to be published.
55. S. E. Dann, M. T. Weller, D. B. Currie, M. F. Thomas, and A. D. Al-Rawwas, *J. Mater. Chem.* **3**, 1231 (1993).

56. P. Adler, A. F. Goncharov, and K. Syassen, *Hyp. Interact.* **95**, 71 (1995).
57. T. Mizokawa, H. Namatame, A. Fujimori, K. Akeyama, H. Kondoh, H. Kuroda, and N. Kosugi, *Phys. Rev. Lett.* **67**, 1638 (1991); D. D. Sarma, *J. Solid State Chem.* **88**, 45 (1990).
58. Y. Fujii, M. Kowaka, and A. Onodera, *J. Phys. C* **18**, 789 (1985).
59. S. Sugai, *J. Phys. C* **18**, 799 (1985).
60. M. P. Pasternak and R. D. Taylor, *Phys. Rev. B* **37**, 8130 (1988).
61. A. L. Chen, P. Y. Yu, and M. P. Pasternak, *Phys. Rev. B* **44**, 2883 (1991).
62. M. P. Pasternak, R. D. Taylor, M. B. Kruger, R. Jeanloz, J-P. Itié, and A. Polian, *Phys. Rev. Lett.* **72**, 2733 (1994).
63. M. Van der Heyden, M. P. Pasternak, and G. Langouche, *J. Phys. Chem. Solids* **46**, 1221 (1985).
64. W. Williamson and S. A. Lee, *Phys. Rev. B* **44**, 9853 (1991).
65. G. R. Hearne, M. P. Pasternak, and R. D. Taylor, *Phys. Rev. B* **52**, 9209 (1995).
66. S. S. Hafner, J. Stanek, and M. Stanek, *J. Phys. Chem. Solids* **51**, 203 (1990).
67. Q. Williams, E. Knittke, R. Reichlin, S. Martin, and R. Jeanloz, *J. Geophys. Res.* **95**, 21549 (1990).
68. M. B. Kruger, R. Jeanloz, M. P. Pasternak, R. D. Taylor, B. S. Snyder, A. M. Stacy, and S. R. Bohlen, *Science* **255**, 703 (1992).
69. K. Moorjani and J. M. D. Coey, *Magnetic Glasses*, Elsevier, Amsterdam (1984).
70. J. J. Hauser, *Phys. Rev. B* **22**, 2554 (1980).
71. M. P. Pasternak, R. D. Taylor, and E. Millner, private communication.
72. Y. Syono, A. Ito, S. Morimoto, T. Suzuki, T. Yagi, and S. Akimoto, *Solid State Commun.* **50**, 97 (1984).
73. J. Staun Olsen, C. S. G. Cousins, L. Gerward, and H. Jhans, *Physica Scripta* **43**, 327 (1991).
74. M. P. Pasternak, S. Nasu, K. Wada, and S. Endo, *Phys. Rev. B* **50**, 6446 (1994).
75. H. Mao, T. Takahashi, W. Bassett, G. L. Kinsland, and L. Merrill, *J. Geophys. Res.* **79**, 1165 (1974).
76. F. Koch and J. B. Cohen, *Acta Cryst. B* **25**, 275 (1975).
77. R. M. Hazen and R. Jeanloz, *Rev. Geophysics Space Phys.* **22**, 37 (1984).
78. G. J. Long and F. Grandjean, *Adv. Solid-State Chem.* **2**, 187 (1991).
79. C. A. McCammon, *J. Magn. Magn. Mat.* **104-107**, 1937 (1992).
80. Q. Williams, R. Jeanloz, J. Bass, B. Svendsen, and T. J. Ahrens, *Science* **236**, 181 (1987).
81. D. Bancroft, E. L. Peterson, and S. Minshall, *J. Appl. Phys.* **27**, 291 (1956); P. M. Giles, M. H. Longenbach, and A. R. Marder, *J. Appl. Phys.* **42**, 4290 (1971); H. G. Drickamer, *Rev. Sci. Instrum.* **41**, 1667 (1970).
82. R. D. Taylor, G. Cort, and J. O. Willis, *J. Appl. Phys.* **53**, 8199 (1982).
83. R. D. Taylor, M. P. Pasternak, and R. Jeanloz, *J. Appl. Phys.* **69**, 6126 (1991).
84. N. von Bargen and R. Boehler, *High Pressure Res.* **6**, 133 (1990).
85. M. Ross and A. K. McMahan, in *Physics of Solids Under Pressure*, J. S. Schilling and R. N. Shelton, eds., p. 161, North-Holland, New York (1981).
86. N. Sakai, K. Takemura, and K. Tsuji, *J. Phys. Soc. Japan* **51**, 1811 (1982).
87. K. Takemura, S. Minomura, O. Shimomura, and Y. Fuji, *Phys. Rev. Lett.* **45**, 1881 (1980).
88. O. Shimomura, K. Takemura, and K. Aoki, in Proc. of the Eighth AIRAPT Int. Conf, Upsalla, Sweden, 1981, C. M. Blackman, T. Johannisson, and L. Tegner, eds., Arkitekkopia, Upsalla (1982).
89. M. Pasternak, A. Simopoulos, and Y. Hazoni, *Phys. Rev.* **140**, A1892 (1965).
90. M. Pasternak, J. N. Farrell, and R. D. Taylor, *Phys. Rev. Lett.* **58**, 575 (1987).
91. J. Moser, G. M. Kalvius, and W. Zinn, *Hyp. Interact.* **41**, 499 (1988). See also A. Gleissner, W. Potzel, J. Moser, and G. M. Kalvius, *Phys. Rev. Lett.* **70**, 2032

(1993) and H. J. Hesse and G. Wortmann, *Hyp. Interact.* **93**, 1505 (1994) for ^{151}Eu Mössbauer spectroscopy in a diamond anvil cell studies on $EuAl_2$ and EuM_2Ge_2, respectively.
92. R. D. Taylor and J. N. Farrell, *J. Appl. Phys.* **63**, 4108 (1988).
93. M. M. Abd-Elmeguid and R. D. Taylor, *Phys. Rev. B* **42**, 1048 (1990).
94. H. G. Zimmer, K. Takemura, K. Syassen, and K. Fisher, *Phys. Rev. B* **29**, 2350 (1984).
95. D. DiMarzio, M. Croft, N. Sakai, and M. W. Shafer, *Phys. Rev. B* **35**, 499 (1987).
96. M. Matlak and W. Nolting, *Z. Phys. B* **55**, 103 (1984).
97. J. N. Farrell and R. D. Taylor, *Phys. Rev. Lett.* **58**, 2478 (1987).
98. G. Parthasarathy and W. B. Holzapfel, *Phys. Rev. B* **37**, 8499 (1988).
99. U. Ladewig, B. Perscheid, and G. Kaindl, *Hyp. Interact.* **10**, 957 (1981).
100. M. Pasternak and A. L. Spijkervet, *Phys. Rev.* **188**, 574 (1969).
101. J. M. Hastings, L. M. Corliss, and M. Pasternak, *Phys. Rev. B* **31**, 3209 (1970).
102. J. Akella, G. S. Smith, and A. P. Jephcoat, *J. Phys. Chem. Solids* **49**, 573 (1988).
103. R. Coehoorn, unpublished results.
104. Z. Fisk and M. B. Maple, *J. Alloys Comp.* **183**, 303 (1992).
105. S. Nasu, *High Pressure Research*, in press.
106. H. F. Grünsteudel, H. J. Hesse, A. I. Chumakov, H. Grünsteudel, O. Leupold, J. Metge, R. Rüffer, and G. Wortmann, in Proc. of the Int. Conf. on Hyperfine Interactions, Leuven, Belgium (1995).
107. U. van Bürck and G. V. Smirnov, *Hyp. Interact.* **90**, 313 (1994) and references therein.
108. U. van Bürck, D. P. Siddons, J. B. Hastings, U. Bergmann, and R. Hollatz, *Phys. Rev. B* **46**, 6207 (1992).
109. R. Röhlsberger, E. Gerdau, W. Sturhahn, E. E. Alp, and R. Rüffer, in ICAME-95 proceedings, to be published in Il Nuovo Cimento.

9

Mössbauer Spectral Studies of Natural Substituted Spinels

Umberto Russo, Susanna Carbonin, and Antonio Della Giusta

1. INTRODUCTION

1.1. Mineralogical Aspects

Oxide minerals are widespread in a large variety of terrestrial and lunar rocks, and in meteorites.[1] They generally occur as minor components, that is, only as a few percent of the rock weight, but they may also be concentrated to form large ore deposits in the form of hematite, chromite, magnetite, and corundum. They are usually grouped according to structural symmetry as reported in Table 1. Modifications of coexisting oxides following changes in temperature, pressure, and oxygen fugacity, fO_2, during rock evolution have a direct influence on the magnetic properties of the earth. A fundamental parameter controlling their distribution, abundance, and composition is the level of initial fO_2 and its variations upon cooling. For instance, at magmatic temperatures, for high fO_2 values, iron oxides crystallize in preference to iron-rich silicates, whereas, with decreasing fO_2, iron is partitioned between oxide and silicate phases. A full comprehension of these phenomena is still unavailable, because of the difficulties arising from the large number of phases, solid, liquid, and gaseous, usually occurring in natural systems at various stages of their evolution.

Oxide minerals simultaneously record information on both the ambient temperature and the chemical potential of oxygen. This follows because oxygen is a component of the chemical system containing the minerals. Consequently, for several pairs of coexisting oxides, we may write a stoichiometric relationship, such as

$$O_2 = 6\ Fe_2O_3 - 4\ Fe_3O_4,$$

which defines the oxygen chemical potential at given temperature and pressure conditions. Important work has been carried out to show why only certain pairs of oxide minerals, such as rutile and hematite, hematite and magnetite, ilmenite and rutile, ilmenite and magnetite, coexist under specific well-defined conditions of temperature, pressure, fO_2 and, obviously, chemical composition. Changes in the oxygen fugacity cause strong exsolution and inter-

U. Russo • Dipartimento di Chimica Inorganica, Metallorganica ed Analitica, Università di Padova, Via Loredan 4, I-35131 Padova, Italy. *Susanna Carbonin and Antonio Della Giusta* • Dipartimento di Mineralogia e Petrologia, Università di Padova, I-35100 Padova, Italy.

Mössbauer Spectroscopy Applied to Magnetism and Materials Science, Volume 2, edited by Gary J. Long and Fernande Grandjean. Plenum Press, New York, 1996.

Table 1. Common Oxide Minerals Grouped According to Symmetry

Rhombohedral	Orthorhombic	Tetragonal	Cubic Monoxides
Hematite, Fe_2O_3	Pseudobrookite, Fe_2TiO_5	Rutile, TiO_2	Periclase, MgO
Corundum, Al_2O_3	Armalcolite, $Fe_{0.5}Mg_{0.5}Ti_2O_5$	Anatase, TiO_2	Wüstite, $Fe_{1-\delta}O$
Ilmenite, $FeTiO_3$	Brookite, TiO_2		Manganite, MnO
			Bunsenite, NiO
		Cubic Spinels	
Spinel, $MgAl_2O_4$	Galaxite, $MnAl_2O_4$	Mg-chromite, $MgCr_2O_4$	Magnetite, Fe_3O_4
Chromite, $FeCr_2O_4$	Ghanite, $ZnAl_2O_4$	Mg-ferrite, $MgFe_2O_4$	Trevorite, $NiFe_2O_4$
Franklinite, $ZnFe_2O_4$	Hercynite, $FeAl_2O_4$	Maghemite, $Fe_{2.66}O_4$	Ulvöspinel, Fe_2TiO_4

growth phenomena which may be observed optically especially in iron-titanium oxides; their study allows petrologists to trace the evolution of the oxide minerals and hence of the host rock.

1. 2. Crystallographic Aspects

Most oxide structures consist of nearly cubic or hexagonal closely-packed oxygen frameworks, with cations filling part of the four or six coordinate interstices. Iron, titanium, and manganese are common elements in oxide minerals; the existence of unpaired 3d electrons gives these ions net magnetic moments. Oxides consequently play a very important role in the origin of the earth's magnetic field.

The structure of spinels, which was first solved[2] by Bragg in 1915, consists of 32 symmetry equivalent oxygen dianions in a cubic unit cell whose lattice parameter, a_0, is about 0.8 nm, and whose symmetry is Fd3m. Sixteen of the 32 six-fold interstices, the M sites, and eight of the 64 four-fold interstices, the T sites, are filled by cations. The T site has perfect tetrahedral symmetry, point symmetry $\bar{4}3m$, whereas the M site has an irregular octahedral symmetry, $\bar{3}m$, with six equal bond distances, which is elongated or compressed along its unique three-fold axis. The T and M sites have coordinates fixed by symmetry, whereas oxygen has one degree of freedom, having lattice parameter coordinates, x, x, x, usually defined[3,4] as u or the "oxygen parameter."

The cations filling T and M sites of natural spinels are mainly Mg^{2+}, Fe^{2+}, Al^{3+}, Cr^{3+}, and Fe^{3+}, but at least twenty other cations with charges from one to six, e.g., Ag^+, Co^{2+}, Cu^{2+}, Zn^{2+}, Ga^{3+}, In^{3+}, Rh^{3+}, Si^{4+}, Sn^{4+}, Ti^{4+}, V^{5+}, and W^{6+}, may enter this structure. A few of the cations have strong site preferences, such as Cr^{3+} and Ti^{4+} for the M site, or Si^{4+} for the T site, but the majority of the cations can enter either site. Most natural spinels contain divalent and trivalent cations, as well as Ti^{4+} and Si^{4+}, which usually are the only high valence cations. Normal spinels are those, such as $MgAl_2O_4$, in which divalent cations order mainly on the T site. Inverse spinels are those with trivalent cations mainly on the T site, such as magnetite. The number of divalent cations on the M site, which must be equal to the number of trivalent cations on the T site, is called the inversion parameter, i. Delocalization of electrons between the T and M sites has been documented extensively through the semiconductor behavior of the inverse spinels. Because magnetic, elastic, dielectric, and optical properties depend on the cation distribution between the T and M sites, many investigations are devoted to unraveling this distribution. This is an important problem in the geological sciences, because the cation distribution strongly depends upon the cooling history of the host rock.

Extensive miscibility normally occurs within the spinel group, but some miscibility gaps exist, mainly in the binary solutions in which one end-member is a normal spinel and the second end member is an inverse spinel, such as the $FeAl_2O_4$-Fe_3O_4 system. Naturally occurring iron and titanium spinels commonly form non-stoichiometric oxides, such as maghemite, $Fe_{2.67}\square_{0.33}O_4$ or γ-Fe_2O_3, where $\square$ represents a cationic vacancy, and titanomaghemite, a formation usually associated with highly oxidizing conditions. In the laboratory, at very high temperatures, spinels with an excess of trivalent cations, and with vacancies replacing the divalent cation, as in $Al_{2.67}\square_{0.33}O_4$, or γ-Al_2O_3 easily crystallize.

1. 3. Analytical Techniques

The structural features of spinels, related to the short-range ordering and to the electronic and magnetic properties become increasingly complex as the number of different cations increases. Thus every available technique must be used to obtain information on the chemical and physical characteristics of these spinels. However each technique may provide only a partial understanding of these properties because it may consider only one, or at most a few aspects, of the very complex problem of spinel characterization. For instance, x-ray diffraction studies provide reliable values for the cubic lattice parameter, a_0, the oxygen parameter, u, and the average bond distances. Quantitative analysis of experimental T and M site electron densities also provides an accurate estimate of the tetrahedral and octahedral sites occupancies, and, hence, of the inversion parameter, i. These fundamental values are, in any case, only the starting point for the full characterization of a spinel. It is also possible to draw an accurate electron density map for the entire unit cell and to investigate electron distribution between the oxygen dianions and the metallic cations.[5] However, this technique has some severe limitations. For instance, many transition elements have similar scattering factors, so that it is sometimes difficult to discriminate between different cations. Moreover, x-ray diffraction determined electron densities are practically insensitive to the oxidation states of the various cations. However, these oxidation states may often be distinguished by their different sizes and bond distances. Another important characteristic is that x-ray diffraction averages over an entire crystal and cannot reveal the compositional variations in a crystal resulting from zoning phenomena related to a continuously changing environment during crystal growth or resulting from subsequent diffusion. Consequently only an average structure may be revealed by x-ray diffraction.

Fortunately, electron microprobe analysis can give quantitative and qualitative information on the chemical composition of small portions of an appropriately prepared crystal surface.[6] Such an analysis has to be repeated many times at different places of a crystal to determine the homogeneity of a crystal and to obtain average results for use in subsequent studies. Electron microprobe analysis cannot distinguish different oxidation states, but the cation distribution between the T and M sites in spinels and the different oxidation states, may be obtained through a minimization procedure that takes into account the information obtained by both microprobe and diffraction techniques. This procedure usually assumes that cations with strong site preferences occupy those sites as far as possible, such that Cr^{3+} and Ti^{4+} are presumed to be located only on the M sites, whereas Si^{4+} occupies only the T sites, at least when present in small amounts. All the other cations are initially equally distributed between both the T and M sites and their relative populations are thus the variable parameters. Moreover, a set of calculated metal-to-oxygen bond distances is required. These values may be obtained from the literature,[7-9] and some typical values are given in Table 2, and this choice is normally crucial for the development of a complete model of the spinel in question. The results must be self-consistent and, whenever possible, should be confirmed by other independent measurements.

Table 2. Various Spinel Metal to Oxygen Bond Distance Models, in Å

Model	Al(III) T	Al(III) M	Mg(II) T	Mg(II) M	Fe(II) T	Fe(II) M	Fe(III) T	Fe(III) M	Ref.
a	1.767	1.909	1.964	2.095	1.996	2.138	1.887	2.025	7
b	1.767	1.909	1.965	2.095	1.996	2.138	1.891	2.020	8
c	1.770	1.910	1.965	2.095	1.995	2.120	1.865	2.025	9

Magnetic susceptibility measurements[10] provide valuable information on the magnetic properties of the spinels, i.e., on the magnitude of the magnetic moments in the spinel and on how the moments interact. These magnetic results are important for technological applications and also provide information on the history of the earth's magnetic field. Indirect information on cation distributions between the T and M sites can also be obtained from the total magnetic moment.

Neutron diffraction[11] studies yield detailed information on both the crystal structure and the magnetic properties of spinels. Indeed, neutron scattering distinguishes between elements, such as magnesium and aluminum, that are almost identical in x-ray diffraction, and unambiguously determines the location and occupancy of interstitial species. It can also measure the average ordered magnetic moment at each crystallographic site, as well as the moment orientation. Measurements can be made from a few millikelvin up to 230 K, and in high pressure environments of up to several kilobars. Because much of this information can be independently obtained from other techniques, neutron scattering can often confirm other results.

Transmission electron microscopy[12] can be used in spinels to detect chemical and structural defects, such as exsolutions, chain-width disorder, stacking faults, clustering, and inclusions, defects that are beyond the resolution of other techniques. As electron microprobe analysis, transmission electron microscopy can investigate a small portion of a sample, a spot as small as 50 nm in diameter.

It was only with the introduction of magic angle spinning and the possibility of investigating spectra obtained with nuclides with nuclear quadrupole moments, that nuclear magnetic resonance spectroscopy became a useful technique[13] at the disposal of mineralogists and petrologists studying spinels. It permits the study of the structure of amorphous or poorly crystalline materials, and order/disorder phenomena both in the solid state and in the melt. The first nuclide to be studied was silicon-29, and, as a result, silicate minerals have been the most extensively studied. Relationships have been found between the chemical shift and the degree of polymerization or the number of aluminum atoms in the silicon second coordination sphere. Perhaps even more interesting for the study of spinels is aluminum-27 nuclear magnetic resonance spectroscopy. Although it is less sensitive than silicon-29, it gives useful information on the aluminum coordination in crystalline and amorphous environments. It is also important because it can distinguish Mg^{2+} from Al^{3+}, cations that have very similar x-ray scattering factors; NMR measurements thus help in the determination of the inversion parameter in spinels rich in both magnesium and aluminum.[14]

The electron spin resonance spectra of Cr^{3+} present on the M site as an impurity have been used to determine the near neighbors of the M site.[15] In this way, the inversion of $MgAl_2O_4$ has been measured with a claimed accuracy of 10 percent. Thus, as reported in

Table 3. Spinel Information Obtainable from Various Analytical Techniques

Technique	Qual. Analysis	Quant. Analysis	Bond Length	Cation Valence	Coord. Number	Magnetic Moment
X-ray Diffraction			×		×	
Microprobe	×	×				
Magnetism				×	×	×
Neutron Diffraction	×	×	×		×	×
ESR		×			×	
TEM	×	×			×	
NMR				×	×	
Mössbauer Effect	×	×		×	×	×

Table 3, each analytical method gives a unique contribution to the study of spinels, and a reliable model can be proposed only on the basis of the mutual confirmation of several techniques.

2. MÖSSBAUER SPECTROSCOPY

Together with x-ray diffraction and microprobe analysis, iron-57 Mössbauer spectroscopy is surely the most commonly used technique for mineralogical and petrological investigation of rocks, thanks to the ubiquitous presence in rocks of iron, even though often only in trace amounts.[16,17] One of its greatest advantages is that it is insensitive to any element other than iron, and so no interference is possible from other cations. As a nuclear technique, it records transitions between nuclear levels, the difference of which is perturbed by the electron density. This perturbation contains extensive information about the environment in which the iron nucleus is embedded, and is a function of the density of the valence shell electrons, their symmetry, and the orientation of their magnetic moments. The final spectrum is the sum of the contributions of each iron nucleus, rather than the average, as is the case for x-ray diffraction and, provided it is possible to separate these contributions in the spectral deconvolution, the spectrum contains information on each nucleus. By means of a careful interpretation, a Mössbauer spectrum can yield valence state, coordination number, site symmetry, and magnetic properties for each iron-containing site present in the absorber. In particular, a well established correlation between the electron density and the isomer shift, δ, permits a simple identification of valence state and coordination number. Typical isomer shifts reported for spinels are given in Table 4. In ideal geometries, characterized by spherically symmetric electron distributions, and thus by zero electric field gradients, iron gives rise to a single line Mössbauer spectrum. The appearance of a quadrupole split doublet is indicative of electronic distortions, distortions which may result from the lattice, as is predominately the case for Fe^{3+}, or from the valence electrons, as is predominately the case for Fe^{2+}. The latter contribution makes the quadrupole interaction, ΔE_Q, for Fe^{2+} strongly temperature dependent. Finally, the appearance of a six line spectrum indicates the presence of a long range magnetically ordered phase, the characteristics of which can be studied in detail.

Another parameter important for the full characterization of a spinel is the Mössbauer spectral linewidth. Component spectral linewidths, which are much larger than the natural linewidth of 0.195 mm/s for iron-57, are indicative of a continuous distribution of the electronic environment. The presence of various metals bonded to oxygen gives rise to configurations with probabilities proportional to the molar fractions of the individual

Table 4. Typical Room Temperature Spinel Isomer Shifts[a]

Coordination Number	Approximate Isomer Shift, mm/s		
	Fe^{2+}	$Fe^{2.5+}$	Fe^{3+}
4	1.0	0.7	0.15
6	1.1	0.8	0.3
8	1.3	–	–

[a]Relative to room temperature α-iron.

elements, each of which is characterized by slightly different hyperfine parameters, the so-called next nearest neighbor effect.[18]

Because every iron ion makes its own contribution to the overall Mössbauer spectrum, the area of each spectral component is approximately proportional to the number of ions in a particular environment. This permits a quantitative determination of the iron content in each site if the recoil free fractions, f, for each of the various iron sites are known. These recoil free fractions are characteristic of a particular iron site because they are a function of the strength by which the metal is bound in its equilibrium position in a solid. However one can assume, at least for similar compounds, that f depends only on the coordination geometry. Thus just two f-factors are necessary, one for the tetrahedral site and one for the octahedral sites. These two values have been found to be similar[19] and their ratio approximates one at low temperatures. Thus, for example, measurements at 78 K give quite accurate values for the iron distribution among different valence states and coordination geometries, provided that no phase change has taken place on lowering the temperature.

Texture effects, due to a preferential orientation of the microcrystals in the sample, may cause asymmetry in the peak intensity of a quadrupole doublet. This is particularly serious in complex spectra, in which some of the absorption peaks are completely unresolved. Texture most frequently occurs when crystals have an easy cleavage plane, as is the case for some sheet silicates, but texture is unusual with spinels. Texture effects may easily be avoided by careful grinding of the sample, possibly under acetone to avoid overheating, to yield a fine, homogeneous powder, and then dispersing it in an inert suspending material, or by measuring the spectrum at the magic angle.

Fitting the Mössbauer spectra of natural spinels is generally not straightforward, as the spectra reflect the complex nature of this class of compounds. The simultaneous presence of iron in different oxidation states and coordination geometries gives rise to an often unknown number of unresolved peaks that have to be deconvoluted to give internally consistent reasonable chemical, physical, and mineralogical parameters. At the end of the fitting procedure, the results must satisfy a model consistent with the contributions of all available experimental techniques.

A common feature in natural spinels is the presence of many different cations on both the tetrahedral and octahedral sites, such that the environments, even of crystallographically equivalent iron sites, may actually vary. Thus each of these iron sites becomes different from a Mössbauer spectral point of view and gives slightly different spectra and, as a consequence, the overall spectrum shows broad, asymmetric lines. Because the iron-oxygen bond distances are normally not significantly affected by iron next nearest neighbors, it is valid to assume that the isomer shift will remain virtually constant. The electronic symmetry at the iron nucleus is certainly influenced by the various next nearest neighbors, so that the quadrupole splitting changes somewhat from one site to another. In such cases, the best approach is to assume a distribution of iron sites with varying quadrupole splittings and a constant isomer shift.

Mössbauer spectral measurements as a function of temperature may be of great help because the isomer shift and the total spectral absorption area are expected to increase with decreasing temperature. In contrast, the quadrupole splitting, ΔE_Q, for iron(III) sites remains virtually constant and that for iron(II) increases substantially. Thus, some absorption peaks become evident upon cooling and the f factors approach each other, making the relative component areas more reliable for quantitative analysis.

3. EXAMPLES

3. 1. Chromite, $FeCr_2O_4$

The interpretation of the Mössbauer spectra of natural chromium containing spinels is normally quite complex and controversial. The results obtained by studying synthetic chromites often may only be utilized as a starting point, as the presence of other metals in natural spinels makes the spectra much more complex. Pure chromite, $FeCr_2O_4$, is a very simple compound containing only iron(II) in the tetrahedral site and its Mössbauer spectrum at room temperature is characterized by a single narrow line[18,20] with the parameters reported in Table 5. The iron site has cubic symmetry, and is surrounded by four oxygen anions in the first coordination sphere and by twelve chromium cations in the next nearest neighbor sphere and, as a consequence, no valence or lattice contribution to the electric field gradient is observed.

The addition of aluminum creates an increasing distortion in the iron environment due to the next nearest neighbor effect, a distortion which lowers the electronic symmetry at the tetrahedral iron site. The singlet is thus split into a doublet of increasing quadrupole splitting with increasing aluminum content, whereas the isomer shift remains virtually constant. In the same way, chromites in which chromium is partially substituted by iron(III), giving rise to spinels with the general formula $Fe^{2+}Cr_{2-x}Fe_x{}^{3+}O_4$, yield spectra that depend upon the degree of iron substitution.[21] This substitution causes a complete rearrangement in the cation distribution among the various valence states and coordination sites, depending upon the re-

Table 5. Crystallographic and Mössbauer Spectral Parameters for Synthetic Chromites[a]

Compound	a_0, Å	δ, mm/s	ΔE_Q, mm/s	H_{int}, kOe	Area, %	Site	Cation Assignment
$FeCr_2O_4$	8.378	0.936	0.00	-	100	T	Fe^{2+}
$Fe_{1.25}Cr_{1.75}O_4$	8.388	0.939	0.244	-	77	T	Fe^{2+}
		0.391	0.463	-	23	M	Fe^{3+}
$Fe_{1.5}Cr_{1.5}O_4$	8.401	0.938	0.316	-	64	T	Fe^{2+}
		0.376	0.441	-	36	M	Fe^{3+}
$Fe_{1.75}Cr_{1.25}O_4$	8.408	0.95	0.35	-	57	T	Fe^{2+}
		0.38	0.42	-	43	M	Fe^{3+}
Fe_2CrO_4[b]	8.379						
$Fe_{2.38}Cr_{0.62}O_4$	8.382	0.279	c	477	33	T	Fe^{3+}
		0.68	c	420	66	M	$Fe^{2.5+}$
$Fe_{2.5}Cr_{0.5}O_4$	8.383	0.277	c	482	33	T	Fe^{3+}
		0.72	c	431	66	M	$Fe^{2.5+}$
$Fe_{2.75}Cr_{0.25}O_4$	8.388	0.277	c	488	33	T	Fe^{3+}
		0.69	c	445	66	M	$Fe^{2.5+}$
Fe_3O_4	8.393	0.283	c	494	33	T	Fe^{3+}
		0.67	c	461	66	M	$Fe^{2.5+}$

[a]Room temperature data obtained from reference 21. [b]This spinel shows a broad Mössbauer spectral absorption band attributed to tetrahedral Fe^{2+}, $Fe^{2.5+}$, and Fe^{3+}. [c]Value not reported.

dox properties and geometric preferences of each cation. Up to $x < 0.68$, all substitutional iron(III) is found in octahedral sites, and there is no change in the molecular structure. The resulting Mössbauer spectra show two doublets due to the distorted tetrahedral iron(II) and octahedral iron(III) cations. This substitution is characterized by the increase in the cubic unit cell lattice parameter, an increase which is approximately linear in x. Compounds with x between 0.68 and 1.38 yield Mössbauer spectra with a single, very broad, asymmetric peak that, with some ambiguity, can be fit with three separate absorptions. The model on which this fitting is based involves the displacement of some tetrahedral Fe^{2+} cations to octahedral sites, leading to the onset of electron delocalization between the iron(II) and iron(III) present on these sites. On the Mössbauer time scale, this electron delocalization is fast enough to give rise to a singlet with an isomer shift corresponding to an average valence state of 2.5. The remaining portion of the spectrum is due to octahedral iron(III) and tetrahedral iron(II). Chromites with x values higher than 1.38 give spectra with two magnetic sextets, clearly indicating that the structure is now very similar to that of magnetite, Fe_3O_4.

Natural chromites normally have very complex compositions. The CR5 chromite from South Africa is particularly rich in chromium, aluminum, magnesium, and iron. In the computation of its site distribution, chromium was assumed to occupy only the octahedral site, whereas aluminum, magnesium, and iron were assumed to occupy both octahedral and tetrahedral sites.[22] The results obtained by using three different sets of metal-to-oxygen bond distances, see Table 2, are given in Table 6. This is a very good example, because the T-O and M-O bond distances require that all the available Mg^{2+} be placed in the T site and Al^{3+} in the M site. Consequently, the cation distribution in CR5 chromite is not very sensitive to the set of bond distances used. Table 6 indicates that the main difference is found in the displacement of some M-site Fe^{3+} to the T site, with an identical, but opposite, displacement of Fe^{2+}. In this situation a large next nearest neighbor effect is expected, and, indeed, complex highly temperature dependent Mössbauer spectra, as shown in Figure 1, are observed. The simplest model reported in the literature to fit these spectra is based[18] on the presence of two iron sites, a tetrahedral site occupied by iron(II), and an octahedral site occupied by iron(III). Support for this model comes from the values of the observed spectral linewidths and, above all, from the good agreement of the Fe^{3+}/Fe_{total} ratio determined by Mössbauer spectroscopy, electron microprobe analysis, or chemical analysis.

An alternative model for understanding the CR5 chromite spectra considers the existence of two tetrahedral iron(II) and one octahedral iron(III) sites.[23] The crystallographic equivalence of the iron(II) sites is removed by the simultaneous presence of various cations around the octahedral interstices. This change strongly affects the electronic symmetry at the iron(II) site, whereas it has a negligible effect on the electron density at both the iron(II) and iron(III) sites. As a consequence, the iron(II) doublets have the same isomer shift and different quadrupole splittings and the iron(III) site remains unsplit. With this model, the Fe^{2+}/Fe^{3+} ratios calculated on the basis of the spectral absorption areas are not in good agreement with those obtained by microprobe analysis. A related model has also been applied to two natural spinels from Sukinda ultramafites,[24] but, in this case, even small variations in the chemical composition led to different spectral hyperfine parameters and, thus, to different site and valence state assignments. The spectra of one of these samples were fit with a tetrahedral iron(III) doublet, an octahedral iron(II) doublet, and a tetrahedral iron(II) doublet. However, the hyperfine parameters for the octahedral iron(II) site are somewhat questionable in that the small quadrupole splitting remains virtually constant, increasing only from 0.34 to 0.39 mm/s upon cooling from 297 to 110 K. In contrast, the isomer shift increases from the normal value of 1.08 mm/s at 297 K to the unusually high value of 1.74 mm/s at 110 K. Two tetrahedral iron(III) and one octahedral iron(II) sites were used to fit the spectrum of the second sample, which was consequently assumed to have crystallized in conditions of high oxygen fugacity. Only later did the first sample crystallize,

Table 6. CR5 Site Populations Determined by X-ray Diffraction and Microprobe Analysis[a]

Model	Fe(II) T	Fe(III) T	Fe(II) M	Fe(III) M	Fe(II)/ Fe	Fe(III)/ Fe	Mg(II) T	Al(III) M	Cr(III) M	Ti(IV) M
a	0.509	0.000	0.008	0.187	73.5	26.5	0.482	0.556	1.226	0.015
b	0.503	0.004	0.014	0.183	73.4	26.6	0.484	0.555	1.227	0.015
c	0.508	0.000	0.008	0.186	73.5	26.5	0.483	0.558	1.225	0.015

[a]Data obtained from references 7, 8, and 9.

when the oxygen fugacity had decreased and Al^{3+} began to be incorporated into the octahedral sites of the spinel structure. A similar model with three different iron sites has been used to fit[22] the spectra of CR5. The 80 K spectrum, shown in Figure 1, reveals the presence of a symmetric doublet due to an iron(III) component and at least two different iron(II) components. The parameters of these components were used to fit the room temperature spectrum by taking into account the expected temperature dependence. Reasonable fits, see Table 7, were obtained, in good agreement with the crystallographic results. As expected neither iron(III) was found on the tetrahedral site nor iron(II) on the octahedral site.

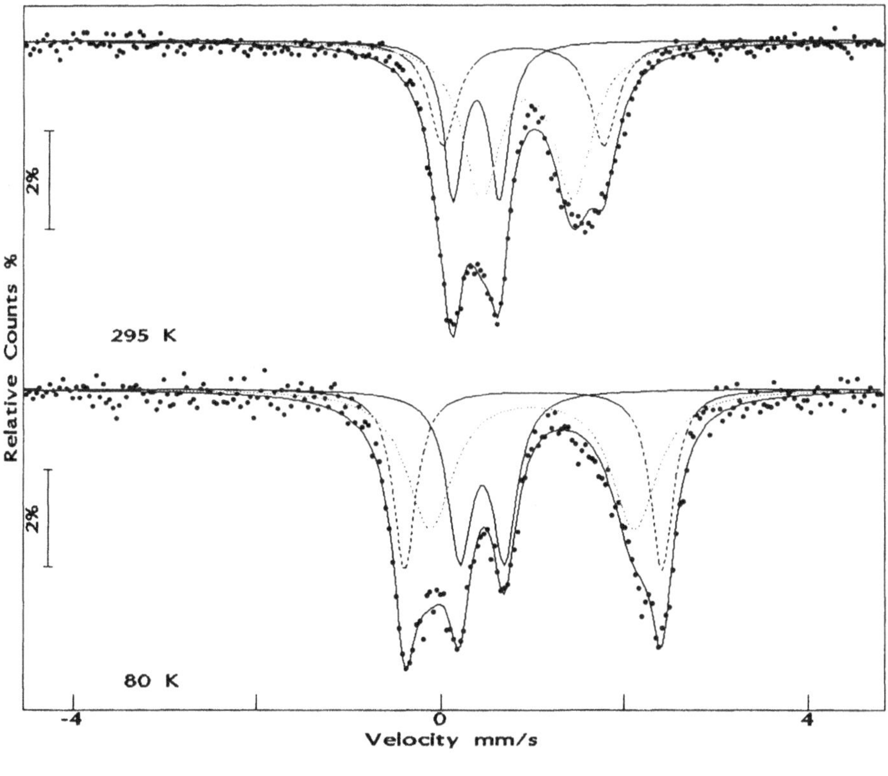

Figure 1. Mössbauer spectra of CR5 spinel collected at 80 and 295 K.

Table 7. Mössbauer Spectral Parameters for the CR5 Spinel

T, K	δ,[a] mm/s	ΔE_Q, mm/s	Γ, mm/s	Area, %	Assignment Site	Cation
297	0.39	0.50	0.25	25	M	Fe^{3+}
	0.94	1.00	0.50	50	T	Fe^{2+}
	0.90	1.76	0.36	25	T	Fe^{2+}
80	0.45	0.49	0.32	26	M	Fe^{3+}
	0.99	2.23	0.62	46	T	Fe^{2+}
	1.01	2.80	0.30	28	T	Fe^{2+}

[a] Relative to room temperature α-iron.

3. 2. Hercynites, $FeAl_2O_4$

Pure hercynite, which is essentially a normal spinel, gives a very simple Mössbauer spectrum characterized by a doublet with an isomer shift of 0.91 mm/s and a quadrupole splitting of 1.57 mm/s, values typical for a tetrahedral iron(II) site.[25] The quadrupole interaction has been ascribed either to a non-cooperative Jahn-Teller distortion[26] or to a slight inversion of the cation distribution.[27,28] Synthetic hercynites yield rather complex Mössbauer spectra which were fit[29] with two doublets, due to both tetrahedral and octahedral iron(II), and a third doublet, due to small amounts of iron(III). The iron distribution among these sites is strongly dependent on the annealing temperature during preparation.

The above model has been applied to an extensive series of natural spinels[30] derived from many different geological and tectonic settings. Apart from large amounts of the hercynite component, these samples usually contain Mg^{2+}, Cr^{3+}, and Fe^{3+}. However, the Mössbauer spectra reveal the presence of two different tetrahedral sites, which, because they are not evident in the x-ray diffraction pattern, have been ascribed to the next nearest neighbor effect. The variation in the quadrupole splitting for these sites, as the composition changes, is due to the different occupancies of the octahedral sites. Fe^{3+} is present in the tetrahedral site, but a partial occupancy of the octahedral site cannot be excluded as its contribution to the Mössbauer spectrum could be hidden by the low velocity iron(II) components. Such an occupancy should, however, be favored by a high Cr^{3+} content. This model is further supported by the extremely good agreement between the Fe^{3+}/Fe_{total} ratios revealed by both Mössbauer spectroscopy and electron microprobe analysis. A more complex model, requiring the presence of four different iron sites, has been used to fit the Mössbauer spectra of a sample of aluminum-rich spinel, TS2, from the Italian Alps.[22] This material is rather inhomogeneous, and not only did the various crystals give different crystallographic parameters, but even the microprobe data changed inside the same crystal, due to zoning effects, so that only average data could be obtained.

Unlike the cation distribution in the CR5 chromite, the cation distribution in TS2, determined by x-ray and microprobe analysis, is very sensitive to the bond distances, because the interatomic distances and site electron densities can be satisfied by various distributions of the four cations entering the tetrahedral and octahedral sites. Table 8 reveals that the third set of metal-to-oxygen bond distances gives results which are rather different from those obtained using the other models. According to models a and b, Fe^{2+} is practically all on the tetrahedral site and Fe^{3+} on the octahedral site, whereas model c places only Fe^{3+} on the tetrahedral site and distributes Fe^{2+} on both sites.

The room temperature Mössbauer spectrum, shown in Figure 2 and fit with the parameters reported in Table 9, is rather complex and only a comparison with the results obtained at 80 K leads to a logical fitting model. Only one site is occupied by iron(III), most likely with octahedral coordination, and accounts for about 20 percent of the total spectral

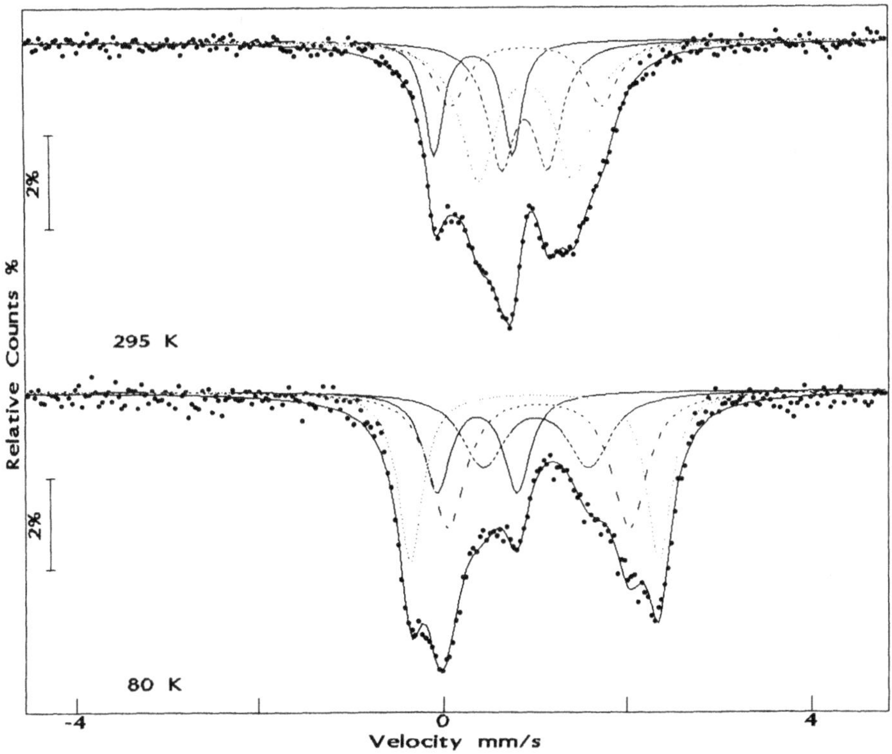

Figure 2. The Mössbauer spectra of TS2 spinel measured at 80 and 295 K.

absorption area. The large quadrupole splitting is attributed to the presence of a large number of possible cation distributions in the second coordination sphere, which is composed of six tetrahedral and six octahedral sites. For a statistical distribution of cations on the available sites the tetrahedral site should contain Mg^{2+}, Al^{3+}, and Fe^{2+}, in large amounts, cations which will give rise to a strong next near neighbor effect. The remaining part of the spectrum is fit with three strongly overlapping doublets. In spite of the great uncertainty in the Mössbauer spectral parameters, the percent distribution of iron is in good agreement with that obtained from x-ray diffraction and electron microprobe analysis.

Table 8. TS2 Site Populations Determined by X-ray Diffraction and Microprobe Analysis[a]

Model	Fe(II) T	Fe(III) T	Fe(II) M	Fe(III) M	Fe(II)/ Fe	Fe(III)/ Fe	Al(III) T	Mg(II) T	Al(III) M	Mg(II) M
a	0.221	0.000	0.003	0.058	79.5	20.5	0.120	0.655	1.816	0.115
b	0.220	0.000	0.003	0.058	79.3	20.7	0.122	0.652	1.813	0.118
c	0.164	0.059	0.060	0.000	79.2	20.8	0.083	0.689	1.812	0.081

[a]Data obtained from references 7, 8, and 9.

Table 9. Mössbauer Spectral Parameters for TS2

T, K	δ,[a] mm/s	ΔE_Q, mm/s	Γ, mm/s	Area, %	Assignment Site	Cation
295	0.34	0.85	0.24	17	M	Fe^{3+}
	0.89	0.50	0.35	25	T	Fe^{2+}
	0.91	1.03	0.47	40	T	Fe^{2+}
	0.92	1.64	0.42	17	T	Fe^{2+}
80	0.37	0.87	0.34	18	M	Fe^{3+}
	1.01	1.11	0.52	19	T	Fe^{2+}
	1.04	1.96	0.47	34	T	Fe^{2+}
	1.00	2.70	0.32	29	T	Fe^{2+}

[a]Relative to room temperature α-iron.

A very simple spectrum, shown in Figure 3, is that of the classical spinel, $MgAl_2O_4$, containing some substitutional divalent iron. The spectrum is from a Swedish mineral referred to as VASA[22] in Table 10. It can be easily fit with two iron(II) doublets with the spectral parameters given in Table 11. One of these doublets, by far the largest in area, is

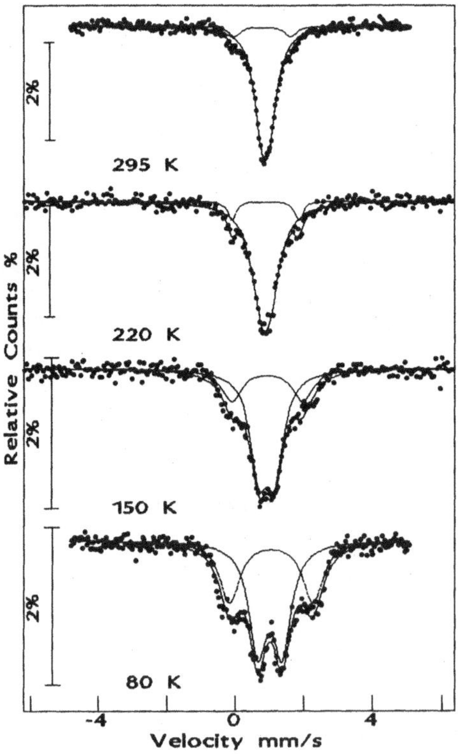

Figure 3. The Mössbauer spectra of VASA spinel measured between 80 and 295 K.

Table 10. VASA Site Populations Obtained from X-ray and Microprobe Analysis[a]

Cation	Site	Unquenched Population	Quenched Population
Al(III)	T	0.125	0.263
Fe(II)	T	0.073	0.073
Fe(III)	T	0.000	0.000
Mg(II)	T	0.793	0.654
Al(III)	M	1.874	1.737
Fe(II)	M	0.001	0.000
Fe(III)	M	0.002	0.001
Mg(II)	M	0.121	0.261

[a]Based on bond distance model b.

characterized by a very small quadrupole splitting. This indicates the high symmetry of the site, a site which appears to have tetrahedral symmetry. The other iron site is in a similar, but much more distorted environment, as is indicated by its large quadrupole splitting. Although the temperature dependence of the hyperfine parameters is quite normal, the behavior of the relative area is somehow unexpected, as the two species present a very different temperature dependence. This has been interpreted as the consequence of different bonding for the two types of iron.

Heating at 1000° C for 10 minutes with subsequent fast quenching to room temperature causes dramatic changes in the Mössbauer spectra of VASA, as is shown in Figure 4 and in Table 12. Both iron sites are strongly affected by the cation rearrangement upon quenching and this is reflected in the quadrupole splitting values that now indicate the presence of much more highly disordered second coordination spheres. At the same time, the isomer shift remains virtually constant indicating a retention of the tetrahedral bonding at both iron sites. This distribution is consistent with that obtained from x-ray diffraction and electron microprobe analysis, see Table 10, which suggests only tetrahedral Fe^{2+} in both unquenched and quenched materials. In this case, as for the CR5 spinel, because of the almost total absence of the iron(III) variable parameters, changing the set of bond distances leads to almost identical results, and thus only the data for set b are reported in Table 10.

The ML2 spinel also has a composition similar to that of TS2, with a somewhat higher iron content,[31] as is reported in Table 13. In this case, iron is more randomly distributed, and, as a consequence, comparable amounts of Fe^{3+} and Fe^{2+} are found on the tetrahedral and octahedral sites. Thus ML2 gives a more complex Mössbauer spectrum, because several

Table 11. Iron(II) Mössbauer Spectral Parameters of the VASA Spinel

T, K	δ,[a] mm/s	ΔE_Q, mm/s	Γ, mm/s	Area, %	Assignment Site	Assignment Cation
295	0.89	0.22	0.53	90	T	Fe^{2+}
	0.79	1.65	0.42	10	T	Fe^{2+}
220	0.89	0.27	0.67	89	T	Fe^{2+}
	0.86	1.93	0.40	11	T	Fe^{2+}
150	0.92	0.46	0.61	71	T	Fe^{2+}
	0.94	2.07	0.70	29	T	Fe^{2+}
80	1.03	0.71	0.63	60	T	Fe^{2+}
	1.06	2.44	0.72	40	T	Fe^{2+}

[a]Relative to room temperature α-iron.

Table 12. Mössbauer Spectral Parameters for the VASA Spinel After Annealing

T, K	δ,[a] mm/s	ΔE_Q, mm/s	Γ, mm/s	Area, %	Assignment
295	0 91	1.90	0 47	45	Fe^{2+}
	0.93	1.18	0.52	55	Fe^{2+}
240	0.99	2.19	0.52	45	Fe^{2+}
	0.99	1.38	0.65	55	Fe^{2+}
200	1.01	2.34	0.48	45	Fe^{2+}
	1.01	1.32	0.74	55	Fe^{2+}
160	1.05	2.58	0.42	52	Fe^{2+}
	1.03	1.81	0.68	48	Fe^{2+}
120	1.06	2.70	0.41	56	Fe^{2+}
	1.08	1.94	0.73	44	Fe^{2+}
80	1.06	2.83	0.35	56	Fe^{2+}
	1.07	2.20	0.70	44	Fe^{2+}

[a]Relative to room temperature α-iron.

slightly different second coordination spheres become equally probable. The 295 K spectrum shows two broad, asymmetric absorption peaks, clearly due to the overlap of various doublets. As the 295 K spectrum does not show clear features, additional spectra were col-

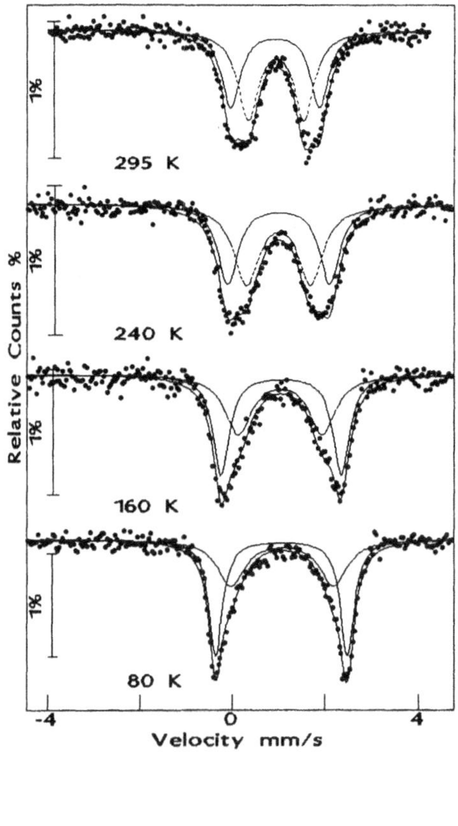

Figure 4. The Mössbauer spectra of VASA spinel after annealing.

Table 13. ML2 Site Populations Determined by X-ray Diffraction and Microprobe Analysis[a]

Model	Fe(II) T	Fe(III) T	Fe(II) M	Fe(III) M	Fe(II)/ Fe	Fe(III)/ Fe	Al(III) T	Mg(II) T	Al(III) M	Mg(II) M
a	0.255	0.109	0.088	0.069	65.8	34.2	0.132	0.496	1.641	0.180
b	0.278	0.085	0.065	0.093	65.8	34.2	0.149	0.480	1.624	0.196
c	0.316	0.045	0.026	0.133	65.8	34.2	0.165	0.465	1.608	0.211

[a]Data obtained from references 7, 8, and 9. The Ti(IV) octahedral site population is 0.018 in all three models.

lected at low temperatures and are shown in Figure 5. At low temperature a weak absorption on the right side of the low-velocity component, an absorption which is diagnostic of an iron(III) species, becomes apparent, but no change occurs in the high velocity absorption. The spectra are thus fit by assuming a continuous distribution of iron(II) sites, with the same isomer shift and varying quadrupole splittings. The resulting parameters are listed in Table 14. In this way, although all details on the distribution of Fe^{2+} are lost, a good estimate of the total area due to Fe^{2+} is obtained. Two more doublets, with parameters which are typical of Fe^{3+}, are then required to fit the remaining absorption area. The resulting fit is in good

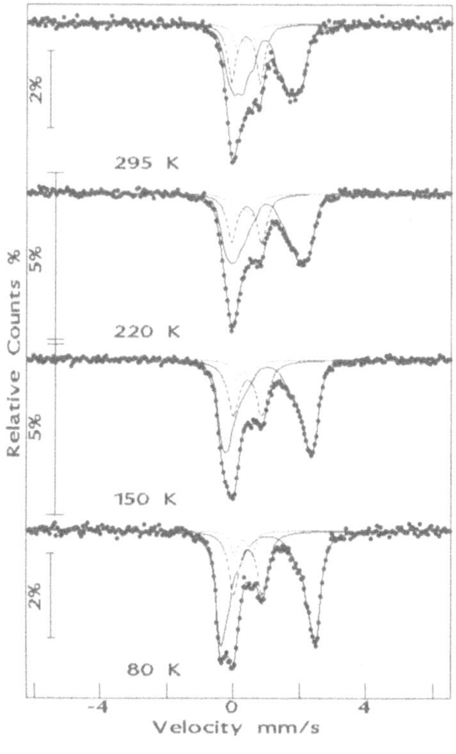

Figure 5. The Mössbauer spectra of the ML2 spinel.

Table 14. Mössbauer Spectral Parameters for the ML2 Spinel

T, K	δ,[a] mm/s	ΔE_Q, mm/s	Area, %	δ,[a] mm/s	ΔE_Q, mm/s	Area, %	δ,[a] mm/s	ΔE_Q, mm/s	Area, %
295	0.36	0.89	24.5	0.29	0.40	10.9	0.95	1.46	64.7
220	0.43	0.90	24.1	0.37	0.43	10.2	1.02	1.94	64.7
150	0.44	0.91	26.7	0.36	0.49	8.7	1.06	2.43	64.5
80	0.49	0.89	26.6	0.37	0.52	7.6	1.11	2.80	65.7

[a]Relative to room temperature α-iron.

agreement with the x-ray diffraction results, as far as the Fe^{2+}/Fe_{total} ratio is concerned. Moreover, the isomer shift values for the two Fe^{3+} components indicate a distribution of the metal between the tetrahedral and octahedral sites with an area ratio that agrees with the one calculated using model c in Table 13.

4. CONCLUSIONS

The Mössbauer spectra of spinels provide a determination of the various iron sites, their coordination and symmetry, the valence state of the iron ions, and a quantitative estimate of the percentage of iron present in each site. It is clear from the chemical and crystallographic complexity of spinels that only in the simplest cases can Mössbauer spectra be deconvoluted with acceptable confidence. Much more often, comparisons among spectra obtained at various temperatures are indispensable as a starting point for obtaining reliable results. Detailed information must sometimes be sacrificed to obtain other parameters with better accuracy. Best fits are achieved only when, with reasonable values for the χ^2 or the misfit parameters, the hyperfine parameters are self-consistent, that is, when they have acceptable values and follow the expected changes with temperature, and when they provide a model which is meaningful on a chemical, physical, and crystallographic basis. Most of all, the results must be in full agreement with those obtained using other independent techniques. Although it is worth noting that in all reported examples the Fe^{2+}/Fe^{3+} ratio measured by Mössbauer spectroscopy agrees well with the values calculated from stoichiometry, it must be remembered that Mössbauer spectral results alone would, in many cases, supply only limited information.

REFERENCES

1. Spectroscopic Methods in Mineralogy and Geology, in *Reviews in Mineralogy*, F. C. Hawthorne, ed., Vol. 18, Mineralogical Society of America (1988).
2. W. H. Bragg, *Phil. Mag.* **30**, 305 (1915).
3. S. Hafner, *Schw. Miner. Petr. Mitteilungen* **40**, 208 (1960).
4. R. J. Hill, J. R. Craig, and G. V. Gibbs, *Phys. Chem. Minerals* **4**, 317 (1979).
5. V. G. Tsirel'son, Ye. L. Belokoneva, Yu. Ž. Nozik, and V. S. Urusov, *Geochem. Intern.* **24**, 124 (1987).
6. S. J. B. Reed, *Electron Microprobe Analysis*, Cambridge University Press, Cambridge (1975).
7. S. Carbonin and A. Della Giusta, unpublished results.
8. A. Della Giusta, S. Carbonin, and G. Ottonello, *Miner. Mag.* in press.
9. H. St. C. O'Neill and A. Navrotsky, *Amer. Mineral.* **68**, 181 (1983).
10. J. A. Lodya, H. Pollak, J. Nell, and A. Albers, *Hyp. Interact.* **90**, 1789 (1994).
11. *International Tables for Crystallography*, Vol. C, Kluwer Academic Publisher, Dordrecht (1992).
12. E. A. Ferrow, *Hyp. Interact.* **90**, 121 (1994).

13. Spectroscopic Methods in Mineralogy and Geology, in *Reviews in Mineralogy,* F. C. Hawthorne, ed., Vol. 18, Chapter 9, Mineralogical Society of America (1988).
14. G. C. Gobbi, R. Christoffersen, M. T. Otten, B. Miner, P. R. Buseck, G. J. Kennedy, and C. A. Fyfe, *Chem. Lett.* 772 (1985).
15. N. Schmocker and F. Waldner, *J. Phys. C: Solid State Phys.* **9**, L235 (1976).
16. N. N. Greenwood and T. C. Gibb, *Mössbauer Spectroscopy*, p. 258, Chapman and Hall, London (1971).
17. E. Murad and J. H. Johnston, in *Mössbauer Spectroscopy Applied to Inorganic Chemistry,* G. J. Long, ed., Vol. 2, p. 507, Plenum, New York (1987).
18. G. M. Bancroft, M. D. Osborne, and M. E. Fleet, *Solid State Comm.* **47**, 623 (1983).
19. G. A. Sawatzky, F. van der Woude, and A. H. Morrish, *Phys. Rev*. **183**, 383 (1969).
20. M. D. Osborne, M. E. Fleet, and G. M. Bancroft, *Contrib. Mineral. Petrol.* **77**, 251 (1981).
21. M. Robbins, G. K. Wertheim, R. C. Sherwood, and D. N. E. Buchanan, *J. Phys. Chem. Solids* **32**, 717 (1971).
22. S. Carbonin, U. Russo, and A. Della Giusta, *Miner. Mag.* in press.
23. Y. L. Chen, B. X. Xu, J. G. Chen, and Y. Y. Ge, *Phys. Chem. Miner.* **19**, 255 (1992).
24. S. Mitra, T. Pal, and T. Pal, *Miner. Mag.* **55**, 535 (1991).
25. L. Larsson, H. St. O'Neill, and H. Annersten, *Eur. J. Mineral.* **6**, 39 (1994).
26. R. E. Vandenberghe and E. DeGrave, in *Mössbauer Spectroscopy Applied to Inorganic Chemistry,* G. J. Long and F. Grandjean, eds., Vol. 3, p. 87, Plenum, New York (1989).
27. M. Tanaka and T. Tokoro, *J. Phys. Soc. Japan* **21**, 262 (1966).
28. C. M. Yagnik and H. B. Mathur, *J. Phys. C: Solid State Phys.* **1**, 469 (1968).
29. J. C. Waerenborgh, M. O. Figueiredo, J. M. P. Cabral, and L. C. J. Pereira, *Phys. Chem. Miner.* **21**, 460 (1994).
30. B. J. Wood and D. Virgo, *Geochim. Cosmochim. Acta* **53**, 1277 (1989).
31. P. Nimis, Ph. D. Thesis, University of Padua (1994).

10

Iron-zinc Intermetallics in Commercial Galvanneal Steel Coatings

Desmond C. Cook and Richard G. Grant

1. INTRODUCTION

The need to improve the corrosion resistance of sheet steel, particularly for use within the automotive industry, has led to a dramatic increase in the use of coated steels in place of cold-rolled sheet steel. Consequently, industrial interest in the processing of zinc alloy coatings has risen over the past decade.[1-4] Hot-dip galvanized and galvannealed sheet steel are two such products in use today. These processes involve the use of zinc and zinc-alloy coatings to protect the steel through the sacrificial or galvanic mechanism,[5] and are an economic way to apply the zinc. Subsequently, today's continuous coating lines are capable of producing materials having well controlled coating thickness and uniformity. In contrast to galvanized steel, galvanneal steel results from the post-dip annealing of the zinc-coated sheet steel, in which iron and zinc are interdiffused to form an iron-zinc alloy coating. Galvanneal can also be produced by heat-treating sheet steel onto which zinc has been electrodeposited.[6-9] Within the galvannealed coating, four iron-zinc phases may be present. They are commonly referred to as the Zeta, Delta, Gamma-1, and Gamma phases.

Manufacture of the most suitable galvannealed coating requires identifying the iron-zinc phases which form during the galvannealing process, having an understanding of the properties of each phase, and knowing how to control the formation of any particular phase or phases in order to obtain optimum material performance. Positive identification of each phase and the fraction present in a galvannealed coating is very difficult, particularly in the on-line situation where it is desirable to know these parameters quickly in order to control the anneal temperature and time. The primary reason for this problem is the lack of high quality data on the crystal structure and the related microstructure of the separate iron-zinc phases. Another reason is the lack of knowledge on the relationship between the pure iron-zinc phases and the phases present in different galvannealed coatings. X-ray diffraction, for example, should be able to separately identify each phase but has only proven to be moderately successful in identifying the Zeta and Gamma phases. In particular, the crystal structures of the Delta and Gamma-1 phases are not well understood and x-ray diffraction has consistently failed to sepa-

D. C. Cook • Department of Physics, Old Dominion University, Norfolk, VA 23529, USA. *R. G. Grant* • Department of Physics, Hollins College, Roanoke, VA 24020, USA.

Mössbauer Spectroscopy Applied to Magnetism and Materials Science, Volume 2, edited by Gary J. Long and Fernande Grandjean. Plenum Press, New York, 1996.

rately identify each, not only in coatings but in standard laboratory produced iron-zinc alloys.

1. 1. Galvannealed Steel

Galvannealed steel is one product that has resulted from post-dip annealing of zinc-coated sheet steel. Galvannealed coatings are comprised of iron-zinc alloy phases which have better corrosion resistance, particularly after painting, and are easier to weld than galvanized coatings. These and other improved properties which have resulted from coating steel, have led to a dramatic increase in their use over the last decade. To meet the automakers' demands for increased and improved galvanneal, the number of continuous hot-dip and electrolytic galvanizing lines has doubled since 1985 to its present number of nearly four hundred.[3] Over 40 million tons of zinc and zinc alloy coated sheet steel are produced worldwide annually.[4] This is evenly divided between the United States, Europe, and Japan. Approximately one-third is purchased by the automotive industry.

In general, the coatings produced by the hot-dip method offer excellent corrosion protection and paint adhesion. Hot-dip coatings are generally thicker and therefore more corrosion resistant than those produced by electroplating. However, the formability, i.e., the bendability of the sheet steel, and coating adhesion suffer as a result of this increased thickness. The mechanical and corrosion properties of galvanneal are controlled by the iron-zinc intermetallics that form in the coating. The corrosion resistance, coating adhesion and ductility, formability, weldability, and paintability, depend on the morphology, composition, and structure of the phases present in the coating.[10-12]

Annealing the zinc-coated sheet steel allows the interdiffusion of zinc and iron which form an alloy coating. The iron content in the coating depends primarily on the anneal temperature and time, with up to four iron-zinc intermetallic phases able to form. They range from the iron rich Gamma phase, which forms a thin layer at the steel surface, to the Gamma-1 and Delta phases, which appear generally as thicker layers in the coating, and finally to the zinc rich Zeta phase, which forms at the coating surface. Other factors which control the composition and properties of the galvannealed coating include zinc bath additives, such as aluminum, as well as the composition and type of steel coated.[13-19] These factors modify the iron-zinc reactivity and diffusion rates and therefore the amount of each phase in the coating.

The presence of each iron-zinc phase in the coating controls the performance of the material in different ways. The Gamma phase is generally a hard thin layer and has the highest iron content of the four phases. It is therefore the easiest phase to weld but the hardest to form. Its corrosion resistance is poor. Increasing the thickness of the Gamma phase leads to a harder coating but it is believed to promote cracking, powdering, flaking, and chipping of the galvanneal. Corrosion resistance and formability improve for the phases containing more zinc. However, adhesion and weldability decrease at the same time. The Delta phase is the most ductile and corrosion resistant of the four phases. Therefore commercial galvanneal containing a majority of the Delta phase is satisfactory for automotive bodies in terms of formability, weldability, and paintability. Present commercial galvannealed coatings normally contain at least 70 percent of the Delta phase with the remaining being the Gamma-1 and Gamma phases.

A significant amount of research has been undertaken recently in order to understand the materials and engineering properties of the separate iron-zinc phases as well as galvannealed coatings. In particular knowledge of hardness, formability, corrosion resistance, weldability, paintability, cracking, bleeding, and fracture behavior have resulted in the need for commercial producers to control the iron-zinc phase composition in galvanneal.[20-27] Correct interpretation of the mechanical properties and the development of models capable of predicting the characteristics and servicibility of newly produced galvannealed coatings, requires accurate identification and characterization of the pure iron-zinc alloys. This has been one intensive research area in which the use of Mössbauer spectroscopy has proven to be an extremely valuable analytical procedure.[28-35]

1. 2. Iron-zinc Alloys

Identification of the different iron-zinc phases, and the fraction of each present in a galvannealed coating is difficult. Typically x-ray diffraction is a simple technique that should allow each phase to be identified. The difficulty in identifying the iron-zinc phases by x-ray diffraction arises because the patterns of three of the phases contain a large number of diffraction lines which are in general broad and overlapping.[36] Only the cubic Gamma phase exhibits a simple pattern and therefore is easily identified. The Zeta phase has also been identified even though its x-ray diffraction pattern contains many broad and overlapping peaks. The diffraction patterns of the Delta and Gamma-1 phases are complex and very similar. In commercial galvannealed coatings, which have been produced under different annealing conditions, and whose mechanical and structural properties are known to be different, the x-ray diffraction patterns are often very similar. Complicating the analysis of the x-ray diffraction data is the immense number of overlapping diffraction lines present in samples containing the Delta and Gamma-1 phases. Hindering the analysis of galvannealed coatings using x-ray diffraction is the fact that the standard powder x-ray diffraction tables, published by the International Center for Diffraction Data,[37] are not complete for any of the four phases. For example, the tabulated data for the Delta phase, $FeZn_7$, was recorded in 1938 before the Gamma-1 phase was known. The need for renewed research on the crystalline and microstructure of high purity iron-zinc alloys has been driven by the use of galvanneal and the need to improve its performance.

This chapter is divided into two sections. The first section deals with the synthesis and characterization of high purity iron-zinc alloys. The preparation of high purity, homogeneous, iron-zinc alloys through a slow diffusion process is discussed. The results of the stoichiometric and homogeneity measurements on the alloys are presented in order to convey the analytical accuracy which should be maintained when calibrating instruments used for galvanneal characterization. The identification and microstructural characterization of the iron-zinc alloys using Mössbauer spectroscopy and x-ray diffraction then follow. The second section of this chapter deals with the identification of the iron-zinc phases in commercially produced galvanneal steel. Using scattering Mössbauer spectroscopy, the phases can now be identified in-situ, without having to strip the coating from the steel substrate. The results of the Mössbauer analysis of commercial galvanneal, produced under a variety of conditions, are presented and compared with some of their mechanical and metallurgical properties.

2. SYNTHESIS OF IRON-ZINC ALLOYS

The ability to prepare and chemically analyze high purity, homogeneous iron-zinc alloys forms the basis for their detailed microstructural characterization. In particular, Mössbauer spectroscopy and x-ray diffraction analyses of the alloys will be discussed. It will be seen that the microstructural properties of at least two of the iron-zinc intermetallic phases are dependent on iron concentration and alloy homogeneity. Therefore the separate identification of the four phases and the determination of the fraction of each in a commercial galvannealed coating can only be performed if the instrumentation is very accurately calibrated. In addition, high quality reference spectra are needed for galvanneal characterization. The availability of high quality iron-zinc alloys permits their use as standards[38] with which commercial producers of galvannealed steel can calibrate their monitoring instruments and more accurately analyze their coatings. The availability and characterization of iron-zinc calibration standards has led to a better understanding of some of the microstructural properties of commercial galvannealed steel. This has resulted in improved coating quality and the ability to tailor galvanneal to suit the environmental conditions to which it will be exposed.

2. 1. Preparation of Iron-zinc Alloys

In order to accurately characterize the iron-zinc alloys, it is necessary to produce high quality samples which are representative of any chosen point on the phase diagram. The most commonly used iron-zinc binary phase diagram,[39] see Fig. 1, shows four intermetallic phases at low iron content. These four phases, which form in galvanneal, are called Gamma, Γ, with 18.0 to 31.0, Gamma-1, Γ_1, with 18.9 to 24.0, Delta, δ, with 8.1 to 13.2, and Zeta, ζ, with 5.9 to 7.1 atomic percent iron. An accurate study requires the samples to be homogeneous and to have an iron content measured to an accuracy of approximately 0.5 atomic percent iron. At the same time, the samples need to be very pure, containing no substitutional or interstitial elements which would effect the crystal structure and general microstructural properties. Several different methods for producing the iron-zinc intermetallics have been reported over the past 30 years.[40,41] However, no one has reported alloy preparations that have resulted in highly homogeneous samples or samples whose iron concentration is accurately known. Most preparation techniques have involved sintering an iron and zinc mixture in vacuum or in an inert atmosphere at either very high temperature or for very long times. Preparation is hampered by the very different melting temperatures of iron and zinc of 1538 and 420°C, respectively. Care must be taken to prevent zinc loss by evaporation as the sintering temperature is increased. Recently the authors have developed a well controlled diffusion technique[42] which produces high quality iron-zinc alloys whose iron content is nearly identical to that of the original undiffused mixture. This accuracy allows the selection of any point on the phase diagram at which to produce and study an alloy. High purity samples can only be produced from high quality starting materials. It is important to prevent iron carbide, nitride, or oxide precipitates from forming, thereby reducing the iron fraction and homogeneity in each phase. To promote uniform diffusion, these materials should be very fine powders which have been thoroughly mixed and pressed into dense tablets.

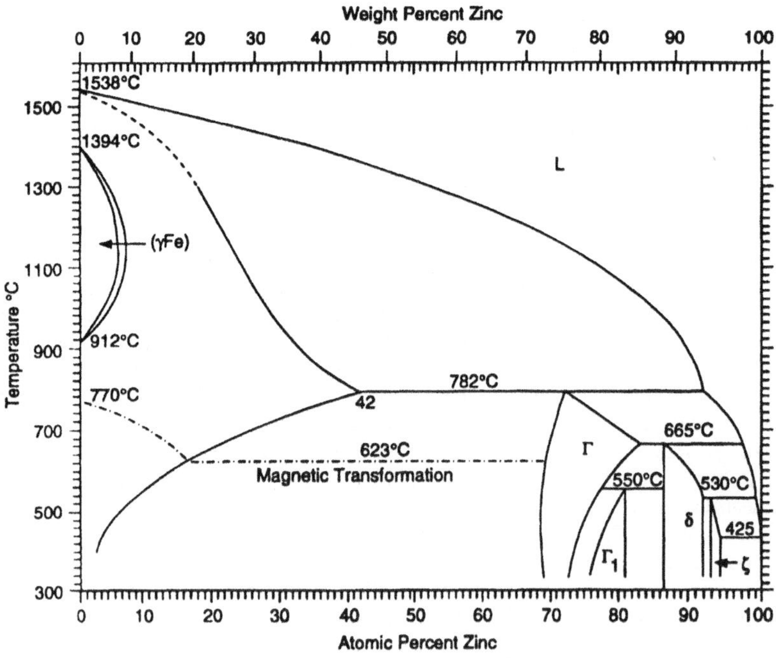

Figure 1. The iron-zinc binary phase diagram as given by Kubaschewski.[39]

2. 1. 1. Materials Used

Many iron and zinc powders are commercially available with different purity and particle size. Generally, the smaller the particle size, the lower the purity. The selection of the iron and zinc powders must be made using different criteria. Because the zinc diffuses into the iron during alloy formation, choosing an iron powder having a small particle size is important even at the expense of the powder purity. Use of fine iron powders containing more than 0.1 weight percent of carbon or oxygen should be avoided due to the formation of iron carbides or oxides.[28] Electrolytic iron micropowder,[43] with 99.7 percent purity, which is commonly used as a mineral supplement in food, is now readily available. It has a particle size of 4 microns and during sample sintering the zinc can easily diffuse through this distance. Of importance is the very low fraction of carbon and nitrogen present in the powder, which should be at most 0.025 weight percent. The particle size for the zinc powder is not as important, because its diffusion is high for alloys prepared above 420°C. Therefore, the highest purity zinc should be chosen. Typically zinc powder having a purity of 99.9 percent and particle size of 20 microns is commercially available.[44] It is important that all powders be purchased in sealed containers and once opened, should be stored in an evacuated container to prevent oxidation and moisture uptake.

2. 1. 2. Alloy Preparation

In order to minimize the zinc loss, the alloys should be prepared in accordance with the iron-zinc binary phase diagram without heating the mixture above the temperature defining the selected point on the phase diagram. Of importance when producing samples at temperatures less than 700°C, is to sinter for sufficiently long times to obtain homogeneity. Using this technique, identical iron-zinc mixtures can be sintered at different temperatures. This allows samples to be produced within the one phase or perhaps spanning several different phases, as is the case for the alloys in the Gamma and Gamma-1 phases containing 19 to 24 atomic percent iron, yet having the same iron concentration.

Samples are prepared by carefully weighing stoichiometric amounts of iron and zinc, 20 to 30 grams total, and then encapsulating and thoroughly mixing the two metals by rotation for 24 hours. All masses should be weighed to an accuracy of 0.0001 grams. A small quantity of each sample, 1 to 1.5 grams, is placed into a one-half inch press tool and subjected to 50 tons per square inch to form a tablet approximately 1 to 2 mm thick. This produces samples with pressed densities greater than 6.5 g cm^{-3}, which is close to the range of alloy densities of 7.1 to 7.4 g cm^{-3}, for the four iron-zinc phases. The tablets are then sealed into an evacuated quartz tube. It is common to place several samples of different iron and zinc ratio into the same quartz tube for preparation at the same temperature. Each tablet should be separated by a small quartz disc to prevent the samples from touching. The samples are then sintered in a computer controlled tube furnace for varying times and temperatures depending on the desired sample. Care must be taken to ensure that the evacuated quartz tube is completely inside the constant temperature hot zone of the furnace, thereby allowing it to reach a uniform temperature which prevents zinc condensation. The sintering temperature is determined from the phase diagram for the particular phase required. In order to minimize thermal stress within the pressed tablet, and to stop zinc loss, the temperature must be increased slowly over a period of about 2 hours until the final sintering temperature is reached. Following sintering for 4 to 10 days, the samples must be quenched in liquid nitrogen to preserve the microstructural properties present at the particular sintering temperature. The tablets are then removed from the quartz tube and crushed finely to a particle size between 5 to 15 μm. Each powdered sample is then individually encapsulated in another evacuated quartz tube and annealed at the same temperature and for the same time as the original sinter. Each sample must be annealed separately during this stage of production.

Fig. 2 shows the high zinc end of the iron-zinc binary phase diagram.[39] This region con-

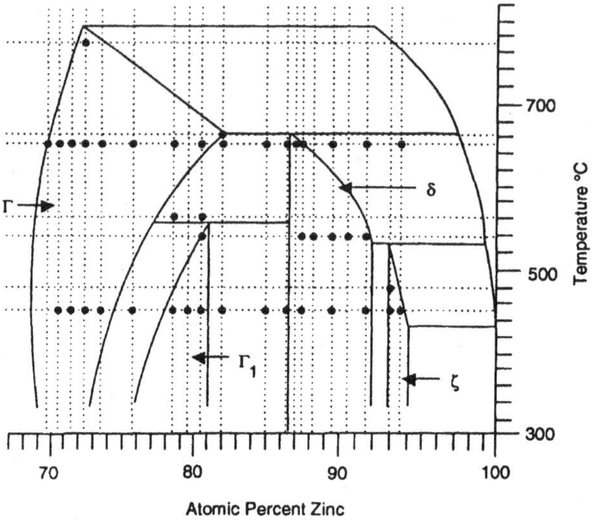

Figure 2. The iron-zinc binary phase diagram showing the atomic percent zinc and the sintering temperatures of the prepared samples.

tains the four main iron-zinc phases, Gamma, Gamma-1, Delta, and Zeta. A total of 31 single phase alloys were produced across the four phases at sintering temperatures of 450, 540, and 650°C, as is shown in Fig. 2. Samples were produced across each of the four phases to investigate the dependence of the crystal structure and hyperfine parameters on the iron content. Fourteen samples were prepared within the Gamma phase because it spans the widest range of iron concentration, i.e., 18 to 30 atomic percent at 650°C. Twelve samples were also produced in the regions between the main phases. To study the effect the sintering temperature has on the crystal structure of the phases, samples were produced at the three different temperatures but with the same iron content.

2. 2. Chemical Analysis of the Iron-zinc Alloys

The chemical composition of all samples was determined by induction coupled plasma spectroscopy and chemical analysis. Additionally, sample homogeneity was determined with an electron microprobe and a scanning transmission electron microscope. The wet chemical analyses were performed on the pure iron and zinc powders, the mixed unsintered powders, and the final sintered and annealed alloys. Once the composition and homogeneity was determined, the samples were microstructurally analyzed using Mössbauer spectroscopy and x-ray diffraction.

2. 2. 1. Induction Coupled Plasma Spectroscopy

Induction coupled plasma spectroscopy was used to measure the iron and zinc fractions of the unsintered iron-zinc powders and the fully annealed intermetallics. By comparing the induction coupled plasma spectroscopy results before and after the sintering process, it was ascertained if there was any iron or zinc lost during the alloying process. Induction coupled plasma analysis was performed with an Applied Research Laboratories, Inc. Model 3410 ICP Spectrometer with a standard plasma minitorch operating with argon gas. Absolute atomic concentrations of iron and zinc were determined using a commercial, custom-made multielement standard,[45] containing 100 mg/l of each of iron and zinc in 2 percent HNO_3. Additional calibration standards were then made using the raw iron and zinc powders used to produce the alloys. A known mass of sample, 100 mg, was dissolved in concentrated nitric acid by heating to approximately 100°C in a fume hood for 30 minutes. A watch glass was placed over the beaker to prevent solution evaporation. Following cooling, the solution was diluted to 2 liters in a volumetric flask using highly purified water. The total atomic concentration was maintained at about 50 ppm so as not to exceed the upper saturation limit of the induction coupled plasma spectrometer. The analysis was performed by setting the spectrometer on the 213.856 and 202.551 nm emission lines of zinc and the 259.940 and 273.955 nm emission lines of iron. The number of iron and zinc atoms per unit volume were measured at the original solution concentrations and also using dilutions of 1 to 1 and 1 to 9 in order to minimize errors due to background counting.

The unmixed and mixed powders, as well as the sintered iron-zinc intermetallics were analyzed by induction coupled plasma spectroscopy. Fig. 3a and b indicates the measured fraction of iron in the mixed powders and in the fully sintered alloys plotted against that expected from the initial masses of iron and zinc before the metals were mixed. Fig. 4a and b

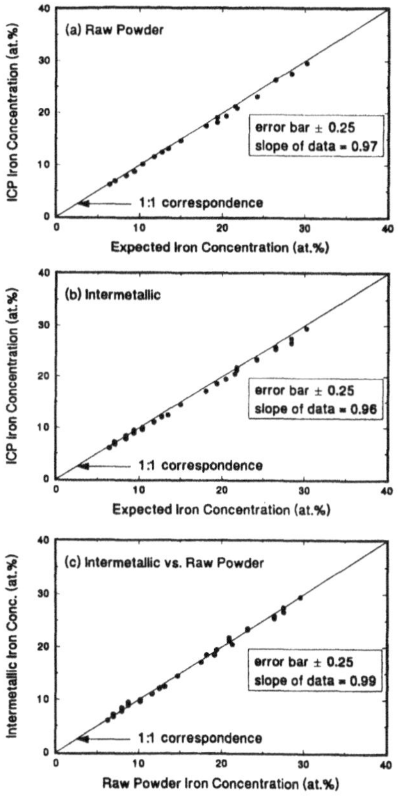

Figure 3. The induction coupled plasma spectral results showing the measured iron concentration compared to the expected value for the raw powders, a, the intermetallic powders, b, and a comparison of the raw and intermetallic powders, c.

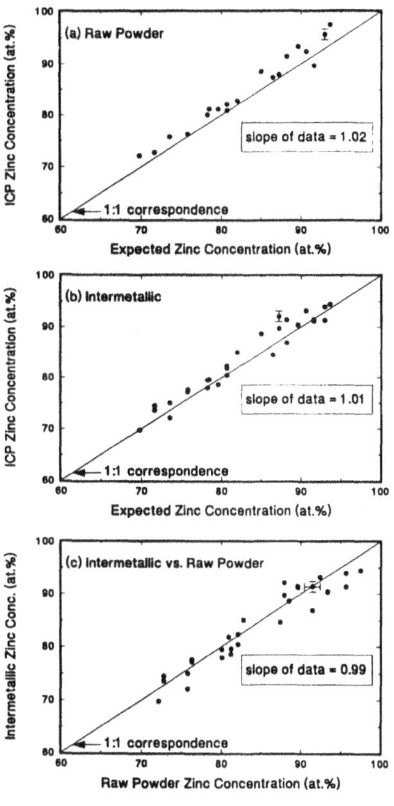

Figure 4. The induction coupled plasma spectral results showing the measured zinc concentration compared to the expected value for the raw powders, a, the intermetallic powders, b, and a comparison of the raw and intermetallic powders, c.

shows the corresponding plots of the induction coupled plasma data for the fraction of zinc in each mixed powder and alloy. For each figure, the iron or zinc fraction is an average of the weight concentration in ppm at the two different iron or zinc absorption wavelengths and at three solution concentrations. On each graph is a solid line showing the 1 to 1 correspondence if the measured induction coupled plasma atomic percent was to agree with the expected value in the original mix before sintering. As can be seen from the figures, the fraction of iron measured in the unsintered mixed powders and fully sintered alloys is less than expected. In contrast, the fraction of zinc is generally higher than expected. Furthermore, Fig. 4a and b indicates that the scatter in the zinc data is much greater than that for the iron. From this data it could be incorrectly concluded that the difference is due to loss of iron during the sintering and annealing processes and that the alloys are deficient in iron. However, in general, a loss of zinc rather than iron would be expected due to the higher vapour pressure of zinc.

A comparison was made between the iron and zinc fractions, measured by induction coupled plasma spectroscopy, in the same samples before and after sintering. These are shown in Figs. 3c and 4c, respectively. It can be seen from the 1 to 1 correspondence line, which is a reasonable fit to the data, that induction coupled plasma spectroscopy measured close to identical iron and zinc concentrations in the sintered and unsintered powders. It is therefore concluded that induction coupled plasma spectroscopy is not able to accurately measure the absolute iron and zinc concentrations to the desired accuracy. The technique underestimates iron concentration and overestimates zinc concentration in the iron-zinc samples. However, induction coupled plasma spectroscopy is able to accurately compare iron and zinc concentrations in different samples and is able to monitor possible changes in the samples due to sintering and annealing. Table 1 summarizes the iron concentrations in one sample from each of the four phases. They can be used to compare the accuracy of each chemical analysis technique discussed.

Table 1. The Iron Content and Homogeneity, H, in Unsintered and Sintered Iron-zinc Alloys[a]

Technique[b]	State	Zeta		Delta		Gamma-1		Gamma	
		% Fe	% H	% Fe	% H	% Fe	% H	% Fe	% H
Expected		7.04		10.37		20.44		29.33	
ICP	Unsint.	6.90(8)		10.2(1)		19.4(2)		28.2(3)	
	Sintered	6.77(5)	99.3	10.0(2)	97.9	19.5(5)	97.5	28.1(7)	97.5
Titration	Unsint.	7.06(9)		10.12(6)		19.9(1)		29.2(2)	
	Sintered	7.20(1)	99.9	10.11(3)	99.7	20.0(1)	99.5	26.7(1)	99.7
μ-probe	Sintered	6.8(1)	98.3	9.4(1)	98.8	21(1)	93.0	25.1(3)	98.8
STM	Sintered	7.6(2)	96.7	9.6(2)	97.8	24(2)	92.6	28(2)	92.3

[a]Iron content given in atomic percent with the standard deviations in the last digit given in parentheses. [b]Expected values from known preparation. Techniques include induction coupled plasma spectroscopy, chemical titration, electron microprobe analysis, and scanning transmission electron microscopy.

2. 2. 2. Chemical Titration

There are several titration techniques available for determining the fraction of iron in materials.[46,47] One such method uses a permanganate solution to oxidize the iron. However, it is important to choose one whose titration end point is not affected by the presence of zinc. Unfortunately, the permanganate solution is not stable in air and because the effect of zinc on the solution is unknown, another oxidation-reduction standard had to be developed[42] with an end point that is not effected by the zinc in the solution.

To accurately determine the fraction of iron in each sample, the following titration technique is recommended. Samples are first reacted with 10 to 15 milliliters of concentrated hydrochloric acid and heated below the boiling point for 20 to 40 minutes. Iron(II) and iron(III) form as products according to the following reactions,

$$2\,Fe + 6\,H^+ \rightarrow 2\,Fe^{3+} + 3\,H_2$$

$$Fe + 2\,H^+ \rightarrow Fe^{2+} + H_2$$

The solution becomes yellow/orange due to the presence of iron(III). During the titration, iron(II) is oxidized to iron(III). Therefore all of the iron(III) must initially be reduced to iron(II). To do this tin(II) chloride, which reduces the iron(III) to iron(II) and changes the solution from yellow/orange to nearly colorless, is used, producing the reaction

$$Sn^{2+} + 2\,Fe^{3+} \rightarrow Sn^{4+} + 2\,Fe^{2+}.$$

Serious errors result if a large excess of tin(II) chloride is added because the tin(II) will also be oxidized during the titration to erroneously give high values for the iron content. Therefore, tin(II) chloride is added dropwise to the yellow/orange solution until it becomes pale green or nearly colorless. A slight excess is added to ensure the complete reduction of iron(III). This slight excess of tin(II) is also oxidized so it does not interfere with the titration. A solution of mercury(II) chloride is added to oxidize the excess tin(II). If a slight excess of tin(II) is used, a white precipitate of mercury(I) chloride forms when the mercury(II) chloride is added. If too much tin(II) chloride is used to reduce the iron(III), a black precipitate of mercury forms when the solution of mercury(II) chloride is added. This difference in precipitates provides an excellent method of checking whether the reduction of iron(III) has been performed correctly through the reactions,

$$Sn^{2+} + 2\,Hg^{2+} + 2\,Cl^- \rightarrow Hg_2Cl_2(s) + Sn^{4+}$$

and

$$Sn^{2+}(excess) + Hg^{2+} \rightarrow Hg^0(s) + Sn^{4+}.$$

If no white precipitate forms, or if the precipitate is black upon the addition of the mercury(II) chloride solution, the sample is discarded. Because air slowly oxidizes iron(II), only one sample is reduced and titrated at a time.

Once the reduction of the iron(III) is complete, sulfuric and phosphoric acids are added to allow for a more complete reaction during the titration and to cause the endpoint of the indicator to be sharper. The indicator, barium diphenylamine sulfonate, is added and the solution is titrated with standardized potassium dichromate solution. The color change at the endpoint proceeds from a blue-green to grey to purple as a consequence of the reaction

$$6\,Fe^{2+} + Cr_2O_7^{2-} + 14\,H^+ \rightarrow 6\,Fe^{3+} + 2\,Cr^{3+} + 7\,H_2O.$$

The procedure described is first used to standardize the titrant, potassium dichromate solution. A pure iron powder sample of known mass is used to measure the molarity of the titrant. Once the volume of titrant needed to oxidize the iron has been measured, the molarity of the titrant can be calculated. Depending on the amount of sample being analyzed, the molarity of the titrant ranges from 0.00667 to 0.01696 mol/l. Once the titrant is standardized, the iron-zinc powders are analyzed. The masses of the mixed raw powders and the iron-zinc intermetallics range from nearly 0.13 to 3 g depending on the theoretical percentage of iron and the amount of the sample available for testing. From the volume of titrant used, the mass of iron in the sample can be determined. Each sample is analyzed three times to reduce titration end point errors.

The concentration of iron in the unsintered mixed powders and the fully sintered alloys,

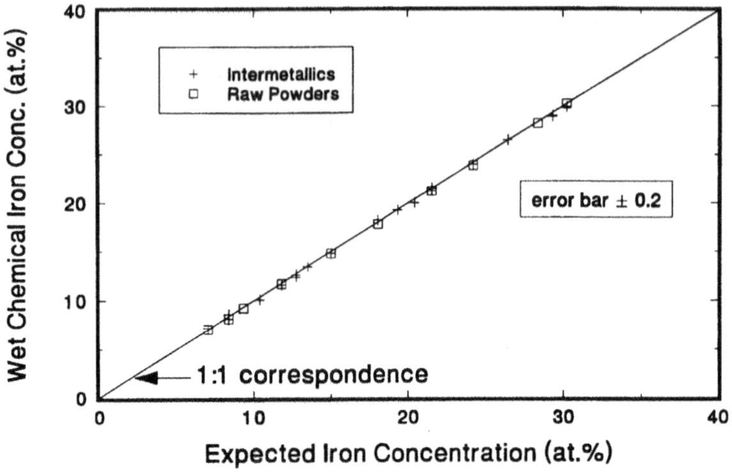

Figure 5. Chemical titration results giving the measured iron concentration as compared to the expected value for the raw powders and the intermetallics.

as determined from chemical analysis, are shown in Fig. 5 plotted as a function of the expected iron fraction. The results show very good agreement between the two fractions. The chemical analysis technique has established that the iron content in iron-zinc samples was within 0.4 weight percent of the expected value. In addition, the majority of samples showed less than 0.1 weight percent difference in iron fraction between the unsintered raw powders and their corresponding fully sintered alloys. Furthermore, repeated chemical analyses using the same sample gave consistent iron concentrations within 0.1 weight percent of each other, establishing that the analytical procedure is very accurate. Table 1 lists the iron concentrations measured in the selected alloy from each phase.

2. 2. 3. Electron Microprobe Spectroscopy

The homogeneity of each sample was measured using an electron microprobe and a scanning transmission electron microscope. An ETEC Autoprobe with a beryllium window, a Kevex energy dispersive system, and a Krisel automation system, was used to measure the iron content in many particles of the same sample in order to monitor the homogeneity between particles. Typical probe volumes were about 10 cubic microns, thus allowing 1 to 2 probes to be made in most particles. Powders were mounted into standard one inch diameter plugs using a high quality epoxy resin. Each plug contained four separated powdered samples which had been finely sprinkled onto the bottom of the mould, which was then evacuated before the epoxy was set and heated to 75°C for 2 hours to speed the hardening. Samples were polished with diamond paste ranging from 600 μm down to 0.25 μm. Finally, the polished samples were coated with a thin layer of carbon to improve the electrical conductivity and reduce charge build-up on the sample while in the microprobe. Twelve pure iron and zinc powders in addition to the sintered alloys were analyzed. X-ray counts were collected for 200 s with a beam current of 215 pA and the data was corrected to weight percent using ZAF corrections. Both the iron and zinc peak positions were determined before each measurement in order to compensate for any electronic drift of the equipment.

Several points within a particle were analyzed, as well as many different particles for each sample. At least eight probes were recorded for each sample analyzed. Pure commercially produced iron and zinc standards, as well as the raw iron and zinc powders, were used for calibration standards because no known iron-zinc intermetallic standards existed at that time. Table 1 includes the microprobe analysis, averaged over each particle in a sample, of each selected alloy of the four iron-zinc phases. The variation in the weight percentage from particle to particle, as indicated by the standard deviation, is a good measure of the homogeneity of the sample. It should be noted that, because pure iron and zinc powders were used for calibration, the absolute weight percentages are not considered to be accurate to better than 1 weight percent. The variation in the weight percentage of iron for each sample is small, thus indicating good sample homogeneity. In general the iron content in each alloy varies from the expected value by less than 0.5 weight percent. Measurements on many particles of the same sample indicated an iron homogeneity greater than 98 percent.

2. 2. 4. Scanning Transmission Electron Microscopy

Scanning transmission electron microscopy was used both to measure the iron content across any one particle and to monitor the homogeneity on a more microscopic scale than is possible with the electron microprobe. In general the resolution of scanning transmission electron microscopy is about 20 nm for a probe volume of ca. 0.02 μm^3. Between 10 to 30 samplings were made across a particle at 30 to 60 nm intervals. To prepare a sample the powdered alloy was embedded in the tip of a size 00 BEEM capsule with low vapor pressure epoxy. The epoxy plug was trimmed to a 20 µm pyramid and then 30 to 50 nm sections were cut dry using a Diatome 45° diamond knife sharpened to a 10 nm edge. The sections were attached on a collodion/Formvar/carbon coated grid[48] which was inserted into a Vacuum Generators HB501 scanning transmission electron microscope for microanalysis. The sample was tilted 10 to 15° towards the detector to minimize absorption and shadowing by the grid bars. The scanning transmission electron microscopy contained an AN 10,000 EDS system equipped with a 30 mm^2 windowless detector having 148 eV resolution.

The composition profile across a particle was determined using a list of tracking points generated prior to automated data acquisition. Fig. 6 shows the scanning transmission electron microscopy particle profiles and scanning electron microprobe photographs of typical particles from a sample prepared in each of the four main iron-zinc phases. The iron concentration and homogeneity for each phase is included in Table 1. Again, because pure iron and zinc powders were used for calibration, the atomic percentages are not considered to be as accurate as the variation in the atomic percentage across a given particle. The atomic percentage remains relatively constant across each particle, demonstrating good homogeneity. At the very edge of some of the particles regions of high oxygen content were observed. This is believed to be due to a very small amount of zinc oxide being present in the initial zinc powder.

Scanning transmission electron microscopy analysis consistently gave iron concentrations within about 0.6 atomic percent of the expected value for the two low iron phases. However for the two high iron phases, the iron concentration was only accurate to about 2 atomic percent of the expected value. Larger variations in iron homogeneity were also observed across individual particles using scanning transmission electron microscopy. These ranged from 0.2 to 2 percent for the low and high iron phases, respectively.

2. 3. Summary

Using the method of slow diffusion of zinc in iron, high purity, homogeneous alloys of the Zeta, Delta, Gamma-1, and Gamma iron-zinc phases could be prepared. The technique allows standard alloys to be prepared at very precise positions on the iron-zinc binary phase diagram. This is especially important at iron concentrations for which the Gamma-1 and Gamma phases overlap. Wet chemical titration analysis is the most accurate and consistent method for determining the iron content in each alloy phase. The availability of the iron-zinc alloys[38] now permits calibration and correction factors for galvanneal monitoring instrumenta-

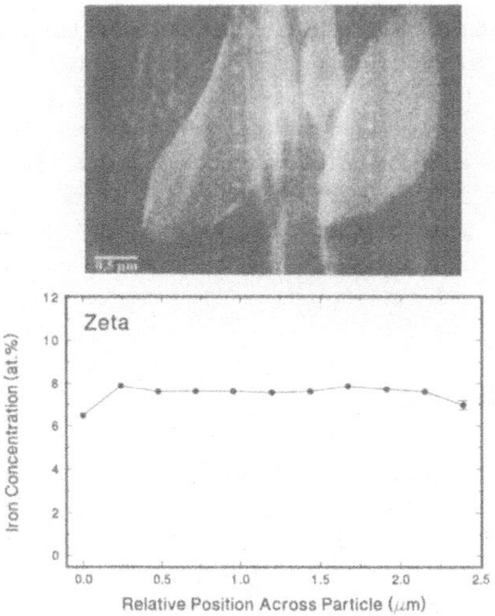

Figure 6A. The scanning electron microscopy photograph, above, and the scanning transmission electron microscopy particle profile, below, of a typical Zeta phase. The + marks in the upper photograph indicate the probe points for the scanning transmission electron microscopy study.

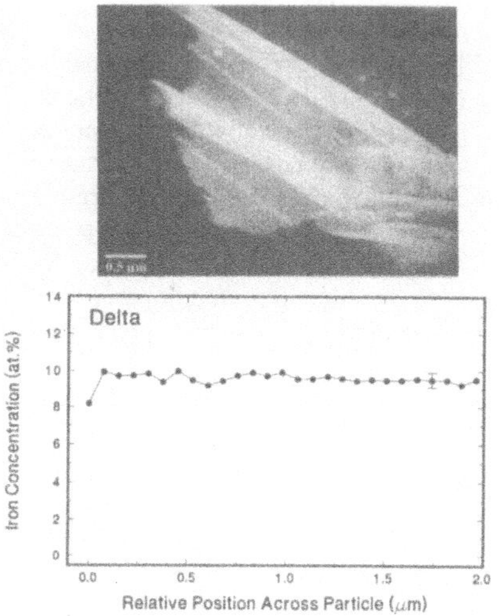

Figure 6B. The scanning electron microscopy photograph, above, and the scanning transmission electron microscopy particle profile, below, of a typical Delta phase. The + marks in the upper photograph indicate the probe points for the scanning transmission electron microscopy study.

tion. Induction coupled plasma spectrometry is one of the popular techniques for determining the average iron content in a galvannealed coating. However the available data shows that induction coupled plasma spectroscopy tends to underestimate the iron and overestimate the zinc fractions in iron-zinc alloys. It has also been determined that the induction coupled plasma procedure is not accurately repetitive.

Table 1 compares the measured iron concentrations in the alloys using the different analytical techniques. Induction coupled plasma spectroscopy is consistently more than 1.5 weight percent low for the Gamma-1 and Gamma phases. Errors of this magnitude made when measuring the total iron weight in a galvannealed coating can lead to incorrect phase fractions being determined. Table 1 also shows that the measured homogeneity decreases as the probe volume decreases. This is not unexpected considering that the chemistry monitors the mean iron fraction in a bulk sample, whereas the electron microprobe and scanning transmission electron microscopy techniques measure much smaller volumes. It is expected that, with additional measurements using the high purity iron-zinc alloys, improved ZAF corrections will allow more consistent iron concentrations to be measured using the different analytical techniques.

The chemical analytical method for measuring the iron concentration is very accurate. Although the procedure is lengthy and repetitive and requires precision it provides accuracy in bulk iron concentration to within 0.5 atomic percent of the expected value. Furthermore, the uncertainty in the chemical measurement is consistently less than ±0.2 atomic percent producing a greater than 99 percent confidence level of being within two standard deviations of the actual bulk iron concentration. When accurate iron concentrations are needed, the titra-

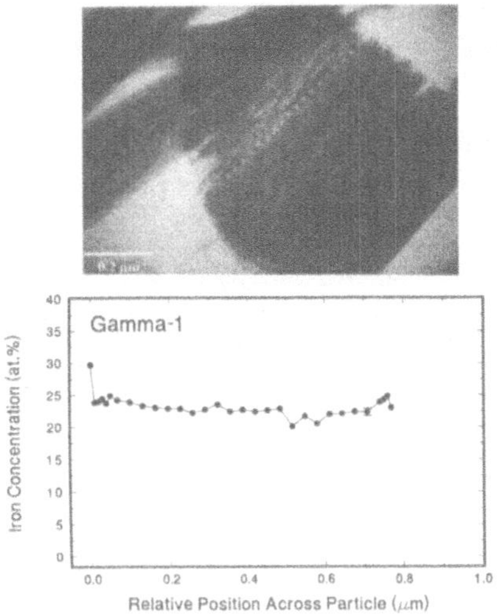

Figure 6C. The scanning electron microscopy photograph, above, and the scanning transmission electron microscopy particle profile, below, of a typical Gamma-1 phase. The + marks in the upper photograph indicate the probe points for the scanning transmission electron microscopy study.

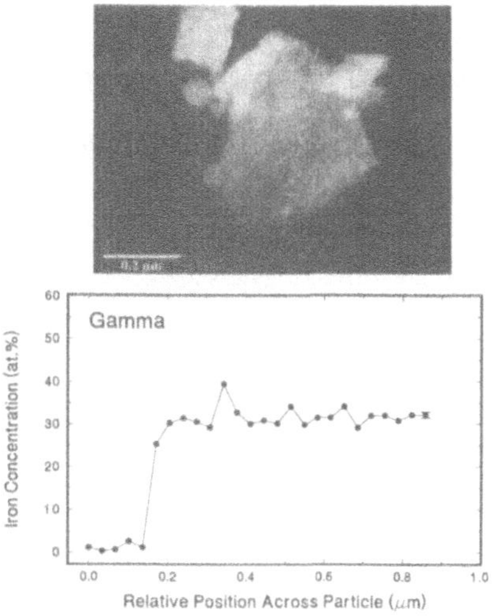

Figure 6D. The scanning electron microscopy photograph, above, and the scanning transmission electron microscopy particle profile, below, of a typical Gamma phase. The + marks in the upper photograph indicate the probe points for the scanning transmission electron microscopy study.

tion method described should be used. As will be discussed in the following section, an accuracy of the iron concentration of better than 0.5 atomic percent is required to allow correct correlation of the parameter changes across each separate phase with iron content.[36]

3. CHARACTERIZATION OF IRON-ZINC ALLOYS

The need to fully characterize the crystalline and microstructural properties of iron-zinc intermetallics has been prompted by the difficulty in identifying the phases formed in galvannealed steel coatings. This results because the two most commonly used industrial probes, x-ray diffraction and x-ray fluorescence, have been reported to be unable to separately identify the Delta, Gamma-1, and Gamma phases in galvanneal. It has been known for several years that Mössbauer spectroscopy can separately identify the four iron-zinc phases.[28-30,49] However, use of the Mössbauer effect to study galvannealed coatings, always results in very complex unresolved spectra comprised of the superimposed signatures from each of the four iron-zinc phases. Deconvolution of the Mössbauer spectrum into the separate phase components has proved to be very difficult, mainly because of the lack of Mössbauer data on the pure iron-zinc alloys. To solve this problem, a detailed study of the 43 iron-zinc intermetallics, listed in Section 2, was undertaken using x-ray diffraction and transmission Mössbauer spectroscopy. The study was performed in order to compile a general database of the crystalline and hyperfine properties of the iron-zinc alloys to permit phase identification in commercial galvanneal. The characterization involved measuring the changes in the lattice and hyperfine parameters across each phase as a function of the iron content and preparation temperature. The samples prepared in the regions between the main phases were similarly analyzed in order to determine their characteristics so that their possible presence in galvannealed coatings could be determined. This section discusses the results of the x-ray diffraction and Mössbauer effect measurements on each of the four iron-zinc phases.

3. 1. Review of Iron-zinc Alloy Properties

3. 1. 1. X-ray Diffraction

X-ray diffraction data previously published on the four iron-zinc phases are very limited and in some cases are over 30 years old. The most recent update to the x-ray powder diffraction files, published by the International Center for Diffraction Data,[37] was for the Zeta phase which references work performed in 1979.[50] However, the reported results for the most complex, Delta phase, were compiled using data recorded in 1928 and 1938.[51,52] This work was performed well before the discovery of the neighboring Gamma-1 phase in 1974.[53] Controversy has existed over the crystal structure of the iron-zinc phases and as a result, the iron-zinc binary phase diagram has been modified several times since 1930. The present Delta phase was originally reported[54] to be separated into a low temperature component, called Delta_1, and a high temperature component then called Delta. The phase diagram was altered in 1972 following the report of two different Delta morphologies in hot-dipped galvanized steel.[55] A high iron Delta phase, δ_{1k}, had a compact appearance. The low iron Delta phase, δ_{1p}, had a columnar appearance and was called Palisade. The alterations to the phase diagram were supported by the different physical properties of the two morphologies. In particular, the microhardness was different for each as was the response to etching for samples prepared for scanning electron microprobe analysis. Of greater interest was the reported increase in iron content at the δ_{1p}/δ_{1k} boundary in the galvanized coatings.[55-58] The Delta phase was split into two distinct phases separated by a mixed phase region. The Delta region of the phase diagram was again altered to its present single Delta phase[39] following the work of Gellings et al.[41] and Bastin et al.[40] Both groups used x-ray diffraction to determine that the lattice parameters and cell structure were not discontinuous across the present iron concentration range of the Delta phase. The investigation by Bastin et al.[40] showed that the crystal structure remains hexagonal between the two iron concentration limits of 8 to 13 atomic percent, and that there is only a small, continuous, change in the lattice parameters over this range. They concluded that, if two Delta phase morphologies exist, x-ray diffraction could not separately identify them. There are also conflicting results concerning the atomic site occupancies by iron and zinc in the Gamma phase. These will be discussed in detail following the Mössbauer spectral analysis of the pure iron-zinc alloys.

Because the pure phases are very difficult to produce and because the use of film to record the early patterns cannot resolve the many diffraction peaks that have now been identified, it is apparent that most of these International Center for Diffraction Data patterns are still incomplete. They do not provide enough information to allow phase identification in galvannealed steel coatings. This discrepancy is also due in part to the complex x-ray patterns for the phases and to the difficulty in characterizing the phases using metallographic cross-section analysis with scanning electron microscopy. Table 2 gives a summary of the crystallographic data presently available on the four iron-zinc phases. Included are the results of the x-ray diffraction analysis to be discussed in this section.

3. 1. 2. Mössbauer Spectroscopy

Only a few papers involving the studies of iron-zinc phases by Mössbauer spectroscopy have been published. Jones and Denner,[59] studied the phases which formed on mild steel during galvanizing. They reported observing the Zeta, Delta_1, and Gamma phases, each of which produced very different spectra. The Delta_1 phase has since been renamed as Delta. Their spectra were recorded in transmission geometry after thinning the steel substrate. Unfor-

Table 2. Structural Information on Four Iron-zinc Phases

	Zeta	Delta	Gamma-1	Gamma
Stoichiometry	$FeZn_{13}$	$FeZn_{10}$	Fe_5Zn_{21}	Fe_3Zn_{10}
Crystal symmetry	monoclinic	hexagonal	fcc	bcc
Space group symmetry	C2/m	$P6_3mc$	$F\bar{4}3m$	$I\bar{4}3m$
Reference pattern[a]	34-1314	13-578	32-478	33-697
Iron, atomic %	7.04	9.09	19.24	23.08
Iron range, atomic %	5.9 - 7.1	8.1 - 13.2	18.9 - 24.0	18.0 - 31.0
Iron range, weight %	5.2 - 6.1	7.0 - 11.5	16.6 - 21.2	15.8 - 27.7
Atoms per unit cell	28	555	408	52
Lattice parameters				
a, Å	10.880(3)	12.790(5)	17.977(3)	9.015(3)
b, Å	7.611(1)			
c, Å	5.079(1)	57.23(5)		
β, deg	100.83			
V, Å^3	413	8107	5810	733
References	36, 37, 50, 62	36, 37, 40, 41, 51, 52, 65	36, 37, 53, 66	36, 37, 53, 67-69

[a]The pattern reference number in the powder diffraction file published by the International Center for Diffraction Data, ref. 37.

tunately, the spectra were dominated by the signal from the steel substrate and the unresolved contribution from the coating was difficult to accurately interpret. In addition, they had to polish the coating down from the top in an attempt to isolate the separate phases. As a result, the accuracy of their data is not sufficient to provide the details now required for accurate analysis of each phase in galvanneal. Graham et al.[60] used conversion electron Mössbauer spectroscopy to also study the iron-zinc phases on galvanized steel. They reported the presence of the Gamma and Zeta phases, as well as two separate Delta phases. Their spectra also lacked resolution which prevented determination of accurate Mössbauer spectral parameters. Of importance was their reported observation of two different Mössbauer spectra for the Delta phase. They attributed this to the presence of the Compact and Palisade morphologies, a conclusion which has not been confirmed.

Recently, three Mössbauer spectral studies were undertaken on separate iron-zinc phases.[28,30,31] Gu et al.[30] undertook a Mössbauer effect study of one sample from each of the four iron-zinc phases and compared the results to their electrodeposited zinc coatings. Cook[28] used Mössbauer spectroscopy to show that, in iron-zinc alloys prepared using iron powder containing greater than 0.1 weight percent carbon, iron carbides readily formed, resulting in mixed and inhomogeneous alloy phases. The following study of four high purity iron-zinc alloys by Grant and Cook[31] showed that multiple iron sites were present in three of the four phases.

3. 2. Experimental Procedure

The x-ray diffraction patterns of the high purity iron-zinc alloys discussed in Section 2, were recorded using a Siemens Model D-5000 diffractometer. In order to upgrade the International Center for Diffraction Data files, the patterns were recorded with and without internal calibration using both copper and molybdenum x-rays. Patterns were recorded using general goniometer scans in the ranges 10 to 160° and 2 to 10° in two-theta and with step increments of 0.020° and a sample time of 1 s. When loading the powder onto the spinner, extreme care was taken to ensure that the top of the sample was exactly the same as that of the holder, thereby minimizing displacement errors which, if not accounted for, will lead to inaccurate identification of the iron content in the Delta and Gamma phases. The patterns were analyzed and lattice parameters determined using commercially available software for peak searching and pattern fitting. Only patterns recorded using the copper K_α x-rays are discussed and illustrated in this section.

Transmission Mössbauer spectra of the iron-zinc alloys were recorded at 300 K. The powdered samples were pressed with a boron nitride binder to form a thin tablet 1 cm in diameter. Most spectra were recorded over the small velocity range of ± 1.2 mms^{-1} using a 50 mCi, ^{57}Co/Rh source. Data collection times ranged from two days to two weeks depending on the iron content of the sample and the desired spectral resolution. The spectra were analyzed using a standard least-squares fitting routine[61] incorporating the transmission integral to account for absorber thickness. All isomer shifts are referenced to α-iron at 300 K.

3. 3. Iron-zinc Phase Analysis

3. 3. 1. Zeta Phase

The Zeta phase spans a narrow iron concentration range between 5.9 and 7.1 atomic percent or 5.2 and 6.1 weight percent iron, as is shown in Fig. 2. It has a peritectic melting point at 530°C and its crystal structure is monoclinic,[62] with space group C2/m. The accepted stoichiometry is $FeZn_{13}$. Atomic structure of the Zeta phase shows each iron atom surrounded by twelve zinc atoms at the vertices of a slightly distorted icosahedron. These icosahedrons then link together to form chains which run parallel to the c-axis. The chains group together to form an approximate hexagonal array, as is shown in Fig. 7.

Three samples were prepared at two different iron concentrations of 6.40 and 7.04 atomic percent within the ζ phase and analyzed using x-ray diffraction and Mössbauer spectroscopy. The x-ray diffraction pattern recorded for the alloy containing 7.04 atomic percent iron using the copper K_α x-rays, is shown in Fig. 8A. Over 100 diffraction peaks were indexed from the

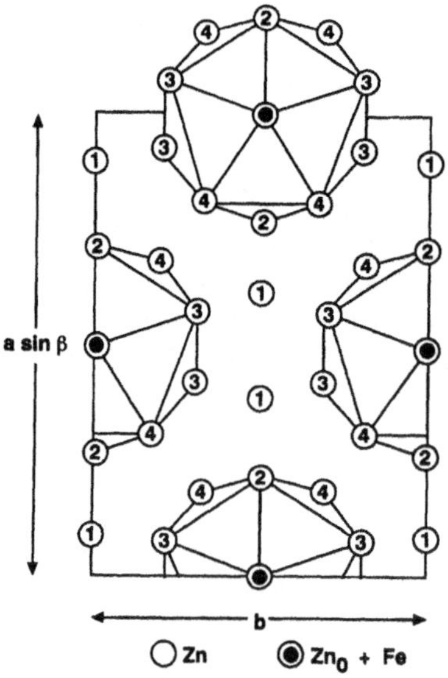

Figure 7. The structure[62] of the Zeta phase viewed along [001].

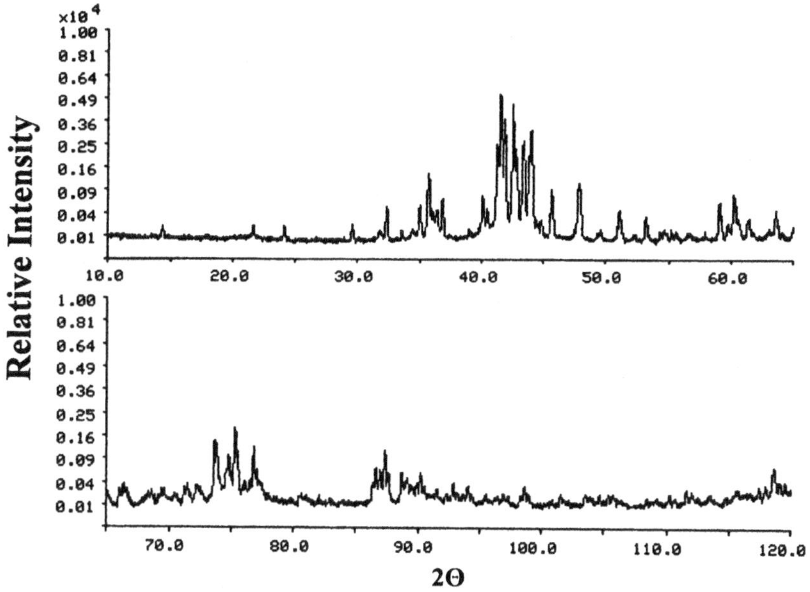

Figure 8A. The x-ray diffraction pattern of the Zeta phase containing 7.04 atomic percent iron.

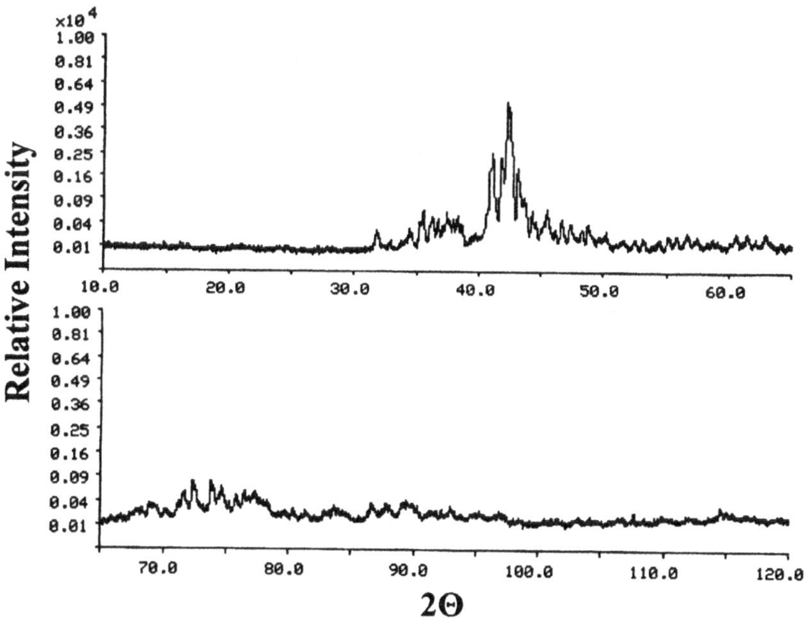

Figure 8B. The x-ray diffraction pattern of the Delta phase containing 8.41 atomic percent iron.

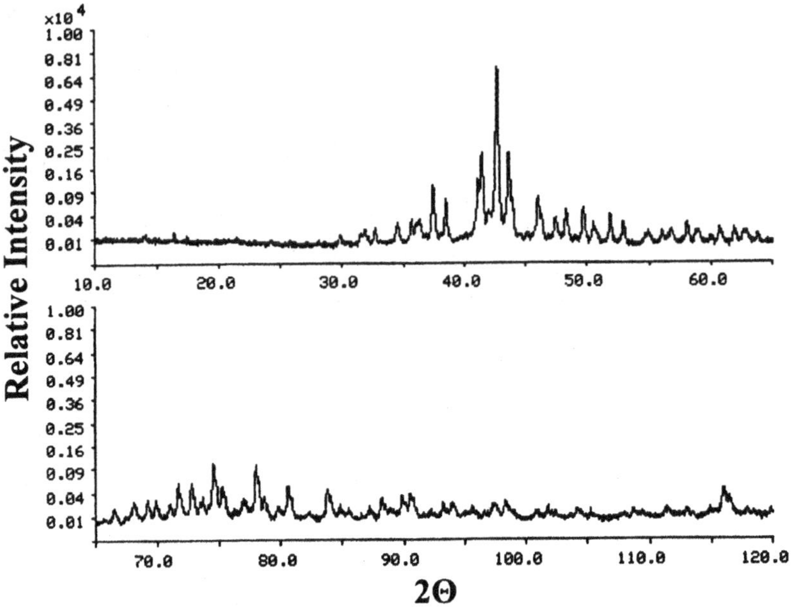

Figure 8C. The x-ray diffraction pattern of the Gamma-1 phase containing 21.54 atomic percent iron.

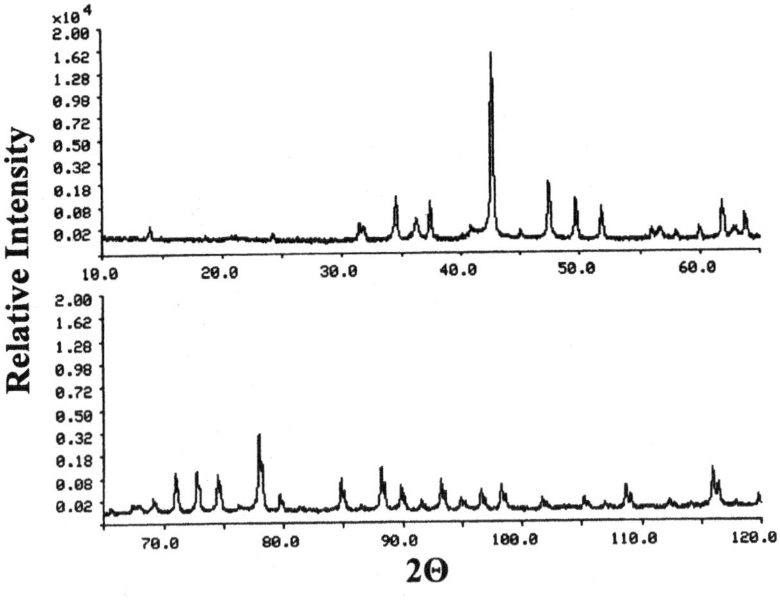

Figure 8D. The x-ray diffraction pattern of the Gamma phase containing 27.63 atomic percent iron.

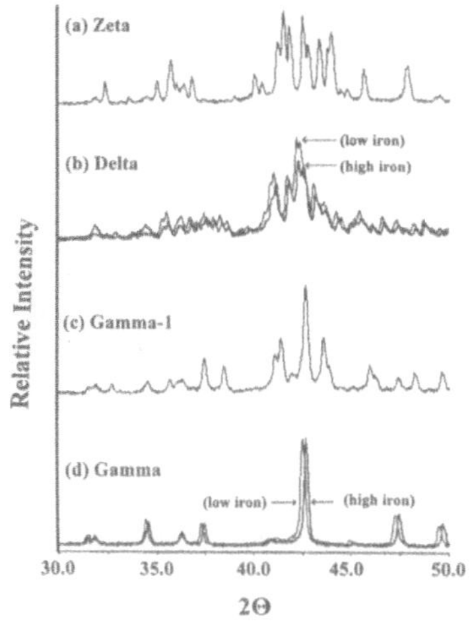

Figure 9. Partial x-ray diffraction patterns for the Zeta, a, Delta, b, Gamma-1, c, and Gamma, d, binary iron-zinc alloys.

pattern analysis.[63] This pattern is also shown over the range 30 to 50° in Fig. 9a. On this scale the individual peaks are better observed and useful in comparative studies with the other phases. The lattice parameters are given in Table 2 and are found to be in close agreement with the present International Center for Diffraction Data pattern reported by Gellings et al.[50]

The Mössbauer spectrum of the Zeta phase with 7.04 atomic percent iron, is shown in Fig. 10a. It is clear that this phase is paramagnetic at 300 K. It clearly exhibits a quadrupole doublet with a very small splitting, in contrast to several early studies which reported the spectrum as a broad single line.[59,60] The quadrupole splitting is expected to be small due to the nearly cubic symmetry of the nearest neighbor zinc atoms. The fit parameters are listed in Table 3. The isomer shift, δ, is 0.489 mms^{-1}, and quadrupole splitting, ΔE, is 0.153 mms^{-1}, values which are similar to those reported by Gu et al.[30] Within experimental uncertainty, the x-ray diffraction and Mössbauer results were found to be independent of iron content and alloy preparation temperature. This is probably due to the narrow iron concentration range of the Zeta phase.

Table 3. Mössbauer Parameters and Iron Site Occupancies for the Four Iron-zinc Phases

Phase	Site	Isomer Shift, mm/s	Quadrupole Splitting, mm/s	Linewidth, mm/s	Relative Area, %
Zeta		0.489	0.153	0.244	100
Delta	A	0.468	0.508	0.275	56.7
	B	0.474	0.216	0.295	37.6
	C	0.275	0.372	0.285	5.7
Gamma-1	A	0.470	0.079	0.298	46.0
	B	–0.042	0.555	0.294	6.2
	C	0.042	0.224	0.394	47.8
Gamma	A	0.365	0.156	0.276	64.0
	B	0.126	0.503	0.293	36.0

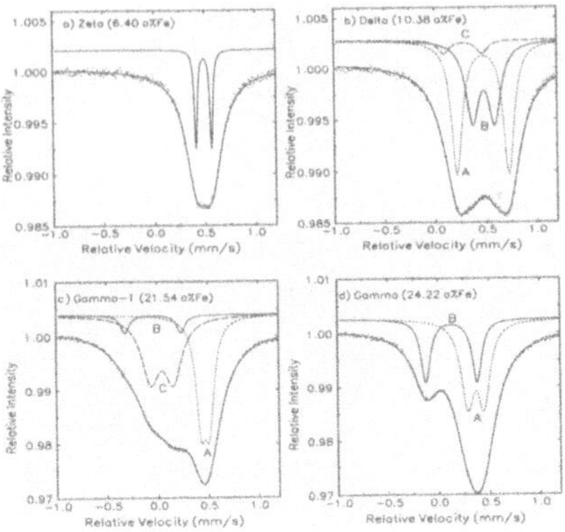

Figure 10. The Mössbauer spectra for the Zeta, a, Delta, b, Gamma-1, c, and Gamma, d, binary iron-zinc alloys.

3. 3. 2. Delta Phase

The Delta phase forms with an iron concentration between 8.1 and 13.2 atomic percent or 7.0 and 11.5 weight percent. It has a peritectic reaction temperature of 672°C and crystallizes in the hexagonal structure with space group $P6_3mc$.[52] The nominal stoichiometry is $FeZn_{10}$, although it can vary between $Fe_{0.89}Zn_{10.11}$ and $Fe_{1.45}Zn_{9.55}$. The unit cell has been reported to be very large, containing 555 ± 8 atoms.[52] However, the atomic arrangement within the unit cell is not known as yet.

A total of ten different samples were analyzed within the Delta phase. These samples spanned six different iron concentrations from 8.41 to 13.01 atomic percent and were prepared at three different temperatures, as is shown in Fig. 2. Analysis of the pure alloys showed that the lattice parameters and hyperfine parameters depend on iron content but not on preparation temperature. Fig. 8B shows the typical x-ray diffraction patterns of the low iron Delta phase containing 8.41 atomic percent iron. Once again over 100 diffraction peaks have been indexed.[63] Fig. 9b shows the pattern plotted between 30 and 50°. Also included is the x-ray diffraction pattern for the high iron Delta phase containing 12.76 atomic percent iron. It can be seen that the positions of the diffraction peaks shift slightly with iron content. Over the iron concentration range of 8.4 to 13.0 atomic percent, the peaks shift about 0.2 degrees in 2θ. Accuracy in sample mounting is very important because displacement errors[64] can easily shift the diffraction peaks by this amount. As seen by the large number of diffraction peaks, the Delta phase is the most complex of the iron-zinc alloy phases and careful analysis is required to index the peaks accurately. For the Delta phase given by the generally accepted stoichiometry $FeZn_{10}$ and containing 9.09 atomic percent iron, the lattice parameters are a = 12.790(5) Å and c = 57.23(5) Å. This unusually long columnar cell has a volume of approximately 8100 Å^3. Both lattice parameters decrease linearly and continuously from a = 12.795(5) to 12.759(4) Å and c = 57.26(5) to 57.10(5) Å as the iron content increased from 8.41 to 13.01 atomic percent. This behavior does not agree with the non-linear relationship between the lattice parameters and the iron content, reported by Bastin et al.[40,65] The lattice parameter ratio, c/a, is independent of iron content, indicating that the contraction of the unit cell with increasing iron content is isotropic. The unit cell volume was found to decrease linearly from 8118 to 8050 Å^3 with increasing iron content.

The results presented for the high purity Delta phase alloys prepared at the three different temperatures of 450, 540, and 650°C, cover the regions designated to the Compact and Palisade morphologies by Ghoniem and Löhberg.[55] The data show continuity of the lattice parameters, cell volume, and crystal structure with iron content. This eliminates the possibility of the Compact and Palisade morphologies existing with different crystal structures. The alloys studied by Bastin et al.[40] were prepared along the boundary of the Delta phase at different temperatures between 530 and 660°C and iron concentrations between 8 and 13 atomic percent. This does not include the regions designated as the Compact and Palisade morphologies on the old phase diagram. Nevertheless, in agreement with Bastin et al. we concluded that pure, laboratory produced, powders of the Delta phase have only one morphology.

Nine of the Delta phase samples were studied by Mössbauer spectroscopy. By fitting the set of spectra in a consistent manner, it was determined that there were three distinct iron sites which have been labelled A, B, and C. Each iron site exhibits a quadrupole splitting but no magnetic hyperfine field at 300 K. This is the first time that more than one iron site has been identified in the Delta phase. However, past research[59,60] did show that some of the Mössbauer spectra were asymmetric doublets. Even though these earlier spectra were fitted to only one iron site, the asymmetry should have warranted more detailed analysis. The Mössbauer spectrum of a typical Delta phase with 10.38 atomic percent iron, is shown in Fig. 10b, and the general fit parameters are included in Table 3. The six Mössbauer spectra recorded at iron concentrations between 8.41 and 13.01 atomic percent are shown in Fig. 11. There are significant changes in the spectra as the iron content is varied. There are three iron sites in each sample although the C site is only marginally populated with iron at low total iron concentrations.

Fig. 12 shows the variation in isomer shift, quadrupole splitting, and relative area of the three sites as a function of total iron content. The isomer shifts of the three sites decrease gradually, but continuously, with increasing iron content. The isomer shifts of the A and B sites are nearly identical. Furthermore, the quadrupole splittings of the A and B sites increase only marginally, in contrast to the rapid increase at site C, as the iron content increases. This indicates that the crystal symmetry of the A and B sites remains relatively constant across the Delta phase, but that of the C site changes continuously. This is supported by the x-ray diffraction data which indicate a uniform expansion of the hexagonal lattice with decreasing iron content. Of particular interest is the change in relative area of the three subspectra. Under the assumption that the recoilless fraction of each iron site is the same, the subspectral relative areas are direct measures of the site occupation by iron. Research is underway at the present time to determine the recoilless fraction, i.e., the Mössbauer temperature, of each site in this and the other phases. Therefore, the iron occupancies at the two main sites, A and B, decrease

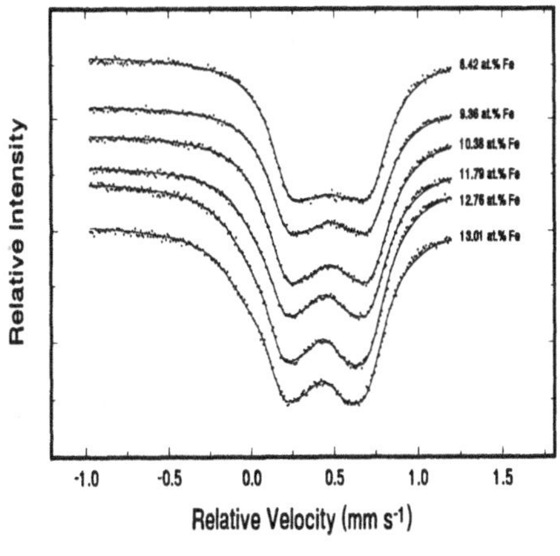

Figure 11. The Mössbauer spectra of the Delta phase alloys containing between 8.42 and 13.01 atomic percent iron.

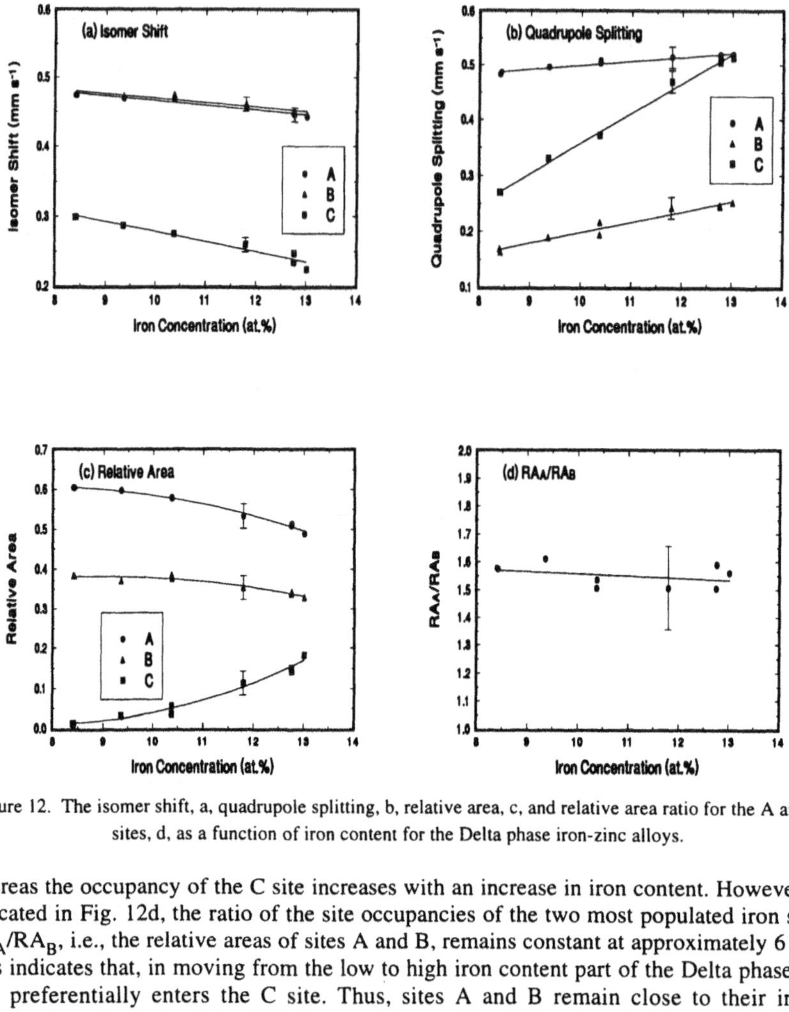

Figure 12. The isomer shift, a, quadrupole splitting, b, relative area, c, and relative area ratio for the A and B sites, d, as a function of iron content for the Delta phase iron-zinc alloys.

whereas the occupancy of the C site increases with an increase in iron content. However, as indicated in Fig. 12d, the ratio of the site occupancies of the two most populated iron sites, RA_A/RA_B, i.e., the relative areas of sites A and B, remains constant at approximately 6 to 4. This indicates that, in moving from the low to high iron content part of the Delta phase, the iron preferentially enters the C site. Thus, sites A and B remain close to their initial occupancies. Analysis of the Mössbauer spectra of the Delta phase has not revealed any information concerning the possible formation of the Compact, δ_k, and Palisade, δ_p, morphologies.

3. 3. 3. Gamma-1 Phase

The Gamma-1 phase was first reported in 1974 by Bastin et al.[53] It forms between the iron concentrations of 18.9 and 24.0 atomic percent, or 16.6 and 21.2 weight percent, has a peritectic reaction temperature of 550°C and the nominal stoichiometry is Fe_5Zn_{21}. The crystal structure is cubic containing 408 atoms per unit cell, and has space group $F\bar{4}3m$. Single crystal x-ray diffraction analysis showed the crystal structures of the Gamma-1 and Gamma phases to be related. If eight unit cells of the Gamma phase are stacked together thereby doubling the lattice dimensions, then the large Gamma-1 unit cell, also with space group $F\bar{4}3m$, is formed.[66,67] It is this relationship between the two crystal structures that leads to overlap of the x-ray diffraction peaks from each phase and complicates the separate identification of the

Gamma-1 and Gamma phases in galvannealed steel coatings. Koster and Schoone[66] determined the atomic positions and site occupancies of the Gamma-1 phase and proposed four site symmetries which they labelled as cubooctahedral, CO, octahedral, OH, outer-tetrahedral, OT, and inner-tetrahedral, IT.

Four high purity alloys in the Gamma-1 phase having three different iron concentrations of 19.35, 20.44, and 21.54 atomic percent, were studied by x-ray diffraction and Mössbauer spectroscopy. Figs. 8c and 9c show the typical x-ray diffraction pattern of the Gamma-1 phase. Over 100 diffraction peaks of the fcc crystal structure have now been indexed and the lattice parameter is 17.977 ± 0.003 Å. Although there appears to be a marginal increase in the lattice parameter with iron content, the change is comparable to its experimental error and the lattice parameter is independent of iron concentration pending further investigation.

X-ray diffraction analysis of the sample prepared in the Gamma-1 phase at 540°C and with an iron concentration of 19.35 atomic percent, showed the sample not to be single phase but to contain a mixture of the Delta and Gamma-1 phases. Chemical analysis confirmed the expected iron concentration of 19.35 atomic percent for this sample. It is likely that the phase diagram at the top of the Gamma-1 phase boundary is not accurate.

Fig. 10c shows the Mössbauer spectrum of a typical sample within the Gamma-1 phase with 21.54 atomic percent iron. Analysis shows the presence of three quadrupole split doublets and the three iron sites have been labelled A, B, and C. The hyperfine parameters are listed in Table 3. Within experimental error they were found to be independent of iron content. The relative iron populations, however, varied at two of the three sites, the A and C sites, with the total iron content in the phase. At a low iron concentration of 19.35 atomic percent, the A and C sites contain 50 and 45 percent of the total iron, respectively. However at a higher iron concentration of 21.54 atomic percent, both sites have equal iron populations of 47 percent. The visible changes in the Mössbauer spectra are minor and probably will prevent the determination of the iron content in the Gamma-1 phase of a galvannealed steel coating. The low atomic density of the Gamma-1 crystal structure is most likely responsible for the measured stability in the lattice and hyperfine parameters as the iron content increases across the Gamma-1 phase.

3. 3. 4. Gamma Phase

The Gamma phase forms between the iron concentrations of 18.0 and 31.0 atomic percent, or 15.8 and 27.7 weight percent. It has a peritectic reaction temperature of 782°C, the highest of the four iron-zinc phases. The nominal stoichiometry is Fe_3Zn_{10}, however, it is quite often written as Fe_5Zn_8 for samples at the high iron side of the phase diagram because the Gamma phase crystallizes in the cubic γ-brass structure with space group $I\bar{4}3m$. For the stoichiometry of cubic γ-brass, the fraction of iron would be 38.5 percent which is greater than allowed by the boundary of the phase diagram. Each lattice point on the bcc Bravais lattice contains a 26 atom cluster. Therefore, there are 52 atoms per unit cell.[68] Fig. 13 shows the cluster of atoms which is situated at the body centered position. The clusters at the eight corner positions have been omitted for simplicity. Each cluster is identical and contains four crystallographically unique sites. They are labelled inner-tetrahedral, IT, outer-tetrahedral, OT, octahedral, OH, and cubooctahedral, CO. These four sites contain 8, 8, 12, and 24 atoms per cluster, respectively. It was expected that, because more iron is observed to dissolve in this intermetallic than is found in other γ-brass binary alloys, the remaining iron would be randomly distributed onto the four sites. Two previous neutron and x-ray diffraction studies[68,69] have been unable to conclusively ascertain which of the four sites are occupied by the iron atoms. Brandon et al.[68] produced a sample in the low iron end of the Gamma phase and determined that the iron occupied only the inner tetrahedral and outer tetrahedral sites. In contrast, Johansson et al.[69] produced a sample in the high iron concentration region of the Gamma phase and proposed that the iron occupied the inner tetrahedral and octahedral sites. Table 4 compares the atomic positions and site occupancies reported by Johansson et al. and

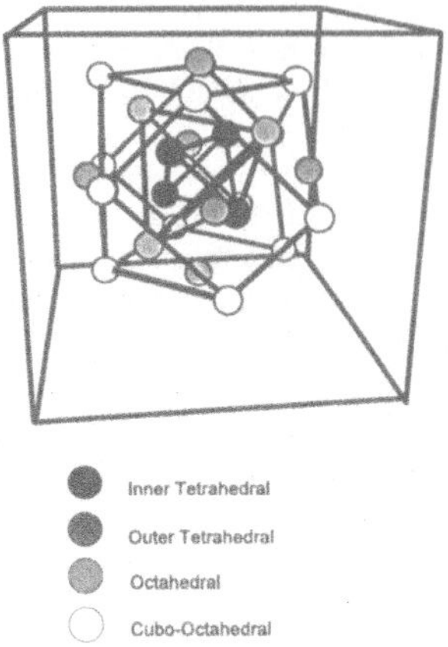

Figure 13. The bcc structure of the Gamma phase showing the 26 atom cluster at the body centered lattice point. This cluster is also repeated at the 8 corner positions giving 52 atoms per unit cell.

Brandon et al. Although both agree with the site labeling, there is disagreement on which sites are occupied by iron and/or zinc. A Mössbauer spectral analysis, presented below, on pure alloys provides information concerning the sites occupied by iron.

Fourteen different samples were prepared covering nine iron concentrations and three temperatures. The iron concentration ranged from 18.04 to 30.24 atomic percent for samples sintered at 650°C, thereby covering the full iron concentration range possible for this phase. Furthermore, because part of the Gamma phase overlaps the Gamma-1 phase in iron content, see Fig. 2, it was possible to produce Gamma samples with the same iron content as the Gamma-1 phase but at a different sintering temperature. Figs. 8d and 9d show the typical x-ray diffraction patterns of the Gamma phase. The lattice parameter of the bcc Gamma phase was found to depend on iron content. As can be seen from Fig. 9d, the diffraction peaks shift to higher angles by about 0.5 degrees as the iron concentration increases from 21.77 to 29.33 atomic percent. Over 70 peaks for the bcc structure have been indexed.[37] The lattice parameter is independent of sample preparation temperature but increases non-linearly from 8.974 to 9.025 ± 0.003 Å as the iron concentration decreases from 30.24 to 18.04 atomic percent. As the low iron content end of the Gamma phase is approached, the lattice parameter converges toward a constant value of 9.025 Å. The lattice parameter of the Gamma phase is nearly one-half of that for the Gamma-1 phase. The site symmetries in both phases are also similar. Therefore it is possible that the atomic cluster geometry for the Gamma-1 phase is very similar to that of the Gamma phase, as is shown in Fig. 13.

Fig. 10d shows a typical Mössbauer spectrum of the Gamma phase with 24.22 atomic percent iron. Its hyperfine parameters are listed in Table 3. Analysis of the fourteen alloys showed that, of the four inequivalent atomic sites, only two, A and B, are occupied by iron in

Table 4. Structural Information for the Iron-zinc Gamma Phase[a]

Site	Sym.	x^b	z^b	Occupancy[b]	x^c	z^c	Occupancy[c]
IT	8c	0.0973(15)		Fe	0.1028(5)		4 Fe + 4 Zn
OT	8c	0.1638(13)		Zn	0.1673(4)		Fe
OH	12e	0.3551(15)		8 Fe + 4 Zn	0.3538(4)		Zn
CO	24g	0.3029(8)	0.0508(12)	Zn	0.3045(2)	0.0491(3)	Zn

[a]Standard deviations in the last digit of the positional parameters are given in parentheses. [b]Data obtained from reference 69. [c]Data obtained from reference 68.

agreement with neutron diffraction studies.[68,69] The Mössbauer spectra of the eight samples prepared at 650°C, with iron concentrations between 19.36 and 30.24 atomic percent, are shown in Fig. 14. The vertical pair of parallel lines on the figure provide a reference to observe the shift in the peaks with iron content. The variations in the fit parameters are plotted against total iron content in Fig. 15. The isomer shift and quadrupole splitting of both sites decrease linearly with increasing iron content. As a result the main spectral peak at positive velocity shifts to lower velocity as the iron content increases. This is primarily due to the corresponding decrease in both the quadrupole splitting and isomer shift of the B site. The other peak does not shift to any appreciable extent. The spectral changes are large enough to allow the iron content of the Gamma phase in galvanneal to be determined.

Important changes in the relative areas allow the actual sites occupied by iron to be identified and have provided information concerning the conflicting reports of Johannson et al.[69] and Brandon et al.[68] Fig. 15c shows that, as the iron content in the Gamma phase increases, the area ratio of the A and B sites approaches 6 to 4. As the iron concentration varies across the Gamma phase, no discontinuity is observed in the Mössbauer spectral parameters. In all spectra the same two iron sites are observed. Therefore, the iron must continuously occupies the same two atomic sites at all times. The relative areas of the two sites approach the limit of 6 to 4 and, in comparison with the allowed atomic positions for each site, see Table 4, it can be concluded that the octahedral and most likely the inner-tetrahedral sites must contain iron. The Mössbauer data agree with the results of Johansson et al. It should be noted that the actual iron ratio of the two sites does not reach 6 to 4 in the Gamma phase because the stoichiometry of cubic γ-brass, Fe_5Zn_8, is never achieved.

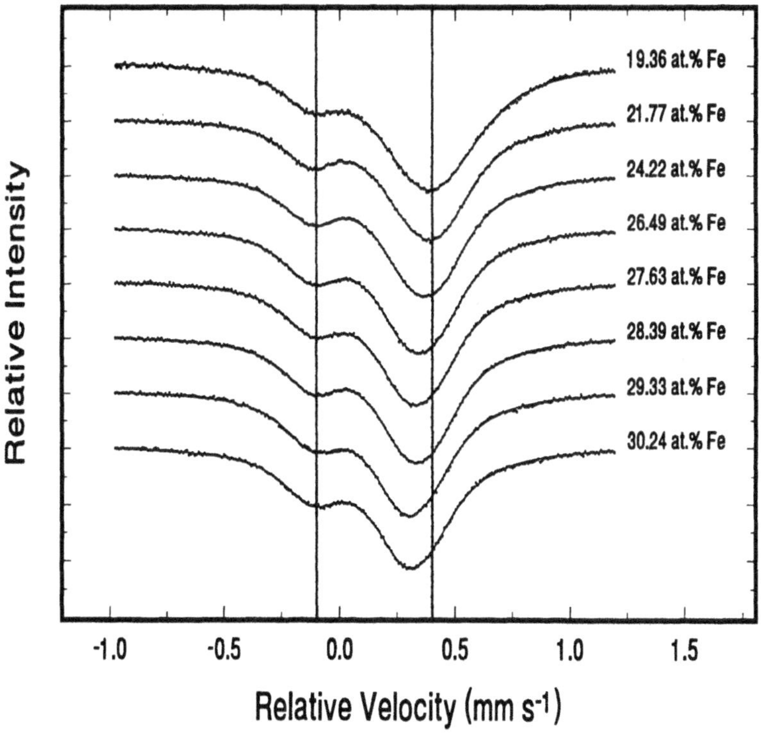

Figure 14. The Mössbauer spectra of the Gamma phase alloys containing between 19.36 and 30.24 atomic percent iron.

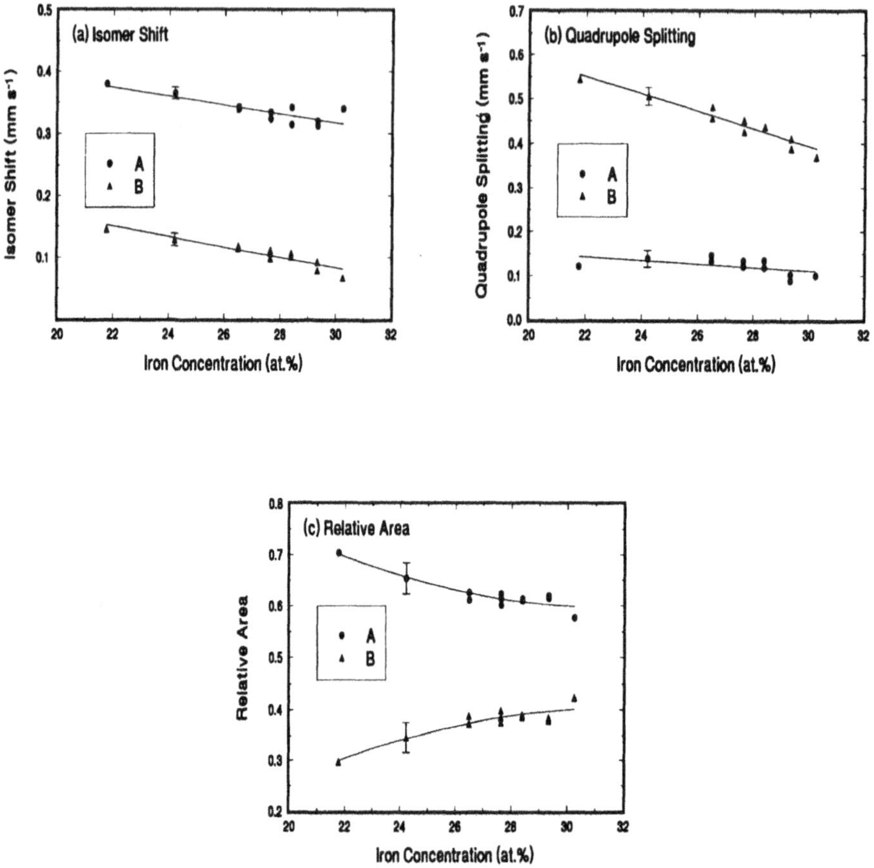

Figure 15. The isomer shift, a, quadrupole splitting, b, and relative area, c, as a function of iron content for the Gamma phase iron-zinc alloys. Solid lines are least squares fits to the data.

The similarity between the Gamma and Gamma-1 phases not only appears in the x-ray diffraction data but also in the Mössbauer spectral parameters. Although not the same numerically, the isomer shifts and quadrupole splittings for the A and B sites in both phases are similar enough to predict that the A, B, and C iron sites in the Gamma-1 phase are the octahedral, inner-tetrahedral, and outer-tetrahedral sites, respectively.

3. 3. 5. Mixed Phases

The iron-zinc intermetallics prepared in the "mixed phase" regions between the main iron-zinc phases were also analyzed by x-ray diffraction and Mössbauer spectroscopy. This was undertaken in order to determine their parameters so that their possible presence in commercial galvannealed coatings could be identified. Five samples were prepared at 540 and 650°C in the Γ - δ region, one sample was prepared at 450°C in the Γ - Γ_1 region, and three samples were prepared at 450°C in the Γ_1 - δ region. The x-ray and Mössbauer spectral results showed that each sample was comprised of a mixture of two iron-zinc phases corresponding to the region in which it was produced. The fraction of each phase was directly related to the proximity of the sample to the pure phase regions. One interesting observation was made from the lattice and Mössbauer fit parameters. For the samples prepared in the mixed phase regions containing the Gamma phase, the fit parameters for that phase component always corresponded to the iron content at the right-hand side of the Gamma phase boundary for the particular sample preparation temperature. That is, the lowest possible iron content was found in the Gamma phase component for the preparation temperature. In addition, for the mixed phase samples containing the Delta phase, the fit parameters always corresponded to the highest iron concentration, that being 13.2 atomic percent. This information will prove to be important for the correct analysis of the phases in galvannealed steel.

3. 4. Summary

X-ray diffraction and Mössbauer spectroscopy have shown their capability of identifying the four main iron-zinc alloy phases. Phase identification by x-ray diffraction does present some difficulties due to the large number of diffraction peaks and the accuracy with which samples need to be mounted. However an important outcome of the alloy characterization by x-ray diffraction has been the recording of new reference patterns which will hopefully soon allow mixed phase samples, e.g., galvanneal, to be successfully analyzed. At the present time the data have proved accurate for the separate identification of the two phase mixtures of ζ - δ and δ - Γ_1. For example, to identify the δ - Γ_1 mixture, a specific angular range between 40 and 45° in two-theta needs to be analyzed. As can be seen in Fig. 16, the Delta phase has a characteristic double peak at 42.2° and the Gamma-1 phase has a well defined peak at 41.4°. The Gamma-1 phase also has a peak at 42.6° but this coincides with the Gamma phase and cannot be used. It should be mentioned that all of the diffraction peaks for the Gamma phase overlap with half of those of the Gamma-1 phase. This results because the same lattice spacings are produced in a fcc structure by doubling the lattice parameter and Miller indices of a bcc structure. Therefore, it is not possible to identify the presence of the Gamma phase in a sample containing both Gamma and Gamma-1 phases based solely on peak location. Separate identification of the two phases must be made by comparing the relative intensities of the Gamma-1 peaks to those of the sum of the Gamma-1 and Gamma peaks. This is difficult in a galvannealed coating where the intensities of diffraction peaks vary with coating preparation.

Mössbauer spectroscopy provides better phase identification than x-ray diffraction because the spectrum of each phase is very different. This will be shown to be valuable for separate identification of the Gamma-1 and Gamma phases in galvanneal. The fitting of the Mössbauer spectra of mixtures of the four pure phases has been shown to give phase fractions

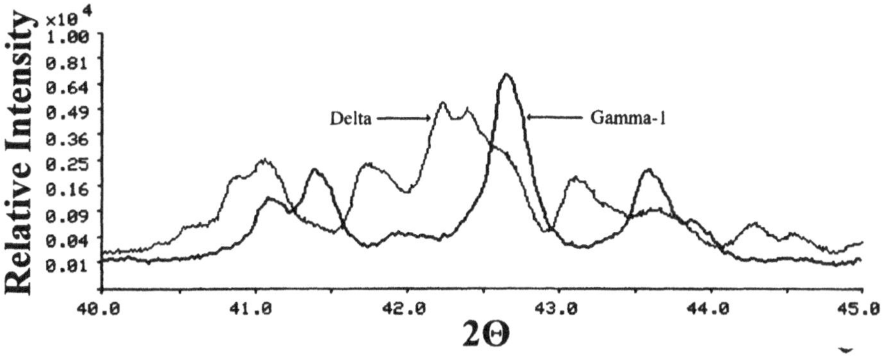

Figure 16. Overlap of the x-ray diffraction patterns for the Delta and Gamma-1 phases.

which are in very good agreement with their known proportions. In general the rough fits are made by constraining the Mössbauer parameters and allowing the relative Mössbauer subspectral areas to vary. Following this, the Mössbauer parameters of the Delta and Gamma phases are separately unconstrained to allow accurate fits which permit the iron concentration in each phase to be measured.

The preparation and characterization of high purity, homogeneous iron-zinc alloys now permit their use as calibration standards[38] for analytical instruments which are used to monitor or characterize commercially produced galvanneal. These commonly include x-ray diffraction and fluorescence equipment as well as atomic absorption and induction-coupled plasma spectrometers, electron microprobes, and scanning electron and scanning transmission electron microscopes. As will be discussed in Section 4, the compiled database of the crystallographic and hyperfine parameters of the iron-zinc intermetallics will facilitate the identification of the iron-zinc alloys in commercial galvannealed coatings.

4. GALVANNEALED SHEET STEEL

The increasing use of metallic coated sheet steel by the automotive industry has led to the demand for materials which have excellent corrosion resistance, formability, weldability, and paintability. In addition, the surface quality must be very good for exterior panels. Galvannealed sheet steel is one product meeting all the requirements and is now widely used by many automobile manufacturers. Galvanneal is also being used in the construction industry for door and window frames as well as in reinforced concrete.

For economic production of galvannealed steel, it is necessary to identify and control the iron-zinc intermetallics which form during the annealing cycle. Phase identification in galvannealed coatings has proved to be a difficult task as was discussed in Section 2. However characterization of the iron-zinc alloy standards discussed in Section 3, now makes it possible to attempt to identify each phase in a coating.

The second part of this chapter deals with the identification of the iron-zinc phases in commercially produced galvannealed sheet steel. Section 4 describes the processing and general properties of galvanneal. The corrosion resistance, formability, paintability, and weldability of galvannealed steel are discussed with particular emphasis on the effect of coating composition on these mechanical properties. In Section 5, the use of Mössbauer spectroscopy and x-ray diffraction to analyze galvanneal is discussed. Also discussed are the Mössbauer results which show positive identification of the iron-zinc phases in each coating of some commercial galvannealed sheet steel produced at different anneal temperatures. The results are compared with x-ray diffraction data and scanning electron metallographic cross-section photographs.

4. 1. Processing of Galvannealed Sheet Steel

Galvannealing is an on-line process involving the diffusion of the iron in a steel substrate and a molten zinc coating at about 500°C. The diffusion results in the formation of a coating containing iron-zinc intermetallics which are generally layered parallel to the steel surface. When diffusion is complete, the iron concentration decreases towards the coating surface and the four phases discussed in Section 3, see Fig. 2, are formed. The quality of the galvannealed coating is mainly influenced by the total iron content and the amount of each phase. These in turn are controlled by many factors including the annealing temperature and time, the chemistry of the zinc bath, and the type of steel used. Each of the iron-zinc phases show different mechanical properties, some of which are not desirable in high quality galvanneal. Therefore being able to identify the phase composition and then control their formation is a major interest to galvanneal producers.

Prior to the galvannealing process, steel sheet is produced through various hot and cold rolling techniques. The resulting coils of sheet steel range from 0.5 to 2.5 mm thick. The coils are thoroughly cleaned, usually by acid pickling, in order to remove any oxide layer which may have formed in the milling process.[1] Once the sheet leaves the payoff coil it undergoes an extensive cleaning which may involve chemical as well as heat treatment. It is then passed through a pre-annealing high temperature furnace, which is typically operated at 600°C or above.[2] A protective atmosphere, typically consisting of hydrogen and nitrogen, is maintained in the furnace to prevent the steel sheet from oxidizing. Pre-annealing allows the steel to regain the ductility lost during the rolling process. Once annealed, the sheet steel enters the molten metal pot and upon emergence passes through gas knives, or wipers, where excess coating metal is wiped away. For galvanized and galvannealed coatings the bath consists of zinc, usually saturated with 0.03 weight percent iron, and often containing small amounts of additional elements, such as 0.1 to 0.2 weight percent aluminum, which helps control or inhibit iron-zinc phase formation and improve the surface finish. Fig. 17 shows the galvannealing process including the several sensors which are presently used.[70] The zinc coated steel passes through the galvannealing furnace where it is heated for up to 10 seconds at temperatures ranging from 500 to 570°C. This allows the interdiffusion of the iron and zinc and forms an alloy coating typically 6 to 11 microns thick. Following the galvannealing furnace, the sheet is cooled and post treated before being coiled at the end of the process. Critical parameters are the annealing temperature and time which can affect the coating thickness, coating iron content, and the quantities of the iron-zinc phases formed. Temperature sensors and total iron content gauges provide some information concerning possible coating composition.

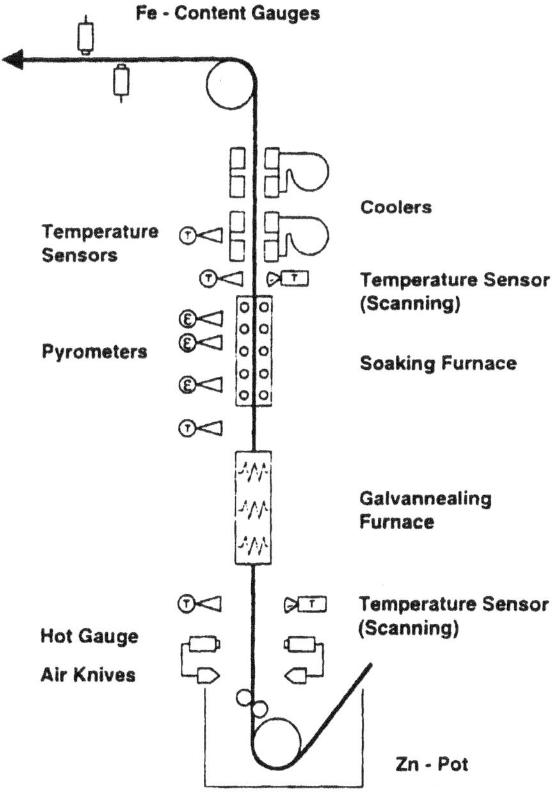

Figure 17. A current standard hot-dip galvannealing line showing the zinc bath, annealing furnace, cooling section, and sensors.[70]

4. 2. Properties of Galvannealed Sheet Steel

The presence of each phase in the coating controls the performance of the material in different ways. The Gamma phase is generally a hard thin layer with high iron content. It is easier to weld but more difficult to form than the other phases. The high iron content also results in the poor corrosion resistance of the Gamma phase. Increasing the thickness of the Gamma phase leads to a harder coating but promotes cracking of the entire coating which can lead to further reduced resistance to corrosion. The formability and corrosion resistance improve for the phases containing more zinc. However, this improvement compromises the weldability. The Delta phase is generally believed to be the most ductile and corrosion resistant of the four phases. The following sections summarize the general characteristics of zinc and zinc-alloy coated steel sheet and discuss the metallurgical properties which are of most concern to the automotive industry. Generally, these metallurgical properties are controlled by the phases present and their relative abundances. However, the primary problems presently facing steel producers involve not only controlling the phase formation but also identifying the phases present within a coating.

4. 2. 1. Corrosion Resistance

The major advantage of galvannealed coatings is their superior resistance to corrosion when used in automotive environments. This is particularly evident if the coatings are painted. Salt spray testing, as well as atmospheric exposure tests, have shown that galvanneal steel coatings have a greater resistance to corrosion than do galvanized coatings.[21] Several factors contribute to the exceptional corrosion resistance of galvannealed coatings. First, the presence of a coating produces a barrier between the corrosive environment and the reactive steel surface. The larger the coating weight or thickness, the larger the protective barrier. This is also true for galvanized steel surfaces. Second, protection through the galvanic or sacrificial mechanisms is observed in both galvanized and galvanneal products. However, it has been shown that the iron-zinc alloys in galvannealed coatings are less electrochemically active than the pure zinc coating in galvanized steel. These factors have contributed to the increased use of galvanneal and the demand for higher quality coatings.

4. 2. 2. Formability

During the production of automobile body panels, sheet steel is subjected to various deformations which may stretch the steel and increase its surface area. In the case of coated sheet steel the problem is further complicated because the coating must also stretch and conform to the bends of the steel base if product quality is to be maintained. Generally, the steel composition, coating process, coating thickness, and the iron-zinc phases present can all affect the formability of galvanneal steel sheet. The adhesion between the coating and the steel substrate is perhaps the most important concern. During the deformation of galvannealed coatings the coating can powder, i.e., the coating can crumble or disintegrate into a fine powder, producing a region where the substrate is exposed to the atmosphere. This may reduce the corrosion resistance of the coating and inhibit the paintability of the final product. However, by keeping coating thicknesses and iron contents low, the powdering loss is minimized. Another parameter affecting the formability is the frictional coefficient between the sheet and the tool surfaces. The frictional coefficient can vary from coating to coating, nevertheless, problems due to friction can be controlled through the proper lubrication and operation of the fabricating tools. Finally, the ductility of a galvannealed coating is an important factor influencing its formability. Coatings containing high amounts of the Gamma phase are generally more brittle and more likely to crack. In contrast, galvanized coatings, primarily containing pure zinc, are more ductile.

4. 2. 3. Paintability

When coated steel sheet is used for exterior automotive body panels, paintability is an aesthetic concern. The cathodic electrophoretic process is the most common technique used in the automotive industry for the application of the primer coat of paint. During this process craters in the paint film may appear. Craters are small defects in the otherwise smooth surface. In general, galvannealed coatings are less resistant to cratering than galvanized coatings. Although galvannealed coatings have a smoother surface, they are more microscopically rough and porous than galvanized coatings. This leads to better paint adhesion and a smoother, more desired appearance.

4. 2. 4. Weldability

Spot welding is the primary method for joining together pieces of steel sheet within the automotive industry. This involves the overlapped sheets being pressed between two copper electrodes and an electrical current passed through them fusing the two sheets together. However, this procedure is affected by the presence of a zinc or zinc-iron coating on the steel surface. The pure zinc surfaces of galvanized coatings tend to alloy with the copper electrodes reducing the life span of the electrodes. Furthermore, spot weld currents must be increased in order to sufficiently fuse galvanized steel sheets. In contrast, galvannealed coatings are generally easier to weld than galvanized steel. This is primarily because the iron-zinc phases have a higher electrical resistivity than pure zinc. Therefore, sufficient resistance heating and fusion of the steel sheets can be achieved at lower currents than in the case of galvanized steel. In turn, this extends the life span of the copper weld tips.

The presence of the iron-zinc phases in the galvannealed steel is normally determined using metallographic cross-sectional scanning electron microprobe analysis. Although the phases cannot be positively identified by this method, alloy layering is commonly observed. Fig. 18 shows the typical cross-sectional micrograph of a galvannealed coating. Color etching the polished sample enhances visual phase separation. The layers generally have different thickness and are labelled from the steel substrate to the surface. A thin bottom layer always forms on the steel substrate. The second and third layers form the bulk of most coatings. It is often difficult to identify the interface between these two layers. The fourth layer is generally rough or jagged and often displays distinct structure indicative of oriented crystals. In many coatings the fourth surface layer is absent. The Mössbauer analysis to be presented in Section

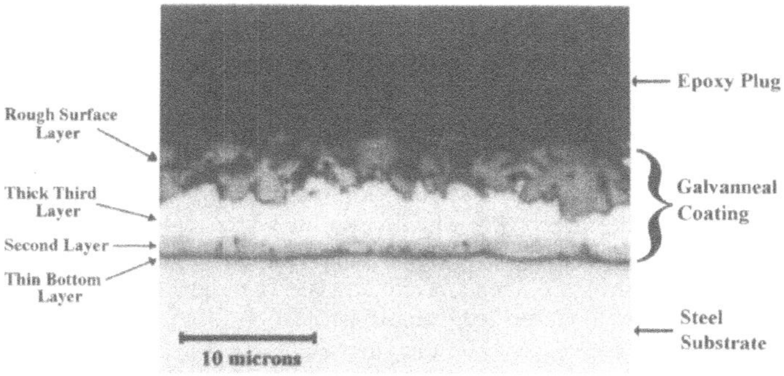

Figure 18. A cross sectional micrograph of a typical galvannealed coating. Note the layering of the iron-zinc intermetallic phases within the galvanneal. The black region is the epoxy plug used to mount and polish the sample and the light region below the coating is the steel substrate.

5 positively identifies each layer as the four iron-zinc phases discussed in Sections 2 and 3. From substrate to surface they are a thin Gamma phase followed by the Gamma-1, Delta, and Zeta phases.

4. 3. Summary

Successful application of galvannealed coatings has provided a means by which steel companies can economically meet the environmental demand to improve the corrosion resistance of steel sheet. In order to tailor the properties of galvanneal to specific applications, it is necessary to understand and control the phase formation during the galvannealing process.

Positive identification and the amount of each phase present in a galvannealed coating has been difficult. As presented in Section 3, Mössbauer spectroscopy and x-ray diffraction are now able to separately identify the pure laboratory produced iron-zinc phases. However, when analyzing galvannealed coatings, separate phase identification using x-ray diffraction is difficult. Furthermore, the presence of the thick steel substrate prevents the use of conventional transmission Mössbauer spectral techniques. Nevertheless, scattering Mössbauer spectroscopy has demonstrated the ability to separately identify the iron-zinc phases without having to remove them from the steel substrate. These results will be presented in Section 5.

5. ANALYSIS OF GALVANNEALED COATINGS

Producing galvannealed sheet steel for specific applications requires the ability to identify the phases formed in the coating during the galvannealing process. An understanding of the microstructural properties of the phases is required to control their formation as well. The ability to now accurately identify the four iron-zinc alloy phases[71-72] by Mössbauer spectroscopy and x-ray diffraction has lead to the use of both techniques to characterize the coatings on commercial galvanneal.

Steel sheet used for galvannealing is normally thicker than 0.5 mm with the galvannealed coating itself being about 10 μm thick. Therefore it is not possible to study galvanneal in-situ using transmission Mössbauer spectroscopy. Transmission spectra can be recorded if the steel is thinned sufficiently to allow gamma-ray penetration. Such investigations were performed by Jones and Denner[59] with only limited success. Their Doppler velocity was too large to allow sufficient resolution of the coating spectrum which was dominated by the substrate signal. Transmission Mössbauer spectrocopy can also be used if the coating is removed from the steel. In doing so however it is uncertain if the complete coating is being detached. It is important to maintain the integrity of the coating because Mössbauer spectroscopy is a non-destructive probe. Therefore studying the galvannealed coating in-situ is advantageous in situations where removing it from the substrate is either difficult or detrimental to its microstructure.

In scattering Mössbauer geometry, detection of the re-emitted radiation following resonant absorption of a gamma-ray, is useful in studying intact coatings and thin films. Because gamma-rays, x-rays, and conversion electrons have different electronic attenuation in materials, detecting the three types of radiation in scattering geometry provides depth dependent information on the sample. This is important in galvannealed coatings where the different iron-zinc phases are generally layered parallel to the steel substrate. The low energy gamma-rays and x-rays undergo only minor attenuation in coatings about 10 μm thick. Scattering Mössbauer spectroscopy is therefore well suited to the study of galvannealed coatings. In addition, by recording the Mössbauer spectrum using the emitted conversion electrons, the surface regions of coatings are able to be studied to a depth of about 0.5 μm. This is a useful method for monitoring the presence of near surface phases such as the Zeta or Delta phases.

5. 1. Sample Preparation and Experimental Procedure

One factor complicating the analysis of coatings in scattering Mössbauer geometry is the direct radiation from the source. The rectangular or cylindrical gas flow proportional detectors[73,74] commonly used for backscattering analysis, are excellent for conversion electron Mössbauer spectroscopy. They have limited use for x-ray Mössbauer spectroscopy, and no use for backscattered gamma-ray Mössbauer spectroscopy. In order to shield the detector from direct source radiation, a toroidal proportional detector should be used.[75-78] Although difficult to manufacture, the toroidal detector has been found to provide excellent resolution and efficiency, equivalent to commercial proportional counters used for transmission Mössbauer spectroscopy.[79]

Commercial galvannealed coatings were studied in-situ using a cylindrical[74] conversion electron Mössbauer spectroscopy detector attached to the toroidal gamma and x-ray Mössbauer detector.[79] The two-stage detector allowed simultaneous collection of the three types of scattering spectra. Both were operated in gas-flow mode using 90 percent He/CH_4 and 90 percent Kr/CH_4, respectively.[80] One interesting feature when recording γ-ray scattering Mössbauer spectra is the increased resonant signal obtained when counting on both the 14.4 keV re-emitted gamma-ray and the 1.8 keV escape peak from krypton which is due to the resonant emission of gamma-rays.

Galvannealed coatings were studied[81] in-situ using scattering Mössbauer spectroscopy incorporating the simultaneous recording of γ-ray, x-ray, and conversion electron Mössbauer spectra at 300 K using a 50 mCi ^{57}Co in Rh source. Some coatings were also studied in transmission geometry by fracturing the coating from the steel substrate. For these samples the conversion electron Mössbauer spectra of the underside of the fractured coating, the side close to the substrate, as well as the surface of the steel substrate from which the coating was removed, were also recorded. One advantage of studying the fractured coating using transmission Mössbauer spectroscopy is to investigate possible crystalline texture due to the iron-zinc diffusion during phase formation. The spectra are recorded before and after crushing the fractured coating. Conversion electron Mössbauer spectra of the underside of the fractured coating accurately identify the iron-zinc phase which forms at the steel substrate, or at the fracture interface. Analysis of the steel substrate using conversion electron Mössbauer spectroscopy identifies whether or not coating material remained attached following fracture. In general, for iron-zinc phase identification in a galvannealed coating, the in-situ recording of only the gamma-ray and conversion electron Mössbauer spectra have been found to be necessary.[80,81]

To prepare a galvannealed coating for the in-situ analysis, the steel sheet was cut to be no larger than 25 × 25 cm to facilitate mounting on the detector. Most samples however were smaller than this, ranging down to 2.5 × 2.5 cm in size to allow mounting inside the conversion electron Mössbauer spectroscopy detector. The coating was then degreased for 5 seconds in acetone. If a galvannealed coating was to be studied in transmission geometry, it was removed from the substrate by shear fracture. To do this, a small strip of the galvanneal approximately 1 cm wide and 5 cm long was cut. Epoxy was used to join a piece of stainless steel sheet to the coating. The coating was scored through to the substrate with a razor blade and pulled apart as shown in Fig. 19. Without exception, all coatings were easily stripped from their substrate, leaving little or no intermetallic material on the steel.[82] With the coating and epoxy still fixed to the stainless steel, the strip could be bent, popping off the epoxy and coating. The epoxy was thinned to approximately 0.5 mm with 600 grit sandpaper making sure not to damage the attached coating. The final result was a piece of galvannealed coating, approximately 1 cm^2, removed from its substrate and attached to a thin layer of epoxy. The fractured coating was analyzed both with transmission Mössbauer spectroscopy through its entire thickness, including the epoxy, and with conversion electron Mössbauer spectroscopy on the now exposed underside of the coating.

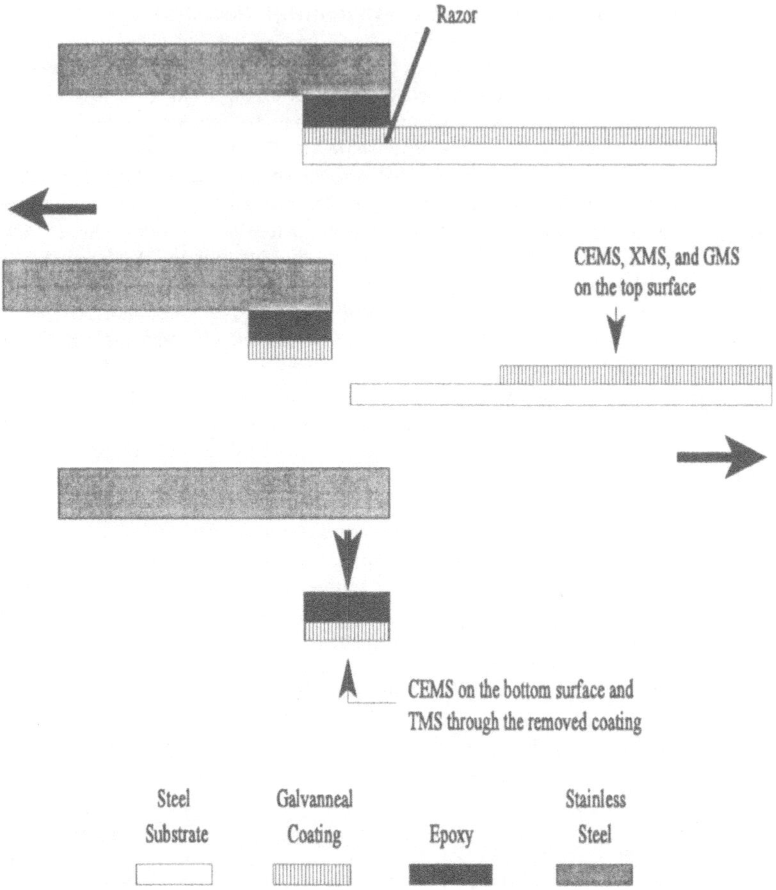

Figure 19. A schematic diagram of the fracturing technique used to cleanly strip a galvannealed coating from the steel substrate. Also shown are the experiments performed on the intact and fractured coatings.

The x-ray diffraction patterns were recorded using a Philips model APD3720 automated powder diffractometer with a copper x-ray tube. Small strips of the galvannealed sheet were mounted directly into the goniometer without the need for the standard holder used for powder samples.

Metallographic analysis was undertaken to determine the coating weight and iron content. The coating weights were measured by the weigh-strip-weigh method. In this technique, a standard disc of a galvannealed coating was weighed. Next the coating was stripped from the substrate using a solution of hydrochloric acid and an organic inhibitor. The remaining substrate was once again weighed and the coating weight calculated. The acid solution containing the dissolved coating was analyzed for iron content using wet chemistry or induction coupled plasma spectroscopy. Scanning electron microscopy photographs were recorded at different magnifications using polished and color etched galvanneal cross-sections.

γ-ray Mössbauer spectroscopy was always found to probe through the entire galvannealed coating and into the steel substrate. This resulted in the superposition of the α-iron base signal from the steel substrate with the coating spectrum. To facilitate the analysis, the base signal was computer subtracted leaving only the γ-ray Mössbauer spectrum of the galvannealed coating. Following the evaluation of the galvanneal using x-ray diffraction and Mössbauer spectroscopy, the relative phase fractions were compared with the thickness of each layer measured from the cross-sectional scanning electron micrographs. The average thickness of the coatings and the layers within the coatings were determined by measuring each layer at ten points across the scanning electron micrograph using a Vernier calliper. The errors correspond to the standard deviations in these measurements.

5. 2. Phase Identification in Galvanneal

Three commercial galvannealed samples were produced at different annealing temperatures to investigate the temperature dependence of phase formation. The same steel substrate, annealing time, and bath aluminum content were used for each sample. The phases identified by Mössbauer spectroscopy were compared with those observed in the x-ray diffraction patterns. A study of the fractions of each iron-zinc phase in a coating was made by comparing the Mössbauer subspectral areas with the thickness of each phase measured from the scanning electron microscopy photographs. Each coating was produced using very low carbon steel and a lead-free, 0.14 weight percent aluminum-containing zinc bath. Table 5 lists the annealinf temperatures and measured metallographic data.

Table 5. Properties and Phase Compositions of Galvannealed Steel Samples

Sample	Anneal. Temp., °C	γ-ray Mössbauer Spectral Phase Fraction, %				Scanning Electron Microsc. Phase Fraction, %				Coating Weight, g/m^2	Weight Percent Iron
		ζ	δ	Γ_1	Γ	ζ	δ	Γ_1	Γ		
FZ1	499	37	43	16	4	42	40	12	6	66.7	8.0
FZ2	552	0	75	19	6	0	85	11	4	59.5	8.7
FZ3	591	0	81	11	8	0	70	16	14	71.7	12.0

Fig. 20 shows three of the Mössbauer spectra of the coating of sample FZ1. The Mössbauer fit parameters to the phases correspond to those reported for the pure iron-zinc alloys in Section 3. Table 5 lists the relative subspectral areas of each phase determined from the fit to the gamma-ray Mössbauer spectrum. The conversion electron Mössbauer spectrum of Fig. 20a shows that the coating surface consists of only the Zeta phase. The thickness of this phase is therefore expected to be greater than 0.5 µm. Conversion electron Mössbauer spectral

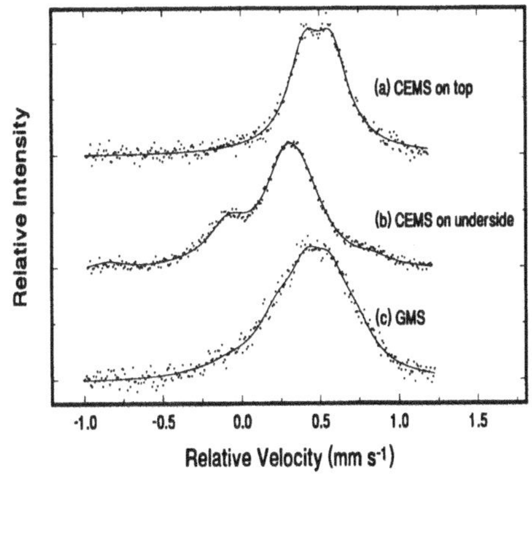

Figure 20. The conversion electron Mössbauer spectra of the surface, a, and the underside, b, and the γ-ray Mössbauer scattering spectrum of the surface, c, of the FZ1 coating, recorded at 300 K.

analysis of the underside of the fractured coating, illustrated in Fig. 20b, shows the presence of both the Gamma and the Gamma-1 phases. The relative fraction of each phase in the conversion electron Mössbauer spectrum was determined to be 97 and 3 percent, respectively. The Mössbauer fit parameters showed that the Gamma phase has a 31 atomic percent iron content. The detection of both the Gamma and Gamma-1 phases by conversion electron Mössbauer spectroscopy indicates that perhaps the Gamma phase is thin, < 0.5 μm, and some of the conversion electron Mössbauer signal from the Gamma-1 phases is able to penetrate through the Gamma layer. Alternatively, if the Gamma layer is not uniformly thick, and the two phases are not uniformly layered, the coating fracture process may have exposed a small fraction of the Gamma-1 phase. The gamma-ray Mössbauer spectrum of this coating, following the subtraction of the substrate signal, is shown in Fig. 20c. Spectral analysis shows the presence of four iron-zinc phases. Subspectral areas show the coating to be comprised of 36.8 percent Zeta, 43.4 percent Delta, 16.0 percent Gamma-1, and 3.8 percent Gamma. The Mössbauer fit parameters to the Delta phase indicate it has a low iron content of 8.1 atomic percent. In addition, with only 4 percent of the gamma-ray Mössbauer spectrum corresponding to the Gamma phase, it is concluded that this phase is in fact very thin, in agreement with the conversion electron Mössbauer spectral analysis. The x-ray diffraction analysis of the coating, shown in Fig. 21, confirms the presence of the predominantly low iron Delta and Zeta phases. It should be noted that in this pattern, the diffraction peak at 42.4° is due to the superposition of the low iron Delta peak at 42.2° and the Zeta peak at 42.5°. Other isolated peaks from the Delta phase confirm the low iron content.

The scanning electron metallographic cross-section photograph for FZ1 is shown in Fig. 22. Color etching of the polished sample has enhanced identification of the intermetallic layers. The bottom layer is a very thin dark layer next to the steel substrate. The average thickness of this layer is 0.6 μm, comprising approximately 6 percent of the total coating thickness. This layer is followed by a lighter layer with an average thickness of 1.2 μm making up nearly 12 percent of the coating. A very light colored layer, having 40 percent of the coating thickness,

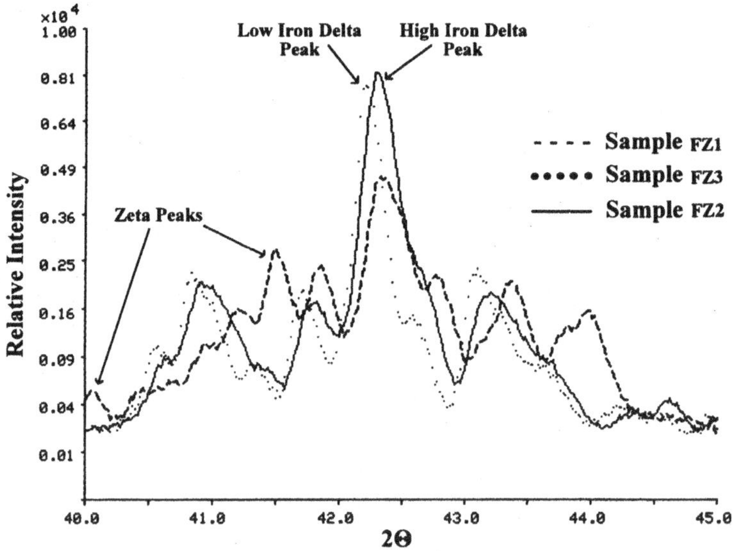

Figure 21. Overlap of the x-ray diffraction patterns of the three galvannealed coatings between 40° to 45° in 2θ. FZ1 shows a large amount of the Zeta phase. FZ2 shows the presence of a low iron content Delta phase, and FZ3 shows the presence of a high iron content Delta phase.

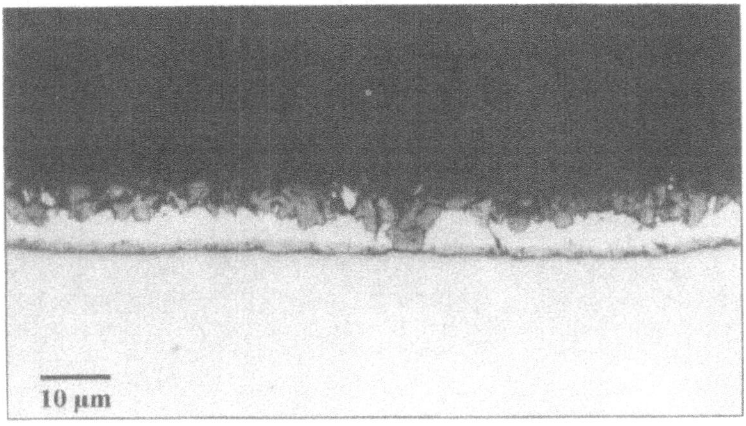

Figure 22. Cross sectional scanning electron micrograph of FZ1.

is next followed by the dark top layer which forms a rough jagged layer on the surface of the coating. A comparison of the layer thicknesses with the Mössbauer phase fractions indicates that the Gamma, Gamma-1, Delta, and Zeta phases form from the substrate to the coating surface, respectively. This is the trend observed in all galvannealed coatings analyzed. The Mössbauer parameters of the Gamma phase correspond to the high iron content end of the phase. The Mössbauer parameters of the Delta phase correspond to the low iron content end of that phase. This is expected in coatings containing large amounts of the Zeta phase because the total iron content in these coatings is generally low.

The galvannealed coating of sample FZ2 is significantly different from that of FZ1. The conversion electron Mössbauer spectrum of the coating surface, Fig. 23a, shows only the presence of the Delta phase with the fit parameters indicating the iron content to be low at 8.7 atomic percent. No Zeta phase was detected. For the fractured coating, analysis of the conversion electron Mössbauer spectrum of the underside, Fig. 23b, shows the presence of a high purity Gamma phase as well as 3 percent of the Gamma-1 phase. The γ-ray Mössbauer

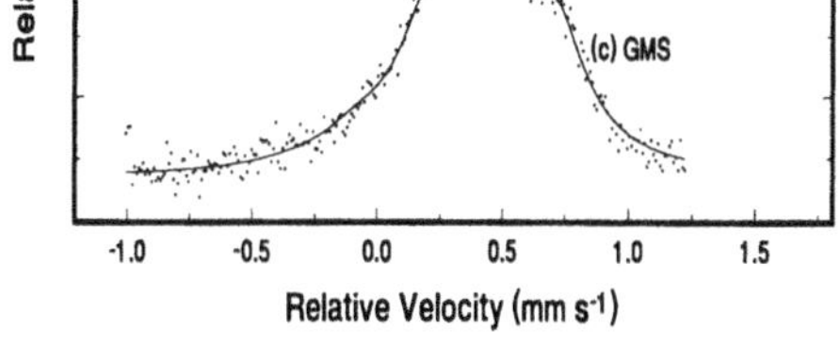

Figure 23. The conversion electron Mössbauer spectra of the surface, a, and the underside, b, and the γ-ray Mössbauer scattering spectrum of the surface, c, of the FZ2 coating, recorded at 300 K.

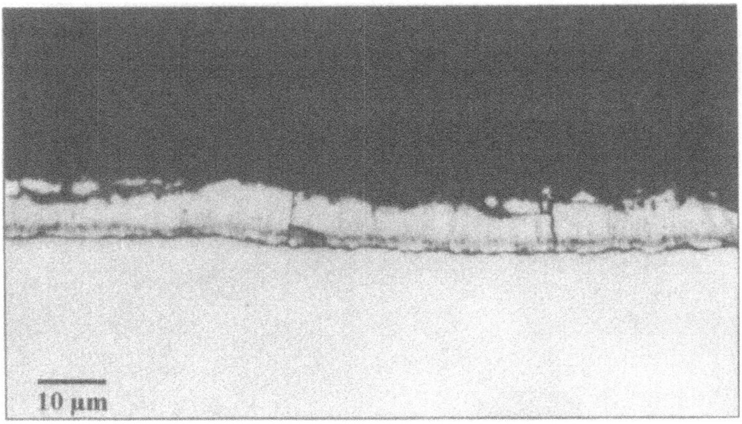

Figure 24. Cross sectional scanning electron micrograph of FZ2.

scattering spectrum of the intact coating subspectral areas, see Fig. 23c and Table 5, shows that the coating is comprised of 75.2 percent of the low iron content Delta phase, 19.1 percent of the Gamma-1 phase, and 5.7 percent of the Gamma phase. The low amount of the Gamma phase once again leads to the prediction that this layer is very thin. The x-ray diffraction pattern, shown in Fig. 21, confirms the presence of a large fraction of the low iron content Delta phase.

The metallographic cross-section of sample FZ2 is shown in Fig. 24. There are some similarities to that of sample FZ1. First, the bottom layer is a very thin, dark, 0.3 µm layer, next to the steel substrate. Second, this layer is followed by a lighter layer which has an average thickness of 0.9 µm and makes up about 11 percent of the coating thickness. A very light layer follows this and makes up the bulk of the coating. However, this coating differs from sample FZ1 in that no rough, jagged fourth layer, the Zeta phase, is observed on the coating surface. The Mössbauer spectral and x-ray diffraction analyses show that both this and sample FZ1 contain the low iron content Delta phase.

The Mössbauer spectra of sample FZ3 are shown in Fig. 25. It can be immediately seen that the conversion electron Mössbauer spectrum of the coating surface, Fig. 25a, is very different from that of FZ1 and slightly different from that of FZ2. Analysis of this conversion electron Mössbauer spectrum shows the surface to contain only the Delta phase. The Mössbauer parameters show this phase to have a high iron content of 13 atomic percent. No

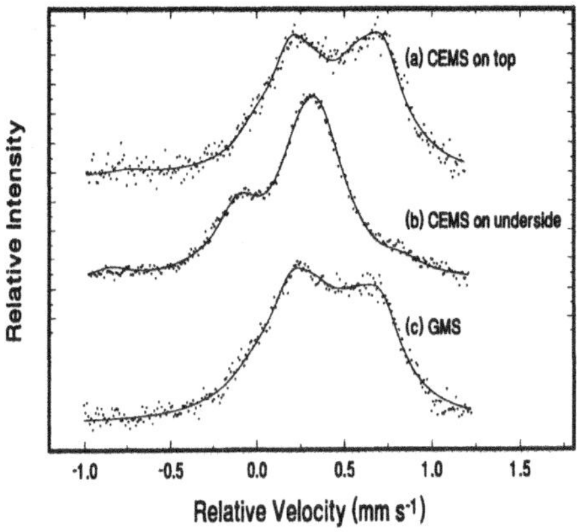

Figure 25. The conversion electron Mössbauer spectra of the surface, a, and the underside, b, and the γ-ray Mössbauer scattering spectrum of the surface, c, of the FZ3 coating, recorded at 300 K.

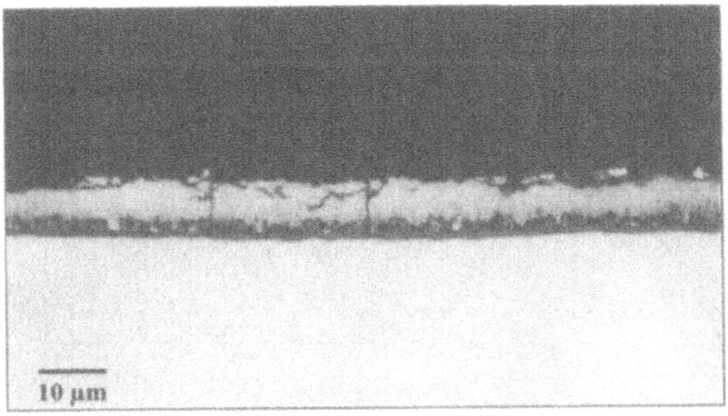

Figure 26. Cross sectional scanning electron micrograph of FZ3.

Zeta phase is present. The conversion electron Mössbauer spectral analysis of the underside of the fractured coating, Fig. 25b, shows only the high iron content Gamma phase with 30 atomic percent iron. This indicates both that the Gamma phase is thicker than in the previous two samples and that the conversion electron Mössbauer spectral signal from the adjoining phase, expected to be Gamma-1, is attenuated and not detected. The γ-ray scattering Mössbauer spectrum of the entire coating is shown in Fig. 25c. Once again this spectrum is very different from that of FZ1 and indicates the predominance of the Delta phase. The Mössbauer fit parameters showed the majority of the coating is the high iron content Delta phase. Uncorrected subspectral areas show the coating composition to be 81.0 percent Delta, 11.0 percent Gamma-1, and 8.0 percent Gamma, as given in Table 5. X-ray diffraction analysis, see Fig. 21, confirms the coating to consist of predominantly the high iron content Delta phase with no Zeta phase being detected. The x-ray diffraction findings are therefore in agreement with the Mössbauer spectral analysis.

The metallographic cross section of sample FZ3 is shown in Fig. 26. The dark bottom layer is a much thicker layer than in the other two samples, averaging 1.2 μm or 14 percent of the total coating thickness. This layer is followed by another patchy dark layer of vertically oriented crystal-like columns having an average thickness of about 1.4 μm. A very thick, light layer is the third and final layer in this coating. No fourth layer is observed on the coating surface as was seen on sample FZ1.

5. 3. Iron Coating Weight

To determine if scattering Mössbauer analysis is able to estimate the iron coating weight, a comparison was made between the Mössbauer spectral area and the measured weight of iron in many different galvannealed coatings. γ-ray scattering Mössbauer spectroscopy probes the entire coating thickness, including the underlying steel substrate. The signal from the substrate is superimposed on the coating signal to produce the typical spectrum shown in Fig. 27. By comparing the total area of the spectrum with the area due to the substrate alone, a measure of the iron content in the coating is obtained. Fig. 28 shows the plot of the relative spectral area due to the coating against the weight of iron in the coatings measured by chemical titration. The linear correlation shows the potential use of Mössbauer spectral analysis as an alternate method to determine the weight of iron in a galvannealed coating without having to strip the coating from the substrate.

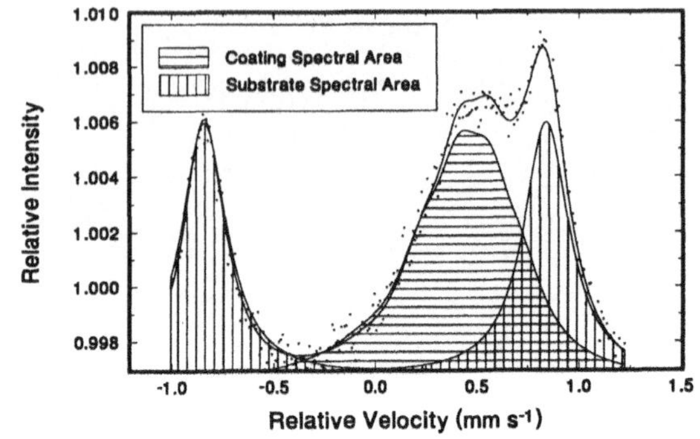

Figure 27. The γ-ray scattering Mössbauer spectrum of a galvannealed coating illustrating the spectral area from the coating and the underlying steel substrate.

5. 4. *Effect of Aluminum in the Galvannealing Bath*

Aluminum is commonly added to hot-dip batch galvanizing zinc baths to inhibit iron-zinc phase formation, improve the luster of the coating, and reduce the atmospheric oxidation of the bath. Although there has been some work on the effect small additions of aluminum to the zinc bath have on the phase formation[16,83-86] the results vary from one study to another. It is known that, upon the initial submergence of the sheet steel into an aluminum containing zinc bath, an aluminum-iron layer forms which inhibits the interdiffusion of the iron and zinc. The kinetics of this inhibition layer are not well understood. Furthermore, little work has been performed on the effect this layer has on alloy formation during the annealing process. It is believed that, because the inhibition layer serves to delay the iron-zinc alloy formation, the resultant galvannealed coating would form in the low iron content end of the phase diagram. That is, assuming that the thickness of the coatings were the same, less of the Gamma-1 and Gamma phases would be formed in a zinc bath containing aluminum.

In order to investigate the effect aluminum has on the phase formation, two coatings were analyzed with Mössbauer spectroscopy and x-ray diffraction. The two samples were prepared under the identical preparation conditions except for the addition of 0.13 percent aluminum to one of the iron saturated zinc baths. The steel was degreased with acetone before being anneal-

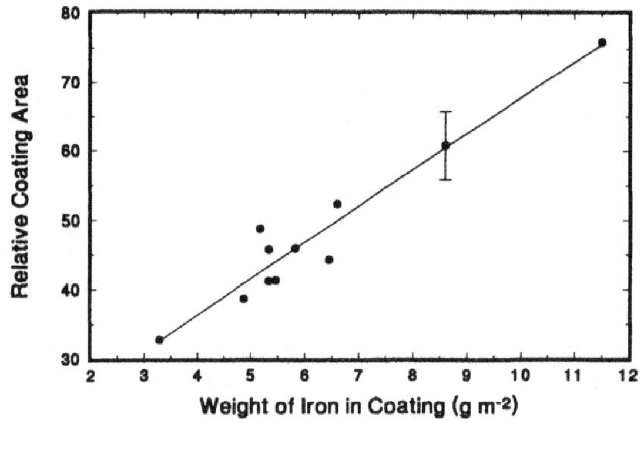

Figure 28. A comparison of the coating spectral area with the weight of iron in galvannealed coatings. The weight of iron in a coating is the product of the coating weight and the coating iron content.

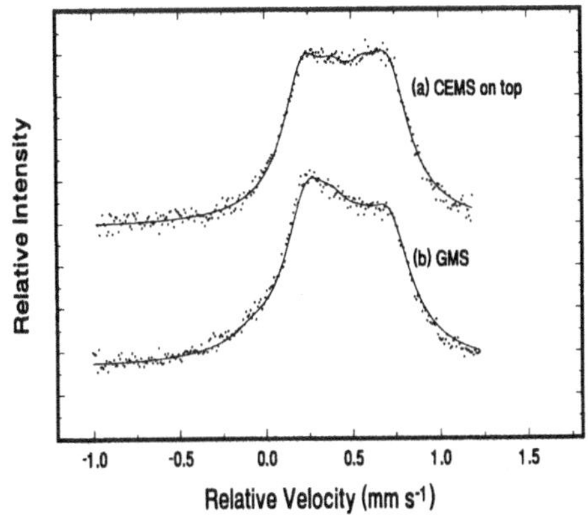

Figure 29. The conversion electron Mössbauer spectrum of the surface, a, and the γ-ray scattering Mössbauer spectrum of the surface, b, of the aluminum-free coating, recorded at 300 K.

ed at 750°C for 2 minutes in a 5 percent H_2 in N_2 atmosphere. Following the annealing, the steel sheet was submerged in an iron saturated zinc bath for 6 seconds. The molten metal bath was maintained at 465°C. The coated steel was then annealed for 10 seconds at 530°C, furnace cooled for 20 seconds, and finally quench cooled.

The Mössbauer spectra of the intact coating of the aluminum free sample are shown in Fig. 29. The conversion electron Mössbauer spectrum of the coating, Fig. 29a, reveals only the pure Delta phase on the surface. The Mössbauer spectral fit parameters indicate the Delta phase to have a low iron concentration of 8.2 atomic percent. The gamma-ray Mössbauer spectrum, see Fig. 29b, shows the presence of three iron-zinc phases, namely the low iron Delta phase, and the Gamma-1 and Gamma phases. No Zeta phase was observed which is in agreement with the conversion electron Mössbauer spectral analysis. The Delta phase makes up approximately 85 percent of the coating composition whereas the Gamma-1 and Gamma phases make up 2.4 and 12.6 percent, respectively. The conversion electron Mössbauer spectral analysis of the underside of the fractured coating showed the presence of only pure high iron content Gamma phase. This agrees with the gamma-ray Mössbauer spectrum which predicts that the Gamma phase has a thickness > 0.5 μm. X-ray diffraction confirmed the absence of the Zeta phase and the presence of the low iron content Delta phase.

The in-situ scattering Mössbauer spectra of the sample containing aluminum are shown in Fig. 30. It can immediately be seen that both the conversion electron and gamma-ray Mössbauer spectra are different from those of the aluminum-free sample. Conversion electron Mössbauer spectral analysis, see Fig. 30a, again shows that only the Delta phase is present on

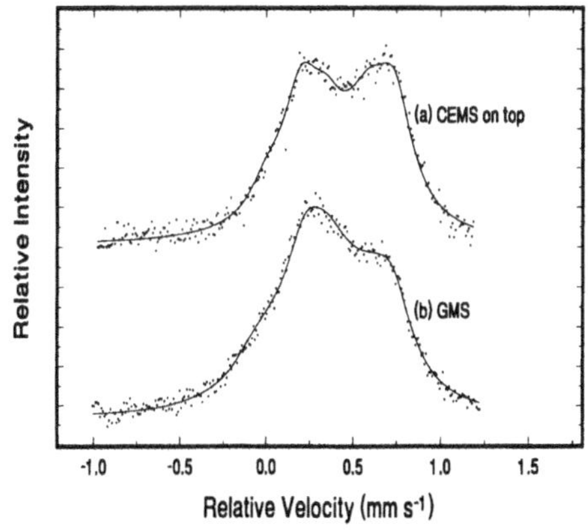

Figure 30. The conversion electron Mössbauer spectrum of the surface, a, and the γ-ray Mössbauer scattering spectrum of the surface, b, of the aluminum containing coating, recorded at 300 K.

the coating surface. The Mössbauer parameters indicate the phase to be the high iron content Delta phase containing 13 atomic percent iron. No Zeta phase was detected. X-ray diffraction analysis confirmed the Mössbauer analysis. The gamma-ray Mössbauer spectrum, see Fig. 30b, shows that the coating is also composed of 67.5 percent of the Delta phase, 13.1 percent of the Gamma-1 phase, and 19.4 percent of the Gamma phase. Conversion electron Mössbauer spectral analysis once again indicates that the pure high iron content Gamma phase is located on the underside of the coating in contact with the steel substrate.

A small addition of aluminum to the zinc bath resulted in an increase in the percentage of iron in the coating. That is, the aluminum added sample was found to contain a greater relative amount of the Gamma-1 and Gamma phases in addition to the increase in the iron content of the Delta phase. This would seem to be in contrast to the expected results. The aluminum-iron inhibition layer is thought to delay the iron-zinc alloy formation. However, comparing the relative coating spectral areas from the gamma-ray Mössbauer spectra, see Fig. 28, it can be determined that the aluminum free coating contains approximately 11 g m^{-2} of iron and the aluminum containing coating approximately 8 g m^{-2} of iron. Therefore, less alloying has resulted from the addition of aluminum to the zinc bath, consistent with the expected results.

5. 5. Summary

Scattering Mössbauer spectroscopy has been shown to be capable of uniquely identifying the iron-zinc intermetallics that form in commercial galvannealed coatings. At the present time it is the only technique that can perform the identification in-situ. A great deal of research now needs to be carried out in order to fully correlate the fraction of each phase identified in the coatings with the large number of variable preparation conditions.

The general findings of the Mössbauer analysis on the herein reported, and many additional, galvannealed coatings are now summarized. Conversion electron Mössbauer spectral analysis of the underside of fractured coatings consistently detects the presence of the relatively high iron content Gamma phase at the bottom of the coating next to the steel substrate. The Gamma phase generally forms a uniform layer, ranging in thickness from 0.3 to 1.2 μm, depending on the preparation conditions. Even in coatings containing only a small amount of the Gamma phase, the thin dark layer next to the steel substrate is always observed in the cross sectional micrographs.

The correspondence between the layer thickness observed in the cross sectional micrographs and the phase abundances determined by Mössbauer spectroscopy shows that the Gamma-1 phase forms next to the Gamma phase in either a light layer or in dark vertical crystal-like columns. It is suspected that these columns start at the Gamma boundary and grow into the Delta phase. In coatings containing less than 10 percent total iron, the Gamma-1 phase forms as a light layer and the Gamma phase forms a very thin layer. In samples containing more than 10 percent iron, a much thicker Gamma layer was found next to the steel substrate followed by very dark vertical crystals of Gamma-1. Generally a medium to high iron content Delta layer is also observed in these coatings. Those samples, which contain a rough, jagged layer on the coating surface, were found to contain a large amount of the Zeta phase on the coating surface. This indicates that the Zeta phase forms on the coating surface in a porous and often non-uniform layer. Many coatings contain only islands of the Zeta phase at the surface, a surface which is predominantly covered by the Delta phase. Furthermore, if a large amount of Zeta is observed, then a low iron content Delta layer is also present. Coatings found to have a Delta layer in the medium to high iron content range, generally show thicker Gamma and Gamma-1 layers. Finally, coatings that contain large amounts of the Zeta phase or a low iron content Delta layer on the surface always possess only a thin layer of the Gamma phase at the coating-steel interface.

Following the fracture of coatings, conversion electron Mössbauer spectral analysis never recorded any iron-zinc intermetallic phases remaining on the steel substrate. In addition, all conversion electron Mössbauer spectra recorded on the underside of the fractured coatings, showed the presence of the Gamma phase whose thickness varied from sample to sample and whose iron content was always high at close to 30 atomic percent. The results indicate that under shear the steel-coating interface is possibly responsible for the adverse powdering and chipping properties of galvanneal.

There appears to be a clear trend between the galvannealing temperature and the fraction of each iron-zinc phase in a coating. The data presented above for the three coatings show that at low annealing temperatures the Zeta phase readily forms, accompanied by a majority of the low iron content Delta phase. As the annealing temperature is increased less, and perhaps none, of the Zeta phase is formed. The Delta phase forms at the surface with a low or medium iron content. With increasing annealing temperature, the iron content in the Delta phase also increases. At the same time more of the Gamma-1 and Gamma phases form in the coating.

Coating iron weight can be determined in-situ using γ-ray scattering Mössbauer spectral analysis. The ratio of coating spectral areas to the total spectral area shows a linear relationship with the measured iron weight in the coatings.

Finally, the effect that aluminum addition to the zinc bath has on the iron-zinc phase formation leads to some important initial findings. Two coatings were prepared under similar conditions except for the addition of 0.13 percent aluminum to one of the iron saturated zinc baths. X-ray diffraction and Mössbauer spectroscopy both showed that the addition of aluminum seems to suppress the iron-zinc alloying resulting in a coating with less total weight of iron. This was observed even though the percentage of iron in the aluminum containing coating was greater. More research in this area is warranted.

6. CONCLUSIONS

There has been much progress over the past 10 years in the understanding of the physical and mechanical properties of zinc and zinc alloy coated sheet steel products. The ability to understand the mechanical properties and production parameters of the steel has succumbed in many cases to the need for basic scientific and materials characterization. Research on iron-zinc alloys, the basic materials that make up galvanneal, has been difficult but rewarding. Convincing galvanneal producers that fundamental research was required has been, and often still is, a monumental task, perhaps more difficult than the actual research itself. However, many close collaborative partnerships have now resulted between university research scientists and galvanneal producers and users in order to allow the product to reach its expected longevity and marketability.

The fundamental research has resulted for the first time in accurate identification of the four iron zinc alloys in galvannealed coatings. The ability to now accurately calibrate monitoring and analytical instrumentation has resulted from the preparation of high purity alloys.

The tailoring of galvannealed coatings for specific applications requires the ability to identify the phases and their relative quantities which form during the galvannealing process. With the compilation of the microstructural properties and hyperfine parameters of the pure iron-zinc phases and the use of scattering Mössbauer geometry for in-situ analysis, phase identification in commercial galvannealed sheet steel is now a reality. The scattering technique also permits the surface phase to be identified and, in the case of the Delta phase, its actual iron content can be determined. It is anticipated that the use of Mössbauer spectroscopy will now enable correlation of production parameters with iron-zinc phase fractions in galvannealed coatings. Furthermore, correlation of the phase fraction with the mechanical and metallurgical properties, such as the formability, weldability, paintability, and corrosion resistance, can also now be investigated.

The door has now been opened for many research scientists to begin new studies on the properties of zinc coated steels. Research on the control of the alloy growth and the reactivity of zinc with the steel, both with and without inhibitors, is still in its infancy. It is one of the most important areas which will be investigated in the next few years. Preventing coating delamination, improving powdering and chipping resistance, as well as formability, all depend on accurately determining the properties of each phase. Finally, fast monitoring techniques are necessary to control the quality of the annealed coating. At the present time the required monitor is not available. The challenge is to find the acceptable on-line monitoring technique which will accurately and quickly measure the fraction of each phase in the coating. The adverse conditions and the enormous scale on which galvanneal steel is produced are generally in conflict with standard analytical instrumentation used in the research laboratory.

ACKNOWLEDGMENTS. The authors are grateful to the many colleagues who provided us with samples, helpful analysis of some of the coatings, and who have worked closely with us for many years. These include Drs. Herbert Townsend, Vijay Jagannathan, Rebecca Tuszynski, and Waldemar Furdanowicz of Bethlehem Steel Corporation, Mr. Alan Jenkins of British Steel Corporation, Mr. Todd Bensinger of National Steel Corp., Dr. Paul Bucklin of Chrysler Corporation, Mr. Jean-Claude Farge of Noranda, Quebec, Canada, Dr. Nai-Yong Tang of Cominco Ltd., Ontario, Canada, and Dr. Bruno De Cooman of OCAS, Belgium. We are most grateful for the continued enthusiasm provided by Dr. Frank Goodwin of ILZRO, and for the research and technical results provided by Mr. Trevor Coddington and Ms. Patricia Cook of Old Dominion University. Finally we are indebted for the financial support provided to us by the International Lead Zinc Research Organization, Inc., under Grant No. ZM-403 and the Virginia Center for Innovative Technology, under Grant Nos. MAT-92-007, 93-018, and 94-151.

REFERENCES

1. H. E. Townsend, L. Allegra, R. J. Dutton, and S. A. Kriner, *Materials Performance* **25**, 36-46 (1986).
2. H. E. Townsend, *NACE Annual Conference and Corrosion Show*, Paper No. 416 (1991).
3. Y. Suemitsu, in *Proc. Galvatech'95: The Use and Manufacture of Zinc and Zinc Alloy Coated Sheet Steel Products into the Twenty-first Century*, p. 3-12, Iron and Steel Society, Inc., C. E. Slater Pub. (1995).
4. T. Hada, in *Proc. Galvatech'95: The Use and Manufacture of Zinc and Zinc Alloy Coated Sheet Steel Products into the Twenty-first Century*, p. 217-224, Iron and Steel Society, Inc., C. E. Slater Pub. (1995).
5. J. S. Kim and J. H. Chung, in *Proc. Galvatech'95: The Use and Manufacture of Zinc and Zinc Alloy Coated Sheet Steel Products into the Twenty-first Century*, p. 135-138, Iron and Steel Society, Inc., C. E. Slater Pub. (1995).
6. T. Ichida, in *Proc. Galvatech'95: The Use and Manufacture of Zinc and Zinc Alloy Coated Sheet Steel Products into the Twenty-first Century*, p. 359-369, Iron and Steel Society, Inc., C. E. Slater Pub. (1995).
7. Y. Li, S. Li, Y. Xu, W. Wu, and P. Auric, *Hyp. Interact.* **69**, 577-580 (1991).
8. M. Gu, M. R. Notis, and A. R. Marder, in *Proc. Galvatech'89: First International Conference on Zinc and Zinc Alloy Coated Sheet Steel*, p. 462-469, Iron and Steel Institute of Japan (1989).
9. T. Adaniya, T. Hara, M. Sagiyama, T. Homa, and T. Watambe, *Plating Surface Finishing*, 52 (1985).
10. B. Dauby, P. Vamkeleffe, and S. Feron, in *Proc. Galvatech'95: The Use and Manufacture of Zinc and Zinc Alloy Coated Sheet Steel Products into the Twenty-first Century*, p. 717-722, Iron and Steel Society, Inc., C. E. Slater Pub. (1995).
11. G. J. Harvey and P. N. Richards, *Met. Forum* **6**, 232 (1984).

12. H. Smith and W. Batz, *J. Iron Steel Inst. Japan* 895 (1972).
13. C. Allen and J. Mackowiak, *Corrosion Sci.* **3**, 87-97 (1963).
14. G. J. Harvey and P. D. Mercer, *Met. Trans.* **14**, 619 (1973).
15. I. P. Biele, in *Proc. Galvatech'95: The Use and Manufacture of Zinc and Zinc Alloy Coated Sheet Steel Products into the Twenty-first Century*, p. 69-75, Iron and Steel Society, Inc., C. E. Slater Pub. (1995).
16. M. Guttmann, Y. Lepretre, A. Aubry, M. J. Roch, T. Moreau, P. Drillet, J. M. Mataigne, and H. Baudin, in *Proc. Galvatech'95: The Use and Manufacture of Zinc and Zinc Alloy Coated Sheet Steel Products into the Twenty-first Century*, p. 295-307, Iron and Steel Society, Inc., C. E. Slater Pub. (1995).
17. C. Coffin and S. W. Thompson, in *Proc. Galvatech'95: The Use and Manufacture of Zinc and Zinc Alloy Coated Sheet Steel Products into the Twenty-first Century*, p. 121-131, Iron and Steel Society, Inc., C. E. Slater Pub. (1995).
18. C. S. Lin and M. Meshii, in *Proc. Galvatech'95: The Use and Manufacture of Zinc and Zinc Alloy Coated Sheet Steel Products into the Twenty-first Century*, p. 335-342, Iron and Steel Society, Inc., C. E. Slater Pub. (1995).
19. M. Lamberigts, C. De Mare, and B. C. De Cooman, in *Proc. Galvatech'95: The Use and Manufacture of Zinc and Zinc Alloy Coated Sheet Steel Products into the Twenty-first Century*, p. 33-38, Iron and Steel Society, Inc., C. E. Slater Pub. (1995).
20. L. M. Bernick and W. C. Sievert, "Processing of Galvannealed Steel," in *Mechanical Working and Steel Processing*, p. 123, Gordon and Breach, New York (1969).
21. F. W. Salt, J. F. Stanners, and K. O. Watkins, *Brit. Corrosion J.* **1**, 5 (1965).
22. J. Chang and H. Wei, *Corrosion Sci.* **30**, 831-837 (1990).
23. J. P. Landriault and F. W. Harrison, in *Proc. CIM/ICM Twenty-fifth Annual Conference of Metallurgists*, Paper No. 29-4 (1986).
24. S. Nomura, T. Akiyama, J. Lee, M. Yamaguchi, and K. Higashi, *Trans. Iron Steel Inst. Japan* **23**, 930 (1983).
25. S. A. Gedeon and T. E. Eager, *Met. Trans. B* **17**, 879 (1986).
26. K. Chou and V. Jagannathan, "Corrosion and Powdering Resistance of Galvanneal for Automotive Applications," Paper No. 375, Corrosion' 91, NACE, Cincinnati, OH (1991).
27. M. Gu and A. R. Marder, *Powder Diffraction* **6**, 89-94 (1991).
28. D. C. Cook, *Mössbauer Effect Study of Four Metal Powder Galvanneal Standards*, Bethlehem Steel Corporation, Report No. ODUDCC0191 (1991).
29. D. C. Cook, *Mössbauer Analysis of the Zinc-Iron Phases on Galvannealed Steel*, Bethlehem Steel Corporation, Report No. ODUDCC0289 (1989).
30. M. Gu, M. R. Notis, and A. R. Marder, *Met. Trans. A* **21**, 273-277 (1990).
31. R. G. Grant and D. C. Cook, *Hyp. Interact.* **94**, 2309 (1994).
32. M. Ghafari, M. Saito, Y. Hirose, and Y. Nakamura, *Hyp. Interact.* **69**, 447-450 (1991).
33. A. Laggoun, A. Hauet, and J. Teillet, *Hyp. Interact.* **54**, 825-830 (1990).
34. A. Hauet, A. Laggoun, and J. Teillet, *Mat. Sci. and Eng. A* **148**, 123-129 (1991).
35. C. Montreuil and B. J. Evans, *J. Appl. Phys.* **49**, 1437-1439 (1978).
36. D. C. Cook and R. G. Grant, *Identification of the Iron-Zinc Phases in Galvanneal Steel Coatings by Mössbauer Spectroscopy and X-Ray Diffraction. Phase I: Characterization of the Fe-Zn Intermetallic Phases*, International Lead Zinc Research Organization, Report No. ODUDCC0193 (1993).
37. *Powder Diffraction File*, International Center for Diffraction Data, Newtown Square, PA, USA.
38. Advanced Metal Coating Analysis, Inc., Virginia Beach, VA, USA.
39. O. Kubaschewski, *Iron-Binary Phase Diagrams*, p. 86-87, Springer-Verlag, Berlin (1982).
40. G. F. Bastin, F. J. van Loo, and G. D. Rieck, *Z. Metallkde* **68**, 359-361 (1977).

41. P. J. Gellings, E. W. de Bree, and G. Gierman, *Z. Metallkde* **70**, 312-314 (1979).
42. R. G. Grant, P. S. Cook, and D. C. Cook, *J. Mater. Res.* **10**, 2454-2462 (1995).
43. A-131 Electrolytic Iron Micropowder, SCM Metal Products, Inc., Research Triangle Park, NC, USA.
44. AS-081 Zinc Powder, OBRON Atlantic Corporation, Painesville, OH, USA.
45. *Induction Coupled Plasma Spectroscopy Standards*, Inorganic Ventures, Inc., Lakewood, NJ, USA.
46. W. E. Harris and B. Kratochvil, *Chemical Separations and Measurement*, Saunders Publishing Co., New York (1974).
47. R. B. Fischer and D. G. Peters, *Basic Theory and Practice of Quantitative Chemical Analysis*, Saunders Publishing Co., New York (1968).
48. W. A. Furdanowicz and K. E. Downey, *J. Microscopy* **174**, 66 (1994).
49. D. C. Cook, R. S. Tuszynski, and H. E. Townsend, *Hyp. Interact.* **54**, 781-785 (1990).
50. P. J. Gellings, E. W. de Bree, and G. Gierman, *Z. Metallkde* **70**, 315-317 (1979).
51. A. Osawa and Y. Ogawa, *Z. Kristallogr.* **68**, 177-188 (1928).
52. H. Bablik, F. Götzl, and F. Halla, *Z. Metallkde* **30**, 248-252 (1938).
53. G. F. Bastin, F. J. van Loo, and G. D. Rieck, *Z. Metallkde* **65**, 656-660 (1974).
54. J. Schramm, *Z. Metallkde* **30**, 131 (1938).
55. M. Ghoniem and K. Löhberg, *Z. Metallkde* **26**, 1026 (1972).
56. H. Bühler, G. Jackl, L. Meyer, and S. Baumgartl, *Microchim. Acta Suppl.* **4**, 75 (1970).
57. P. Gellings, *Corrosion Sci.* **14**, 507 (1974).
58. A. Jena and K. Löhberg, *Z. Metallkde* **73**, 517-521 (1982).
59. R. D. Jones and S. G. Denner, *Scripta Metal.* **8**, 175-180 (1974).
60. M. J. Graham, P. E. Beaubien, and G. I. Sproule, *J. Mater. Sci.* **5**, 626-630 (1980).
61. D. C. Cook and J. D. Cashion, *Hyp. Interact.* **5**, 479-486 (1978).
62. P. J. Brown, *Acta Cryst.* **15**, 608-612 (1962).
63. Submitted for publication to International Center for Diffraction Data, Newtown Square, PA, USA.
64. D. L. Bish and J. E. Post, *Modern Powder Diffraction, Reviews in Mineralogy*, Vol. 20, Mineralogy Society of America (1989).
65. G. F. Bastin, F. J. van Loo, and G. D. Rieck, *Z. Metallkde* **67**, 694-698 (1976).
66. A. S. Koster and J. C. Schoone, *Acta Crystallogr. B* **37**, 1905-1907 (1981).
67. P. J. Gellings, G. Gierman, D. Koster, and J. Kuit, *Z. Metallkde* **71**, 70-75 (1980).
68. J. K. Brandon, R. Y. Brizard, P. C. Chieh, P. K. McMillan, and W. B. Pearson, *Acta Crystallogr. B* **30**, 1412-1417 (1974).
69. A. Johansson, H. Ljung, and S. Westman, *Acta Chem. Scand.* **22**, 2743-2753 (1968).
70. A. Stadlbauer, F. Rubenzucker, K. Zeman, and E. Fuhrmann, in *Proc. Galvatech'95: The Use and Manufacture of Zinc and Zinc Alloy Coated Sheet Steel Products into the Twenty-first Century*, p. 81-85, Iron and Steel Society, Inc., C. E. Slater Pub. (1995).
71. D. C. Cook and R. G. Grant, in *Proc. Galvatech'95: The Use and Manufacture of Zinc and Zinc Alloy Coated Sheet Steel Products into the Twenty-first Century*, p. 497-508, Iron and Steel Society, Inc., C. E. Slater Pub. (1995).
72. R. G. Grant and D. C. Cook, *Il Nuovo Cimento B*, in press.
73. D. C. Cook and E. Agyekum, *Nuc. Instr. Methods B* **12**, 515-520 (1985).
74. D. C. Cook, *Hyp. Interact.* **29**, 1463-1466 (1986).
75. B. Fultz, United States Patent No. 4,393,306 (1983).
76. L. Blaes, H. G. Wagner, U. Gonser, J. Welsch, and J. Sutor, *Hyp. Interact.* **29**, 1571-1574 (1986).

77. P. Schaaf, L. Blaes, J. Welsch, H. Jacoby, F. Aubertin, and U. Gonser, *Hyp. Interact.* **58**, 2541-2546 (1990).
78. U. Gonser, P. Schaaf, and F. Aubertin, *Hyp. Interact.* **66**, 95-100 (1991).
79. R. G. Grant, Ph. D. Thesis, Old Dominion University (1995).
80. D. C. Cook and R. G. Grant, *Identification of the Iron-Zinc Phases in Galvanneal Steel Coatings by Mössbauer Spectroscopy and X-Ray Diffraction. Phase II: Construction and Testing of a Coating Analyzer*, ILZRO, Inc., Report No. ODUDCC0294 (1994).
81. R. G. Grant and D. C. Cook, in *Proc. Galvatech'95: The Use and Manufacture of Zinc and Zinc Alloy Coated Sheet Steel Products into the Twenty-first Century*, p. 509-514, Iron and Steel Society, Inc., C. E. Slater Pub. (1995).
82. D. C. Cook, *Mössbauer Effect Study of Galvanneal Steel Coating Fracture Surfaces*, Bethlehem Steel Corporation, Report No. ODUDCC0192 (1992).
83. A. R. P. Ghuman and J. Goldstein, *Met .Trans. A* **2**, 2903 (1971).
84. H. Smith and W. Batz, *J. Iron Steel Inst. Japan* 895 (1972).
85. T. Q. Coddington and D. C. Cook, in *Proc. Galvatech'95: The Use and Manufacture of Zinc and Zinc Alloy Coated Sheet Steel Products into the Twenty-first Century*, p. 515-517, Iron and Steel Society, Inc., C. E. Slater Pub. (1995).
86. T. Q. Coddington and D. C. Cook, *Il Nuovo Cimento B,* in press.

11

Mössbauer Effect Studies of Materials Prepared by Mechanochemical Methods

S. J. Campbell and W. A. Kaczmarek

1. INTRODUCTION

1. 1. Background

Over the past two decades or so, ball milling has evolved from being a standard technique in powder metallurgy, used primarily for particle size reduction, to its present status as an important method for the preparation of either materials with enhanced physical properties or, indeed, new phases, or new materials. Correspondingly, terms such as mechanical alloying, mechanochemical methods, and mechanosynthesis, or various variants, are becoming increasingly common in the materials science and condensed matter physics literature.

Introductory, pioneering perspectives on mechanical alloying based on milling techniques have been published by Benjamin,[1,2] and a more recent, splendid, overview of the principles and applications of milling has been published by the American Society for Metals Committee on Milling.[3] The extent to which mechanical alloying and related synthesis methods have influenced materials science can be gauged both by the increased number of research publications in recent years, and by the incidence of new journals and international conferences devoted to, or extended in scope to incorporate, mechanical alloying. *NanoStructured Materials*[4] is an example of a new journal in this field. Likewise, the International Conferences on Nanostructured Materials, a series which started in Cancun in 1992, and the International Symposia on Mechanically Alloyed and Nanocrystalline Materials, which developed from the International Symposium on Solid-State Reactions and Mechanical Alloying, Grenoble, 1990 and the International Symposium on Mechanical Alloying, Kyoto, 1991, are now well established conferences in the field. Clearly mechanical alloying and mechanochemical treatments are now well established routes for the preparation of new materials. This is further indicated by the wide range of materials which have been treated and synthesized in this way, including metal-metal oxide nanocomposites,[5,6] 3d transition metal based compounds,[7] nano-

S. J. Campbell • School of Physics, University College, University of New South Wales, Australian Defence Force Academy, Canberra ACT 2600, Australia. *W. A. Kaczmarek* • Department of Applied Mathematics, Research School of Physical Sciences and Engineering, The Australian National University, Canberra ACT 0200, Australia.

Mössbauer Spectroscopy Applied to Magnetism and Materials Science, Volume 2, edited by Gary J. Long and Fernande Grandjean, Plenum Press, New York, 1996.

crystalline materials,[8-10] non-metallic inorganic materials,[11] as well as a wide range of metals and alloys.[12] Although this chapter is concerned primarily with the mechanical treatment of materials using the well established principles of milling, the usefulness and applicability of related techniques, such as repeated rolling and/or pressing and high-strain deformation, in the preparation of new materials should be noted.[12,13] In particular, Shingu and co-workers[14-17] have demonstrated the equivalence of the deterministic approach of repeated rolling and the probabilistic approach of ball-milling in forming super-saturated solid solutions of Ag-Cu powders and mechanically alloyed Al-Fe. Likewise, high-strain deformation has been used to prepare submicrometer grain structures.[18]

The advantages of the Mössbauer effect as a microscopic probe of the local atomic environments in condensed matter are well known[19-22] and the measurement of the hyperfine interactions is of primary interest. These hyperfine interactions include the electric monopole interaction, i.e., the isomer shift, the electric quadrupole interaction, i.e., the quadrupole splitting, and the magnetic dipole or nuclear Zeeman interaction, i.e., the magnetic hyperfine splitting. These interactions often provide detailed insight into the structural and magnetic environment of the Mössbauer isotope. Indeed, four decades after its discovery, Mössbauer spectroscopy still continues to develop as a sophisticated scientific technique, both because of its intrinsic interest and because of its scope for ever wider and more incisive applications in the investigation of materials. This is borne out, for example, by the increasing use of nuclear resonant diffraction[23] and scattering[24] of synchrotron radiation, as synchrotron sources become more readily accessible to the scientific community. As expected, Mössbauer spectroscopy has become one of the main techniques used to understand the structural and magnetic properties of the new phases and materials prepared by mechanochemical synthesis, particularly those containing iron. Indeed, Mössbauer spectroscopy is often the most effective way of characterizing the range of structures, phases, and metastable states which can be produced and which sometimes co-exist in milled products.

In addition to phase and site identification, Mössbauer spectral characterization of materials can provide information on particle size and surface effects, as well as on fine particle dynamic properties.[25] However, it is important to note that Mössbauer spectroscopy alone rarely provides sufficient insight into a material for a full and proper understanding of its properties. Rather, the sometimes unique information available from Mössbauer spectroscopy complements the information obtained by other techniques, such as x-ray or neutron diffraction, thermal analysis, electron microscopy, and magnetization measurements.

Milled Cu-Fe. The extent to which Mössbauer spectroscopy complements and extends the information available from other techniques in the investigation of mechanically alloyed materials is nicely illustrated by using the Cu-Fe alloy system as an introductory example. Indeed this system illustrates the potential of mechanical alloying in the preparation of new phases. Supersaturated solid solutions, which extend well beyond the solubility limit of ca. 4.4 atomic percent iron in copper at the melting point[26] of 1094 °C, may be prepared by mechanical alloying.[27-31] Interest in the preparation of metastable Cu-Fe alloys by mechanical alloying has also been stimulated by the enhancement in magnetic properties reported for related granular and nanocomposite magnetic materials.

The Cu-Fe alloy system is of long standing interest to Mössbauer spectroscopists, both because of the Kondo effect exhibited by dilute **Cu**-Fe alloys, and because of the range of iron states and phases, particularly the antiferromagnetic fcc γ-iron phase, which can be stabilized in the coherent fcc copper matrix.[32-34] Indeed, interest in the various iron states of Cu-Fe alloys continues, with recent studies characterizing the high temperature behavior of pairs of iron atoms.[35] Macri et al.[31] have recently reported the production, upon mechanically alloying powder mixtures of copper and iron for 16 h., of both fcc copper-like phases of $Cu_x Fe_{100-x}$, with x > ca. 40, and bcc iron-like phases, with x < 30. These results were found to be in good agreement with earlier work.[27,36] By using a combination of x-ray diffraction and neutron diffraction, Macri et al.[31] were able to examine in detail the relative enhancement of scattering from fcc copper-like and bcc iron-like phases, respectively. The extra insight which Mössbauer

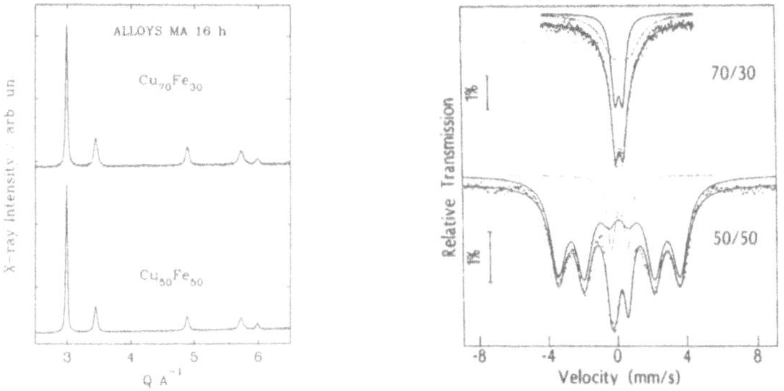

Figure 1. X-ray diffraction patterns and room temperature Mössbauer spectra for mixtures of $Cu_{50}Fe_{50}$ and $Cu_{70}Fe_{30}$, milled for 16 hours. Reproduced with permission from Macri et al.[31]

spectroscopy brings to such an investigation is shown in Fig. 1, in which the x-ray diffraction patterns of $Cu_{70}Fe_{30}$ and $Cu_{50}Fe_{50}$, mechanically alloyed for 16 hours, reveal a single fcc solid solution for both samples, whereas the corresponding room temperature Mössbauer spectra reveal significant differences in the local environments of the iron atoms. Although details of the subspectral components have yet to be resolved, the essentially doublet-like nature of the $Cu_{70}Fe_{30}$ spectrum, compared with the predominantly broad distribution of magnetic hyperfine splittings observed in the $Cu_{50}Fe_{50}$ spectrum, see Fig. 1, appears to reflect differences in the local magnetic environments of the iron atoms in these two, otherwise structurally similar, alloys.[31] It should be noted that further clarification of the subspectral components present in the spectra shown in Fig. 1 could be obtained from variable temperature Mössbauer spectral measurements. In particular, low temperature spectra would establish the extent to which relaxation effects, associated with fine particles resulting from the milling process, are exhibited by the spectra.

1. 2. Scope of the Chapter

As noted above, Mössbauer spectroscopy has already been used extensively in the study of materials prepared by milling, mechanical alloying, and related processes. The increased incidence, over the past five to ten years, of Mössbauer spectral research in the field is indicated by several review articles. Campbell and Gleiter[12] have provided an overview of Mössbauer spectral studies of nanostructured materials, including systems prepared by mechanical alloying. The contributions which Mössbauer spectroscopy have made in understanding the mechanosynthesis of nanocrystalline materials have also been reviewed by Le Cäer and Matteazi.[10] More recently Würschum[37] has reviewed the contributions which various nuclear techniques, including Mössbauer and nuclear magnetic resonance spectroscopy, perturbed angular correlation, positron annihilation, and muon spin rotation measurements, have made to the study of nanocrystalline metals and alloys.

The goal of this chapter is both to provide an introduction to and a broad overview of ball milling and mechanochemical synthesis and to point out the nature of the special insights which Mössbauer spectroscopy can provide in contributing to a fuller understanding of the solid state reactions and mechanisms which take place during milling and related treatments. The discussion, however, is not restricted to Mössbauer spectral measurements and related analyses, but also includes relevant results obtained with other techniques, such as x-ray and neutron diffraction, scanning electron microscopy, and differential thermal analysis, results which contribute to an improved understanding of materials prepared using mechanical treatment methods.

In addition to a brief overview of the milling devices commonly used in industrial and research applications, Section 2 includes a discussion of important milling parameters and environments, and an outline of some of the approaches adopted so far in modelling the milling process. An outline of the principal processes and mechanisms which occur during mechanochemical treatments is given in Section 3. The wide range of topics to which Mössbauer spectroscopy has already contributed prevents a comprehensive review of all of these applications. Rather the approach adopted in Section 4 is to present, through selected examples, some of the unique insights afforded by Mössbauer spectroscopy, followed by a discussion of some aspects of the technologically important iron carbides, nitrides, and oxides, materials which have been the subject of considerable research activity in recent years. The concluding remarks incorporate a brief prognosis for the future contributions of Mössbauer spectroscopy to the study of mechanochemically prepared materials.

Milled Iron. As an example by which to conclude this introductory section, it is appropriate to consider the milling of iron powder. Iron, as we all know, is an element of great technological importance in materials. In addition, because of the favorable properties of the iron-57 isotope, iron is an almost ideal probe for Mössbauer effect measurements, with over 80 percent of all Mössbauer spectral studies being carried out with this isotope.[38] Because of these favorable properties, a number of Mössbauer spectral investigations, which incorporate the milling of iron using a variety of milling devices, operating conditions, and milling environments, have been reported.[39-47]

In the context of this chapter, which considers the effect that different milling environments have on milled products, see Section 2. 2, a pertinent feature of the spectra shown in Fig. 2 is the behavior they reveal when iron powder is milled in a range of environments, i.e., in a vacuum, Fig. 2b, in helium, Fig. 2c, with a surface active agent, the surfactant DDAA, $C_{28}H_{59}O_2N$, Fig. 2d, and in nitrogen, Fig. 2e. The range of small but significant differences revealed by the analysis of these spectra, and shown by the subspectral components of Fig. 2, reveals the important role that the milling environment can play in determining the nature of milled iron. The decrease in mean grain sizes, $<\delta>$, from ca. 95 to 45 nm, and further to ca. 8 nm, as the iron powder is milled in a vacuum, in helium, and with a surfactant, respectively, is indicated by the increased contributions of 6, 7, and 19 percent, respectively, for the subspectral component areas associated with the interfacial regions, the regions which separate the nanocrystalline grains. This behavior is similar to that observed in nanocrystalline iron prepared by the gas condensation technique.[48] Milling in nitrogen, however, is found[41] to additionally result in a bcc iron nitride phase containing ca. 4 ± 1 atomic percent nitrogen, as is shown in Fig. 2e. Although similar spectral features were observed by Nasu and Shingu[39] and Tanaka et al.[40] upon milling iron in argon, the additional subspectral component, of reduced magnetic hyperfine field compared with that of the nanocrystalline α-iron grains, which they observed, was linked to the boundary regions of the grains and the strain fields resulting from the milling. There was also some evidence for slight contributions from finely dispersed iron oxides. In contrast, no evidence of a subspectral component due to iron atoms in grain boundaries was reported by Schaaf et al.[45] in their study of iron powder milled in argon, but rather an additional subspectrum was obtained[49] with hyperfine parameters consistent with the presence of chromium contamination originating from the milling vial and balls.

A brief comparison between similar studies on milled iron powder,[39-41,45] although indicating the sensitivity of Mössbauer spectroscopy for such studies, also illustrates several problems. First, although similar spectral features are observed for iron milled in vacuum, helium,[41] and argon,[39,40] the relative contributions of the interfacial regions and the grain boundaries, or the effects due to strain fields, have yet to be resolved. This situation can be clarified by measurements based on controlled grain growth during low temperature annealing, and by variable temperature Mössbauer effect measurements. Second, the nature of the milling

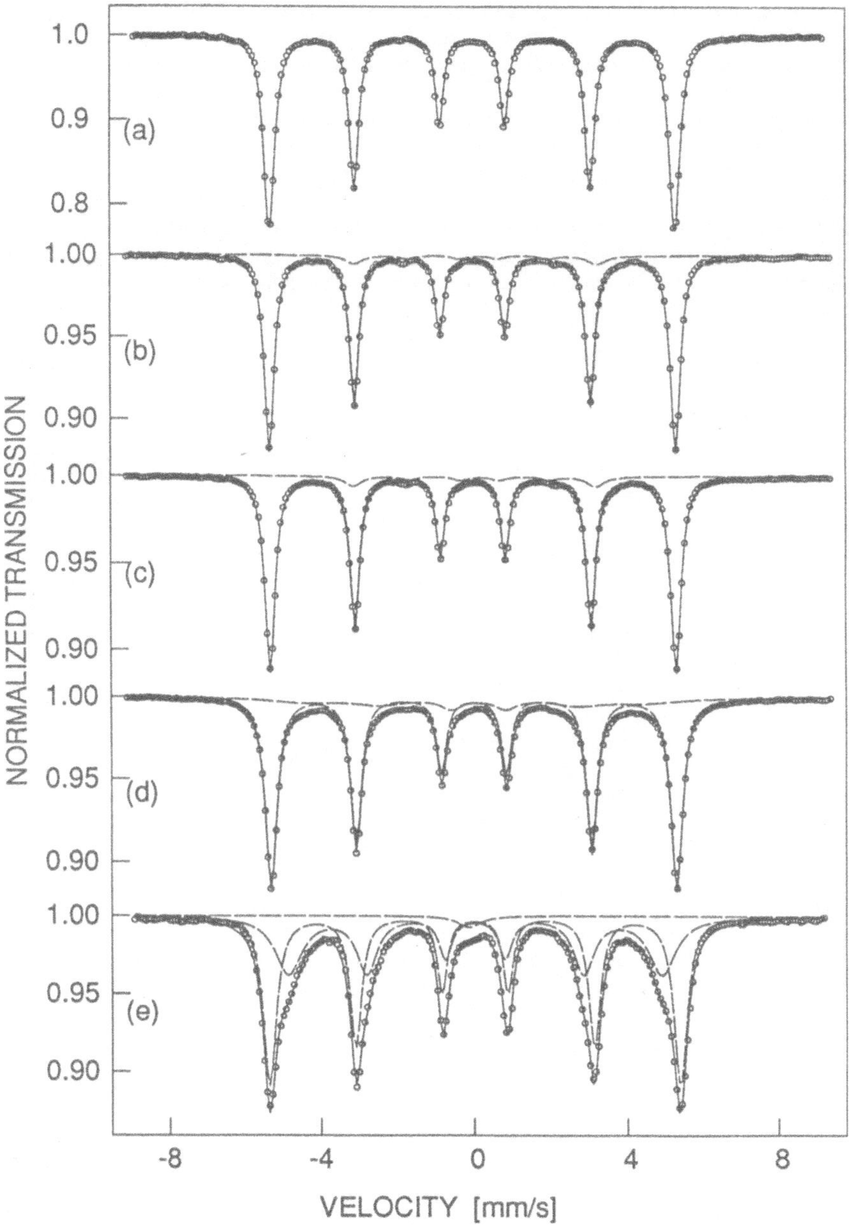

Figure 2. Room temperature Mössbauer spectra of (a) the starting iron powder, (b) iron ball milled in a vacuum for ca. 610 hours, (c) iron ball milled in helium for ca. 620 hours, (d) iron ball milled with a surfactant, DDAA, for ca. 250 hours, and (e) iron ball milled in nitrogen for 264 hours.

process intrinsic to mechanochemical treatments will, depending upon the relative hardness of the mill vessel, balls, and charge, result in wear and contamination. For example, in the work reported by Schaaf et al.,[45] chromium from the mill and balls led to contamination of the milled iron product, see also Section 4. 1. Third, besides such wear-induced impurities, as is so often the case with experimental work, further impurity effects can sometimes occur as the result of a poor vacuum, impure gases, and poor seals. As has already been noted, such problems were found to yield finely dispersed iron oxides when iron was milled in argon.[39,40] While the above points generally indicate problems which have to be further clarified or resolved, a key point in the present discussion is that the sensitivity of Mössbauer spectroscopy permits these problems to be identified, thus contributing to the eventual complete solution and understanding of the reactions and processes which occur during mechanochemical treatments.

2. MILLING

2. 1. Background

As mentioned above, grinding has evolved from its former industrial role, predominantly used for particle size reduction, to a mainstream method used for the preparation of new materials and phases. The mechanical treatment of both inorganic and organic solids remains extremely important to many technological materials, including ceramic materials, metallurgical materials, and ore dressings. In addition to particle size reduction, the deformation of materials at high strain rates by mechanical treatment offers a method for the synthesis of large quantities of amorphous or nanocrystalline powders. These powders can then be cold or hot compacted to the required bulk form. Furthermore, research has revealed that many solid state reactions, which normally occur only at elevated temperatures, can be facilitated or accelerated by high energy milling.[50]

The main features of the formation of amorphous or nanostructured states, are now well established.[51-57] In parallel with the experimental studies, theoretical models for the milling process and the resulting structural transformations have been developed.[58,59] Both as a result of the development of new milling techniques and the exciting results ensuing from mechanical treatment methods, increasingly complex compounds have been mechanochemically treated.[60] Indeed, the field of mechanochemistry has been established to study chemical reactions between otherwise nominally "non-reactive" materials. Efforts to establish a theoretical framework for understanding the mechanisms of mechanochemical reactions are continuing because their governing principles are not yet fully understood. The main insight into the mechanisms of such reactions has, to date, been derived primarily from empirical experimental studies.

2. 2. Milling Conditions

The milling conditions are obviously very important in determining the processes which occur during mechanical treatment. A summary of the main factors which should be considered in carrying out a ball milling experiment is presented schematically in Fig. 3. It is well known that the properties of the milled products, such as the particle size distribution, the degree of disorder or amorphization, and the final stoichiometry, depend on the milling conditions[61] and, as such, the more complete the control and monitoring of the milling conditions, the better. In particular it is important to track the microstructural changes which take place during milling. Only recently, however, have systematic studies on the standardization of the milling conditions and their description been proposed.[62,63] This standardization is important because of the availability of a wide range of different types of mills, each with their own individual characteristics and different milling conditions. Information about the main parameters to be considered in the milling process is presented in Section 2. 3. 5.

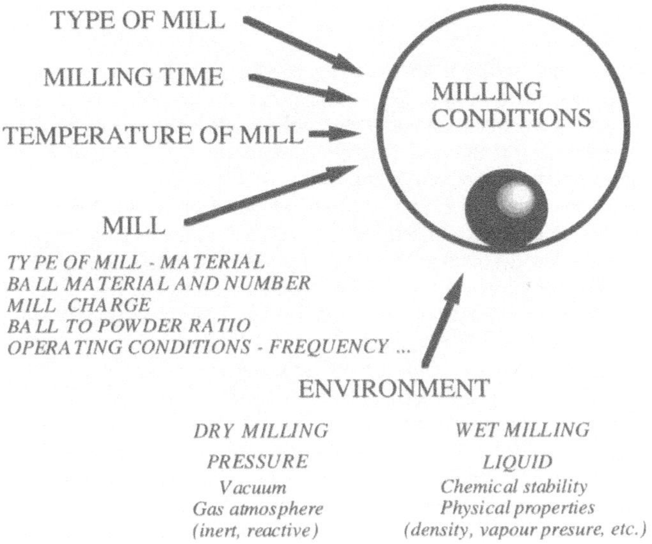

Figure 3. An overview of the main factors to be considered in a ball milling experiment.

2. 3. *Milling Devices*

The first overall description and analysis of the different types of mills used in mechanical alloying was presented by Maurice and Courtney.[62] The majority of the mills used for experimental research and for the production of small batches of materials are variants of the ball mill. The high energy associated with the heavy balls impacting on the powdered material produces, for example, the necessary cold welding needed for either the formation of alloys, or, in the case of brittle materials, the fracture required for particle size reduction. The main types of mills currently used in the field of mechanical alloying or mechanochemistry include stirred ball mills or attritors, mixer mills, planetary or vertical type mills, and rotary or horizontal ball mills. Typical capacities of the various commercial mills[64] are given in Table 1.

Reflecting the increasing interest in mechanical alloying and the use of different types of mills, several investigations[58,59,62] into the fundamental mechanics and physics of the milling process have been carried out. Various factors, such as the ball mass, impact velocity, impact angle, and impact duration, as well as the geometrical proportions of the milling chamber, the ratio of the ball to powder mass, and the filling volume, influence the performance of the mill and affect the resulting material.

Table 1. Typical Capacities of Different Types of Mills

Mill type	Sample weight
Vibratory mills	up to 200 g
Planetary mills	up to 4 containers each 250 g
Vertical mills	up to 4 containers each 250 g
Attritors	0.5 to 100 kg
Horizontal mills	up to several tons

2. 3. 1. Vibratory Mill

Vibratory or mixer mills, as shown in Fig. 4, are generally used for small scale sample production. They impart a three-dimensional motion to the ball charge and, because they operate at high speed, a greater amount of energy is transferred to the material being processed. Experimental results and theoretical predictions have shown[61,65] that milling times in mixer mills are much shorter than in other types of mills.

The milling takes place in a totally closed container which is cylindrical with curved or hemispherical ends. The processes occuring in this type of mill are characterized by the repeated impact of the balls on the charge material. The impact speed, together with the impact frequency, are the most important parameters in determining the rate of mechanical energy transferred to the charge and thus the overall kinetic behavior.[62] A typical example of this kind of mill is the Spex 8000 mixer or vibratory mill. Such mills are used widely as high energy mills for mechanical alloying and for the production of nanophase materials. Further modifications of this type of mill incorporate forced cooling, to permit extended milling times for powder mixtures with exothermic reactions, e.g., the formation of metal carbides or nitrides, and operation in an inert or reactive gas atmosphere, such as argon, nitrogen, or ammonia. Recently, a new approach to the construction of a high capacity, high energy mill, based on a similar vibratory action mechanism was proposed.[66] This new mill was found to exhibit improved control of the impact velocities, with higher velocities, and correspondingly higher energies.[65]

2. 3. 2. Planetary Mill

Planetary ball mills, as shown in Fig. 4, are the next most commonly used mills for mechanical alloying. Milling is performed in sealed bowls into which the appropriate ball char-

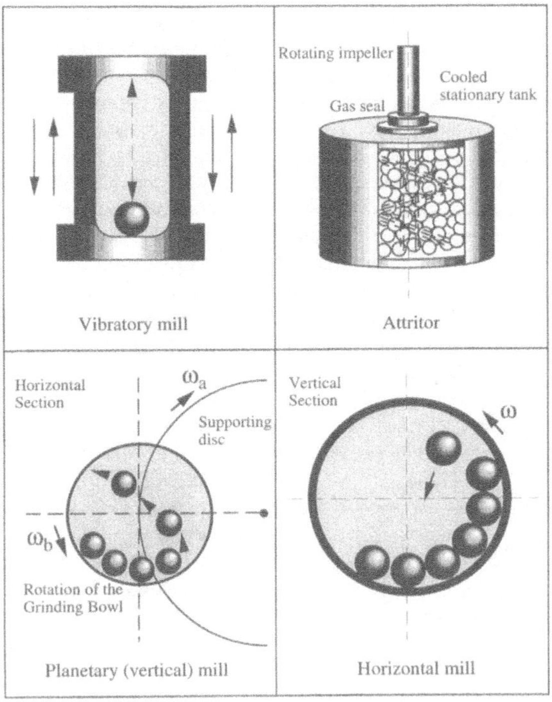

Figure 4. Schematic representations of the different types of milling devices.

ge is placed together with the constituent powders. Centrifugal forces caused by the rotation of the supporting disc and the autonomous turning of the grinding bowl act on the medium and the grinding container. Because the turning directions of the supporting disc and the bowl are opposed, the centrifugal forces alternately are synchronized and opposed. Thus the grinding balls and the charge alternatively roll on the inner wall of the bowl, and are lifted and thrown across the container at high speed. The milling process takes place mainly as a result of impact and friction. A typical example of this type of mill is the Fritsch planetary ball mill, P-5. An overview of the characteristics and applications of planetary mills in mechanical alloying and mechanochemistry has recently been published.[59,67]

2. 3. 3. Vertical Mill or Attritor

The attritor is essentially a vertical mill. Such devices were widely used for laboratory investigations during the 1970s, the first decade of major developments in the field of mechanical alloying. As shown in Fig. 4, the usually sealed cylindrical tank contains numerous grinding balls that are activated by a vertical shaft having a set of horizontal impellers which are set progressively at right angles to each other. The impellers energise the ball charge, causing powder size reduction as a result of impacts between the balls and the container wall, the agitator shaft, the impellers and the balls themselves. Mechanical energy is transferred to the powder by impact, friction, and interparticle collisions. Based on geometrical considerations, it is clear that the greatest number of collisions in this system occur between the balls and the powder. These mills, made of various materials, are available in a wide range of sizes with variable shaft speeds and with temperature control.[68]

The first description of the application of an attritor in mechanical alloying was published by Benjamin.[1,2] A detailed study of the effects of different ball sizes on the efficiency of the attritor has been reported by Cook and Courtney.[69] In this study, based on cinematographic techniques for observing the ball movements, a kinetics approach was employed to analyse the particle sizes and the evolution of the microstructure. It has been found that, for differently sized balls, the particle welding rate was increased and a greater deformation of the powder particles was observed. This study covered the important topic of the effect of ball size on the mechanical alloying of metallic powders. Additional studies are required in order to describe the influence of the milling parameters on non-metallic materials, materials for which crystal structure defect formation and propagation mechanisms are important.

2. 3. 4. Horizontal Mill

A conventional horizontal ball mill, see Fig. 4, is a relatively large device which can cope with large volumes of powder over long milling times. Only recently has this basic design been adapted to smaller scale requirements.[70] The Anutech Uni-Ball-Mill has been specially designed for mechanical alloying and mechanochemical applications. Because most of the results presented in this chapter were obtained on powders prepared with this type of mill, details of the Uni-Ball-Mill construction and its milling characteristics will be described in detail.

As mentioned above, three types of milling devices are generally used for mechanical alloying, vibrating mills, attritors, and planetary mills. However, none of the commercially available devices were designed with the specific needs of the mechanical alloying process in mind. Indeed, this factor became particularly evident in recent years as the dependence of the final milled product on the type of milling device and energy regime has been clearly demonstrated, e.g., for the formation of an amorphous Ni-Zr phase.[53,71] Further, studies of the amorphization process for Fe-Ni at various milling intensities,[72] showed that the location on the phase diagram for Fe-Ni mixtures, prepared by mechanical alloying, is directly related to the milling intensity. The milling intensity corresponds to the actual processing temperature of the individual particles during impact. In this respect, the kinetics of the amorphous phase formation, as well as the extent of the glass-forming region, were shown to be crucially depen-

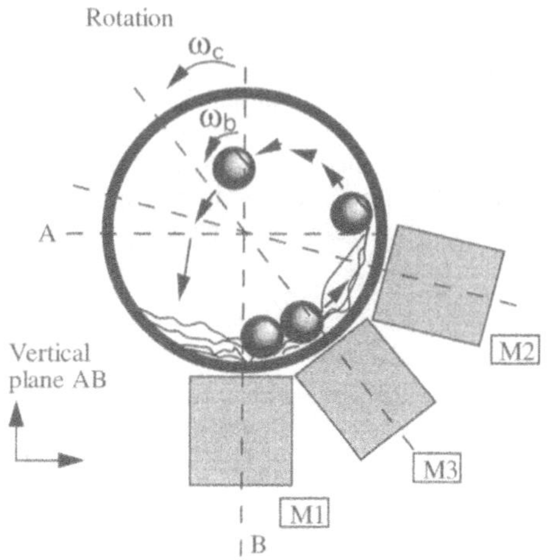

Figure 5. The basic design of the Uni-Ball-Mill showing the stainless steel cell and balls. M1, M2, and M3 indicate the range of possible positions of the Nd-Fe-B permanent magnets corresponding to different energy configurations, as discussed in the text.

dent on the type of milling device, the ball movement pattern, and the milling energy.[73] In addition, relatively little is known at present about the milling conditions to be selected in order to produce final products with *a priori* selected physico-chemical properties. With such factors in mind, a mill should be simple and offer a wide range of milling conditions, including those encountered in commercial devices, and further it should permit the careful control of milling parameters.

The basic design of the Uni-Ball-Mill is shown[70] in Fig. 5. The commercially available Uni-Ball-Mill II is equipped with four milling containers, comprising stainless steel cells with the approximate dimensions of 50 mm width and 140 or 250 mm diameter. The ball charge can be up to four 25 mm diameter Cr-hardened steel balls, whose movement is restricted in the vertical plane by the cell walls. Additionally, the motion of the balls is controlled by a magnetic field using Nd-Fe-B type permanent magnets, magnets whose field intensity and direction can be adjusted externally. By adjusting the spatial and/or temporal profiles of the magnetic field, the trajectories, impact energy, and shearing energy can be varied. The three general patterns of ball movement that can be achieved with this device are shown in Fig. 6. In the case illustrated in Fig. 6a, the magnetic field holds the balls mainly at the bottom of the cell which is rotating with the frequency, ω_c. Friction causes the balls to rotate in the same direction with the frequency, $\omega_b = \omega_c R/r$, where R is the radius of the cell and r is the radius of the ball. Periodically, the outer ball on the right-hand side is released, completes most of the circle while being pushed against the cell wall by the centrifugal force, and then hits the left most ball at the bottom of the mill. In this mode of operation the powder is worked both by impact and by shearing. As is indicated in Fig. 6b, the balls may be confined to the bottom part of the cell all of the time, either by increasing the intensity of the magnetic field, or by decreasing the frequency of rotation of the cell. In this mode, the balls both rotate and oscillate around the equilibrium position at the bottom of the cell, and the powder is fragmented mostly by shearing. In the third case, as illustrated in Fig. 6c, by decreasing the cell rotation frequency and by relocating the magnet, a situation may be achieved whereby the ball, which is released from the bottom, is not fully pinned to the wall by centrifugal force, and can either hit one of the remaining balls, or the opposite wall of the cell. Finally, it should be noted that, depending upon the distance between the balls and the magnets, the effective mass of the balls can be increased by a factor of ca. 80, from 60 g to 5 kg. Hence, by changing the position of the magnet, the type of energy transferred to the milled material can be varied, yielding a range of energy modes from the same milling assembly. Thus, for the magnet positions, M1, M2, and M3, indicated in Fig. 5, we find that, for M1 the energy transfer is mainly frictional, for M2 the energy transfer is mainly kinetic, and for M3 both types of transfer occur. In this way a high

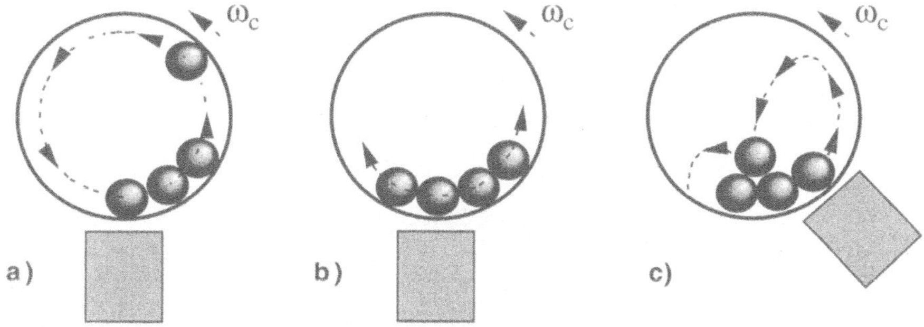

Figure 6. The three main patterns of ball movements achieved with the Uni-Ball-Mill by control of the rotation frequency and the magnet positions. The different energy configurations are discussed in the text.

energy mode can be arranged by choosing magnet positions between M1 and M3 or M1 plus M2.

In summary, a key feature of this milling device is the freedom it provides for studying the same material under a range of energy transfer conditions. The relatively small scale of this device, coupled with the relatively low value of the maximum energy attainable, means that extended milling times may be required to attain the same milled products for a given system. An example is the Fe_3C phase in milled Fe-C mixtures, see Section 4. 2. 1, which can be obtained in shorter times with other milling devices.[74,75] As mentioned in Section 4. 2. 2, Calka and coworkers[76,77] have extended the basic design of this mill by incorporating either heating or cooling of the mill and charge, and by adding a means for monitoring pressure changes in the cell during milling. This allowed Millet et al.[76,78] to investigate the formation of GaN by mechanical alloying through a liquid-gas reaction which required heating the mill to ca. 100 °C, i.e., above the melting point of gallium. Likewise, the study of iron milled in nitrogen or ammonia at different temperatures led to an increased nitrogen content upon milling at ca. 77 K as compared with milling at higher temperatures, see Section 4. 2. 2. Schematic designs of these extensions to the basic design of the Uni-Ball-Mill are shown in Fig. 7.

Recent experiments on milling boron and silicon in an ammonia atmosphere, show that, depending on the specific magnet placement, M1 or M3, different phases can be obtained after annealing the milled products.[79] Milling silicon in ammonia initially produced a nanostructured material and then Si_3N_4, products that were obtained under either magnet placement. However, milling boron in ammonia at low energy, predominately by shearing M1 milling, led to the formation of a boron based nanostructured material which crystallized to a hydride or a nitride upon annealing. In contrast, milling boron in ammonia at high energy, predominately by impact M3 milling, led to the formation of an amorphous material which transformed to polycrystalline boron upon annealing. Similar experiments have been carried[73] out on the Al-N system.

2. 3. 5. Modelling the Main Milling Parameters

Because a final milled product is, in most cases, dependent on the milling conditions, information about the milling parameters is of fundamental importance. Almost all milling studies report the time of mechanical treatment along with some of the milling parameters, such as the vibrational amplitude and frequency for vibratory mills, the velocity of rotation for planetary mills, the ball to powder mass ratio, the number of balls, or the filling volume. All of these parameters directly affect the resulting milled material. In practice, however, it is sometimes difficult to compare results obtained in the study of the same or similar materials in different laboratories because of the different types of mills used.

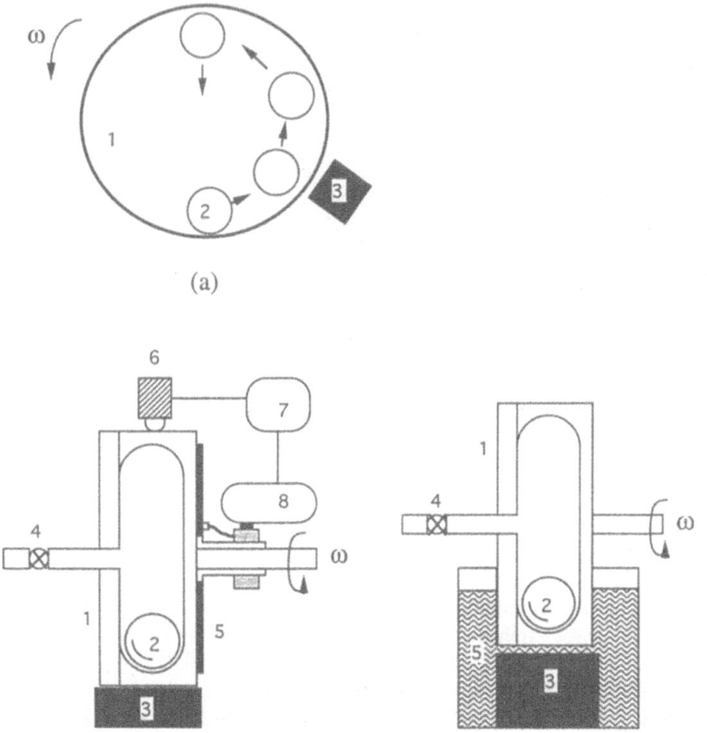

Figure 7. A schematic diagram of the Uni-Ball-Mill, modified to permit either heating or cooling, and monitoring of the cell pressure during milling either (a) at room temperature, 1, the mill, 2, hardened steel balls, 3, magnet, (b) at elevated temperatures, 4, valve, 5, resistance heater, 6, thermocouple, 7, temperature controller, 8, power supply, or (c) at 77 K, 5, liquid nitrogen. Reproduced with permission from Chen et al.[77]

The first paper to address the modelling of the mechanical alloying process was published by Davis et al.[80] The first calculation of the energy transferred to a material in attritor and vibrating mills was reported by Maurice and Courtney.[62] Their calculations considered the impact strains and temperature increases per impact for pure aluminum, copper, iron, and niobium. Further impact modelling studies were concerned mainly with temperature effects during the collisions. Bhattacharya and Arzt,[81] on the basis of an approximation to Hertz's theory of a perfectly elastic impact,[82] calculated the upper limit for the impact temperature. The contact temperature, T_c, of 1260 °C, which they calculated for niobium powder is substantially higher than the values of the bulk temperature increase obtained under similar conditions for an impact velocity of 6 m/s.[62] This difference is readily understood, however, as the high temperature is localized only in areas close to the two contacting surfaces. Similar calculations of the contact temperature for small NaCl crystals of approximately $2.45 \times 2.45 \times 1.5$ mm^3, gave a T_c of 670 °C, a result which should be compared with the measured[81] value of 550 °C. However, the local contact temperature is only one of the many parameters which are required to fully describe the milling conditions.

Table 2. Main Parameters for Analysis of Mills

Parameters		Symbol	Units	Relationships
Mill	Amplitude of vibration	A	mm	
	Time of milling	t	s	
	Frequency of vibration	ν	s^{-1}	
	Ball mass	m_b	g	
	Powder mass	M	g	
	Number of balls	n_b		
Average Energy	Average mill energy intensity	I	W	
	Average transferred energy dose	D	J	$D = I\,t$
	Specific average transferred energy dose	D_M	J/g	$D_M = D/M$
	Specific mill energy intensity	I_M	W/g	$I_M = I/M$
Ball Collisions	Collision frequency	N	s^{-1}	
	Collision duration	τ^*	s	
	Time to second collision with mass m	τ	s	
	Powder mass under one collision	m	g	$m = M/(N\tau)$
Local Energy	Local single collision transferred dose	D^*	J	
	Local energy intensity	I^*	W	$I^* = D^*/\tau^*$
	Specific local dose	D^*_m	J/g	$D^*_m = D^*/m$
	Specific local energy intensity	I^*_m	W/g	$I^*_m = I^*/m$

A summary of the main parameters affecting the mechanical treatment of powders has been developed by Streletskii.[63] As indicated in Table 2, which summarises the important parameters for vibratory and planetary mills, the main features used to describe the milling conditions, in particular the parameters describing the average energy and the ball collisions, can be obtained directly from simple calculations. However, it should be noted that the local energy distribution parameters are also dependent on the type and structure of the material being milled. Saturation like effects, such as particle size or structure stabilization, are often observed on prolonged milling because the energy is no longer absorbed by the fine powder grains. In addition, the physical properties of the milled material evolve during the milling process and the final amount of energy transferred locally to the powder grains will be a function of the milling time. For very fine powders, the energy transferred to the material is minimal, and this observation was confirmed theoretically by Maurice and Courtney.[62] Their calculations indicate that the plastic work consumes only a small fraction of the available elastic energy.

As mentioned above, several theoretical studies[58,59,62,65,69] have followed from the extensive experimental research carried out during the earlier years. Such studies are directed towards both determining how the media dynamics are altered when different grinding balls or mills are used, and modelling the mechanical alloying process itself. Given the complex and multifaceted nature of the mechanical alloying process, it is important to remember that no single model can provide a complete and precise description of the behavior of a system with such diverse elements involving mechanical motion, heat transfer, thermodynamics, and kinetics. A broad overview of the recent developments in modelling the mechanical alloying process has been published by Maurice and Courtney.[58] These authors have extended the scope of the subject by developing computer modelling programs, programs which incorporate the physics of powder deformation, fragmentation, and coalescence in the process.[83,84]

A more detailed perspective on the changes which occur at the local level and the evolution of powder microstructure, particularly the atomic structures of grain boundaries and their effect on the properties of nanocrystalline materials, has been published by Phillpot et al.[85] The physical properties of porous solids are usually correlated with the volume fraction of the pore space, i.e., the porosity, of the solid. The properties of porous solids can be approximated by the simple relation, $\sigma/\sigma_s = C\phi^n = C(\rho/\rho_s)^n$, where σ is the physical property of interest, ϕ is the porosity, ρ is the density, n and C are phenomenological correlation coefficients, and the subscript s designates a property of the solid material.[86] In order to avoid the well-known experimental difficulties associated with the elimination of pores during the preparation of fully dense materials, all structural and physical property oriented models assume a fully dense and "relaxed ground state" material. Recent analyses of the macroscopic properties of model disordered materials have resulted in excellent agreement with experimentally measured properties.[87] Nonetheless, it is clear that a full description of the processes occuring during mechanochemical treatments, and their link with the complex structures of the milled products, remains as a formidable analytical challenge.

2. 4. Characterization of Milled Powders - Experimental Techniques

In essence most, if not all, of the mainstream experimental techniques available in condensed matter physics and materials science have been used to examine milled samples. These include x-ray and neutron diffraction, transmission electron microscopy, scanning electron microscopy, elemental analysis by energy dispersive x-ray analysis, atomic absorption spectroscopy, Rutherford backscattering, differential scanning calorimetry, differential thermal analysis, and thermogravimetric analysis. Additional methods of sample characterization are covered by microhardness, powder specific surface, shape, and size analysis. Standard a.c. magnetic susceptibility or d.c. magnetization measurements, as a function of magnetic field and temperature, are also frequently used. In general, several experimental techniques should be combined in order to fully characterize the changes which occur in a material during milling. Indeed, because of the ultra fine scale of the materials, all these techniques are useful and necessary in order to fully understand their structural features. Although limited to the Mössbauer effect isotopes, such as iron-57 and tin-119, the importance of these elements, particularly iron, means that Mössbauer spectroscopy, with its ability to provide detailed information on the local atomic environment in a fairly routine and straightforward manner, is the microscopic technique most widely used in the study of milled materials. As mentioned in Section 1. 2, an overview of the study of nanocrystalline metals and alloys by nuclear spectroscopic techniques, including perturbed angular correlation and nuclear magnetic resonance, has been published recently.[37] Nonetheless, there is a need for further work in order to obtain microchemical analysis on the requisite fine scale. Certainly atom-probe field-ion microscopy and scanning tunnelling microscopy offer the required lateral resolution for such chemical mapping.

3. MATERIALS PROCESSING

An impression of the broad range of materials which can be synthesized and treated by mechanical grinding, mechanical alloying, and mechanochemistry is presented in Fig. 8. The approximate ranges of particle sizes resulting from these processes are also shown, along with the main regions of structural and chemical changes which occur under the conditions indicated, see also Fig. 3. The relatively simple grinding process is usually used for particle size reduction, whereas the more refined techniques of mechanical alloying and mechanochemistry are used to produce new structural states and new reactions.

3. 1. Amorphization

The first amorphous alloys were formed by vapor condensation of metals on a cooled substrate.[88] Solidification by quenching rapidly from the melt was introduced a few years

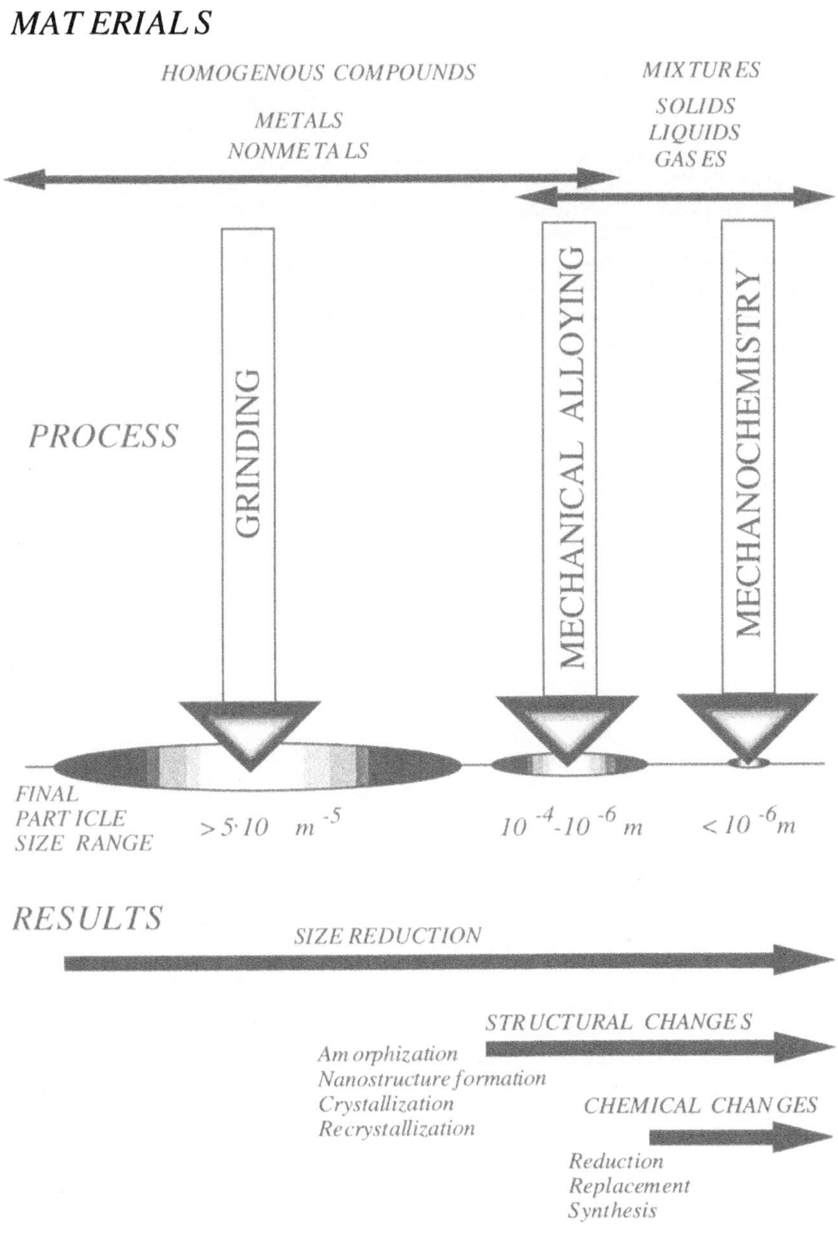

Figure 8. A schematic impression of the range and type of materials which can be treated and synthesized by mechanical grinding, mechanical alloying, and mechanochemistry. The approximate ranges of particle sizes and the structural and chemical states and changes of the milled products are also shown.

later.[89] Amorphous alloys attracted considerable technical interest from the very outset, especially for industrial applications in high-strength materials. Early studies demonstrated that mechanical deformation can induce amorphization.[90,91] Schultz[92] concluded that amorphization by mechanical alloying is based on a solid state reaction and, neglecting the existence of intermetallic phases, occurs under a metastable thermodynamic equilibrium. The formation of an alloy by mechanical treatment is believed to involve the same interdiffusion reaction as that occuring in nanocrystalline materials.[93] The first qualitative description of the mechanical alloying process was based on the transfer of energy during ball-powder-ball or ball-powder-wall impacts.[1,2,94] Many industrial methods were proposed by White[95] and later developed by Koch and coworkers[96] for producing bulk quantities of amorphous alloys and intermetallics which could not be produced by rapid quench liquid phase technology.[7,97] As outlined recently by Yavari et al.,[28] developments in the mechanical treatment of materials have permitted the production of alloys of immiscible elements.

The formation of amorphous alloys by solid state mechanical alloying and milling has been described fully by Weeber and Bakker,[51] Schwarz,[52] Koch,[54] and Enzo et al.[55] The two main categories of amorphization by ball milling are, first, the mechanical alloying of elemental or dissimilar alloy powders involving material transfer between the components and, second, the mechanical milling or grinding of a single component, e.g., an intermetallic compound, such that no material transfer occurs. Mechanical alloying and mechanical milling can lead, respectively, to the vitrification of elemental crystalline metallic powder mixtures and intermetallic compounds. Intermixing of the different atomic species occurs on an atomic scale with a large negative heat of mixing,[56] and this can lead to the destabilization and amorphization of the crystalline lattice.[98,99] In both mechanical alloying and mechanical milling, the Gibbs free energy of the crystal has to be increased above that of the amorphous phase. Details of the thermodynamic transformations and the processes occuring during amorphization by mechanical treatment are presented in a number of reviews.[52,54]

3. 2. Formation of Nanostructures

The first systematic investigations of nanocrystalline or nanostructured materials were carried out by Gleiter.[100,101] In their initial studies, disorder was introduced heterogeneously into a crystal by incorporating a high density of lattice defects, such as dislocations or grain boundaries.[100] However, later nanostructured metals and alloys have been made primarily by evaporation of the metals in a high purity inert gas, followed by condensation onto a liquid nitrogen cooled cold finger, and then compaction in an ultra-high vacuum. Other preparations[12,101] include gas-phase synthesis, laser ablation, thermal pyrolysis, chemical condensed-phase synthesis, and high-energy ball milling. Composites of amorphous and nanocrystalline phases can have superior properties when compared with those of the amorphous material.[57] The slow crystallization of metallic glasses on heating is a well established preparation for nanostructured materials with excellent soft[102] and hard[103] magnetic properties. An effective method for producing exciting new partially crystalline, partially amorphous materials, directly from the melt, has recently been proposed.[104]

In the case of ball milling, the intense cold working imposed on the metallic or ceramic powder decreases the thermodynamic stability of the material by increasing the number of extended imperfections, such as dislocations and interfaces. The grain size reduction results in a microstructure which is similar to that found for nanocrystalline metals produced by gas condensation.[105] Materials milled for a prolonged period consist of small crystals separated by high-energy grain boundaries in which the atomic arrangement lacks any long or short range order. The enthalpy stored through the extensive ball milling is considerably higher than for other known cold working methods. In the latter case, the enthalpy stored during severe cold working of metals and alloys rarely exceeds 2 kJ/mol and is only a small fraction of the heat of fusion,[106] ΔH_f. However, for ball milling the enthalpy is considerably larger and can reach values typical of the crystallization enthalpies[106] of a metallic glass, i.e., of the order of 0.4

times the heat of fusion. Such enthalpy storage, in the form of dislocations and grain boundaries, is not achieved in conventionally processed materials, and the grain boundary energy of ball milled nanocrystalline solids seems to be considerably larger than the grain boundary energy of fully equilibrated grain boundaries.[107]

Grain size reduction generally takes place as follows. Depending on the crystal structure of the material, plastic deformation proceeds by slip and twinning at low and moderate strain levels, whereas at high strain levels, the deformation of shear bands, which consist of dense networks of dislocations, becomes the dominant deformation mechanism. The shear bands in which the deformation is localized, have a typical width of ca. 0.1 to 1 μm. The local shear instability of a crystal lattice may be triggered by material inhomogeneities and enhanced by thermoplastic instabilities, instabilities which result from a nonuniform heat transfer during the deformation process. In the early stages of ball milling, the average atomic level strain increases due to the increasing dislocation density. At a certain dislocation density within these heavily strained regions, the crystal disintegrates into subgrains which are initially separated by low-angle grain boundaries with offset angles[108] of less than 20°. The formation of subgrain structures results in a decrease in the atomic level strain. During further processing, deformation occurs in shear bands located in previously unstrained parts of the material, such that further subgrain size reduction occurs, and the relative orientation of the final grains with respect to each other ultimately becomes random. The nanocrystalline grains themselves are relatively dislocation free.[109]

The basic ideas of nanostructured materials apply to many different types of solids, metals, alloys, semiconductors, high T_C superconductors, diluted and concentrated magnetic materials, as well as mixed carbide, nitride, or oxide based ceramics. These nanostructured materials incorporate many defects, such that the fraction of atoms situated in the cores of these defects become comparable to the fraction of atoms in the residual crystals. Depending on which type of defect, e.g., dislocations, grain boundaries, or interphase interfaces, is incorporated, different kinds of nanostructured materials, with different physical properties, may result.[57,110]

3. 3. Reactive Ball Milling - Mechanochemistry

Mechanochemistry is an exciting and rapidly developing area of materials science. The topic of mechanochemistry and the mechanical activation of solids has been reviewed recently by Boldyrev.[60] In mechanochemistry, interest is focused not just on the physical mass transport or diffusion through defects, or on phase transformations, but also on the stoichiometry of the milled product. Indeed, over the past decade, the increased application of milling techniques in the preparation of materials has drawn attention to the chemical reactions which may occur as a part of the milling process. As outlined by Boldyrev,[60] many different chemical reactions must be taken into account when the milling of solids is carried out either in a liquid environment or when the starting material comprises a mixture of elements and/or compounds.

The temperature increase which accompanies the mechanical treatment of solids forms the basis of a theory of combustion, and such theories for exothermic reactions are useful in understanding the synthesis of metal carbides.[111] A further approach is based on the changes expected in the reactivity of solids due to mechanical treatment as a result of the increase in the specific surface area of the solid. In support of this approach, it is believed that extended chemisorption and a lowering of the surface potential are among the main requirements for a faster reaction.[60] Such effects are particularly important for the mechanical treatment of solids dispersed in liquids,[112] or reactive gases.[113] The formation of new surfaces during mechanical treatment is believed to be one of the main sources of the reactivity. However, the grinding experiments of Schrader and Hoffmann,[114] demonstrated that a correlation does not always exist between the reactivity and the increase in the surface area. Indeed, our recent findings[115,116] on the mechanochemical reduction of haematite show a similar trend in that the rate at which haematite is transformed to magnetite decreases as the particle size decreases, see Section 4. 3. 1. In addition, the effect that the crystal defects have on the reactivity has been considered,[114] and accounts for only a small part of the overall increase in reactivity.[117] Also it has been found that many processed materials have particular or characteristic chemical reactions.[11,118] Overall it is clear that the reactions which take place under mechanical treatment are complex, although three main categories, as outlined below, can be considered, solid-solid, solid-liquid, or solid-gas reactions.

3. 3. 1. Solid-solid Reactions

The introduction of a large number of crystal defects upon mechanical treatment of a mixture of powders provides the conditions required for the promulgation of solid state reactions, reactions which lead to mechanical alloying of the two solids.[111,119] As discussed by Johnson,[107] in this kind of reaction the new product is obtained from the low temperature region of a phase diagram, i.e., the solid state, rather than from the high temperature region, i.e., the liquid state. An intermediate amorphous phase is often found on milling, so that additional thermal treatment may be required to produce the required stable phase.[120] In addition, the type of mill and the associated milling energy also affect the phases which can be produced, as has been demonstrated for the preparation of $MoSi_2$ by Lee et al.[121] The formation of solid phases from metal-metal and metal-metalloid mixtures is well described by intergrain diffusion.[122,123]

As is well known, the mechanical treatment of mixtures of metals can generate a range of morphological, structural, or stoichiometric changes.[124] In the case of intermetallics, the physical processes connected with the energy stored in the intergrain boundaries are the main contribution to the structural changes, and amorphous, nanocrystalline, quasicrystalline, or new stable crystalline phases can be obtained.[62,125] If a chemically reactive element or compound is present in the starting mixture, then chemical reactions as well as structural changes can take place. These chemical reactions are mainly chemical reduction or chemical synthesis. Phase separation can occur in very stable chemical compounds, e.g., complex oxides, as a result of mechanical activation.[126] This type of processing has been studied in detail for simple iron oxides by Matteazzi and Le Caër,[127] who observed a partial reduction of haematite on milling with graphite. Chemical synthesis activated by mechanical treatment has been reviewed by Lin and Nadiv,[11] Kosmac and Courtney,[118] and, more recently, by Uchrin et al.[67]

Mechanochemical treatment has already been used to prepare a wide range of new materials and compounds, including borides, carbides, nitrides, silicides, and structured nanocomposite materials. This method clearly offers a new perspective for the synthesis of inorganic materials. Full details about materials synthesis by mechanical treatment are given in the reviews by Fecht[128] and Matteazzi et al.[5,129] Several examples of the solid-solid reactions which take place as a result of milling treatment, and the insight provided by Mössbauer spectroscopy, are presented in this chapter. In addition to the introductory example of the Cu-Fe system presented in Section 1. 2, we also consider, in Section 4, Cr-Fe, W(Fe-57), Co-Fe-Si-B, alumina-tin nanocomposites, and the reduction of $FeCl_3$, Nd-Fe-B, Fe-C, and iron milled with pyrazine, $C_4H_4N_2$. A concluding example of milled W-C powders is given in Section 5.

3. 3. 2. Solid-liquid Reactions

The concept of wet milling is well established in ball milling technology,[3] particulary in industrial applications in which nominally non-reactive liquids are used during the powder processing to aid in particle size reduction. Many different types of simple or complex liquids have been used for powder particle dispersion. In addition, many different anti-agglomeration agents, which prevent the build up of multiparticle aggregates and prevent the powders from sticking to the mill surface, have been used. As a consequence, wet milling generally results in a more rapid decrease in particle size than does dry milling. In addition, a more uniform distribution of particles sizes is observed for many ceramic or magnetic powders. In general, for short milling times and low milling energies, the influence of the dispersing liquid used during milling can be neglected. However, for long milling times and/or high milling energies, the liquid-solid interaction can play an important role in determining the stoichiometry and structure of the final product. Indeed, nominally non-reactive liquids can become reactive under the continuous energy transfer taking place in a mill. As a result of physical and/or chemical adsorption, there is direct contact between the molecules of the liquid and the atoms on the surface of the milled particles, and thus different types of chemical reactions can be expected.

Milling materials with surface active substances, or surfactants, is becoming increasingly important in mechanochemistry. The main chemical reaction occurs as a result of the decomposition of the liquid molecules and a subsequent interaction with the solid material due to the milling action. Two main types of surface active substances, either water based or organic solvent based, are used in wet milling. With water hydrolysis takes place and, depending on the surface electrostatic potential and pH, cations or anions can be adsorbed onto the surface of the grains.[130] With organic solvents the situation is more complex as their decomposition depends on their bond energies as well as on the chemical and structural properties of the solid being milled. Liquid hydrocarbons consisting of linear chain or ring containing molecules, are used as dispersing agents. For example, for niobium powder milled in hexane or methanol, the formation of a carbide phase was detected by x-ray diffraction and chemical analysis,[112] and the carbon content was found to increase linearly with milling time. In addition, the presence of hydrogen in the solvents is expected to influence the morphology and properties of the final product. Furthermore, for both types of liquids, continuous mechanical treatment provides autocatalytic processes by which a fresh surface can supply the catalyst for continuous adsorption, diffusion, and chemical bonding.

Kaczmarek and Ninham[131] have recently carried out a systematic investigation of the wet milling of barium ferrite with cationic and anionic surfactants. An indication of the extra advantages which surfactants can bring to the mechanical treatment of materials, through changing the bulk and surface properties of the milled powders, is shown in Fig. 9. In this figure, the typical experimental situations for two different surfactants, the cationic surfactant, dihexadecyldimethylammonium acetate, DHDAAc, and the anionic surfactant, sodium didodecylsulfosuccinate, SDDS, are depicted, along with barium ferrite powder, $BaFe_{12}O_{19}$, dispersed in water. As is indicated in Fig. 9, the surface potential of the barium ferrite powder is influenced by the pH. For a $pH < 5$, the surface of the powder is positively charged, whereas for a $pH > 5$ the surface is negatively charged. The experimental conditions can clearly be modified by changing the surfactant and the pH of the dispersion medium. Several examples of the influence of a liquid environment on the mechanical treatment of solids are presented in Section 4. 3. These include the reduction of haematite to magnetite by wet milling in a vacuum, and the milling of barium ferrite in the presence of surfactants DHDAAc and SDDS.

3. 3. 3. *Solid-gas Reactions*

Very fine powders have relatively large surface areas and thus are highly reactive not just with oxygen but also with other gases, such as hydrogen or nitrogen. Normally two features are associated with an increase in reactivity, first, a decrease in the particle size and, second, an increase in the temperature. Mechanical treatment of materials provides both of these features through a continuing decrease in the particle size and an increase in the local temperature. As outlined in Section 4. 2, metal nitrides, particulary those with 3d transition metals, have been successfully synthesized by milling the metal powder in both nitrogen[41,132] and ammonia.[113,133,134] Ammonia is the simplest amine and is easily chemisorbed onto a metal surface. Because of the catalytic properties of transition metal surfaces and the continuous mechanical agitation during milling, ammonia decomposes by the Temkin-Pyzhev reaction,[135] or, in simple form, by the reverse Haber-Nernst reaction, $2\ NH_3 \Leftrightarrow N_2 + 3\ H_2$, through a multistep process involving the absorption of the NH_3, NH_2, NH, and N species onto the metal surface. Additional factors, such as pressure or gas volume, influence the final product.[73]

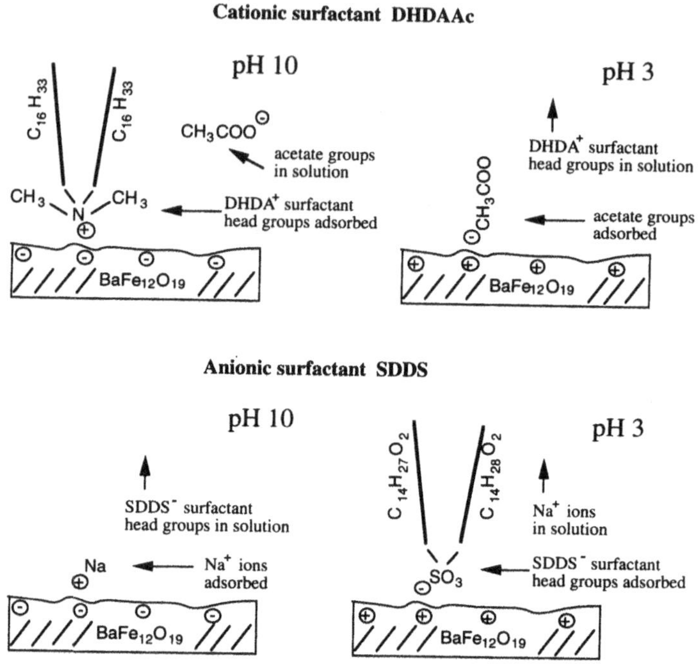

Figure 9. Schematic representation of the interactions between the molecules of cationic and anionic surfactants and the surface of $BaFe_{12}O_{19}$ powder for different pH values of the water dispersion.[131]

4. MÖSSBAUER EFFECT STUDIES

As noted in Section 1. 2, Mössbauer spectroscopy has been used extensively in the study of a range of materials prepared by mechanical attrition and related methods. A consideration of all of these studies is beyond the scope of this chapter. Rather, a selection of examples will be presented with the aim of illustrating the diversity of topics relevant to mechanochemical synthesis which can be examined using Mössbauer spectroscopy. These examples include, the determination of grain boundary widths in nanophase materials, the amorphization of materials, the formation of alumina-tin nanocomposites, the preparation of ultrafine metallic powders by mechanochemical reduction, and the formation of mechanically alloyed rare earth permanent magnet materials. Wear of the mill and balls during milling can contaminate the resulting materials, as is also illustrated in several cases. The mechanical treatment of the technologically important iron carbides, nitrides, and oxides are illustrated in detail in Sections 4. 2 and 3, respectively.

4. 1. Selected Examples

Cr-Fe grain boundaries and Debye temperatures. Fultz et al.[46] have carried out a systematic study of the average widths of the grain boundaries in nanophase alloys prepared by mechanical attrition. Their comprehensive investigation of 18 binary iron alloys revealed features in the Mössbauer spectra which they associated with iron-57 atoms at or near grain boundaries. An outline of the approach which they adopted, based on a detailed analysis of nanometer-scale Cr-Fe crystallites synthesized by mechanical alloying follows.[136,137]

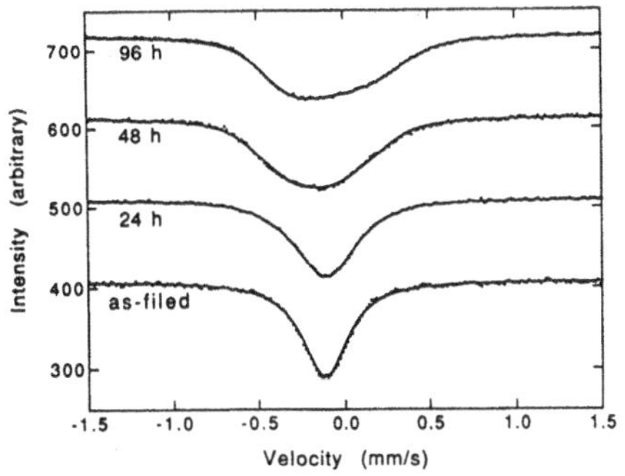

Figure 10. The room temperature Mössbauer spectra of a chromium-18.5 atomic percent iron alloy after ball milling for the times indicated. Reproduced with permission from Kuwano et al.[136]

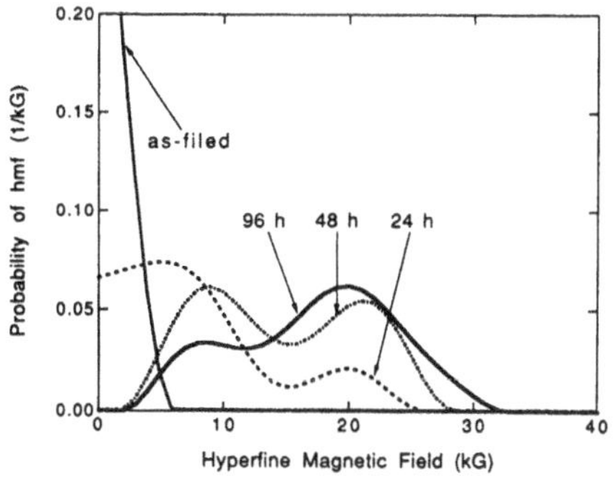

Figure 11. The probability distribution of the magnetic hyperfine fields obtained from the fits to the Cr-18.5 atomic percent Fe spectra of Fig. 10. Reproduced with permission from Kuwano et al.[136]

The key point in the analysis of the room temperature Mössbauer spectra, obtained from the alloy resulting from milling 18.5 atomic percent iron with chromium for various times, see Fig. 10, is the pronounced bimodal distribution of the magnetic hyperfine fields observed in all but the original unmilled alloy, see Fig. 11. The high field component shows a systematic relative increase in its fractional contribution to the spectra with decreasing grain size as the milling proceeds. In contrast, the high field component shows a decrease in its fractional contribution with grain size increase, increase which can be brought about by annealing at 400 °C for successively longer times. Both of these changes provide strong evidence that this high field component originates with iron-57 at the grain boundaries.[136] Fig. 12 shows the volume fraction of atoms at the grain boundaries obtained by Kuwano et al.[136] for a series of Cr-Fe alloys as a function of the crystallite size. The full lines in Fig. 12 represent the results of Monte Carlo simulations used to create a microstructure having a crystallite size distribution qualitatively similar to that of the ball milled powder. The best description of the data was obtained with the '9nn' model, where '1nn' denotes the grain boundary volume fraction obtained by counting the atoms with at least one neighbor in another crystallite, the '2nn' grain boundary volume fractions were obtained by counting all atoms that had either first or second neighbors in another crystallite, and so on. This in turn led to an average grain boundary width of about 1.2 nm, a value which is similar to the 1 nm value obtained by using a simple geometrical model based on spherical particles, but somewhat larger than that reported for nanocrystalline iron and palladium prepared by gas condensation.[136] Overall, a systematic investigation[46] of a set of binary Fe-X nanocrystalline alloys, with grain sizes in the range of 3 to 19 nm, revealed that the smallest grain sizes of 3 to 6 nm were obtained for the bcc Cr-Fe, Mo-Fe, and Fe-Ti alloys and for the fcc Fe-Mn and Fe-Ni alloys. In turn it was shown that these bcc alloys had grain boundaries widths of ca. 1 nm, whereas the fcc alloys had smaller grain widths of ca. 0.5 nm. In addition, the isomer shifts for the iron-57 atoms at the grain boundaries in the Cr-Fe, Fe-Ti, and Fe-Mn alloys were found to be positive with respect to the spectral component of the nanocrystalline grains.[46] This behavior, shown in Fig. 13 for Cr-Fe, is consistent with a reduced density at the grain boundaries, in agreement with the trend exhibited by the Debye temperatures of the crystalline and grain boundary components of Cr-Fe, as will be outlined below.

The separation of the crystalline and grain boundary components in the Mössbauer spectra of Cr-Fe alloys, see Fig. 11, permits the Debye temperatures, θ_D, of the two spectral hyperfine field components in milled Cr-18.5 atomic percent Fe sample to be determined from an analysis of the temperature dependence of their isomer shifts. As shown in Fig. 13, fits based on the usual expression for the second order Doppler shift gave good agreement with the experimental

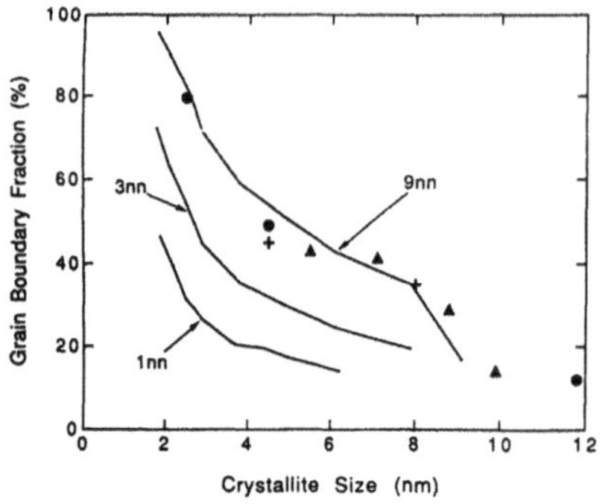

Figure 12. The volume fraction of the atoms at grain boundaries as a function of crystallite sizes for a series of milled Cr-Fe alloys, •, Cr-18.5 at% Fe, +, ball milled Cr-18.5 at% Fe followed by annealing, Δ, chromium powder ball milled to produce Cr-22 at% Fe followed by annealing. The full lines indicate the results of calculations based on a Monte Carlo simulation of a microstructure similar to that of the ball milled powder, as is discussed in the text. Reproduced with permission from Kuwano et al.[136]

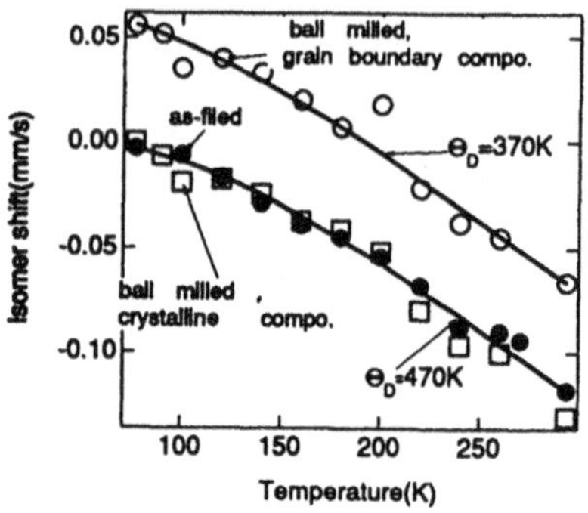

Figure 13. The isomer shifts of the crystalline and grain boundary components of milled Cr-18.5 at% Fe powder as a function of temperature. The behavior of the original as-filed powder is also shown for comparison. Reproduced with permission from Kuwano et al.[137]

data and θ_D values of 470 K for both the original, as-filed, alloy and the crystalline component of the ball milled alloy, and 370 K for the grain boundary component.[137] The former value is in good agreement with the 467 K value of bulk α-iron, and the latter value compares well with the 345 K value obtained for nanocrystalline iron prepared by inert gas condensation followed by high pressure vacuum compaction.[48] In summary, it is clear that Mössbauer spectroscopy can provide new information on the behavior of nanocrystalline alloys. The reduced Debye temperature of 370 K for the grain boundary component is most likely a result of the reduced number of near-neighbor atoms at the grain boundaries.[137] Such behavior is in accord with the reduced atomic density and the corresponding enhancement of the specific heat reported for the interfacial components of nanocrystalline materials.[101]

Contamination. A further point of interest in the study of Cr-Fe alloys is that the iron present in the milled Cr-22 atomic percent Fe powder examined by Kuwano et al.[136] resulted from iron contamination by the steel balls and vial upon milling the initially pure chromium filings for 96 hours. Clearly, in this case the extensive iron contamination was used to advantage. Correspondingly, Kuwano et al.[136] demonstrated that, after milling for 96 hours, a starting alloy of nominal concentration of 18.5 atomic percent iron in chromium had reached an iron concentration of ca. 48 atomic percent.

Grain boundaries in tungsten containing iron-57. A different use of Mössbauer spectroscopy in characterising milled materials, and particularly the grain boundaries, was adopted by Rixecker et al.[49] in their investigation of non-magnetic tungsten milled with ca. 1 atomic percent iron-57 as a dilute Mössbauer spectral probe. This study has the advantage that only local electric hyperfine interactions are present. The study also provides signal enhancement from the grain boundaries because of the preferential segregation of the solute iron-57 atoms into the grain boundaries. A detailed analysis of the distribution of the isomer shifts and quadrupole splittings observed in the Mössbauer spectra of the as-milled and annealed $W_{97.5}(Fe\text{-}57,Co)_{2.5}$ samples revealed components attributable to iron located at substitutional lattice sites, at grain boundaries, and in the vicinity of lattice defects. Based on a simple geometric model of monodispersed spherical particles, a grain boundary thickness of ca. 1.8 nm was obtained for the milled tungsten sample, a thickness which is similar to that reported above for nanocrystalline Cr-Fe.

Amorphization. As outlined in Section 3. 1, the preparation of amorphous materials by mechanical alloying has been extensively investigated and the the main processes which occur during amorphization are now more or less fully understood.[52,54,92,138-140] Indeed, the extent to which mechanical alloying was manifestly successful in the preparation of amorphous materials has contributed significantly to the widespread acceptance of mechanical alloying and related methods as a novel alternative means for preparing a wide range of new materials.

The class of amorphous materials which had previously been prepared through melt spinning or vapor quenching is of particular interest, because, as mentioned above, mechanical alloying permits the preparation of amorphous materials in powder form. The amorphous powder can then be consolidated into the required bulk form, thus extending the potential range of technological uses beyond that of the sometimes restrictive ca. 50 μm thin ribbon form available from melt spinning. The combination of useful magnetic and physical properties exhibited by magnetic amorphous alloys prepared by rapid quenching from the melt, and their potential for technological applications, are well known.[141,142] Amorphous alloys of $Fe_{100-x}B_x$ prepared by liquid quenching,[143] for x values of 11 to 28 atomic percent, or by vapor quenching,[144] for x values of 30 to 90 atomic percent, have been thoroughly investigated. The Fe-B system has also been studied extensively by mechanical alloying techniques, with both nanostructured and amorphous states being obtained, depending on the composition of the starting powder mixtures.[145] Summaries of the Mössbauer spectral studies of mechanically alloyed Fe-B have been published,[10,12] and interest in its solid state amorphization continues.[146] Here we consider the example of amorphous soft ferromagnetic Co-Fe-Si-B alloys which, for selected compositions, exhibit magnetostriction values close to zero and low magnetic hysteresis losses over a wide range of frequencies.

By combining melt spinning of amorphous Co-Fe-Si-B ribbons, followed by their crystallization by heat treatment, and dry[147] or wet[148] ball milling, amorphous Co-Fe-Si-B can be prepared in the powder form. Fig. 14 shows the x-ray diffraction patterns of a melt spun $Co_{70.4}Fe_{4.6}Si_{15}B_{10}$ ribbon and of the same material following a crystallization treatment at 600 °C in vacuum for 10 min and dry milling in helium for 50 hours.[149] The essentially disordered amorphous-like nature of the ball milled sample, as evidenced by the similarity of its x-ray diffraction pattern with that of the melt spun ribbon, is further supported by the similarity of the 4.2 K Mössbauer spectra of the two samples, see Fig. 15. Although the mean magnetic hyper-

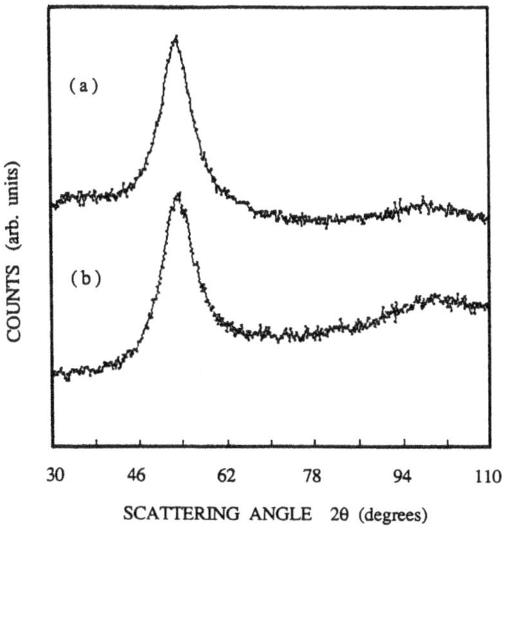

Figure 14. The x-ray diffraction patterns of $Co_{70.4}Fe_{4.6}Si_{15}B_{10}$ as (a) a melt spun ribbon and (b) after ball milling in helium for 50 hours following crystallization of the melt spun ribbon by heat treatment at 600 °C for 10 min.[149]

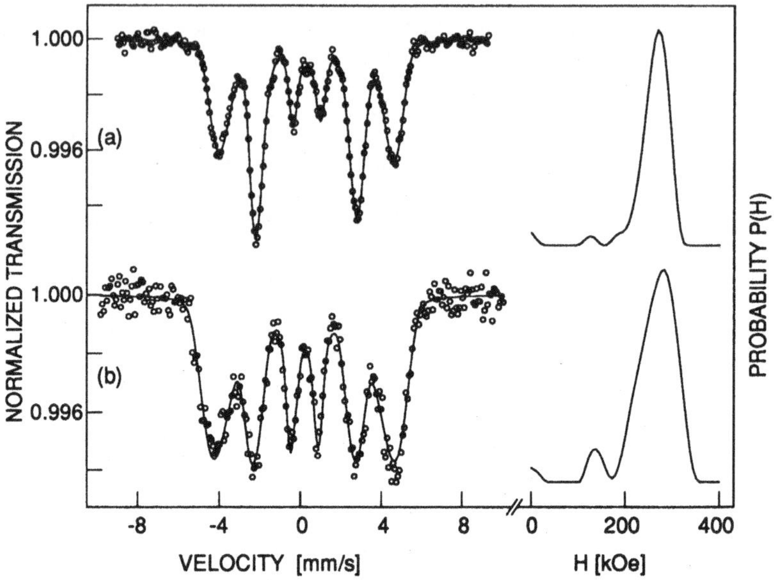

Figure 15. The 4.2 K Mössbauer spectra (a) of melt spun ribbons of $Co_{70.4}Fe_{4.6}Si_{15}B_{10}$, and (b) of a sample ball milled for 50 hours following crystallization of the melt spun ribbon by heat treatment at 600 °C for 10 min. The probability distributions of the magnetic hyperfine fields are also shown.[149]

fine field value is 25.8 T in both spectra, the width of the probability distribution of the fields, $\Delta P(B_{hf})$, increases from ca. 4.2 T in the melt spun ribbon to ca. 5.5 T in the milled sample. This increase indicates a less well defined local atomic environment with increased compositional variations in the ball milled sample as compared with the melt spun sample. This change in the local atomic environment is also indicated by the lower value of the anisotropy field, B_a, of ca. 0.3 T obtained from the fit to the spectra of the ball milled sample as compared with the B_a value of ca. 0.9 T obtained for the melt spun sample.

The investigation of the Co-Fe-Si-B alloys nicely illustrates the importance of the ball-to-powder mass ratio in such studies. For example, Matyja et al.[150] found only a contribution from an amorphous phase in the recrystallized amorphous ribbons of $Co_{78-x}Fe_xSi_{13}B_y$, with $x \leq 8$, which had been milled with a ball to powder mass ratio of 5, whereas ca. 80 percent of an amorphous phase was found[147] upon milling a recrystallized $Co_{70.3}Fe_{4.7}Si_{15}B_{10}$ ribbon for 90 hours with a ball to powder mass ratio of ca. 50. Because the results[149] shown in Figs. 14 and 15, yielded ca. 100 percent of the amorphous phase on milling $Co_{70.3}Fe_{4.7}Si_{15}B_{10}$ for 50 hours with a ball to powder mass ratio of ca. 84 and, by allowing for the different starting compositions of the Co-Fe-Si-B alloys and the different mills, it appears that the increased fraction of the amorphous phase is linked primarily to the increased ball to powder mass ratio.

The effect of wet milling Co-Fe-Si-B in a range of anionic and cationic surfactants has also been examined.[148,151] The main point revealed by comparison of the room temperature and 4.2 K Mössbauer spectra of the resulting products is that relaxation effects occur as a result of milling in the presence of certain surfactants.[151] This is linked with the improved capacity for particle size reduction exhibited by wet milling and is indicated by the 4.2 K spectra of $Co_{70.3}Fe_{4.7}Si_{15}B_{10}$ wet milled in hexane with a series of different surfactants, see Fig. 16. By comparison, the x-ray diffraction patterns of these samples were similar to the pattern of the dry milled sample, see Fig. 14b, a similarity which is consistent with a disordered amorphous-like or nanostructured state.

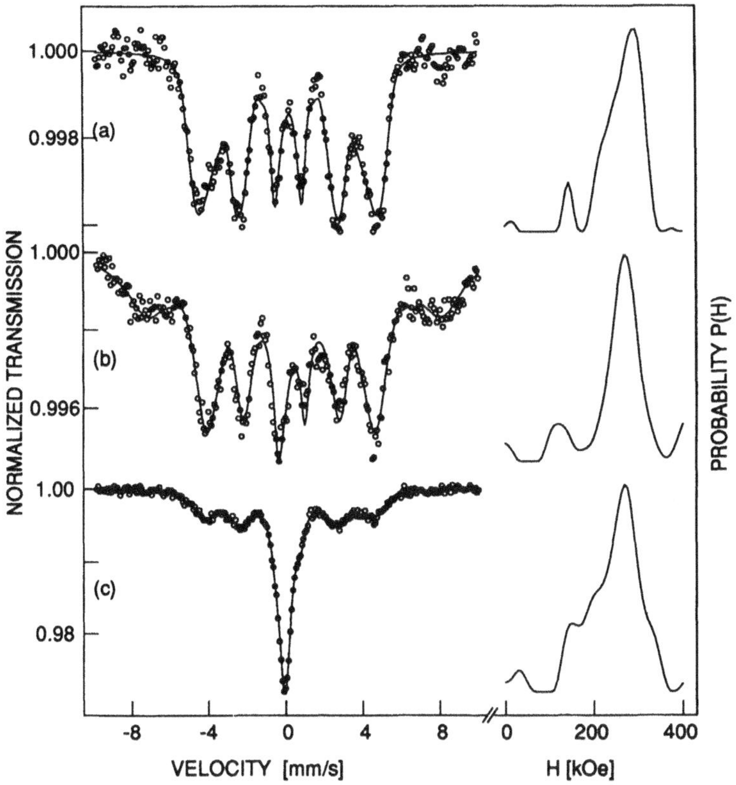

Figure 16. The 4.2 K Mössbauer spectra of $Co_{70.4}Fe_{4.6}Si_{15}B_{10}$ wet milled in hexane with (a) the surfactant, $NaC_{28}H_{28}O_7S$, for 262 hours, (b) the surfactant, $C_{28}H_{59}O_2N$, for 180 hours, and (c) the surfactant, $C_{26}H_{56}NBr$, for 180 hours. The probability distribution curves of the magnetic hyperfine fields represent the behavior of only the disordered phase.[151]

In summary, it is clear that Mössbauer spectral studies of Co-Fe-Si-B and related alloys, combined with x-ray diffraction measurements, clarify the extent to which disordered states are produced by milling. In addition, relaxation effects, consistent with a superparamagnetic-like behavior of the magnetic moments are observed on milling in the presence of certain surfactants. This relaxation behavior is linked primarily to the reduced particle sizes and the improved shape homogeneity obtained for samples wet milled with surfactants. In addition to the care needed to prevent the degradation of the magnetic properties as a result of the particle size reduction, care must also be taken during milling with surfactants to prevent the formation of iron oxides and to avoid other chemical reactions which produce new, unwanted, phases.[151]

Contamination. As an interesting corollary to the above study of amorphous metallic glasses prepared in the powder form by ribbon melt spinning, recrystallization, and then ball milling, Bansal et al.[152] demonstrated that the opposite effect - crystallization of a metallic glass, an Fe-B-Si Allied Metglas, 2605S-2, occurs during ball milling. Given that severe mechanical deformation is expected to increase the chemical and topological disorder in metallic glasses, this crystallization[152,153] is initially surprising. However, in addition to the effects that oxygen uptake during milling may have on increasing the kinetics of the crystallization,[154] Bansal et al.[152] have shown that the observed crystallization is consistent with chemical changes in the powder caused by contamination during milling. This is indicated by the Mössbauer spectra shown in Fig. 17, and the corresponding P(H) curves for the Fe-B-Si metallic glass in various states. The as-received Fe-B-Si ribbon, Fig. 17a, exhibits a broad, featureless, distribution of magnetic hyperfine fields, a distribution which is characteristic of a metallic glass. In contrast, the spectra and related P(H) curves for both the sample milled for 6 hours, Fig. 17b, and for the milled sample after splat-quenching, Fig. 17c, show evidence for both a high field, iron rich, component and a low field, iron poor, component, a component which is consistent with the presence of a small amount of Fe_2B. Given that all of the mechanical history of the milling process, but not the chemical contamination resulting from the milling, is lost as a result of the melt prior to the splat-quench, their spectra and analyses,[152] shown in Fig 17, strongly support the chemical composition data and indicate an increase in iron content with increasing milling time. In addition, the x-ray diffraction patterns of the samples which were first milled and then splat-quenched from the melt, revealed an increasing amount of a crystalline phase in the splat, a phase that increased with milling time.

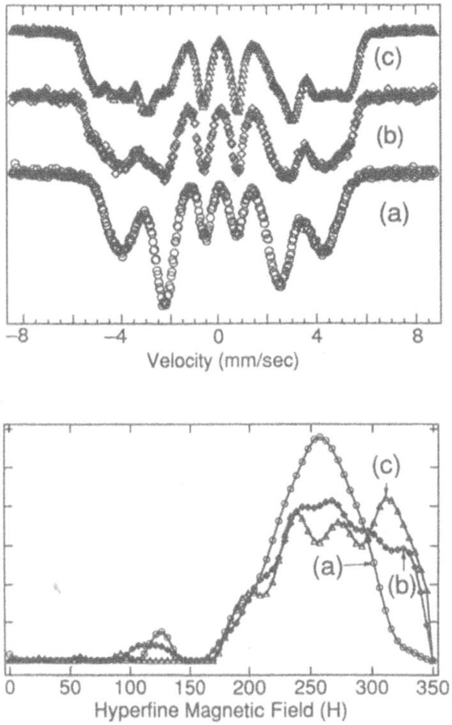

Figure 17. The room temperature Mössbauer spectra and the corresponding magnetic hyperfine field distributions for Fe-B-Si melt spun ribbons (a) as received, (b) after milling for 6 hours, and (c) after milling for 6 hours followed by splat quenching from the melt. Reproduced with permission from Bansal et al.[152]

Tin-119 Mössbauer spectroscopy. After iron-57, tin-119 is the most widely used Mössbauer isotope and ca. 10 percent of the Mössbauer spectral literature reports work with this isotope.[38] Tin-119 has been used very effectively in the study of a number of tin containing systems subjected to milling or mechanochemical treatment.[155-159] Not unexpectedly, the Fe-Sn system has been extensively studied,[39,155,156,159] and various features, such as the extension of the range of solid state solubility, phase formation, and the nanostructured and amorphous states of mechanically alloyed Fe-Sn, have been studied from the perspective of the hyperfine properties of both iron-57 and tin-119. Indeed, Kientz et al.,[159] in their studies of nanocrystalline Fe-Sn solid solutions, have noted that the hyperfine interactions of tin-119 make it particularly well suited for studies of both structural and magnetic disorder at grain boundaries. First, the tin atom does not have a magnetic moment, but the tin nuclei do experience a transferred magnetic hyperfine field from their surrounding magnetic neighbors. Second, the tin-119 magnetic hyperfine field depends sensitively on both the number and the spatial arrangement of the neighboring magnetic atoms.

Mechanosynthesis has been shown to be useful for the large scale production of a wide range of nanophase materials, including metal carbides, nitrides, fluorides, aluminides, silicides, and semiconducting compounds, as well as metal-oxide and metal-sulfide composites, see, for instance, Matteazzi et al.[160] Because of their potential for practical applications as cermets, such nanocomposites as α-Al_2O_3-M, which are produced by the reactive milling of aluminum with metal oxides, M_xO_y, have been investigated either by direct grinding of α-Al_2O_3 and β-tin powder mixtures or by the reactive milling of aluminum and SnO_2 mixtures.[157] The series of tin-119 Mössbauer spectra obtained at 77 K for these two investigations are shown, for various milling times, in Figs. 18a and b, respectively. Analysis of the grinding experiments, shown in Fig. 18a, experiments which were carried out with a starting mixture of two parts of Al_2O_3 powder and three parts of tin powder, indicated three main components, an Sn^{4+} doublet, a β-tin singlet, and an Sn^{2+} doublet. The systematic behavior of the subspectra corresponding to these components with milling time, and particularly the increase to saturation of both the isomer shift and the quadrupole splitting of the Sn^{4+} and Sn^{2+} components, combined with the changes in the x-ray diffraction patterns, indicated that the Sn^{2+} components come mainly from a SnO_x phase, with x of ca. 1.1, a phase which becomes increasingly disordered with increased grinding time. The reactive milling experiments, see Fig. 18b, with mixtures of four parts of aluminum powder and three parts of SnO_2 powder, show a gradual decrease with milling time of the fractional area of the SnO_2 component, which has an isomer shift of ca. 0 mm/s. This change is commensurate with increases in the subspectral areas of both the β-tin singlet, with an isomer shift of ca. 2.5 mm/s, and the doublet, with an isomer shift of ca. 3.0 mm/s and a quadrupole splitting of ca. 2.1 mm/s, associated with SnO_x, where x is again ca. 1.1. The disordered SnO_x phase is observed to form in both preparations and, in the case of the grinding experiments shown in Fig. 18a, the SnO_x is considered to form during grinding at the interface between the alumina and tin grains. Although complementary x-ray diffraction and NMR measurements reveal different structural characteristics for the alumina present in both experiments, tin-119 Mössbauer spectroscopy has provided clear evidence for the formation of alumina-tin nanocomposites. The resulting spectra permit the contributions from the other tin-based phases to be determined.[157]

Mechanochemical reduction: fine powders. The possibility of chemical reduction of metal oxides by mechanical alloying in the presence of a reducing agent has been demonstrated by Schaffer and McCormick[161] who produced copper by milling CuO with calcium. Similarly, metallic zinc was obtained by milling ZnO with calcium, with the β-brass alloy forming upon the coreduction of CuO and ZnO with calcium. This combination of mechanical treatment with oxidation/reduction or with solid state displacement processes, has been applied successfully to the synthesis of a range of materials, including amorphous molybdenum and tungsten, crystalline metals,[162] rare-earth permanent magnet alloys,[163] and ultrafine powders.[164,165]

As outlined by Ding et al.[164] in their recent paper on the mechanochemical synthesis of iron

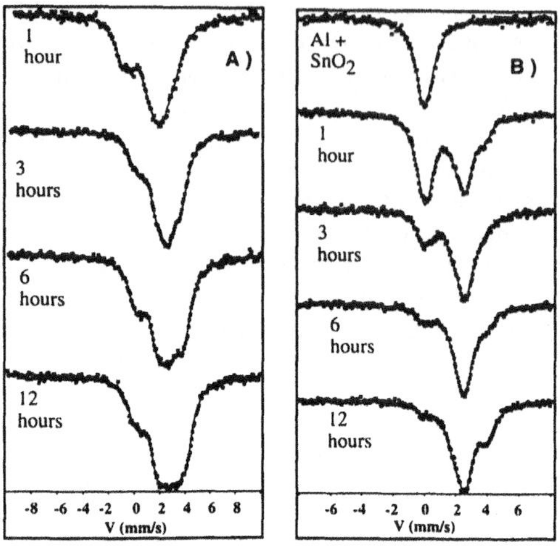

Figure 18. The 77 K tin-119 Mössbauer spectra obtained from (a) milling a starting mixture of two parts of Al_2O_3 powder and three parts of tin powder, subjected to grinding for the times indicated and (b) the reactive milling of a mixture for the times indicated of four parts of aluminum powder and three parts of SnO_2. Reproduced with permission from Le Caër et al.[157]

powder, ultrafine metallic powders have potential applications in a diverse range of fields, including catalysis, magnetic recording, magnetic fluids, and composite materials. In addition this technique offers high yields at relatively low cost as compared with other methods of preparation. By milling anhydrous $FeCl_3$ in the presence of sodium, Ding et al.[164] formed iron powders with average particle sizes, $<\delta>$, of ca. 10 nm, via the solid state displacement reaction, $FeCl_3 + 3\ Na \rightarrow Fe + 3\ NaCl$. Fig. 19 shows the Mössbauer spectra of the as milled powder and the powder after washing with deionized, deoxygenated water and a final rinse with methanol. The spectrum of $FeCl_3$ is also shown in this figure for comparison. In addition to the familiar dominant sextet of the α-iron powder, the spectrum of the as milled powder shows a subspectral component, with a fractional area of ca. 10 percent, resulting from unreacted $FeCl_3$. Both the unreacted $FeCl_3$ and the unreacted sodium were removed by washing. X-ray diffraction studies also confirmed the removal of the NaCl upon washing. Further, an additional spectral doublet was observed in the Mössbauer spectrum of the washed powder, as is shown in Fig. 19. This doublet indicates that ca. 5 to 10 percent of the iron is present as a paramagnetic oxide as a result of the washing process, although this additional doublet component could also be associated with a fine particle superparamagnetic component, or with nanoclusters of iron.[164]

Magnetization measurements on the cold pressed iron samples discussed above, which have been milled and washed, revealed coercivities in the range of 350 to 550 Oe depending on the milling conditions. These coercive fields are much larger than that of bulk α-iron which has a coercive field of a few Oe. Indeed, they are also larger than the coercivities of mechanically milled iron particles of ca. 1 μm diameter, which comprise nanoscale crystallites. This is because demagnetization can take place more readily at the exchange coupled boundaries in such nanoscale crystallites. In contrast, the reduced value of the magnetization of ca. 168 emu/g, measured for the washed sample at 50 kOe, which is about 20 percent lower than that of bulk α-iron, is likely due to either surface oxidation or to superparamagnetic relaxation, as is observed in the Mössbauer spectrum of the washed sample shown in Fig. 19. Overall it is clear that this milling based method has the potential for large scale production of the ultrafine metallic powders required for the many applications outlined above.

Mechanically alloyed permanent magnets. The announcement in 1983 of the discovery of new high performance permanent magnets, based on Nd-Fe-B, understandably generated great attention. Indeed, interest in the preparation, properties, and applications, of

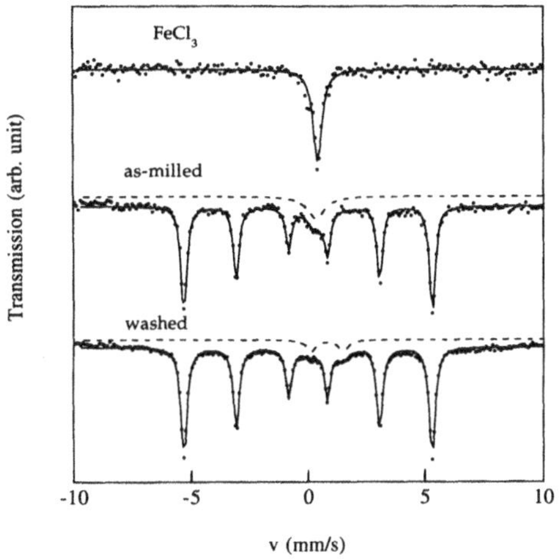

Figure 19. The room temperature Mössbauer spectra of as-milled and washed powder obtained by milling a mixture of $FeCl_3$ and sodium (maximum absorption 10 %). The spectra reveal the predominance of α-iron produced as a result of the milling-induced solid state displacement reaction, $FeCl_3$ + 3 Na → Fe + 3 NaCl. The additional components, indicated by the dashed lines, are discussed in the text. The spectrum of the $FeCl_3$ starting material is shown for comparison. Reproduced with permission from Ding et al.[164]

these and related materials continues[166] unabated to this day. Schultz and coworkers[167-169] have demonstrated the suitability of mechanical alloying as a non-equilibrium, low-temperature processing technique for the preparation of these new permanent magnet materials. The milling first produces a powder with a layered microstructure of iron and neodymium, with the boron atoms being embedded in the Fe-Nd interfaces. Continued milling leads to a refinement of the microstructure with the $Nd_2Fe_{14}B$ hard magnetic phase being formed upon heat treatment of the milled powders.[169] In addition to the continuing strong interest[170-172] in Nd-Fe-B magnets prepared by mechanical alloying or mechanical milling, a range of rare-earth alloys, including $Sm_2Fe_{17}N_{2.6}$, $Sm_2Fe_{17}C_2$, $SmCo_5$, Sm_2Co_{17}, and SmFeTi, have been prepared using these techniques. The reader is referred to McCormick[163] for an overview of this work.

An indication of the phases present in melt-spun and subsequently milled Nd-Fe-B is shown by the series of spectra given in Fig. 20.[171] The Mössbauer spectra of $Nd_2Fe_{14}B$ and related systems have been extensively investigated.[22,173-175] The room temperature spectrum of melt-spun $Nd_{15}Fe_{77}B_8$ shows features[176,177] characteristic of $Nd_2Fe_{14}B$, although a small amount, ca. 7 percent, of α-iron is also discernible in the spectrum. Milling a melt-spun sample in helium for 100 hours results in the spectrum shown in Fig. 20b, a spectrum which is indicative of a multiphase material. Here the $Nd_2Fe_{14}B$ phase forms only ca. 11 percent of the spectral area, whereas ca. 52 percent of the fractional area is α-iron, ca. 34 percent is Fe_2B, with a mean hyperfine field of ca. 23.7 T, and ca. 3 percent is paramagnetic $Nd_{1+\varepsilon}Fe_4B_4$. The result of long term, i.e., ca. 18 months, room temperature aging, see Fig. 20c, is to produce a slight increase, of ca. 3 percent, in the α-iron component and a corresponding decrease in the $Nd_2Fe_{14}B$ component. For this aged sample, see Fig. 20c, the main effect of vacuum heat treatment at 600 °C for one hour is to increase the α-iron fraction, from ca. 55 to 69 percent, at the expense of the $Nd_2Fe_{14}B$ and other phases.

Similar results to those just mentioned have also been obtained by Miao et al.[172] in their comparative study of mechanically alloyed and mechanically milled $Nd_{10}Fe_{84}B_6$. Their Mössbauer spectra and analyses indicated that ca. 85 percent of the iron was in the α-iron phase in the mechanically alloyed powder. In contrast, ca. 51 percent of the iron was in the α-iron phase in the mechanically milled sample. This latter value agrees well with the ca. 52 percent of α-iron observed upon mechanical milling of the melt-spun $Nd_{15}Fe_{77}B_8$ alloy, as is shown in Fig 20b. In an investigation of the effects of heat treatment on the milled samples, Miao et al.[172]

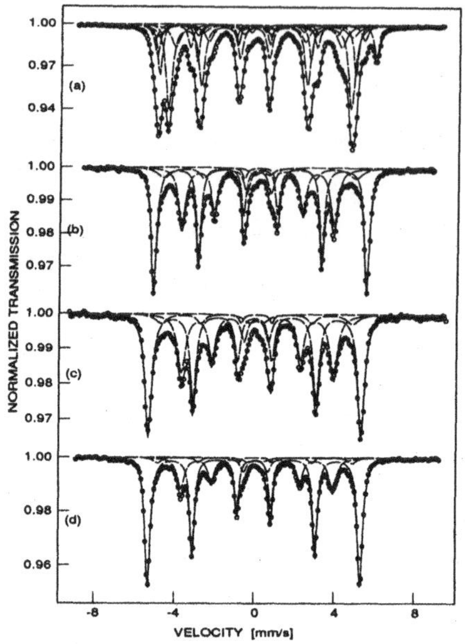

Figure 20. The room temperature Mössbauer spectra of (a) melt spun $Nd_{15}Fe_{77}B_8$, (b) melt spun Nd-Fe-B after ball-milling in helium for 100 hours, (c) the ball-milled state of (b) after room temperature aging for ca. 18 months, and (d) the ball-milled state of (c) after annealing at 600°C in vacuum for one hour.[171]

obtained ca. 22 percent of $Nd_2Fe_{14}B$ upon annealing a mechanically milled sample at 500 °C for 30 min, an amount which increased to ca. 64 percent upon annealing at 850 °C for 30 min. The structural differences between the annealed mechanically milled, and the annealed mechanically alloyed states of $Nd_{10}Fe_{84}B_6$, including the reduced mean grain size and the more uniform grain size distribution of the latter as compared with the former, yields the superior remanent magnetization and coercivity exhibited by the annealed mechanically milled powders. In a related study of mechanically milled $Nd_{2x}Fe_{100-3x}B_x$, where x is 2 to 6, Miao et al.[178] found that the degree of remanence enhancement increased with an increasing fraction of α-iron, whereas the coercivity decreased, as expected for exchange coupled nanocomposite magnets. The optimum structure was found to consist of a relatively uniform mixture of $Nd_2Fe_{14}B$ and α-iron with an average grain size of ca. 20 nm.[178]

4. 2. Carbides and Nitrides

Carbides and nitrides exhibit many useful properties and have many important industrial applications and, as a result, their properties, production, and applications continue to be extensively investigated.[113,179] Because of the potential that milling treatments offer for relatively low cost, large scale production of materials, alloys, and new phases,[2,3,180] mechanosynthesis methods have been used widely in the preparation of a range of carbides[74,181,182] and nitrides.[113,182] Because of the particular importance of the Fe-C system to the steel industry and because of the renewed interest in the magnetic properties of the Fe-N system,[183] both systems have attracted special attention in the preparation of iron carbides and nitrides by mechanochemical means. The equilibrium phases and products of the Fe-C and Fe-N systems have been investigated[184] in detail by Mössbauer spectroscopy with an overview of the nitride phases which are formed when iron is implanted with nitrogen.[185]

4. 2. 1. Iron Carbides

Because steel technology has directed research activity toward the low carbon region of the iron-carbon phase diagram, i.e., below ca. 10 atomic percent carbon, less attention had been

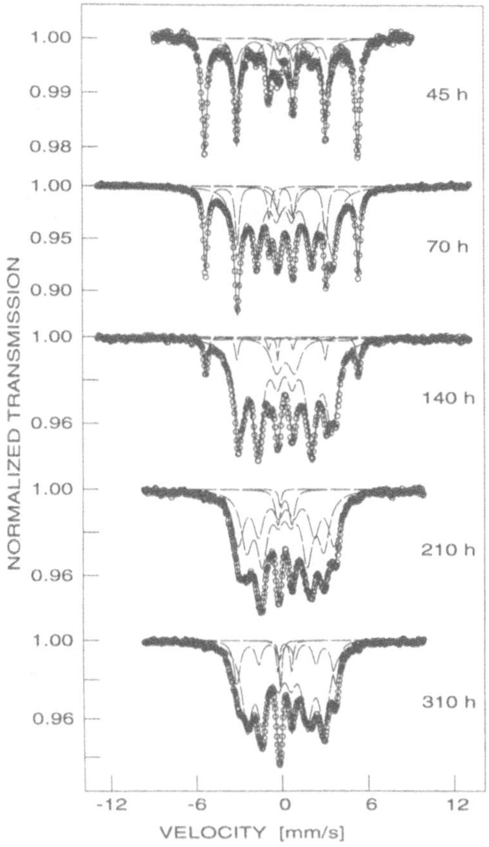

Figure 21. The room temperature Mössbauer spectra of 50:50 atomic percent mixtures of iron and carbon powders milled for the times indicated. The fits to the spectra are discussed in the text.[193]

devoted to the study of high carbon content phases. This is in part due to carbon sublimation and the difficulties[186] in preparing high temperature phases such as Fe_3C and Fe_7C_3. By using ball milling and related methods, however, it has proven relatively straightforward to produce both Fe_3C and Fe_7C_3 albeit, in a metastable, non-equilibrium form.[74,75,186-193] Indeed, a detailed experimental investigation and complementary analysis of the kinetic behavior of the mechanosynthesis of iron carbides of initial composition $Fe_{75}C_{25}$ has only recently been reported.[192]

An indication of the various phases and structural states which are obtained on milling powder mixtures of iron and activated carbon is given by the spectra[193,194] shown in Figs. 21 and 22. Milling a 50:50 atomic percent mixture of iron and carbon powders for 45, 70, and 140 hours produces increasing fractions of ca. 24, 61, and 80 percent of a subspectral sextet associated with the Fe_3C phase, an increase which is accompanied by a corresponding decrease in the area of the α-iron sextet, see Fig. 21. The spectra also reveal features near the center associated with the γ-phase of austenite which has a characteristic singlet and doublet absorption spectrum.[195] The differences in the spectra for the material milled for 140 and 210 hours, however, lead to a fit based primarily on a sextet due to Fe_3C and a second sextet, with a mean isomer shift of ca. 0.2 mm/s and a mean hyperfine field of ca. 17 T. The parameters of the second sextet are similar to those reported for Fe_7C_3 thin films.[196] The fraction of the Fe_7C_3 phase was found to increase from ca. 51 to 74 percent when the milling time was increased

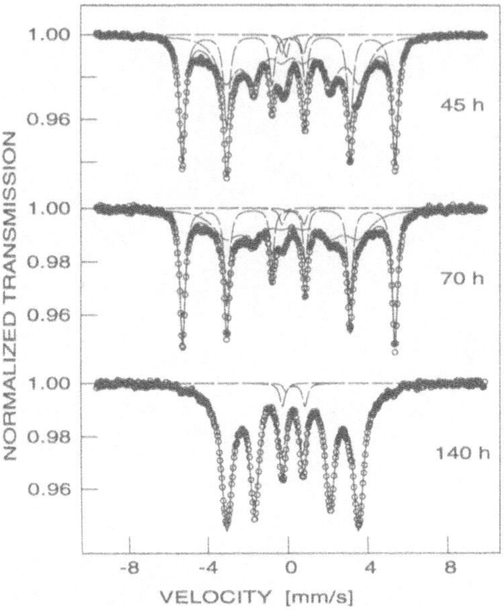

Figure 22. The room temperature Mössbauer spectra of 75:25 atomic percent mixtures of iron and carbon powders milled for the times indicated. The fits to the spectra are discussed in the text.[193, 194]

from 210 to 310 hours and, correspondingly, the fraction of the Fe_3C phase decreased from ca. 42 to 22 percent. Whereas milling a 50:50 atomic percent mixture of iron and graphite powders for 310 hours resulted in a product containing only ca. 74 percent of the Fe_7C_3 phase, appropriate milling and processing conditions can produce a complete reaction of the iron and carbon to form only the orthorhombic Fe_7C_3 phase.[74] By comparison, a 75:25 atomic percent mixture of iron and carbon was found to transform almost entirely to the Fe_3C phase upon milling for 140 hours, see Fig. 22. The trends in the phase changes observed on milling were found to agree with the modelling of the process kinetics reported for powder mixtures with this composition.[192]

A feature of the spectra of the 75:25 atomic percent mixtures of iron and carbon milled for 45 hours, see Fig. 22, is the large linewidths, Γ, of ca. 1.7 mm/s obtained for the broadened sextet whose mean hyperfine parameters are similar to those of Fe_3C. For comparison, a linewidth of ca. 0.7 mm/s is reported for the Fe_3C-type phase in the other milled samples. This behavior agrees with the formation of amorphous Fe_3C reported by Tanaka et al.[186] upon milling 66:34 to 83:17 atomic percent mixtures of iron and carbon and is confirmed by differential scanning calorimetry measurements on the milled samples discussed above.[75,193,194]

Differential scanning calorimetry measurements on the series of 50:50 milled samples, see Fig. 23, show similar overall trends, but with several distinctive features. The pronounced exothermic peak at ca. 700 °C for the material milled for 45 hours, for which the Mössbauer spectrum indicates ca. 24 percent of the Fe_3C phase, is due to the transformation of the remaining milled, but not yet fully reacted, material to the Fe_3C phase.[75] Likewise, the sample milled for 70 hours, which shows ca. 61 percent of Fe_3C, has the same peak at ca. 700 °C but, as expected, of reduced intensity. The absence of the exothermic peak at ca. 700 °C in the sample milled for 140 hours confirms the assignment of the 700 °C peak to the transformation of the milled Fe-C mixture to Fe_3C. This occurs because the sample milled for 140 hours is already predominantly in the Fe_3C state, as is shown by the Mössbauer spectrum in Fig. 21, and thus there is little unreacted material left to transform to Fe_3C. Because the material milled for 210 hours is comprised of ca. 42 percent of Fe_3C and 51 percent of Fe_7C_3, the exothermic

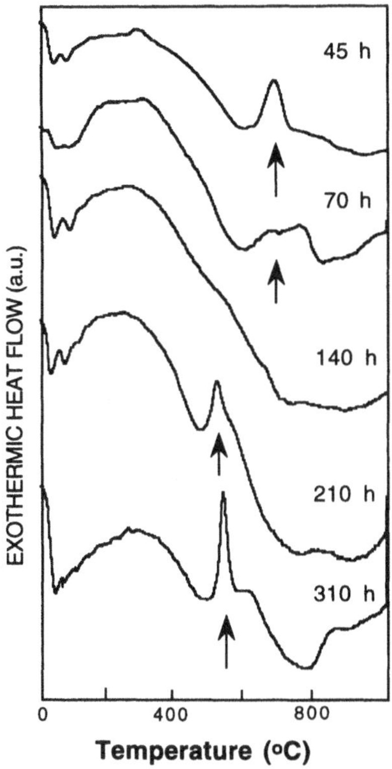

Figure 23. The differential scanning calorimetry measurements of 50:50 atomic percent mixtures of iron and carbon powders milled for the times indicated. The features at ca. 540 and 700 °C are discussed in the text.[193,194]

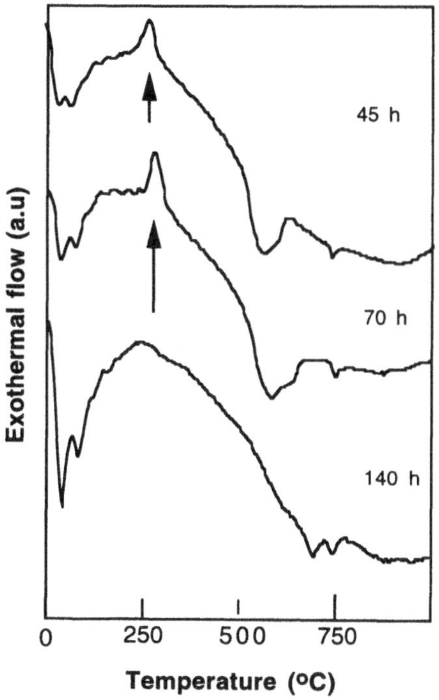

Figure 24. The differential scanning calorimetry measurements of 75:25 atomic percent mixtures of iron and carbon powders milled for the times indicated. The feature at ca. 300 °C is discussed in the text.[193]

peak at ca. 540 °C, combined with the absence of any peak around 700 °C, indicates that the transformation from Fe_7C_3 to Fe_3C takes place at ca. 540 °C. This is confirmed by the differential scanning calorimetry curve obtained for the sample milled for 310 hours, which, with ca. 74 percent of Fe_7C_3, exhibits a more pronounced exothermic peak at ca. 540 °C. As mentioned above, the behavior of the milled 75:25 sample is consistent with the initial formation of an amorphous phase. In agreement with Tanaka et al.,[186] the exothermic peak at ca. 300 °C in the differential scanning calorimetry curves of the 75:25 atomic percent mixtures of iron and carbon powder milled for 45 and 70 hours, see Fig. 24, represents the crystallization of Fe_3C from the amorphous phase. Because this crystallization has occurred at ca. 300 °C, there is only a slight indication of any further crystallization of Fe_3C at 700°C.

The above discussion is confirmed by the differential scanning calorimetry curve obtained for the 75:25 atomic percent mixtures of iron and carbon powders milled for 140 hours, see Fig. 24. The absence of any exothermic features at ca. 300 and 700 °C is consistent with the Mössbauer spectrum shown in Fig. 22, a spectrum which indicates that this material is almost entirely Fe_3C.

In summary, it is clear that Mössbauer spectroscopy, when used with x-ray diffraction and differential scanning calorimetry measurements, has provided details of the milled products in the Fe-C system and has allowed their structural states and transformations to be determined.

4. 2. 2. Iron Nitrides

As mentioned above, in addition to the interest in preparing new iron nitride phases by mechanical treatment, the Fe-N system itself has recently attracted renewed interest because of the high magnetization reported[183,197,198] for the metastable iron nitride, α''-$Fe_{16}N_2$.

A number of different methods have been used to prepare a range of iron nitride phases by milling. El-Eskandarany et al.,[132] for example, obtained Fe_2N in coexistance with α-iron and α'-Fe_4N upon milling iron in a stream of nitrogen. In contrast, supersaturated bcc Fe-N and the high temperature ε-$Fe_{2\text{-}3}N$ phase were obtained by Koyano et al.[133] upon milling iron powder in an ammonia atmosphere. The effects of milling a mixture of α-iron and α'-Fe_4N powders in argon[199] or nitrogen[200] have been investigated, as has the structural transformation of mechanically ground[201] α'-Fe_4N. The milling of mixtures of iron and $Fe_{2\text{-}3}N$ powders has also been studied.[202] Lee et al.[202] have summarized the phases obtained both upon milling and with subsequent heat treatments and found that significant amounts of the strongly magnetic α''-$Fe_{16}N_2$ phase are found in some cases. Chen et al.[77] have introduced a novel development into the synthesis of iron nitrides by investigating the milling of iron powder in nitrogen and ammonia atmospheres at 200 °C, room temperature, and liquid nitrogen temperature, while monitoring the pressure in the milling chamber. Several nitrides, ε-$Fe_{2\text{-}3}N$, α'-Fe_4N, and a supersaturated solid solution of nitrogen in iron, were observed in the resulting x-ray diffraction patterns, with, surprisingly, the highest nitrogen content nitrides resulting at the lowest milling temperature. This is believed to be due to the non-equilibrium nature of the milling process in which additional solute sites at lattice defects, and the reduced atomic motion of nitrogen in iron at low temperature, permit nitrogen contents well above the equilibrium solid solubility of nitrogen in iron. In addition, for the same milling time and temperature, Chen et al.[77] found that nitride powders of higher nitrogen content, ε-$Fe_{2\text{-}3}N$, were formed when ammonia rather than nitrogen was used. This is believed to be due to the catalytic role of the hydrogen in ammonia.

As an extension of the milling of iron powder in nitrogen or ammonia, Kaczmarek and co-workers[203,204] have recently investigated milling iron powder in various amines, such as piperazine, $C_4H_{10}N_2$, pyrazine, $C_4H_4N_2$, and pyrazole, $C_3H_4N_2$. They found that pyrazine and pyrazole were more efficient than piperazine in the preparation of iron nitrides.[203] Figs. 25a-d show the x-ray diffraction patterns of iron powder milled for different times in a vacuum with pyrazine in the stoichiometric ratio of three parts iron to one part nitrogen.[205] The main feature of this work is the emergence of the hexagonal ε-phase, Fe_3N-type structure, with broadened diffraction lines and a relatively high level of rounded background scattering, results which are consistent with the presence of disordered structural and nanostructural inhomogeneities. The Mössbauer spectra of the milled samples, shown in Figs. 26a-d, are generally in good agreement with the x-ray diffraction results. An analysis based on a distribution of magnetic hyperfine fields indicates a predominance of up to 95 percent, depending on the milling time, of ε-Fe_3N_x with an excess of nitrogen, i.e., with $x > 1$. Nonetheless, the broad range of the hyperfine fields is also consistent with small contributions from the nitrogen deficient ε-Fe_3N_y phase with $y < 1$, and with the α'-Fe_4N-type phase, both of which are typified by large $\langle B_{hf} \rangle$ values of ca. 30 to 34 T.[185] The presence of a nitrogen rich phase, tending to the Fe_2N composition, is indicated by the small quadrupole doublet in the fit.[205] In addition to the structural effects noted above, chemical inhomogeneities associated with the milling process, as well as the effects of carbon and hydrogen from the pyrazine, will also contribute to the broad distribution of hyperfine parameters found in the spectra shown in Fig. 26, although, as outlined below, the contribution of the carbon and hydrogen have been clarified by gas chromatography.

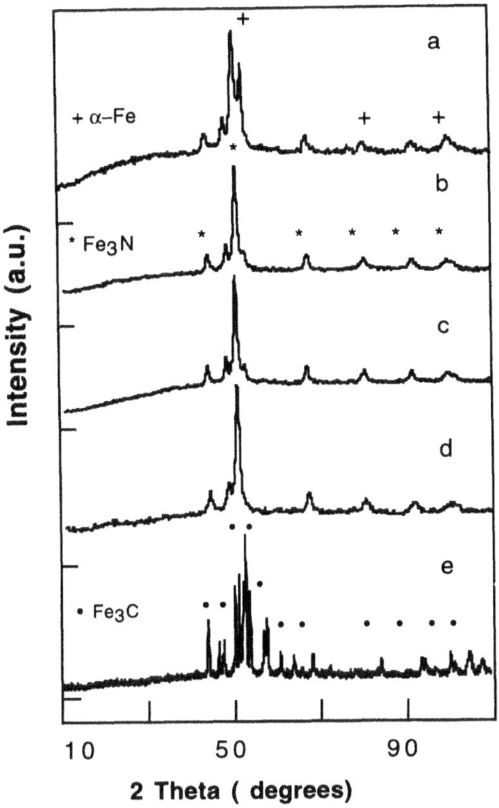

Figure 25. The x-ray diffraction patterns of iron ball-milled with pyrazine in a vacuum for (a) 95 hours, (b) 210 hours, (c) 355 hours, and (d) 500 hours. In (e) the sample milled for 500 hours was subsequently annealed at 550 °C for 30 min.[205]

The weight losses obtained upon heating milled samples, see for example the weight loss[206] of the sample milled for 95 hours in Fig. 27, occur predominantly as a result of the evolution of nitrogen upon the decomposition of ε-Fe_3N_x at ca. 500 °C. This confirms that ε-Fe_3N_x, with <x> values of ca. 1.1 to 1.2, is the main phase present in the milled samples. By comparison, the carbon weight loss was found to remain relatively unchanged upon heating to 650 °C, see Fig. 27. Annealing the sample milled for 500 hours at 550 °C for 30 min, i.e., above the decomposition temperature of ε-Fe_3N_x, revealed clear evidence in both the x-ray diffraction pattern, see Fig. 25e, and the Mössbauer spectrum, see Fig. 26e, for the formation of the Fe_3C phase. This suggests that the carbon remains essentially unchanged during the milling process, probably in a molecular form either on the particle surfaces or in the grains, although some carbon may also be located in interstitial sites in the hexagonal ε-Fe_3N_x structure.[205] Finally, the increased chamber pressure measured during the milling process and the small fraction of ca. 25 percent of hydrogen remaining after milling, combined with the lower energy of the C-H bond as compared with the C-C or C-N bonds, indicates that the hydrogen from pyrazine, $C_4H_4N_2$, is primarily evolved as hydrogen gas during the milling process.

Clearly the mechanical treatment of materials carried out in the presence of heterocyclic compounds, such as pyrazine, $C_4H_4N_2$, and related amines, offers interesting, novel prospects for the mechanochemical synthesis of nitrides and carbides, prospects which should be explored further. However, the additional complexity introduced by the presence of multielement materials may mitigate against the routine use of such milling environments for materials synthesis.

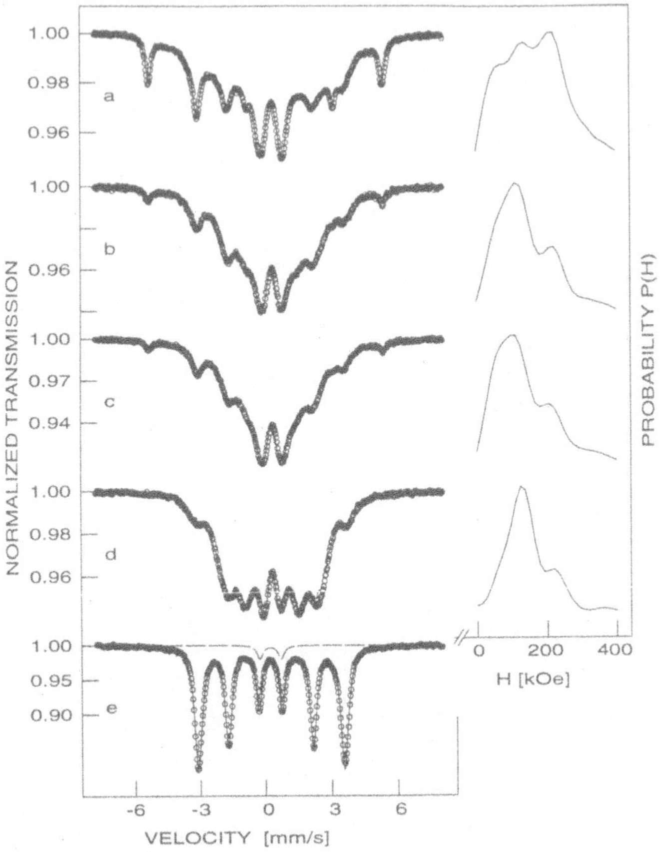

Figure 26. The room temperature Mössbauer spectra of iron ball-milled with pyrazine in a vacuum for (a) 95 hours, (b) 210 hours, (c) 355 hours, and (d) 500 hours. In (e) the sample milled for 500 hours was subsequently annealed at 550 °C for 30 min. The P(H) curves for (a) to (d) represent the behavior of only the distributed component.[205]

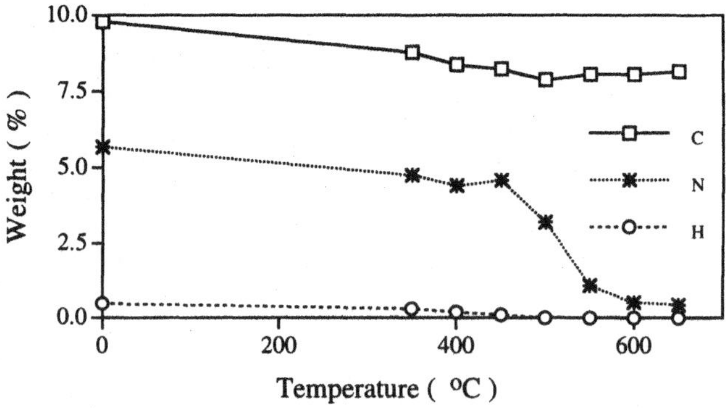

Figure 27. The weight losses for carbon, nitrogen, and hydrogen as determined by gas chromatography carried out on the sample of iron milled in pyrazine for 95 hours.[206]

4. 3. Iron Oxides

Iron oxides are among the most technologically important materials in our society, with many applications, including magnetic separation, magnetic fluid devices, magnetic recording media, and as permanent magnets. Pankhurst and Pollard[207,208] have recently reviewed both the applications and interest in materials such as haematite, α-Fe_2O_3, goethite, α–FeOOH, feroxyhite, δ'-FeOOH, maghemite, γ-Fe_2O_3, magnetite, Fe_3O_4, and barium ferrite, $BaFe_{12}O_{19}$. They have also pointed out the usefulness and relevance of Mössbauer spectroscopy in the study of such magnetic materials. The interested reader should also see Pringle and Long[173] and Bowen et al.[209] for further detailed discussion of the Mössbauer spectral study of such magnetic oxides.

In addition to its traditional use for particle size reduction, ball milling, and related mechanical treatments, have been applied[115,118,127] in recent years to the study of iron oxides, with the overall goal of enhancing their magnetic properties. These interests have been stimulated both by the recent developments in milling techniques, see Section 2, and the potential insights available from milling in the different environments now quite commonly used, see Section 3. 3. Here we consider the effects of milling haematite, α–Fe_2O_3, and barium ferrite, $BaFe_{12}O_{19}$, in a variety of environments. Boldyrev,[60] in his recent review of the mechanochemistry and mechanical activation of solids, has discussed the structural changes which take place upon the mechanical activation of ferrites, spinels, and orthoferrites.

4. 3. 1. Haematite, α–Fe_2O_3

Because of the technological importance of the iron oxides and hydroxides, the preparation and properties of these materials have been thoroughly investigated over the years.[210] A schematic diagram of the iron oxides, and the processes which link and transform them, is given in Fig. 28. In this figure the arrows represent oxidation or reduction reactions.[115,210] The effects of milling magnetite, Fe_3O_4, have been investigated by a number of groups.[6,211-214] Among their interests[212,214] are the reduction of Fe_3O_4 through the displacement reaction, $3\ Fe_3O_4 + 8\ Al \rightarrow 9\ Fe + 4\ Al_2O_3$, and the preparation of magnetic nanocomposite materials.[6,213] The main interest in this section is, however, centred on the fascinating structural changes which take place on milling haematite, changes which can yield a total phase transformation to magnetite under optimal conditions.[115]

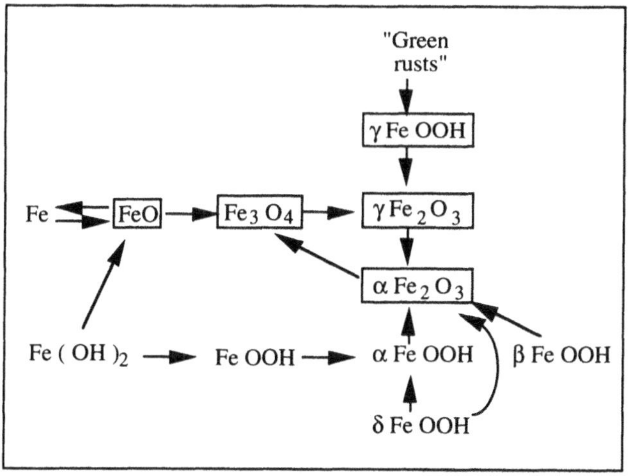

Figure 28. A schematic diagram of the iron oxides. The arrows indicate oxidation or reduction reactions. Reproduced with permission from Burkin[210] as reported by Kaczmarek and Ninham.[115]

An early indication of the structural changes which occur upon ball milling iron oxides was obtained when a mixture of oxides was milled in water and in a variety of polar and non-polar organic media. Uehara[215] found that milling maghemite, $\gamma-Fe_2O_3$, magnetite, Fe_3O_4, and haematite, $\alpha-Fe_2O_3$, in water resulted in a spinel, $\gamma-Fe_2O_3 - Fe_3O_4$, solid solution. More recently Matteazzi and Le Caër,[127,216] in an investigation of the reduction of $\alpha-Fe_2O_3$ by dry ball milling with carbon in an inert atmosphere, obtained nanocrystalline magnetite with an average crystallite size of ca. 11 nm. In addition, Kosmac and Courtney,[118] in their appraisal of the milling and mechanical alloying of inorganic non-metallic systems, noted the influence that iron, resulting from the wear of the grinding media, can have on the phase transformations of haematite during high energy milling. In turn, Kaczmarek and co-workers[115,116] have carried out a systematic study of the mechanochemical treatment of $\alpha-Fe_2O_3$ under a range of environments and milling conditions. Their treatments led to the synthesis of both Fe_3O_4 and $\gamma-Fe_2O_3$ after annealing the milled product at the low temperature of ca. 230 °C, although their main conclusion was that $\alpha-Fe_2O_3$ could be transformed to Fe_3O_4 by the low energy wet milling of haematite in a vacuum. The essence of their comprehensive investigation is shown in Fig. 29 in which I(104), the intensity of the (104) x-ray diffraction peak of $\alpha-Fe_2O_3$, is plotted against milling time for haematite milled in a variety of environments. This graph provides an approximate but convenient measure of the transformation of $\alpha-Fe_2O_3$ to Fe_3O_4 upon milling, with the transformation being found to proceed most readily for wet-milling of $\alpha-Fe_2O_3$ in water in an evacuated vial.[115]

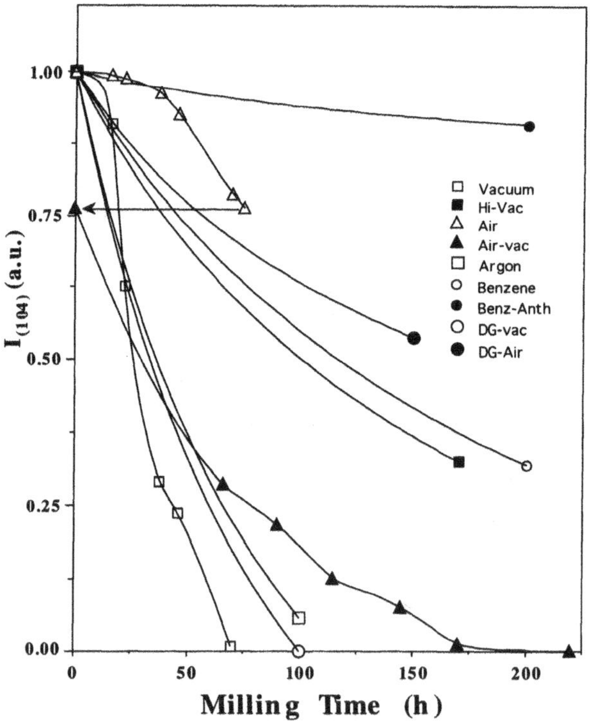

Figure 29. The intensity of the (104) x-ray diffraction peak of $\alpha-Fe_2O_3$ versus milling time for haematite milled in various environments, where Anth is anthracene and DG is diethylene glycol.[115]

Microscopic evidence of the gradual transformation of $\alpha-Fe_2O_3$ to Fe_3O_4 is provided by the room temperature Mössbauer spectra, shown in Fig. 30, of $\alpha-Fe_2O_3$ wet-milled in a vacuum.[217] The fits to the spectra result in a combination of subspectra with mean magnetic hyperfine fields, $<B_{hf}>$, in good agreement with those of $\alpha-Fe_2O_3$ and the tetrahedral A and octahedral B sites of Fe_3O_4, see Fig. 31a. The areas of the subspectral components of the $\alpha-Fe_2O_3$ and Fe_3O_4 phases, see Fig. 31b, show variations in good agreement with the x-ray results presented in Fig. 29. There is a continuous decrease in the amount of $\alpha-Fe_2O_3$ as the transformation to the Fe_3O_4 phase proceeds with increasing milling time. An interesting feature in the Mössbauer spectra of the milled Fe_3O_4 product is the increasingly large linewidth exhibited by the subspectrum of the octahedral B site with increasing milling time, see Fig. 31c. By comparison, the linewidth for the tetrahedral A site remains relatively unchanged at ca. 0.4 mm/s, as does the linewidth of the remaining, unreacted, $\alpha-Fe_2O_3$. This pronounced increase in linewidth for the B site, from ca. 0.4 mm/s after 24 hours to ca. 0.74 mm/s after 144 hours of milling, is due to electron hopping which occurs as a result of the sharing of the octahedral

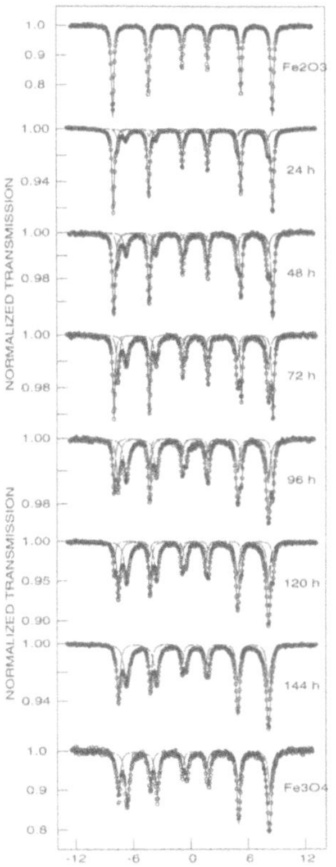

Figure 30. The room temperature Mössbauer spectra of α-Fe_2O_3 wet-milled in a vacuum for the times indicated. The spectra of pure α-Fe_2O_3 and Fe_3O_4 are shown for comparison.[217]

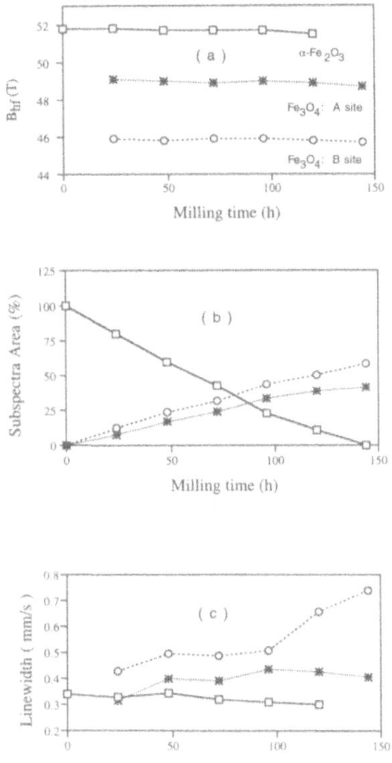

Figure 31. The mean magnetic hyperfine fields, (a), the fractional areas of the sub-spectral components, (b), and the linewidths, (c), resulting from a fit of the Mössbauer spectra of α-Fe_2O_3 measured after milling for various times, where □ is α-Fe_2O_3, ∗ is the A site of Fe_3O_4, and O, is the B site of Fe_3O_4.[217]

sites between Fe^{2+} and Fe^{3+}. The relaxation rate for the hopping increases as the deviation from stoichiometric Fe_3O_4 increases.[218,219] Thus the linewidth, of the octahedral site in the Mössbauer spectrum is correlated with the stoichiometry of milled $Fe_{3-v}O_4$, where v is the vacancy concentration.

Strong supporting evidence for the formation of a vacancy defect $Fe_{3-v}O_4$ product upon milling α–Fe_2O_3 was provided by pulsed neutron scattering measurements carried out on α–Fe_2O_3 subjected to low energy wet milling in a vacuum for about 200 hours.[220] Fig. 32 shows the time of flight neutron diffraction pattern obtained for this sample. The optimal refinement was obtained with ca. 14 percent of the α–Fe_2O_3 phase (*R3c*) and ca. 86 percent of the Fe_3O_4 phase (*Fd3m*). These values agree well with the subspectral component areas determined for these phases from the Mössbauer spectrum of the same sample.[220] The neutron data were analy-

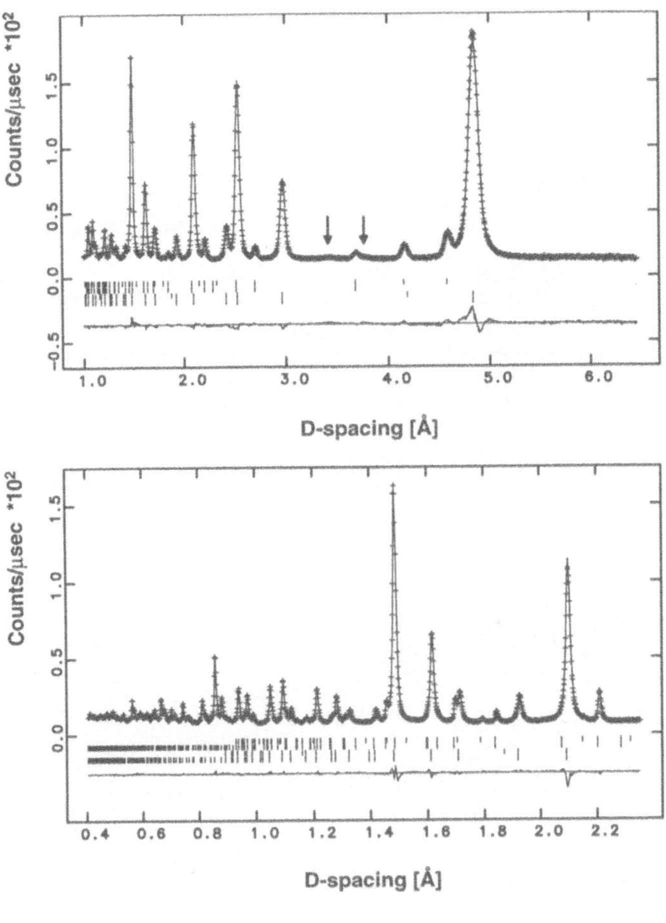

Figure 32. The time of flight neutron diffraction pattern of haematite, α-Fe_2O_3, wet-milled in a vacuum for ca. 200 hours. Top, the low angle detector bank with $\theta = 35°$ and bottom, the backscattering detector bank and $\theta = 145°$. The tick marks are, from top to bottom, magnetic α-Fe_2O_3, α-Fe_2O_3, magnetic Fe_3O_4, and Fe_3O_4. The arrows indicate the diffuse scattering peaks as discussed in the text.[220]

zed using three possible distributions of the vacancies in the magnetite, vacancies restricted to the octahedral sites, vacancies restricted to the tetrahedral sites, or vacancies distributed over both sites. The optimal refinement led to a stoichiometry of $Fe_{2.818}O_4$ for the milled product with v = 0.182 and a vacancy concentration, t = 0.044, for the tetrahedral sites, w/2 = 0.069 for the octahedral sites, where v = t + w, and R_{wp} = 2.50 percent. Such behavior is consistent with the detailed analysis of the applied field Mössbauer spectra measured by Annersten and Hafner,[219] who reported vacancy distributions over both the tetrahedral and octahedral iron sites in off stoichiometric magnetite. In comparison, the refinement for the model with vacancies located solely on the more energetically favorable octahedral sites led to a stoichiometry of $Fe_{2.914}O_4$ with R_{wp} = 2.61 percent. Analysis of the peak broadening and peak shapes led to mean particle sizes, $<\delta>$, of ca. 90 nm with isotropic strain effects present to the extent of ε = 0.43 percent, where ε is the lattice microstress factor, $\varepsilon = \Delta a/a$. These results are consistent with the particle size dimunition and strain present in a mechanochemically treated material. The arrowed features in Fig. 32 highlight two small diffuse scattering peaks, located at interplanar spacings of 3.41 and 3.72 Å, which are evident in the raw data but which are not apparent in the limited space available for Fig. 32. These features correspond well with the superstructure lines observed in some of the series of $Fe_3O_4 - \gamma\text{–}Fe_2O_3$ compounds[219] and are likely to indicate the onset of ordering of the vacancies in the milled spinel, as is observed in fully ordered maghemite, $\gamma\text{–}Fe_2O_3$, which corresponds to off stoichiometric $Fe_{3\text{-}v}O_4$, with v = 1/3.[220]

In confirming the structural transformation of $\alpha\text{–}Fe_2O_3$ to $Fe_{3\text{-}v}O_4$ upon wet-milling in a vacuum, the neutron diffraction analysis has indicated the presence of vacancies predominantly on the octahedral sites. For instance, v is equal to 0.182 for the milled $Fe_{3\text{-}v}O_4$ of Fig. 32. These findings are in good agreement with the evidence for the increased electron hopping relaxation time associated with the vacancies created in the shared Fe^{2+} and Fe^{3+} octahedral sites with increasing milling time, as was discussed above for the Mössbauer spectra shown in Fig. 31c. The effects of milling $\alpha\text{–}Fe_2O_3$ in a range of 0.1 to 2 molar NaCl solutions, yields results in a 0.1 molar solution, i.e., above the ca. 0.078 molar concentration of reduced bubble coalescence in water,[221] which are similar to those obtained in water. Higher concentrations of NaCl, however, led to a reduced reaction rate for the transformation of $\alpha\text{–}Fe_2O_3$ to $Fe_{3\text{-}v}O_4$.[222] This reductive transformation, from $\alpha\text{–}Fe_2O_3$ to $Fe_{3\text{-}v}O_4$ during low energy mechanochemical treatment, is believed to occur primarily as a physical, rather than a chemical process, a process that ruptures the oxide surface layers and permits oxygen to be released. Although significant insight into the structural changes taking place during the reduction of $\alpha\text{–}Fe_2O_3$ to Fe_3O_4 has resulted from a combination of x-ray diffraction, Mössbauer spectroscopy, and neutron diffraction measurements, it is clear that further studies are required to fully identify the factors and mechanisms responsible for this transformation. For example our current conversion electron scattering Mössbauer measurements on wet milled haematite will clarify the extent to which maghemite, $\gamma\text{-}Fe_2O_3$, is present on the surface of milled particles, and contributes to the overall transformation to magnetite.

4. 3. 2. Barium Ferrite, $BaFe_{12}O_{19}$

Barium ferrite remains a very important material in the magnet industry with applications, for example, as a permanent magnet material[223] or, in fine particle form, in high density magnetic recording media.[224,225] Indeed, because of its very small switching field distribution and its propensity for yielding a narrow distribution of particle sizes, combined with its intrinsic magnetic properties, barium ferrite is one of the best materials for flexible recording media.[224] Mössbauer spectroscopy has been especially useful in providing details of the magnetic behavior of barium ferrite and related materials, such as their sublattice site occupancies, particle sizes, spin configurations, and magnetic anisotropies.[173,226-230]

Ball milling has been used for a long time in the preparation of oriented barium ferrite samples. In particular, ball milling can be used to produce ferrite particles of selected mean par-

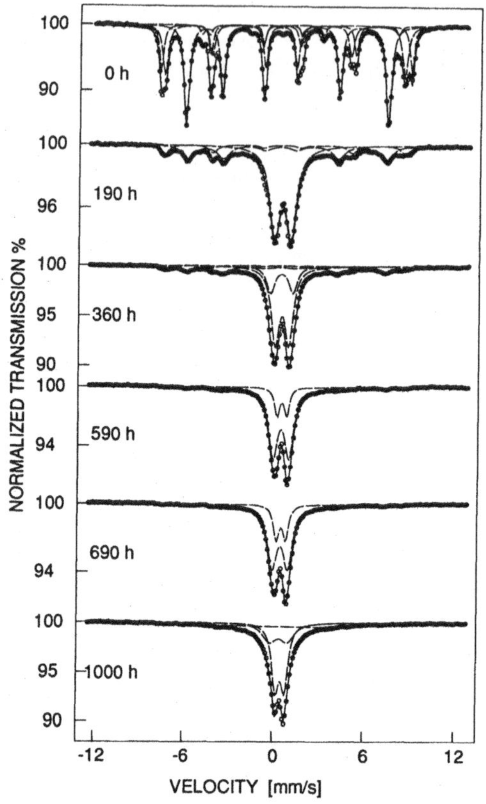

Figure 33. The room temperature Mössbauer spectra of $BaFe_{12}O_{19}$ before, top, and after high energy milling in air for the indicated times.[234]

ticle sizes for compacting in a strong magnetic field before their final sintering.[231] Of particular significance is the increase in coercivity reported[232,233] for milled and annealed barium ferrite. Indeed, with the goal of enhancing the magnetic properties of barium ferrite, we have investigated[233,234] the effects of milling $BaFe_{12}O_{19}$ in a range of environments, such as in a vacuum or in air, and in the presence of surfactants, followed by heat treatment.

Fig. 33 shows a series of room temperature Mössbauer spectra of $BaFe_{12}O_{19}$ measured after high energy milling in air for different times. The spectrum of $BaFe_{12}O_{19}$ agrees well with published spectra,[226] and is well fitted using four sextets to represent the five different crystallographic sites, the three octahedral 12k, $4f_2$, and 2a sites, the tetrahedral $4f_1$ site, and the bipyramidal 2b site. The similarity in magnetic hyperfine field values for the $4f_1$ and 2a sites leads to the use of four rather than five sextets. The scanning electron micrographs of these milled samples reveal the expected steady decrease in particle size with increased milling time, see Fig. 34. In addition, with milling the particles become increasingly homogeneous, such that after milling in air for 1000 hours, the particles have sizes in the range of 0.1 to 0.5 μm.[235] After 360 hours of milling the x-ray diffraction patterns[234] show a pronounced structural disorder with partial decomposition to α–Fe_2O_3 particles and typical mean grain sizes in the range of 15 to 150 nm. The Mössbauer spectra are consistent with superparamagnetic relaxation associated with fine particles of both $BaFe_{12}O_{19}$, mainly during the initial stages of milling, and α–Fe_2O_3 as the milling proceeds. A comparison with the Mössbauer spectra obtained by Kündig et al.[236] in their detailed study of finely dispersed α–Fe_2O_3 particles, indicates that quadrupolar effects, similar to those observed in Fig. 33, were obtained for α–Fe_2O_3 particles

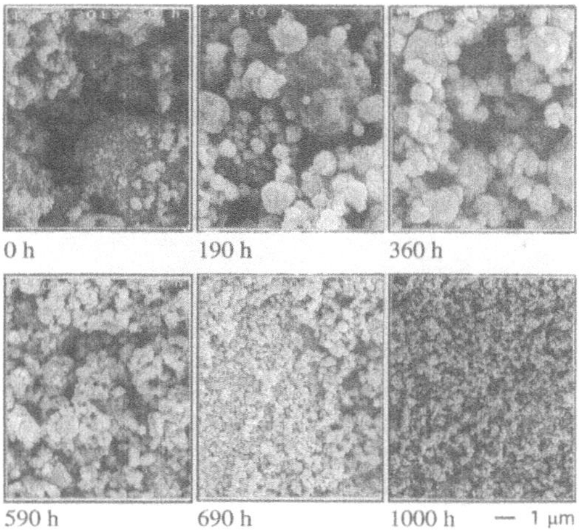

Figure 34. Scanning electron micrographs of $BaFe_{12}O_{19}$ milled in the high energy mode in air for the times indicated.[235]

with average diameters of ca. 10 to 15 nm.[236-238] The superparamagnetic relaxation observed in the Mössbauer spectra, indicative of ca. 10 to 20 nm fine particles, when combined with the similar grain sizes determined from the line broadening observed by x-ray diffraction, indicates that the relatively large particle sizes of 100 to 500 nm found in the sample milled for 1000 hours, and observed in the scanning electron micrographs shown in Fig. 34, comprise smaller individual particles which are clumped together.

As mentioned in Section 3. 3. 2, the effects of milling barium ferrite powder in the presence of cationic or anionic surfactants have also been investigated.[230,233] Fig. 35 shows the Mössbauer spectra obtained after low energy milling of barium ferrite with a cationic surfactant, DHDAAc or $C_{36}H_{75}NO_2$, which has a pH of 3 and is referred to as surfactant C3, and an anionic surfactant, SDDS or $NaC_{28}H_{55}S$, which also has a pH of 3 and is referred to as surfactant A3. Similar room temperature spectra are observed after milling with either surfactant, as is shown in Fig. 35. The resulting spectra exhibit features characteristic of $BaFe_{12}O_{19}$, as well as features characteristic of relaxation. Because of the low energy mode used in the surfactant milling experiments, the particle sizes are larger than those obtained when the equivalent samples are air-milled in the high energy mode, see Fig. 33. In this case only a partial collapse of the magnetic hyperfine splitting, due to relaxation, is observed. Likewise, the 4.2 K spectra of the C3 and A3 samples, Figs. 35c and d, respectively, show the predominant features characteristic of the 4.2 K spectrum[228,239] of $BaFe_{12}O_{19}$, but with an extra subspectral contribution resulting from residual relaxation effects.[230]

The similarity in the behavior of the cationic C3 and anionic A3 milled samples, as shown by their x-ray diffraction patterns and their Mössbauer spectra, see Fig. 35, can, in spite of the nominal difference in the ionic surfactant states, be understood in terms of a simple model of electrostatic attraction and stabilization, with either the head group or the hydrocarbon tail group of the surfactant bound to the ferrite surface, as is shown schematically in Fig. 9. Because a barium ferrite surface has a zero charge at a pH of 5 ±0.5,[240] its surface is positively charged during wet-milling at a pH of 3. With the cationic surfactant DHDAAc, the $DHDA^+$ head group is repelled and goes into solution, and the negatively charged acetate group is adsorbed onto the positively charged surface. Conversely, with the anionic surfactant SDDS, the Na^+ ions are in solution and the DDS^- head group is adsorbed onto the surface.[131] In both cases the $-CH_2-$

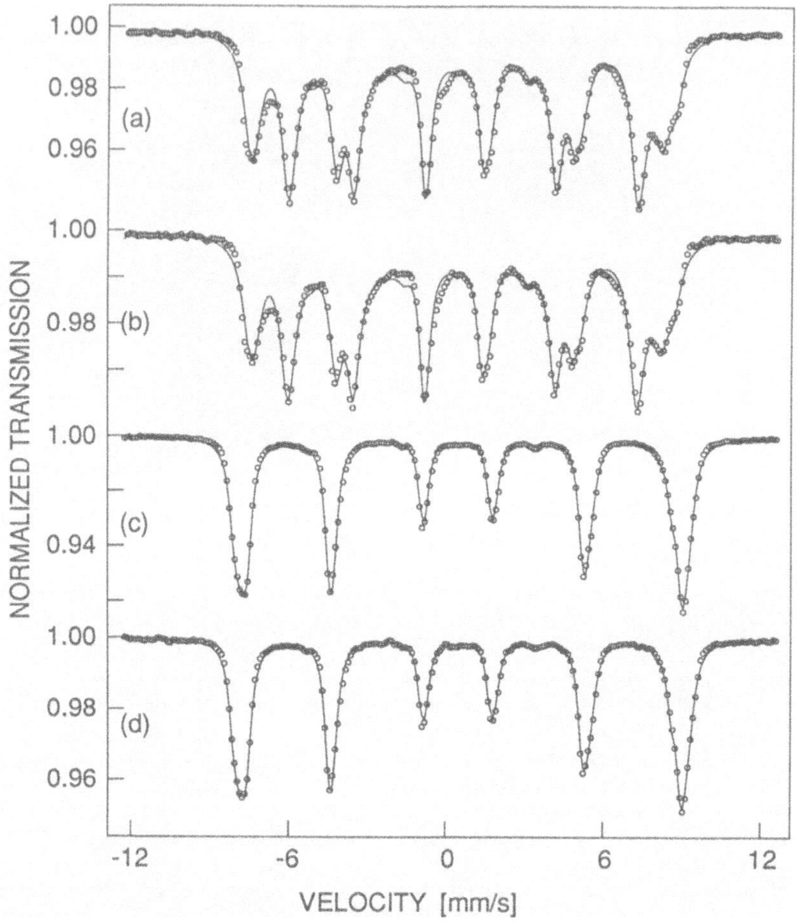

Figure 35. The room temperature Mössbauer spectrum of barium ferrite wet-milled for 310 hours (a) with the C3 cationic surfactant, DHDAAc, and (b) with the A3 anionic surfactant, SDDS. Also shown are the 4.2 K Mössbauer spectra for C3 (c) and A3 (d).[230]

groups are electrostatically bound to the surface. In effect, therefore, at a pH of 3, both the cationic and anionic surfactants have equivalent interfacial properties with the resultant behavior appearing to be governed primarily by electrostatic interactions.

The main effects which occur during ball-milling of barium ferrite in air are outlined schematically in Fig. 36.[235] The reaction at the oxygen-oxide interface becomes increasingly effective with prolonged milling due to the rupture of the iron-oxygen bonds at the surface of the particles. Likewise, decomposition of the complex oxide, $BaFe_{12}O_{19}$, can take place as a result of the diffusion due to the ionic currents, J, associated with charge transport, as is depicted in Fig. 36c. The larger particles observed in the vacuum milled state, with $\langle\delta\rangle$ values of ca. 1 μm after 1000 hours of milling, as compared with a $\langle\delta\rangle$ value of ca. 0.3 μm after air-milling for 1000 hours, are linked with the absence of oxygen during milling. Following the particle size reduction upon initial milling in a vacuum, as is indicated schematically in Fig. 37, the particles are consolidated as a result of a mechanism similar to that occuring in mechanical

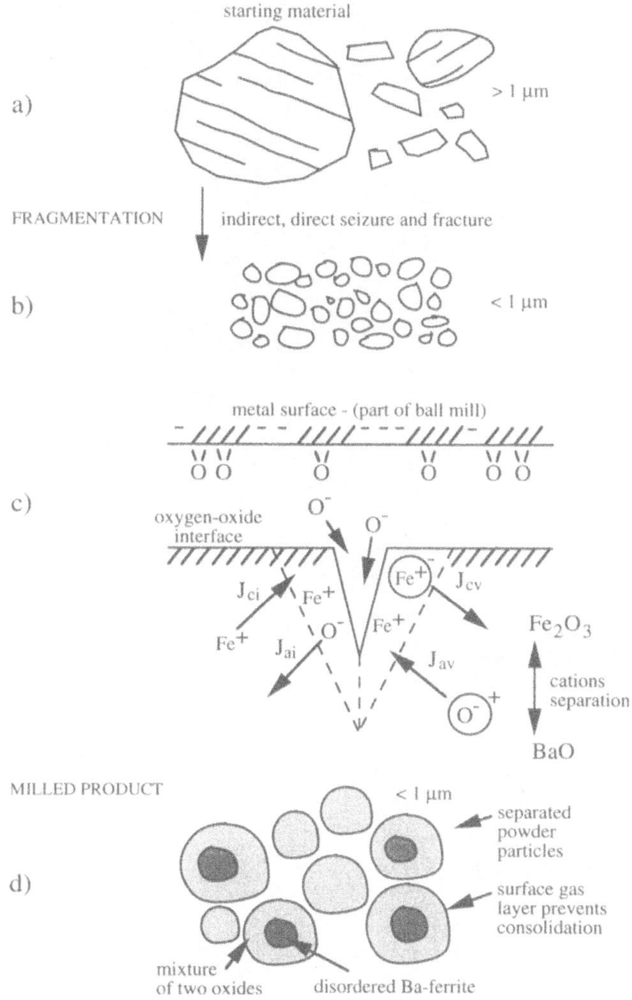

Figure 36. An illustration of the initial fragmentation, a and b, and decomposition mechanism, c, leading to the milled product, d, upon prolonged ball milling of barium ferrite in air.[235] J_{ci} and J_{ai} indicate the cationic and anionic interstitial currents and J_{cv} and J_{av} the cationic and anionic vacancy currents, respectively.

alloying, where simultaneous plastic deformation and fracture of powder particles coexist. The clustering of small particles of ca. 1 μm around large ones of ca. 10 μm diameter, observed in the vacuum milled sample, is likely to be due to magnetic interactions in which the larger particles have a larger magnetic remanence.[235] Of ultimate significance is the overall improvement observed in the magnetic properties when milled barium ferrite samples are annealed at 1000 °C. For example, coercivity values of ca. 390 to 450 kA/m, as compared with ca. 75 kA/m for the unmilled barium ferrite powder, are obtained for both the air-milled and vacuum milled materials. The corresponding saturation magnetization values of $4\pi M_S$ of ca. 340 kA/m, are about 10 percent lower than that of the unmilled barium ferrite.[235,241] While this remarkable five to six fold increase in coercivity can be linked with the fine crystalline grain

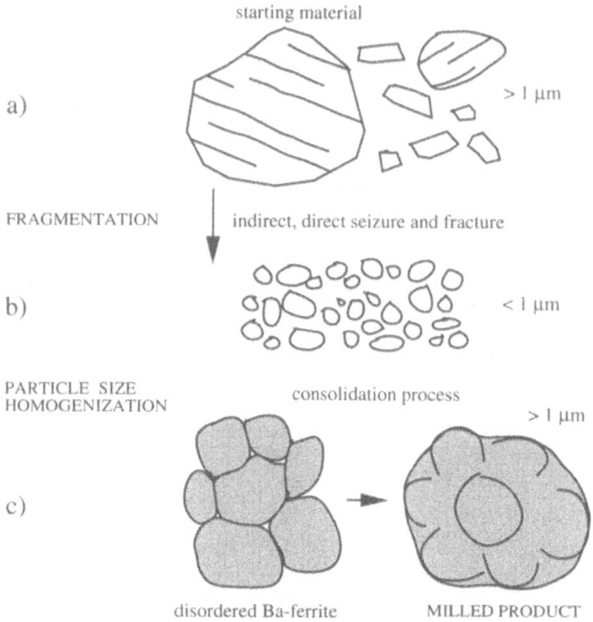

Figure 37. An illustration of the initial fragmentation, a and b, followed by consolidation, c, which occurs upon prolonged ball milling of barium ferrite in a vacuum.[235]

structure of the milled state, it is clear that further investigations are required to fully understand the relationship between the microstructural properties of the milled $BaFe_{12}O_{19}$ and the enhanced magnetic behavior obtained upon annealing.

5. CONCLUDING REMARKS

Ball milling and related techniques, such as mechanical attrition and mechanical alloying, are now well established as important methods for the preparation of new phases and materials, and for enhancing the physical properties of known materials.[13] An overall aim of this chapter has been to illustrate, through examples, the scope, diversity, and applicability afforded by mechanical alloying and mechanochemical treatment, in the preparation of materials. An additional aim has been to demonstrate the significant role which Mössbarer spectroscopy has and will continue to play, in delineating the microscopic details of the local structural and magnetic behavior of the milled products. In addition to the unique insights provided into the behavior of condensed matter by the ready measurement of hyperfine interactions,[19] Mössbauer spectroscopy has proved invaluable as an analytical technique for the ready identification of the phases present in the often multiphase milled products. This application of Mössbauer spectroscopy as a 'finger-print' technique follows, as a result of the reinvestigation, by mechanical treatment, of many established and important systems, including, alloys, carbides, nitrides, and oxides of iron, as was discussed in Section 4. Nonetheless, in spite of its suitability and sensitivity in the study of materials prepared by mechanochemical synthesis, Mössbauer spectroscopy is normally used in conjunction with other experimental techniques, such as x-ray diffraction, electron microscopy, and differential scanning calorimetry, as was outlined briefly in Section 2. 4.

The nominal simplicity of the milling techniques can belie the complexity of the processes taking place in the mechanochemical treatment of materials. Indeed mechanochemical treatment can, to a first approximation, be regarded as a fairly messy business with a myriad of milling conditions and energy modes, reaction processes, and contamination factors, as is outlined briefly below. The degree of reproducibility which is attained on milling the same system under different conditions, e.g., Fe-C as outlined in Section 4. 2, can therefore appear surprising, although the probabilistic nature of the milling process[14,15] ultimately ensures the essential reproducibility of these results. Apart from the geometry and scale of the milling device which essentially governs the energy of the milling process, other important parameters are the rotation or vibration frequency, the number and mass of the balls, the mass of the charge and, correspondingly, the ball to powder mass ratio. Likewise, the milling environment, mill and ball material, the quality of the vacuum and any lubricants present, and surfactants, can significantly affect the reactions and the ultimate milled products. Also, on the practical side, care has to be taken to reduce the risk of combustion associated with the production of ultrafine particles.[164] Recently, increasing attention has been devoted both to controlling the ambient temperature of the mill, typically at values between 77 and 470 K, and to monitoring any pressure changes which occur during the milling.[76,77] A factor which must be taken into account in appraising milled products, is that of contamination. Contamination occurs most commonly as a result of the wear of the mill and ball media, or upon degradation of the milling environment, such as is the case with the complex molecules of some surfactants. Indeed, as discussed in Section 4. 1, significant iron contamination was reported in the study of the Cr-Fe system, to the extent that the iron contamination from the wear on the steel balls and vial was the source of 22 atomic percent iron in the chromium milled by Kuwano et al.[136] Similarly, the crystallization of Fe-B-Si metallic glasses upon milling was found to be consistent with chemical changes in the powder caused by contamination, rather than as a result of any intrinsic process associated with the milling.[152] Also, as an example of the influence of the milling envi-

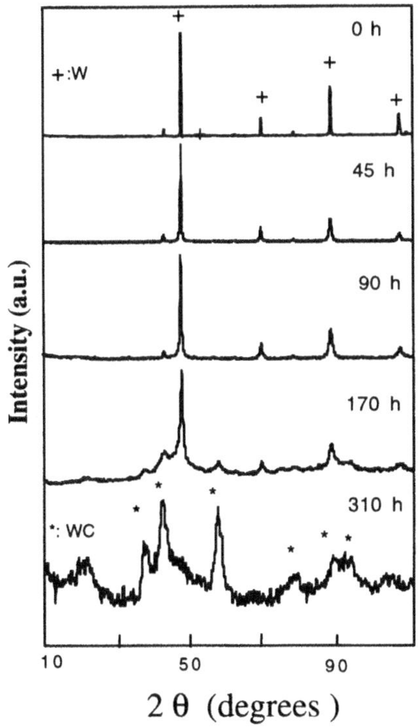

Figure 38. The x-ray diffraction patterns of a 50:50 atomic percent mixture of tungsten and carbon powders milled in a vacuum for the times indicated.[242]

ronment, in the exploratory investigation of the milling of iron in the presence of pyrazine, $C_4H_4N_2$, with the goal of preparing iron carbides and/or nitrides,[203,205] see Section 4. 2. 2, attention had to be paid to the behavior of the hydrogen atoms which result from the breakup of the $C_4H_4N_2$ molecule.

The recent preparation of tungsten carbide by milling tungsten powder and activated carbon in a vacuum[75] provides a further example of iron contamination resulting from the wear of the steel mill and balls. This work also nicely demonstrates the usefulness of iron-57 Mössbauer spectroscopy in the study of such nominally iron-free systems. Evidence for the formation of the extremely hard WC phase in a disordered nanostructured state is provided by the x-ray diffraction patterns of a 50:50 atomic percent mixture of tungsten and carbon powder milled in a vacuum for 170 and 310 hours, see Fig. 38. The formation of the WC phase is clearly established upon annealing the milled samples at 1000 °C for 1 hour, see Fig. 39. After heat treatment at 1000 °C for 20 hours the resulting material consisted almost entirely of the WC phase.[242] The room temperature Mössbauer spectra of the milled samples show evidence of increasing iron contamination with milling time, with the spectral absorption area increasing from ca. 0.003 percent after 45 hours to ca. 3 percent after milling for 310 hours, see Fig. 40. In agreement with the increased production of the very hard WC phase as milling proceeds, the rate of change of the spectral absorption area with time increases significantly after 170 hours, the time at which the WC phase first becomes discernible in the x-ray diffraction patterns, see Fig. 38. The spectrum for the sample milled for 310 hours is similar to the paramagnetic spectrum obtained by Matteazzi and Le Caër[74] upon milling a 1:3:1 atomic percent mixture of iron, tungsten, and carbon. In addition, the spectrum obtained after 210 hours of milled was similar to that obtained for κ-Fe-W-C.[243] The iron may be in the WC(Fe), FeW_3C, or Fe_3W_3C phase, although this point must be clarified by our continuing investigation of the milled Fe-W-C system.

Clearly the question of contamination is an important one which must be considered in all milling and related experiments. It is apparent that Mössbauer spectroscopy can be used to advantage in characterizing the effects of iron contamination. Of course magnetization measure-

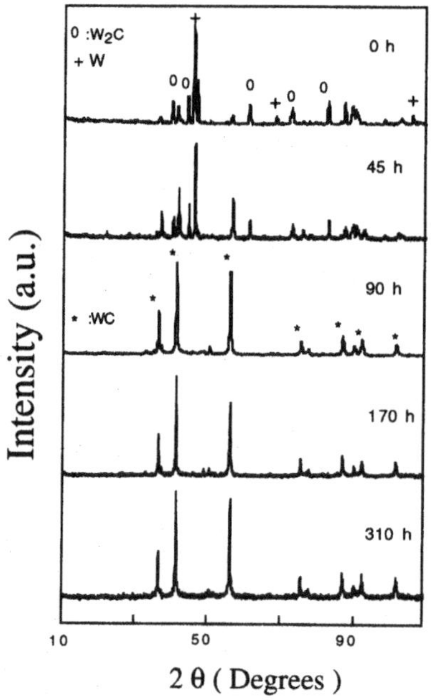

Figure 39. The x-ray diffraction patterns of the milled powders of Fig. 38 following heat treatment in a vacuum at 1000 °C for 1 hour.[242]

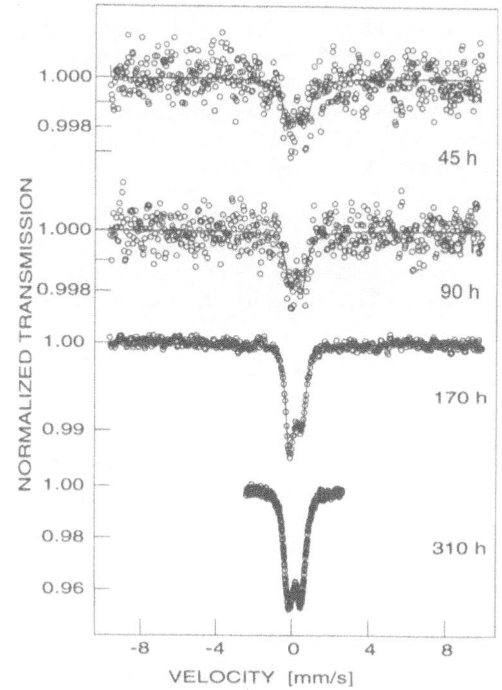

Figure 40. The room temperature Mössbauer spectra of the milled samples of Fig. 38.[242]

ments are also helpful in evaluating iron contamination in milled, nonmagnetic materials.[244] In those cases where even slight contamination is likely to be a significant problem, then mechanical alloying using the more deterministic approach[14] of repeated rolling and pressing, offers an alternative solution.

The diverse range of materials and systems which have already been subjected to milling and mechanical treatment reflects the acceptance by the condensed matter physics and material science communities of the relevance and importance of these techniques in the synthesis of new materials. While the fields of mechanical alloying and mechanochemical synthesis will continue to develop and flourish for the foreseeable future, it appears likely that increasing attention will be devoted to the overall milling conditions via, for example, controlled mill temperatures and continuous monitoring of the mill pressure,[77] and the increasing use of surfactants.[245] Such developments will lead to a better control and an improved understanding of the processes which occur during milling. As indicated by the example of barium ferrite, see Fig. 9 and Section 4. 3. 2, the use of surfactants and similar materials as agents in the milling process clearly offers the potential for more selected control of the structural and morphological, as well as the thermal and magnetic properties, of milled products. Nonetheless, while the use of such chemicals in liquid or solid form offers these novel prospects, the added complexities in the milling process, resulting from the breakdown of these materials, as was discussed in Section 4. 2. 2, may restrict their use during routine materials synthesis by milling. In addition, attention has to be paid to particle size reduction upon milling in surfactants, as this may cause degradation of the magnetic behavior as a result of the associated magnetic relaxation effects.

As mentioned in Section 1. 2, Mössbauer spectroscopy has contributed[10,12,37] significantly to our understanding of mechanically treated systems containing iron and tin. The fundamental importance of iron as an integral element in many systems of technological interest, means that iron-57 Mössbauer spectroscopy will continue to be used extensively in the investigation of milled products. Even in those cases where the resolution of the Mössbauer effect does not permit the unambiguous determination of the hyperfine interaction parameters, such as the distributed hyperfine parameters observed in the Mössbauer spectra of the Cr-Fe alloys prepared by mechanical attrition,[46] see Section 4. 1, the well developed, and in some cases very sophisticated methods of spectral analysis available to Mössbauer spectroscopists,[246-251] can provide detailed insight into the behavior of the materials under investigation. In the case of the mechanically alloyed Cr-Fe, Mössbauer spectral studies led to novel information about both the average width of the grain boundaries and the Debye temperatures of the grains and the grain boundaries.[46,137]

In summary, materials synthesis based on ball milling and related mechanical methods is an expanding and exciting field in materials science and condensed matter physics. In addition to the research challenges which this field of science offers, these milling techniques can be readily adapted for large scale, economic, production of mechanically alloyed[252] ultrafine powders.[164] [See also note added in proof].

ACKNOWLEDGMENTS. Parts of the work presented here were carried out in collaboration with our colleagues at University College, the University of New South Wales, Australian Defence Force Academy, and the Australian National University, and we thank them for their assistance and cooperation. We also wish to thank the secretarial and administrative staff of the University College, and particularly Ms. Ann Kelly, for their typing skill and assistance in the preparation of the manuscript. Aspects of the authors' work presented here have been supported by the ANU-UNSW Research Collaborative Scheme, the Science and Engineering Council, U.K., and the Australian Research Council.

REFERENCES

1. J. S. Benjamin, *Metall. Trans.* **1**, 2943 (1970).
2. J. S. Benjamin, *Scientific American* **234**, 48 (1976).
3. W. E. Kuhn, I. L. Friedman, W. Summers, and A. Szegvari, in *Metals Handbook,* Ninth Edition, Vol. 7, Powder Metallurgy, p. 56, American Society for Metals, 1984.
4. *NanoStructured Materials*, Pergamon Press, New York.
5. P. Matteazzi and G. Le Caër, *J. Am. Ceram. Soc.* **75**, 2749 (1992).
6. L. Takacs, *NanoStruct. Mater.* **2**, 241 (1993).
7. A. Y. Yermakov, *Mater. Sci. Forum* **88-90**, 577 (1992).
8. H. J. Fecht, *NanoStruct. Mater.* **1**, 125 (1992).
9. C. C. Koch, *NanoStruct. Mater.* **2**, 109 (1993).
10. G. Le Caër and P. Matteazzi, *Hyp. Interact.* **90**, 229 (1994).
11. I. J. Lin and S. Nadiv, *Mater. Sci. Eng.* **39**, 193 (1979).
12. S. J. Campbell and H. Gleiter, in *Mössbauer Spectroscopy Applied to Magnetism and Materials Science*, Vol. 1, G. J. Long and F. Grandjean, eds., p. 241, Plenum Press, New York, 1993.
13. J. S. Benjamin, *Mater. Sci. Forum* **88–90**, 1 (1992).
14. P. H. Shingu, K. N. Ishihara, K. Uenishi, J. Kuyama, B. Huang, and S. Nasu, in *Solid State Powder Processing*, A. H. Clauer and J. J. de Barbadillo, eds., p. 21, The Minerals, Metals, and Materials Society, Warrendale, 1990.
15. P. H. Shingu, K. N. Ishihara, and K. Kuyama, in *Proc. of the Thirty-fourth Japanese Congress on Materials Research*, p. 19, Society of Materials Science, Kyoto, 1991.
16. S. Nasu, S. Morimoto, H. Tanimoto, B. Huang, T. Tanaka, J. Kuyama, K. N. Ishihara, and P. H. Shingu, *Hyp. Interact.* **67**, 681 (1991).
17. S. Nasu, S. Imaoka, S. Morimoto, H. Tanimoto, B. Huang, T. Tanaka, J. Kuyama, K. N. Ishihara, and P. H. Shingu, *Mater. Sci. Forum* **88-90**, 569 (1992).
18. R. Z. Valiev, R. R. Mulyukov, and V. V. Ovchinnikov, *Phil. Mag. Lett.* **62**, 253

(1990); R. Z. Valiev, *Nanostructur. Mater.* **2**, 73 (1995).
19. U. Gonser, ed., *Mössbauer Spectroscopy*, Topics in Applied Physics, **5**, Springer-Verlag, Berlin, 1975.
20. P. Gütlich, R. Link, and A. Trautwein, *Mössbauer Spectroscopy and Transition Metal Chemistry*, Springer-Verlag, Berlin, 1978.
21. U. Gonser, ed., *Mössbauer Spectroscopy II*, Topics in Current Physics, **25**, Springer-Verlag, Berlin, 1981.
22. G. J. Long and F. Grandjean, eds., *Mössbauer Spectroscopy Applied to Magnetism and Materials Science*, Vol. 1, Plenum Press, New York, 1993.
23. E. Gerdau and U. van Bürck, in *Resonant Anomalous X-ray Scattering*, G. Materlik, C. J. Sparks, and K. Fischer, eds., p. 589, Elsevier, Amsterdam, 1994.
24. G. V. Smirnov, *Nuclear Resonant Scattering of Synchrotron Radiation*, IAE–5907/9, Russian Research Centre "Kurchatov Institute", Moscow, 1995.
25. A. H. Morrish, *Mat. Sci. Eng.* **B3**, 383 (1989).
26. R. P. Elliott, *Constitution of Binary Alloys*, First Suppl., p. 372, McGraw-Hill, New York, 1965.
27. K. Uenishi, K. F. Kobayashi, S. Nasu, H. Hatano, K. N. Ishihara, and P. H. Shingu, *Z. Metallkunde* **83**, 132 (1992).
28. A. R. Yavari, P. J. Desré, and T. Benameur, *Phys. Rev. Lett.* **68**, 2235 (1992).
29. T. Amborse, A. Gavrin, and C. L. Chien, *J. Magn. Magn. Mater.* **124**, 15 (1993).
30. P. Crespo, A. Hernando, and A. Garcia Escorial, *Phys. Rev. B* **49**, 13227 (1994).
31. P. P. Macri, P. Rose, D. E. Banda, N. Cowlam, G. Principi, and S. Enzo, *Mater. Sci. Forum* **179-181**, 249 (1995).
32. S. J. Campbell, P. E. Clark, and T. J. Hicks, *J. Phys. F: Metal Phys.* **6**, 249 (1976).
33. S. J. Campbell, P. E. Clark, and T. J. Hicks, *Solid State Commun.* **22**, 509 (1977).
34. U. Gonser and M. Ron, in *Applications of Mössbauer Spectroscopy*, R. L. Cohen, ed., Vol. II, p. 281, Academic Press, New York, 1980.
35. I. Spirov, T. Tomov, and S. Asenov, *Proc. Int. Conf. on the Applications of the Mössbauer Effect*, 1996.
36. J. Eckert, J. C. Holzer, C. E. Krill, and W. L. Johnson, *J. Appl. Phys.* **73**, 2794 (1993).
37. R. Würschum, *NanoStruct. Mater.* **6**, 93 (1995).
38. J. G. Stevens, *Hyp. Interact.* **71**, 1537 (1992).
39. S. Nasu and P. H. Shingu, in *Ordering and Disordering in Alloys*, A. R. Yavari, ed., p. 402, Elsevier, Amsterdam, 1992.
40. T. Tanaka, S. Nasu, B. Huang, K. N. Ishihara, and P. H. Shingu, *Nucl. Instr. Meth., Phys. Res. B* **76**, 195 (1993).
41. K. D. Jayasuriya, S. J. Campbell, A. Calka, and J. Jing, *Proc. Second International Conference on the Structural Applications of Mechanical Alloying*, J. J. de Barbadillo, F. H. Fores, and R. Schwarz, eds., p. 27, ASM, Materials Park, Ohio, 1993.
42. L. Daróczi, D. L. Beke, G. Posgay, G. F. Zhou, and H. Bakker, *NanoStruct. Mater.* **2**, 515 (1993).
43. P. Sinha and G. S. Collins, *Hyp. Interact.* **92**, 949 (1994).
44. S. Szabó, I. Brovko, M. K. Varga, D. L. Beke, and G. Posgay, *NanoStruct. Mater.* **6**, 973 (1995).
45. P. Schaaf, G. Rixecker, E. Young, C. N. J. Wagner, and U. Gonser, *Hyp. Interact.* **94**, 2239 (1994).
46. B. Fultz, H. Kuwano, and H. Ouyang, *J. Appl. Phys.* **77**, 3458 (1995).
47. D. C. Cook, T. H. Kim, and J. C. Rawers, *Proc. Int. Conf. on the Applications of the Mössbauer Effect*, 1996.
48. U. Herr. J. Jing, R. Birringer, U. Gonser, and H. Gleiter, *Appl. Phys. Lett.* **50**, 472 (1987).
49. G. Rixecker, R. Birringer, H. Gleiter, and U. Gonser, *Proc. Int. Conf. on the*

Applications of the Mössbauer Effect, 1996.
50. M. Umemoto, S. Shiga, K. Raviprasad, and I. Okane, *Mater. Sci. Forum* **179-181**, 165 (1995).
51. A. W. Weeber and H. Bakker, *Physica B* **153**, 93 (1988).
52. R. B. Schwarz, *Mater. Sci. Eng.* **97**, 71 (1988).
53. J. Eckert, L. Schultz, E. Hellstern, and K. Urban, *J. Appl. Phys.* **64**, 3224 (1988).
54. C. C. Koch, *Ann. Rev. Mater. Sci.* **19**, 121 (1989); C. C. Koch, *J. Non-Cryst. Solids* **117-118**, 670 (1990).
55. S. Enzo, E. Bonetti, I. Sollenta, and G. Cocco, *J. Phys. D: Appl. Phys.* **24**, 209 (1991).
56. H. Bakker, G. F. Zhou, and H. Yang, *Progr. Mater. Sci.* **39**, 159 (1995).
57. H. Gleiter, *NanoStruct. Mater.* **6**, 3 (1995).
58. D. Maurice and T. H. Courtney, *J. Metals* **8**, 10 (1992).
59. M. Abdellaoui and E. Gaffet, *Mater. Sci. Forum* **179-181**, 339 (1995).
60. V. V. Boldyrev, *Solid State Ionics* **63-65**, 537 (1993).
61. P. Pochet, L. Chaffron, and G. Martin, *Mater. Sci. Forum* **179-181**, 91 (1995).
62. D. R. Maurice and T. H. Courtney, *Metall. Trans.* **21A**, 289 (1990).
63. A. N. Streletskii, *Proc. Second International Conference on the Structural Applications of Mechanical Alloying*, J. J. de Barbadillo, F. H. Fores, and R. Schwarz, eds., p. 51, ASM, Materials Park, Ohio, 1993.
64. J. Kerr, *Metal Powder Report* **48**, 36 (1993).
65. D. Basset, P. Matteazzi, and F. Miani, *Mater. Sci. Eng.* **A174**, 71 (1994).
66. D. Basset, P. Matteazzi, and F. Miani, *Mater. Sci. Eng.* **A168**, 149 (1993).
67. J. Uchrin, R. Uchrin, and E. G. Avvakumov, *Mater. Sci. Forum* **179-181**, 425 (1995).
68. H. Zoz, *Mater. Sci. Forum* **179-181**, 419 (1995).
69. T. M. Cook and T. H. Courtney, *Metall. Trans.* **26A**, 2389 (1995).
70. A. Calka and A. P. Radlinski, *Mater. Sci. Eng.* **A134**, 1350 (1991).
71. E. Hellstern and L. Schultz, *Appl. Phys. Lett.* **48**, 124 (1986).
72. C. Kuhrt and L. Schultz, *J. Appl. Phys.* **73**, 1975 (1993).
73. A. Calka and J. I. Nikolov, *NanoStruct. Mater.* **6**, 409 (1995).
74. P. Matteazzi and G. Le Caër, *J. Am. Ceram. Soc.* **74**, 1382 (1991).
75. G. M. Wang, A. Calka, S. J. Campbell, and W. A. Kaczmarek, *Mater. Sci. Forum* **179-181**, 201 (1995).
76. P. Millet, A. Calka, J. S. Williams, and G. H. J. Wantenaar, *Appl. Phys. Lett.* **63**, 2505 (1993).
77. Y. Chen, A. Calka, T. Halstead, and J. S. Williams, *Mater. Sci. Eng.* **206A**, 24 (1996).
78. P. Millet and A. Calka, *Mater. Sci. Forum* **179-181**, 321 (1995).
79. J. I. Nikolov, A. Calka, and J. S. Williams, *NanoStruct. Mater.* **6**, 401 (1995).
80. R. M. Davis, B. McDermott, and C. C. Koch, *Metall. Trans.* **19A**, 2867 (1988).
81. A. K. Bhattacharya and E. Arzt, *Scripta Met.* **27**, 749 (1992).
82. E. H. Love, *A Treatise on the Mathematical Theory of Elasticity*, Fourth Edition, p. 193, Dover, New York, 1944.
83. D. Maurice and T. H. Courtney, *Metall. Mater. Trans.* **26A**, 2431 (1995).
84. D. Maurice and T. H. Courtney, *Metall. Mater. Trans.* **26A**, 2437 (1995).
85. S. R. Phillpot, D. Wolf, and H. Gleiter, *Scripta Metall. Mater.* **33**, 1245 (1995).
86. M. Sahimi, G. Gavalas, and T. Tsotsis, *Chem. Eng. Sci.* **45**, 1443 (1991).
87. A. P. Roberts and M. A. Knackstedt, *J. Mater. Sci. Lett.* **14**, 1357 (1995).
88. W. Buckel and R. Hilsch, *Z. Phys.* **138**, 109 (1954).
89. W. Klement, R. H. Willens, and P. Duwez, *Nature* **187**, 869 (1960).
90. C. A. Pampillo, *Scripta Metall.* **6**, 915 (1972).
91. H. Iwasaki and T. Masumoto, *J. Mater. Sci.* **13**, 2171 (1978).
92. L. Schultz, *Mater. Sci. Eng.* **97**, 15 (1988).
93. J. Horvath, R. Birringer, and H. Gleiter, *Solid State Commun.* **62**, 319 (1987).
94. P. S. Gilman and J. S. Benjamin, *Ann. Rev. Mater. Sci.* **13**, 279 (1983).

95. R. L. White, Ph.D. Dissertation, Stanford University, 1979.
96. C. C. Koch, O. B. Cavin, C. G. McKamey, and J. O. Scarbrough, *Appl. Phys. Lett.* **43**, 1017 (1983).
97. A. Ye. Yermakov, Ye. Ye. Yurchikov, and V. A. Barinov, *Fiz. Met. Metalloved.* **52**, 1184 (1981); *ibid* **54**, 935 (1982).
98. C. Politis and W. L. Johnson, *J. Appl. Phys.* **60**, 1147 (1986).
99. E. Hellstern and L. Schultz, *Phil. Mag.* **B56**, 443 (1987).
100. H. Gleiter, in *Deformation of Polycrystals*, N. Hansen, A. Horsewell, T. Leffers, and H. Lilholt, eds., p. 15, *Proc. Second Risø Int. Symp. Metallurgy and Materials Science*, Roskilde, 1981.
101. H. Gleiter, *Progr. Mater. Sci.* **33**, 223 (1989).
102. G. Herzer, *Mater. Sci. Eng.* **A133**, 1 (1991).
103. R. Coehoorn and J. Duchateau, *J. Mater. Sci. Eng.* **99**, 131 (1988).
104. R. Nagarajan and K. Chattopadhyay, *Acta Metall. Mater.* **42**, 947 (1994).
105. R. Birringer, H. Gleiter, H. P. Klein, and P. Marquardt, *Phys. Lett.* **102A**, 365 (1984).
106. W. L. Johnson, *Prog. Mater. Sci.* **30**, 81 (1986).
107. A. Johnson, in *New Materials by Mechanical Alloying Techniques*, E. Arzt and L. Schultz, eds., p. 354, Informationsgesellschaft Verlag, Calw-Hirsau, 1988.
108. E. Hellstern, H. J. Fecht, C. Garland, and W. L. Johnson, *MRS Symp. Proc.* **132**, 137 (1989).
109. D. A. Rigney, *Ann. Rev. Mater. Sci.* **18**, 141 (1988).
110. F. H. Froes and C. Suryanarayana, *J. Metals* **6**, 12 (1989).
111. G. B. Schaffer and P. G. McCormick, *Metall. Trans.* **22A**, 3019 (1991).
112. T. Kaneyoshi, T. Takahashi, and M. Motoyama, *Scripta Metall. Mater.* **29**, 1547 (1993).
113. A. Calka and J. S. Williams, *Mater. Sci. Forum* **88-90**, 787 (1992).
114. R. Schrader and B. Hoffman, in *Festkörperchemie. Beitrage aus Forschung und Praxis*, V. Boldyrev and K. Meyer, eds., p. 528, Grundstoffindustrie, Leipzig, 1973.
115. W. A. Kaczmarek and B. W. Ninham, *IEEE Trans. Magn.* MAG-**30**, 732 (1994).
116. W. A. Kaczmarek, I. Onyszkiewicz, and B. W. Ninham, *IEEE Trans. Magn.* MAG-**30**, 4725 (1994).
117. V. V. Boldyrev, M. Bulens, and B. Delmon, *The Control of the Reactivity of Solids*, p. 16, Elsevier, Amsterdam, 1979.
118. T. Kosmac and T. H. Courtney, *J. Mater. Res.* **7**, 1519 (1992).
119. P. G. McCormick, V. N. Wharton, M. M. Royhani, and G. B. Schaffer, in *Microcomposites and Nanophase Materials*, D. C. Van Aken, G. S. Was, and A. K. Gosh, eds., p. 65, Warrendale, 1991.
120. T. Takahashi and D. C. Dunand, *Mater. Sci. Eng.* **A192-193**, 186 (1995).
121. P. Y. Lee, T. R. Chen, J. L. Yang, and T. S. Chin, *Mater. Sci. Eng.* **A192-193**, 556 (1995).
122. L. Lu and M. O. Lai, *Materials Design* **16**, 33 (1995).
123. F. H. Froes, C. Suryanarayana, K. Russell, and C.-G. Li, *Mater. Sci. Eng.* **A192-193**, 612 (1995).
124. P. Yu. Butyagin, *Mater. Sci. Forum* **179-181**, 25 (1995).
125. P. Matteazzi, G. Le Caër, and E. Bauer-Grosse, European Patent No. EP0454522A1, (30/10/91) Pr. FR 9004820, (13/4/90).
126. T. Puclin, W. A. Kaczmarek, and B. W. Ninham, *Mater. Chem. Phys.* **40**, 73 (1995).
127. P. Matteazzi and G. Le Caër, *Mater. Sci. Eng.* **A149**, 135 (1991).
128. H. J. Fecht, in *Nanophase Materials*, G. C. Hadjipanayis and R. W. Siegel, eds., p. 125, Kluwer Academic Publisher, 1994.
129. P. Matteazzi, G. Le Caër, and A. Mocellin, in *Ceramics: Charting the Future*, P. Vincenzini, ed., Advances in Science and Technology, p. 1481, TECHNA, Roma, 1995.
130. M. B. Mooiman and K. C. Sole, *J. Metals* **6**, 18 (1994).

131. W. A. Kaczmarek and B. W. Ninham, *Mater. Chem. Phys.* **40**, 21 (1995).
132. M. S. El-Eskandarany, K. Sumiyama, K. Aoki, and K. Suzuki, *Mater. Sci. Forum* **88-90**, 809 (1992).
133. T. Koyano, C. H. Lee, T. Fakunaga, and U. Mizutain, *Mater. Sci. Forum* **88-90**, 809 (1992).
134. K. Majima, M. Itoh, S. Yagyu, S. Katsuyama, and H. Nagai, in *Proceedings of the 1993 Powder Metallurgy World Congress*, Japanese Society of Powder Metallurgy, p. 706, 1993.
135. G. C. Bond, *Catalysis by Metals*, p. 371, Academic Press, London, 1962.
136. H. Kuwano, H. Ouyang, and B. Fultz, *NanoStruct. Mater.* **1**, 143 (1992).
137. H. Kuwano, H. Ouyang, and B. Fultz, *Mater. Sci. Forum* **88-90**, 561 (1992).
138. M. Oehring, Z. H. Yan, T. Klassen, and R. Bormann, *phys. stat. sol. (a)* **131**, 671 (1992).
139. J. Koike, *Phys. Rev. B* **47**, 7700 (1993).
140. Y. Chakk, S. Berger, B.-Z. Weiss, and E. Brook-Levinson, *Acta Metall. Mater.* **42**, 3679 (1994).
141. T. Egami, *Rep. Prog. Phys.* **47**, 1601 (1984).
142. G. K. Nicolaides, M. Pissas, D. Niarchos, R. D. Taylor, and K. V. Rao, *J. Appl. Phys.* **75**, 5853 (1994).
143. U. Köster and U. Herold, in *Glassy Metals I*, H.-J. Güntherodt and H. Beck, eds., p. 225, Springer, Berlin, 1981.
144. C. L. Chien and K. M. Unruh, *Phys. Rev. B* **25**, 5790 (1982).
145. J. Jing, A. Calka, and S. J. Campbell, *J. Phys.: Condens. Matter* **3**, 7413 (1991).
146. J. Balogh, L. Bujdosó, Gy Faigel, T. Kemény, and I. Vincze, *Mater. Sci. Forum* **179-181**, 775 (1995).
147. A. Calka, A. P. Pogany, R. A. Shanks, and H. Engelmann, *Mater. Sci. Eng.* **A128**, 107 (1990).
148. W. A. Kaczmarek, A. Calka, and B. W. Ninham, *phys. stat. sol. (a)* **141**, K1 (1994).
149. S. J. Campbell, K. D. Jayasuriya, A. Calka, and J. Jing, *Nucl. Instr. Meth., Phys. Res. B* **76**, 81 (1993).
150. H. Matyja, D. Oleszak, and J. Latudv, *Mater. Sci. Forum* **88-90**, 297 (1992).
151. K. D. Jayasuriya, S. J. Campbell, A. Calka, and J. Jing, *Nucl. Instr. Meth., Phys. Res. B* **76**, 85 (1993).
152. C. Bansal, B. Fultz, and W. L. Johnson, *NanoStruct. Mater.* **4**, 919 (1994).
153. M. L. Trudeau, R. Schulz, D. Dussault, and A. Van Neste, *Phys. Rev. Lett.* **64**, 99 (1990).
154. M. L. Trudeau, L. Dignard-Bailey, R. Schulz, D. Dussault, and A. Van Neste, *NanoStruct. Mater.* **2**, 361 (1993).
155. S. Nasu, P. H. Shingu, K. N. Ishihara, and F. E. Fujita, *Hyp. Interact.* **55**, 1043 (1990).
156. G. Le Caër, P. Matteazzi, and B. Fultz, *J. Mater. Res.* **7**, 1387 (1992).
157. G. Le Caër, S. Begin-Colin, Y. Calage, P. Matteazzi, and F. Miani, *Hyp. Interact.* **94**, 2229 (1994).
158. A. F. Cabrera, F. H. Sánchez, and L. Mendoza-Zélis, *Mater. Sci. Forum* **179-181**, 231 (1995).
159. M. O. Kientz, G. Le Caër, P. Delcroix, L. Fournes, B. Fultz, P. Matteazzi, and B. Malaman, *NanoStruct. Mater.* **6**, 617 (1995).
160. P. Matteazzi, D. Basset, F. Miani, and G. Le Caër, *NanoStruct. Mater.* **2**, 21 (1993).
161. G. B. Schaffer and P. G. McCormick, *Appl. Phys. Lett.* **55**, 45 (1989).
162. N. G. Danielian, S. K. Janazian, and V. V. Melnichenko, *Mod. Phys. Lett.* **B5**, 1355 (1991).
163. P. G. McCormick, *Proc. First Int. Conf. on Processing Materials for Properties*, H. Henein and T. Oki, eds., p. 675, The Materials Society, 1993.

164. J. Ding, W. F. Miao, P. G. McCormick, and R. Street, *Appl. Phys. Lett.* **67**, 1 (1995).
165. J. Ding, T. Tsuzuki, P. G. McCormick, and R. Street, *J. Alloys Comp.*, in press.
166. *Proc. Int. Workshop on Magnetic Materials,* Perth, Australia, December 1995, *J. Phys. D: Appl. Phys.*, in press.
167. L. Schultz, J. Wecker, and E. Hellstern, *J. Appl. Phys.* **61**, 3583 (1987).
168. L. Schultz, K. Schnitzke, and J. Wecker, *J. Magn. Magn. Mater.* **80**, 115 (1989).
169. L. Schultz, *Mater. Sci. Forum* **88-90**, 687 (1992).
170. T. Nakamura, A. Inoue, K. Matsuki, and T. Masumoto, *J. Mater. Sci. Lett.* **8**, 13 (1989).
171. S. J. Campbell, K. D. Jayasuriya, A. Calka, J. M. Cadogan, J. Jing, and E. Wu, *Proc. Seventh Int. Symposium on Magnetic Anisotropy and Coercivity in Rare-Earth Transition Metal Alloys*, Canberra, Australia, July 1992, p. 68, Hi-Perm Laboratory, Research Centre for Advanced Mineral and Materials Processing, The University of Western Australia, Scott Four Color Print, Perth, 1992.
172. W. F. Miao, J. Ding, P. G. McCormick, and R. Street, *J. Appl. Phys.*, in press.
173. O. A. Pringle and G. J. Long in *Mössbauer Spectroscopy Applied to Inorganic Chemistry*, G. J. Long and F. Grandjean, eds., Vol. 3, p. 13, Plenum Press, New York, 1989.
174. K. Krop, *Physica B* **159**, 66 (1989).
175. M. Fähnle, K. Hummler, and T. Beuerle, *J. Magn. Magn. Mater.* **127**, L278 (1993).
176. J. M. Cadogan and J. M. D. Coey, *Appl. Phys. Lett.* **48**, 442 (1986).
177. G. J. Long, F. Grandjean, and O. A. Pringle, *J. Magn. Magn. Mater.* **125**, L29 (1993).
178. W. F. Miao, J. Ding, P. G. McCormick, and R. Street, unpublished results.
179. T. Ya. Kosolapova, *Carbides, Properties, Production and Applications*, Plenum Press, New York, 1971.
180. J. J. Fischer, *Adv. Mater. Proc.* **138**, 43 (1990).
181. A. Calka and J. S. Williams, *Scripta Metall. Mater.* **27**, 999 (1992).
182. A. Calka, J. I. Nikolov, and B. W. Ninham, *Proc. Second. Int. Conf. on Structural Applications of Mechanical Alloying*, J. J. de Barbadillo, F. H. Froes, and R. Schwarz, eds., p. 189, ASM, Ohio, 1993.
183. J. M. D. Coey, K. O'Donnell, Q. Qinian, E. Touchiais, and K. H. Jack, *J. Phys: Condens. Matter* **6**, L23 (1994).
184. M. Ron, in *Applications of Mössbauer Spectroscopy*, R. L. Cohen, ed., Vol. II, p. 329, Academic Press, New York, 1980.
185. M. Kopcewicz, J. Jagielski, A. Turos, and D. L. Williamson, *J. Appl. Phys.* **71**, 4217 (1992).
186. T. Tanaka, S. Nasu, K. N. Ishihara, and P. H. Shingu, *J. Less-Comm. Metals* **171**, 237 (1991).
187. G. Heinike, *Tribochemistry*, Akademie-Verlag, Berlin, 1984.
188. G. Le Caër, E. Bauer-Grosse, A. Pianelli, E. Bouzy, and P. Matteazzi, *J. Mater. Sci.* **25**, 4726 (1990); F. Miani, P. Matteazzi, D. Basset, and G. Le Caër, *Hyp. Interact.* **94**, 2219 (1994).
189. T. Nasu, K. Nagaoka, N. Itoh, and K. Suzuki, *J. Non-Cryst. Solids* **122**, 216 (1990).
190. T. Tanaka, S. Nasu, K. Nakagawa, K. N. Ishihara, and P. H. Shingu, *Mater. Sci. Forum* **88-90**, 269 (1992).
191. E. P. Elsukov, V. A. Barinov, and L. V. Ovechkin, *J. Mater. Sci. Lett.* **11**, 662 (1992).
192. P. Miani, P. Matteazzi, and D. Basset, *J. Alloys Comp.* **204**, 151 (1994).
193. G. M. Wang, A. Calka, S. J. Campbell, and W. A. Kaczmarek, *NanoStruct. Mater.* **6**, 389 (1995).
194. S. J. Campbell, G. M. Wang, A. Calka, and W. A. Kaczmarek, unpublished results.
195. J. M. R. Genin, O. N. C. Uwakweh, J. Schanen, and P. Bauer, *Hyp. Interact.* **41**, 575 (1988).
196. O. A. Pringle, G. J. Long, J. L. Li, W. J. James, F. Grandjean, and G. C.

Hadjipanayis, *IEEE Trans. Magn.* **28**, 2862 (1992).
197. J. M. D. Coey, *Physics World*, p. 25, August 1993.
198. K. H. Jack, *J. Alloys Comp.* **222**, 160 (1995).
199. T. Koyano, T. Takizawa, T. Fakunaga, and U. Mizutain, *Japan. J. Appl. Phys.* **37**, L1524 (1993).
200. R. S. Figueiredo and J. Foet, *Mater. Sci. Forum* **179-181**, 219 (1995).
201. K.-J. Kim, K. Sumiyama, K. Shibata, and K. Suzuki, *Mater. Sci. Eng.* **A181-A182**, 1272 (1994).
202. P.-Y. Lee, T.-R. Chen, and T.-S. Chin, *J. Alloys Comp.* **222**, 179 (1995).
203. W. A. Kaczmarek and I. Onyszkiewicz, *Mater. Sci. Forum* **179-181**, 195 (1995).
204. W. A. Kaczmarek, B. W. Ninham, and I. Onyszkiewicz, *J. Mater. Sci.* **30**, 5514 (1995).
205. G. M. Wang, S. J. Campbell, and W. A. Kaczmarek, *Hyp. Interact.C* **1**, 472 (1996).
206. G. M. Wang, S. J. Campbell, and W. A. Kaczmarek, unpublished results.
207. Q. A. Pankhurst and R. J. Pollard, *J. Phys: Condens. Matter* **5**, 8487 (1993).
208. Q. A. Pankhurst and R. J. Pollard in *Mössbauer Spectroscopy Applied to Magnetism and Materials Science*, G. J. Long and F. Grandjean, eds., Vol. 1, p. 77, Plenum Press, New York, 1993.
209. L. H. Bowen, E. De Grave, and R. E. Vandenberghe, in *Mössbauer Spectroscopy Applied to Magnetism and Materials Science*, G. J. Long and F. Grandjean, eds., Vol. 1, p. 115, Plenum Press, New York, 1993.
210. A. R. Burkin, *The Chemistry of Hydrometallurgical Processes*, p. 157, Spon, London, 1966.
211. R. H. Torres Sanchez, E. M. Curt, C. Volzono, R. C. Mercader, and A. L. Cavalieri, *Mater. Res. Bull.* **25**, 553 (1990).
212. L. Takacs, *Mater. Lett.* **13**, 119 (1992).
213. M. Pardavi-Horvath and L. Takacs, *J. Appl. Phys.* **73**, 6958 (1993).
214. L. Takacs, H. Dlamini, and H. Pollak, *Proc. Int. Conf. on the Applications of the Mössbauer Effect*, 1996.
215. Y. Uehara, *Bull. Chem. Soc. Japan* **48**, 3383 (1975).
216. P. Matteazzi and G. Le Caër, *Hyp. Interact.* **68**, 177 (1991).
217. S. J. Campbell, W. A. Kaczmarek, and G. M. Wang, *NanoStruct. Mater.* **6**, 735 (1995).
218. J. M. D. Coey, A. H. Morrish, and G. A. Sawatzky, *J. Physique* **32**, C1-271 (1971).
219. H. Annersten and S. F. Hafner, *Z. Krist.* **137**, 321 (1973).
220. M. Hofmann, S. J. Campbell, and W. A. Kaczmarek, *Proc. European Powder Diffraction Conference*, Chester, UK, July 1995, *Mater. Sci. Forum*, in press.
221. V. S. J. Craig, B. W. Ninham, and R. M. Pashley, *J. Phys. Chem.* **97**, 10192 (1993).
222. S. J. Campbell, W. A. Kaczmarek, E. Wu, and G. M. Wang, *Proc. ISMAR-95*, Sydney, Australia, July 1995, *Bull. Magn. Res.*, in press.
223. M. Zenger, *Int. J. Mater. Product Tech.* **9**, 265 (1994).
224. E. Köster, *J. Magn. Magn. Mater.* **120**, 1 (1993).
225. T. Suzuki, T. Oguchi, Y. Kishimoto, and N. Saito, *J. Magn. Magn. Mater.* **120**, 25 (1993).
226. G. Albanese, A. Deriu, E. Lucchini, and G. Slokar, *Appl. Phys.* **26A**, 45 (1981).
227. R. Hollatz, W. Sturhahn, J. Frohn, E. Gerdau, and R. Rüffer, *Hyp. Interact.* **41**, 479 (1988).
228. Q. A. Pankhurst, *J. Phys: Condens. Matter* **3**, 1323 (1991).
229. V. K. Sankaranarayanan, Q. A. Pankhurst, D. P. E. Dickson, and C. E. Johnson, *J. Magn. Magn. Mater.* **125**, 199 (1993).
230. S. J. Campbell, W. A. Kaczmarek, E. Wu, and K. D. Jayasuriya, *IEEE Trans. Magn.* MAG-**30**, 742 (1994).
231. R. S. Tebble and D. J. Craik, *Magnetic Materials*, Wiley, London, 1969.
232. K. Haneda and H. Kojima, *J. Am. Ceram. Soc.* **57**, 68 (1973).

233. W. A. Kaczmarek and B. W. Ninham, *IEEE Trans. Magn.* MAG-**30**, 717 (1994).
234. S. J. Campbell, E. Wu, W. A. Kaczmarek, and K. D. Jayasuriya, *Hyp. Interact.* **92**, 933 (1994).
235. W. A. Kaczmarek, *Mater. Sci. Forum* **179-181**, 313 (1995).
236. W. Kündig, H. Bömmel, G. Constabaris, and R. H. Lindquist, *Phys. Rev.* **142**, 327 (1966).
237. E. Tronc, *Proc. Int. Conf. on the Applications of the Mössbauer Effect*, 1996.
238. P. Prené, E. Tronc, J. P. Jolivet, and J. L. Dorman, *Proc. Int. Conf. on the Applications of the Mössbauer Effect*, 1996.
239. K. Haneda and A. H. Morrish, *IEEE Trans. Magn.* **25**, 2597 (1989).
240. W. A. Kaczmarek, E. Z. Radlinska, and B. W. Ninham, *Mater. Chem. Phys.* **35**, 31 (1993).
241. W. A. Kaczmarek and B. W. Ninham, *J. Appl. Phys.* **76**, 6065 (1994).
242. G. M. Wang, P. Millet, A. Calka, and S. J. Campbell, *Mater. Sci. Forum* **179-181**, 183 (1995).
243. P. Jernberg and R. Wappling, *J. Solid State Chem.* **49**, 123 (1983).
244. W. A. Kaczmarek, A. Calka, and B. W. Ninham, *phys. stat. sol. (a)* **141**, K123 (1994).
245. A. P. Radlinski, A. Calka, B. W. Ninham, and W. A. Kaczmarek, *Mater. Sci. Eng.* **A134**, 1346 (1991).
246. S. J. Campbell and F. Aubertin, in *Mössbauer Spectroscopy Applied to Inorganic Chemistry*, G. J. Long and F. Grandjean, eds., Vol. 3, p. 183, Plenum Press, New York, 1989.
247. G. Le Cäer and R. A. Brand, *Hyp. Interact.* **71**, 1507 (1992).
248. B. Fultz, in *Mössbauer Spectroscopy Applied to Magnetism and Materials Science*, G. J. Long and F. Grandjean, eds., Vol. 1, p. 1, Plenum Press, New York, 1993.
249. R. E. Vandenberghe, E. De Grave, and P. M. A. de Bakker, *Hyp. Interact.* **83**, 29 (1994).
250. J. Y. Ping and D. G. Rancourt, *Hyp. Interact.* **92**, 1203 (1994).
251. E. Yu. Tsymbal, A. M. Afanas'ev, M. Fricke, J. Hesse, and F. Parak, *Z. Phys.* **B94**, 217 (1994).
252. G. M. McColvin and M. J. Shaw, *Mater. Sci. Forum* **88–90**, 235 (1992).

Note added in proof.

The continuing development and expansion of the field of materials synthesis based on mechanical treatment is clearly indicated by the number of papers published on this topic in recent years. For example, using the term 'mechanical alloying' in an OCLC electronic information search of the titles and abstracts of papers in the INSPECT database, the Physics, Electronics, and Computing database created by the Institute of Electrical Engineers, revealed 150 papers published between 1970 and 1974, 230 papers between 1975 and 1979, 260 papers between 1980 and 1984, 430 papers between 1985 and 1989, and 1360 papers between 1990 and 1994. The increasing trend through each decade is clear, with an average of ca. 40 papers per year in the 1970's, 70 in the 1980's, and over 250 per year so far in the 1990's. A more detailed inspection of the figures for the 1990's again reveals an increasing trend, with 170 papers published in 1990, 230 papers in 1991, 290 papers in 1992, 310 papers in 1993, 360 papers in 1994, and 540 papers in 1995. By using 'mechanochemical' or 'mechanochemistry' as the search word, the more recent extension of the synthesis of new materials based on mechanochemical and related methods is revealed by the 41 papers published between 1970 and 1979, the 94 papers published between 1980 and 1990, and the 129 papers published between 1991 and 1996. So far there is a total of 264 such papers, and again, the increasing trend in this burgeoning field of materials synthesis is clear. The INSPECT data base is accessible on the OCLC electronic information search; http://ora.rsch.oclc.org/oclc/ menu/home1.htm.

Indeed, the extent to which research developments are occuring in this field is reflected by the new insight presented in the proceedings of the recent TMS-AIME Symposium on *The Structure and Properties of Nanophase Materials.* Among the papers with particular relevance to topics considered in this chapter are the following. First, Le Caër and Delcroix [*NanoStruct. Mater.* **7**, 127 (1996)] have outlined some recent contributions of Mössbauer spectroscopy to the characterization of nanostructured materials, including the ways in which subtle phenomena in nanophase materials can be studied. The microstructures of a wide range of iron based powder compositions, processed in a low energy ball mill with different alloying elements and milling environments, have been reported by Rawers and Korth [*NanoStruct. Mater.* **7**, 25 (1996)], see Section 1. 2. In addition, Huang et al. [B. L. Huang, R. J. Perez, P. J. Crawford, S. R. Nutt, and E. J. Lavernia, *NanoStruct. Mater.* **7**, 57 (1996)] have synthesized nanocrystalline $Fe_{78}B_{13}Si_9$ by cryogenic high energy ball milling of Metglas, see Section 4. 1. They discussed the different crystallization mechanisms occuring at different stages of milling, and characterized the crystallization kinetics.

An up to date perspective on nanocrystalline and nanophase materials has also been described by Shull [*NanoStruct. Mater.* **7**, 265 (1996)], whose prediction of a saturation in the number of meetings in the general field of nanostructured materials seems to have come true, with 14 major meetings in 1993, 26 meetings in 1994, 26 meetings in 1995, and 11 meetings scheduled for 1996. The current list of meetings can be accessed at "http://cobalt.nist.gov". As explained by Shull, this list will help the coordination between the organisers of conferences in this field.

Author Index

Italicized numbers indicate reference numbers wherein the specified author can be found.

Subject Index

GPSR Compliance

The European Union's (EU) General Product Safety Regulation (GPSR) is a set of rules that requires consumer products to be safe and our obligations to ensure this.

If you have any concerns about our products, you can contact us on ProductSafety@springernature.com

In case Publisher is established outside the EU, the EU authorized representative is:

Springer Nature Customer Service Center GmbH
Europaplatz 3
69115 Heidelberg, Germany

Batch number: 10761599

Printed by Printforce, the Netherlands